i	Effective interest rate per interest period
i_c	Combined interest rate (with inflation included)
i_r	Real interest rate (without inflation included)
I_t	Index of price change $(= I_n/I_k)$ (inflation for period k to period n)
IRR	Internal rate of return
j	Percentage increase in amount each period (geometric series)
M	Number of compounding periods per year
MACRS	Modified accelerated cost recovery system
MARR	Minimum attractive rate of return
N	Number of compounding periods. Also, depreciable life
NPW	Net present worth
P	Present sum of money
$Pr[\cdot]$	Probability of $[\cdot]$
PW	Present worth (method, or equivalent)
PW-C	Present worth-cost (method, or equivalent)
r	Nominal interest rate per year (= annual percentage rate)
R_k	Revenues (cash inflows) for period k
R\$	Real (uninflated, constant worth) dollars
RN	Random number
RND	Random normal deviate
RR	Rate of return
s	Calculated standard deviation
S	Number of standard deviations (for standardized normal distribution)
σ	Population (or estimated) standard deviation
SL	Straight-line method of depreciation
SYD	Sum-of-the-years' digits depreciation method. Also arithmetic operation
t	Effective income tax rate
TI	Taxable income
$U[\cdot]$	Utility of $[\cdot]$
$V[\cdot]$	Variance of $[\cdot]$

*Capital Investment Analysis
for Engineering and Management*

PRENTICE HALL INTERNATIONAL SERIES
IN INDUSTRIAL AND SYSTEMS ENGINEERING

W. J. Fabrycky and J. H. Mize, Editors

SECOND EDITION

Capital Investment Analysis for Engineering and Management

John R. Canada

North Carolina State University
Raleigh, NC

William G. Sullivan

Virginia Polytechnic Institute and State University
Blacksburg, VA

John A. White

University of Arkansas
Fayetteville, AR

Prentice Hall, Inc. Upper Saddle River, New Jersey 07458

Library of Congress Cataloging-in-Publication Data

CANADA, JOHN R.
  Capital investment analysis for engineering and management / John
R. Canada, William G. Sullivan, John A. White. -- 2nd ed.
    p.   cm.
  Rev. ed of: Capital investment decision analysis for management
and engineering. c1980.
  Includes bibliographical references and index.
  ISBN 0-13-311036-2
  1. Capital investments--Evaluation.   I. Sullivan, William G.,
. II. White, John A.,   . III. Canada, John R.
Capital investment decision analysis for management and
engineering.   IV. Title.
HG4028.C4C3. 1996
658.15'242--dc20   95-411

Cover designer: Bruce Kenselaar
Acquistion editor: Alice Dworkin
Copy editor: Kristen Cassereau
Production editor: Judy Winthrop
Manufacturing buyer: Donna Sullivan

 © 1996, 1980 by Prentice-Hall, Inc.,
Simon & Schuster/A Viacom Company
Upper Saddle River, New Jersey 07458

Printed in the United States of America

10   9   8   7   6   5   4

ISBN 0-13-311036-2

Prentice-Hall International (UK) Limited, *London*
Prentice-Hall of Australia Pty. Limited, *Sydney*
Prentice-Hall Canada, Inc., *Toronto*
Prentice-Hall Hispanoamericana, S.A., *Mexico*
Prentice-Hall of India Private Limited, *New Delhi*
Prentice-Hall of Japan, Inc., *Tokyo*
Simon & Schuster Asia Pte. Ltd., *Singapore*
Editora Prentice-Hall do Brasil, Ltda., *Rio de Janeiro*

DEDICATION

to

Our Mentors and Friends

Frank F. Groseclose

Robert N. Lehrer

Contents

2
Computations Involving Interest _____ *19*

3
Equivalent Worth Methods
for Comparing Alternatives _____ *50*

4
Rate of Return Methods
for Comparing Alternatives _____ *69*

5
Estimating for Economic Analyses _____ 103

6
Consideration of Depreciation
and Income Taxes _____ 140

7
Analyses for Government Agencies
and Public Utilities _____ 190

Part Two
Capital Investment Evaluation
Under Risk and Uncertainty

10
Introduction to Risk and Uncertainty _____ 269

11
Sensitivity Analysis _____ 289

12
Analytical and Simulation
Approaches to Risk Analysis ——————— *314*

13
Decision Criteria and Methods
for Risk and Uncertainty ——————— *341*

14
Decision Tree Analysis ——————— *363*

15
Statistical Decision Techniques ——————— *387*

Part Three
Additional Topics in Capital
Investment Decision Analysis ——————

16
*Mathematical Programming
for Capital Budgeting* ———————————— *401*

20
The Analytic Hierarchy Process ——————— *483*

Appendix E
The Standardized Normal
Distribution Function, F(S)

Bibliography

Glossary

Index

Preface

This book is intended primarily for advanced undergraduate or graduate study for students in disciplines emphasizing analytical investment decision methodologies, particularly in engineering and engineering management. The scope and coverage of the book also make it suitable as a text and a reference in industry.

This book is an evolution of *Intermediate Economic Analysis for Management and Engineering* by John R. Canada, published by Prentice Hall in 1971. That was revised with the capable co-authorship of John A White and renamed *Capital Investment Decision Analysis for Management and Engineering,* published by Prentice-Hall in 1980. This, the second edition of that book, slightly renamed, incorporates the co-authorship talents of William G. Sullivan, who provided the primary creative revisions to this edition. It continues as a text and reference that is more concise and yet more advanced than traditional applied works, and is highly lucid—incorporating abundant example problems and solutions.

Part I continues to be a rather succinct summary of basic capital project evaluation techniques, beginning with interest and related equivalency economic comparison methodologies in Chapters 2, 3, and 4. It has been enhanced by very extensive new material on deterministic estimating techniques (Chapter 5) and on consideration of income taxes, updated in view of the 1993 federal law (Chapter 6). It also contains substantial revisions on analyses for public organizations (Chapter 7), replacement analyses (Chapter 8), and capital planning and budgeting (Chapter 9).

Part II continues to include a wide spectrum of useful techniques for including risk and uncertainty in capital investment analyses. Chapter 10 introduces risk and uncertainty concepts with some emphasis on estimating in terms of probabilities. Chapter 11 illustrates tabular and graphical means for exploring sensitivities. Chapters 12 and 13 include a wide range of techniques for considering variability of outcomes, particularly when probabilities can be estimated. Chapter 14 focuses on the use of decision tree analysis as a means of taking into account future (usually probabilistic) outcomes, alternatives, and decisions in determining the best initial choice. Chapter 15 deals with statistical decision techniques, with emphasis on Bayesian concepts for revising probabilities and evaluating the worth of those revisions.

Part III contains additional topics thought useful, generally in more advanced studies of capital investment analysis. Chapter 16 presents several mathematical programming applications. Chapter 17 overviews the topic of activity-based costing as it relates to capital investment analysis. Chapter 18 concerns how to deal with inflation. It might have been included in Part I, but we decided to err on the side of simplification by assuming in either chapters that inflation is not relevant in the comparisons and analyses being made. Chapter 19 and 20 include several widely used multiple attribute decision methodologies, ending with the analytic hierarchy process.

For the use of Part I, only a knowledge of first-year algebra is required, while for much of Parts II and III it is assumed that the student understands the basic analysis procedures of Part I and has a fair knowledge of elementary probability and statistics. Some fundamental probability concepts are explained in the text, but those who need further background will find that the first half of most probability and statistics texts will provide adequate review material. Complete understanding of the application of some specialized quantitative techniques to investment economic analyses will be facilitated by prior exposure to the theory underlying those techniques. For an abbreviated first course on the fundamentals of engineering or project economy, Part I can serve as an applications-oriented text that contains essentially the same breadth of coverage as traditional undergraduate texts. The integration of project economic analysis into the larger picture of capital budgeting within the firm is accomplished in Chapter 9, which contains a major section on procedures and forms used by a large corporation.

For a course in economic evaluation of alternative projects at the advanced undergraduate or initial graduate level, Part I can be used for review purposes as needed with Parts II and III providing the primary study material. Since the chapters in Parts II and III are largely independent of one another, one can include or delete chapters according to the needs of individuals or classes.

Innumerable persons—friends, colleagues, and helpers—have contributed to the development of this work, and so complete acknowledgment is not possible. Once again, the preparation of this book was made much more tolerable than would have otherwise been possible by the extremely competent secretarial services of Mrs. Dot Cupp. Our wives, Wanda, Mary Lib, and Janet, helped by providing encouragement and (usually) good working conditions. Dr. Jerome Lavelle was the primary author of Chapter 17. Special thanks are extended to Dr. Jim Luxhoj for his comments on the cost estimating material in Chapter 5. Ms. Elin Wicks helped to clarify our treatment of AHP in Chapter 20. Suggestions offered by our reviewers—

Dr. William F. Girouard (California Polytechnic University, Pomona), Dr. James S. Noble (University of Missouri, Columbia) and Jerome P. Lavelle (Kansas State University)—were very thoughtful and provided "food for thought" as revisions were being made. To all these, as well as to the authors and publishers providing reprint permissions, and to many others unnamed, we wish to express our gratitude.

John R. Canada
William G. Sullivan
John A. White

PART ONE

Basic Capital Project Evaluation Techniques

1

Introduction and Cost Concepts

This book concentrates on techniques for comparing and deciding between capital investment alternatives on the basis of economic desirability. With the increasing complexity of our industrial technology, economic decision making is becoming more difficult and at the same time more critical. Economic analyses serve to quantify differences between alternatives and reduce them to bases that facilitate project comparison and selection. The importance of use of these methods varies with alternatives under consideration. In general, the use of these techniques is vitally important, for there is much to be saved or lost by virtue of the particular alternative chosen in usual project investment decisions. Indeed, project investment decisions are critically important factors in determining the success or failure of a firm.

1.1 Recognition of a Problem Opportunity

The starting point in any conscious attempt at rational decision making must be recognition that a problem or opportunity exists. In typical situations, problem recognition is obvious and immediate. A broken machine or completely inadequate production capability, for example, causes awareness rather readily. But there usually exist numerous significant opportunities for improvement or alleviation of what would become future problems which are not obvious without search and thought. Once one is aware of the problem or opportunity, action can be taken to solve or take advantage of it.

It has long been acknowledged that economic analysis of complex alternatives should be most valuable when performed as an integral part of the "big picture" of relevant considerations facing the decision maker. This is often called the "systems analysis" approach.

1.2 Systems Analysis

✓ Systems analysis is a coordinated set of procedures that addresses the fundamental issues of *design* and management: that of specifying how people, money, and materials should be combined to achieve a larger purpose. It includes investigation of proper objectives; comparing quantitatively, where possible, the cost, effectiveness, and risks associated with the alternative policies or strategies for achieving them; and formulating additional alternatives if those examined are found wanting.

The five basic elements of a systematic analysis (consistent with the "scientific method") are

1. definition of objectives,
2. formulation of measures of effectiveness,
3. generation of alternatives,
4. evaluation of alternatives,
5. selection.

1.3 Generation of Alternatives

The need for imagination and creativity in the generation of alternatives cannot be overstated, for its lack is a common defect of many analyses. To emphasize the point, no matter how good an analysis and selection among two or more alternatives is made, if there exists some yet unidentified alternative that is superior to any of the alternatives considered, then the solution will be suboptimal—indeed, it may be drastically less than optimal.

The search for alternatives may be thought of as involving two kinds of tasks— the identification of *classes* or functionally different alternatives and the identification of the most attractive variations or subalternatives for each particular class. For example, a crowded plant space problem might be alleviated by classes of alternatives such as building more space, leasing more space, relayout of existing space, subcontracting work, reducing product lines, increasing shift work, etc. Within each of these classes of alternatives there may be any number of variations to be considered. The analyst can create great benefits to the organization by ensuring that the problem and objectives are clearly stated. Such statements facilitate the identification of applicable classes of alternatives. Also, variations for each class must be judiciously selected for analysis so that no significantly superior alternative is overlooked or eliminated from consideration.

1.4 Importance of Estimates in Economic Analyses

Since economic analyses are concerned with which alternative or alternatives are best for future use, they are, by nature, based on estimates of what is to happen in the future. The most difficult part of an economic analysis is estimating relevant quantities for the future, for the analysis is no better than the estimates comprising it. Most estimates are based on past results, and the usual best source of information on past results is the accounting records of the enterprise. Accounting data must frequently be supplemented by statistical, economic, and engineering data as well as good judgment and mature analysis to arrive at valid cost estimates. Chapter 5 focuses on the vital subject of estimation. Cost concepts important to economic analyses in general are described next.

1.5 Cost Concepts

The word *cost* has many meanings in many different settings. The kinds of cost concepts that should be used depend on the decision problem at hand.

There are two facets to estimating for economic analyses: (1) determining the appropriate quantity to estimate, and (2) making the estimate itself. Most estimates to be made are costs and revenues (negative costs).

For economic analyses, the analyst should be concerned with the *marginal* *(incremental)* costs of using resources for the alternative. What actually constitutes the appropriate marginal cost depends on both the scale and timing of the alternative. To determine this quantity, the term "cost" should be defined as follows:

> The cost of a resource is the decrease in wealth that results from committing this resource to a particular alternative; that is, before any of the benefits of the alternative are calculated.

This definition avoids unquestioning use of normal accounting information such as book values and one-time costs and focuses on what is the change in the value of assets effected by a particular action or decision. The following section discusses major cost concepts and issues useful for determining the marginal costs applicable in economic analyses.

1.5.1 Past Versus Future Costs

Costs and other financial events that previously have occurred or have been presumed are tabulated and summarized within the accounting function of an organization. Past costs can be no more than a guide or source of information for the prediction of future costs. Specifically, the analyst should resist the temptation to rely unquestioningly on what may be described as the "actual cost" data from the accounting function. Even if the data are accurate, they will be at best recorded historical costs for similar circumstances.

Future costs often differ significantly from the costs of similar activities in the past. The analyst should be prepared to adjust past costs to reflect probable changes by the time a proposed alternative is to be implemented, taking into account changes both in the prices of the resources and in the amount of each to be required. Prices of resources can be affected by innumerable factors such as climate, geography, and labor regulations.

In economic analyses one is concerned with projecting what is expected to happen in the future as a result of alternative courses of action. Past costs often serve as a useful guide for such projections. It should be remembered that the viewpoints of the accountant and of the economic analyst are generally very opposite—one is an historian and the other is a fortuneteller.

1.5.2 Joint Costs

One of the difficulties in using accounting records is that some of the costs may be recorded in a single category even though they are in fact joint (or common) costs for many different activities. This may be true for labor and material as well as overhead cost items. These joint costs are often allocated to different products, services, or projects using more or less arbitrary formulas.

Any such formula should be scrutinized most carefully by the analyst interested in determining true marginal costs, that is, in the prediction of how total costs are affected by the alternatives under consideration. To determine applicable costs for any alternative, the analyst must both trace out each of the costs unique to the alternative and determine the portion of joint costs that is due to the alternative being examined. It is likely that the results will be very different from cost data readily available from the accounting function.

1.5.3 Usual Accounting Classification of Production Costs

The accounting function of an enterprise keeps records of happenings affecting the finances of the enterprise. Accounting records of production costs normally are separated into three main categories:

1. direct labor;
2. direct material;
3. overhead.

Direct labor costs or *direct materials costs* are those labor or materials costs that can be conveniently and economically charged to products or jobs on which the costs are incurred. Examples are, respectively, the cost of a turret lathe operator and the cost of the bar stock required to produce a large number of a given part.

By contrast, indirect labor cost and indirect materials cost are those costs that cannot be conveniently and economically charged to particular products or jobs on which the costs are incurred. Examples are, respectively, the cost of a janitor serving several departments or products and the cost of tool bits used on different products. Indirect labor and indirect material costs are part of the third category, *overhead costs,* which includes all production costs other than the costs of direct labor and direct material. Examples of other types of overhead costs are power, maintenance, depreciation, insurance, etc. Overhead costs are often referred to as "indirect costs" or "burden." An example of ways in which overhead costs are allocated and should be taken into account in economic analyses is given in Chapter 5.

Appendix 1-A, which appears at the end of this chapter, contains a very brief exposition of accounting fundamentals for those who have had no previous exposure to the subject. While an understanding of accounting fundamentals is not essential to progress in understanding the content of this book, this appendix should be useful to students with no previous exposure, at least to help provide the framework of the economic setting within which business decisions are made.

1.5.4 Fixed and Incremental Costs

Fixed and incremental (also called *variable* or *marginal*) costs can best be defined with respect to changes in the alternatives (e.g., equipment, methods, production levels, etc.) under consideration. As the names imply, fixed costs remain constant and incremental costs change (increment or vary) with respect to the change(s) under consideration.

One classical application of fixed and incremental (variable) cost concepts is in considering the effect of differing levels of production or facility utilization on total production cost. For example, space and equipment costs are often considered fixed costs while direct labor and direct material are often considered variable costs of production. Such categorizations are made difficult because no cost is fixed for all time. What may be considered fixed over one period may be a variable cost for another period. The distinction between fixed and variable costs depends on the decision at hand and the time horizon relevant to them and should be reexamined for every particular decision.

To illustrate the problem, suppose a contractor has a backhoe for which the total costs for a year are the sum of depreciation, fuel, insurance, and operator wages. If the machine is idle for a slack period of the year, then the fuel and wages are the only variable (marginal) costs, since the depreciation and insurance are fixed for that period of time regardless of how much the backhoe is used. Neglecting the effect of pricing during slack periods on prices and businesses that can be obtained during other periods, the decision maker should be willing to sell the services of the backhoe for any price greater than the cost of fuel and wages. If, however, the contractor were offered a yearly rental on the machine at a price that essentially covered only the cost of fuel and wages, he should not accept. For example, the relevant marginal costs over a year should also include the insurance costs that could be eliminated if the backhoe were not to be used for that time. Over an even longer term, even the depreciation expenses should be considered variable since the contractor may choose not to buy a new backhoe. This example illustrates the general principle that the proportion of total costs that may be considered variable decreases as the period of time for the decision under consideration increases.

If one is making an economic analysis of a proposed change, only the variable (marginal) costs need be considered, since only prospective differences between alternatives need be taken into account.

1.5.5 Long- and Short-Run Costs

If an alternative system entails use of otherwise idle capacity, then the immediate opportunity costs and marginal costs will be low compared to the average costs. If, however, the system is already operating near capacity, the opportunity costs of additional output are likely to be high compared to the average costs. In either case, the long-run marginal costs will tend to be less extreme than the short-run marginal costs and will tend to somewhat approach, but not necessarily equal, the average costs.

Long-run marginal costs also depend on the changes in technology associated with shifts in level of production. In many industries marginal costs decrease with increasing production over time as more efficient facilities are placed into operation. In many other industries both the long-run and short-run marginal costs may increase with increasing output as added resources become more scarce.

Determination of long-run marginal costs rests on an adequate identification of the opportunities that will appear and disappear in the future. Although some industries may have been characterized by particular cost trends in the past, these may well not apply in any particular situation projected into the future. Indeed, past trends may well be reversed by future events.

1.5.6 Opportunity Costs

An opportunity (or alternative) cost is the value of that which is foregone (prior to the calculation of any benefits) because limited resources are used in a particular alternative, thereby causing one to give up the opportunity or chance to use the resources for other possible income-producing or expense-reducing alternatives. It is the same as the *shadow price* of a resource in classical economics. The opportunity cost is the usual appropriate measure of the marginal cost of a resource for economic analyses of alternatives. While it can be equal to the price paid for a resource, it is often very different from that actual outlay. Indeed, the use of a resource normally entails an opportunity cost even if the resource were obtained without cost.

As an example, suppose a particular project involves the use of firm-owned warehouse space that is presently vacant. The cost for the space that should be charged to the project in question should be the income or savings that the best perceived alternative use of the space would bring to the firm. This may be much more or much less than the average cost of the space that might be obtained from accounting records.

The opportunity cost of a resource is often fairly nebulous and hard to estimate. It may be the sacrifice of future earnings rather than present cash. In the most general sense, the cost of using a resource on one project is the cost of not having it available for the best alternative, whether that alternative is to sell the resource or to invest it productively in some other alternative that will bring future benefits. Extending the warehouse example from above, for instance, the opportunity cost of the space could be related to either the cash value that could be obtained from the outright sale of the space or the value associated with some other productive use of the space over time.

In estimating opportunity costs, it is useful to distinguish between resources that can be identically replaced, such as loads of sand or pieces of steel, and those that are somehow unique, such as a specific piece of real estate. For identically replaceable resources, for which there is a ready market, the opportunity cost for the resource is merely the market cost of the replacement or, equivalently, the salvage price of the resource if it is already possessed and will not be replaced.

For a resource that is somehow irreplaceable, the opportunity cost for the resource can be estimated as the cost of replacing the unique resource with the least undesirable substitute available. For example, suppose that the unique resource is an engineer who is especially talented at designing improved methods to reduce costs. The opportunity cost of using her on any project X is her particular value on some alternative project(s) Y where she could make the best improvements with the time available. If, however, the assignment of the especially talented engineer to this project X will cause one or more engineers to be hired to substitute for her on the project(s) she would have done otherwise, then the total cost of assigning her to project X is the net savings foregone plus the salaries of the substitute engineer(s).

As another example, consider a student who could earn $30,000 for working during a year and who chooses instead to go to school and spend $12,000 to do so. The total cost of going to school for that year is $42,000: $12,000 cash outlay and $30,000 for income foregone. (*Note:* This neglects the influence of taxes and assumes that the student has no earning capability while in school.)

1.5.6.1 Opportunity cost in determination of interest rates for economic analyses.   A very important use of the opportunity cost principle is in the determination of the interest cost chargeable to a proposed capital investment project. The proper interest cost is not just the amount that would be paid for the use of borrowed money, but is rather the opportunity cost, i.e., the return foregone or expense incurred because the money is invested in this project rather than in other possible alternative projects. Even when internally owned funds rather than borrowed funds are used for investing, the interest cost chargeable is determined by the same opportunity cost principle. In classical economics terminology, the opportunity cost is a measure of the maximum benefit that, for any given situation, can be obtained from an extra unit of capital.

As an example, suppose a firm always has available certain investment opportunities such as expansion or bonds purchases that will earn a minimum of, say, $X\%$. This being the case, the firm would be unwise to invest in other alternative projects earning less than $X\%$. Thus, in computing the cost of various alternatives, the analyst may simply add in $X\%$ of the amount invested for each. Such a cost may be thought of as the opportunity cost of not investing in the readily available alternatives.

As another example, consider the interest cost for investment in a car, with money to be obtained from one of the following three financing alternatives:

(**I**)   borrow and pay 14%;

(**II**)   take out of savings account earning 5%;

(**III**)   cash in a "hotshot investment" that you are confident would earn 30%.

For financing alternative I, the 14% is a cash cost rather than an opportunity cost and should present no conceptual difficulty. For II, the 5% is an opportunity cost because that amount will be given up if the money is not left in savings. Similarly, for III, the 30% is the opportunity cost if one invests in the car rather than keeps the "hotshot investment." The wide range between the 5% and the 30% is not highly unusual, for the opportunity cost of any resource (including money) is very much a function of what alternative(s) is foregone.

In economic studies, it is necessary to recognize the time value of money irrespective of how the money is obtained, whether it be through debt financing, through owners' capital supplied, or through reinvestment of earnings generated by the firm. Interest on project investments is a cost in the sense of an opportunity foregone, an economic sacrifice of a possible income that might have been obtained by investment of that same money elsewhere.

1.5.6.2 Opportunity cost in replacement analyses.

As another illustration of the opportunity cost principle, suppose a firm is considering replacing an existing piece of equipment that originally cost $50,000, presently has an accounting book value of $20,000, and can be salvaged now for $5,000. For purposes of an economic analysis of whether or not to replace the existing piece of equipment, the investment in that equipment should be considered as $5,000; for by keeping the equipment, the firm is giving up the *opportunity* to obtain $5,000 from its disposal. This principle is elaborated upon in Chapter 8.

1.5.7 Sunk Costs

Sunk costs are costs resulting from past decisions and are therefore irrelevant to the consideration of alternative courses of action. Thus, sunk costs should *not* be considered directly in economic analyses.

As an example, suppose Joe Student finds a used car he likes on a Saturday and pays $50 as a "down payment," which will be applied toward the $1,000 purchase price but which will be forfeited if he decides not to take the car. Over the weekend, Joe finds another car that he considers equally desirable for a purchase price of $910. For purposes of deciding which car to purchase, the $50 is a sunk cost and thus should not enter into the decision. The decision then boils down to paying $1,000 minus $50, or $950, for the first car versus $910 for the second car.

A classical example of a sunk cost occurs in the replacement of assets. Suppose that the piece of equipment examined in the last section, which originally cost $50,000, presently has an accounting book value of $20,000 and can be salvaged now for $5,000. For purposes of an economic analysis, the $50,000 is actually a sunk cost. However, the viewpoint is often taken that the sunk cost should be considered to be the difference between the accounting book value and the present realizable salvage value, which is called "book loss" or "capital loss." According to this viewpoint, the sunk cost is $20,000 minus $5,000, or $15,000. Neither the $50,000 nor the $15,000 should be considered in an economic analysis, except for the manner in which the $15,000 affects income taxes, as discussed in Chapter 6.

Often sunk costs and opportunity cost considerations occur together. For another example, consider the plight of Joe and Joan Hapless, who bought a stock for $10,000 four years ago, only to see it decline to $1,000 in value as of one year ago. Since then it has rebounded to be worth $3,000 now. Joan now says, "Sell the dog, we've lost $7,000!" but Joe says, "No, by keeping it for the last year we've made $2,000!" Who is right? The answer is neither, based on the respective rationales stated. Both are guilty of counting those sunk costs—the $10,000, the $1,000, and the $2,000 can all be considered "sunk" because they are all the result of the past. Only the $3,000 the stock is now worth is relevant. It is the opportunity cost if the stock is kept now; so the decision should boil down to what is the best use of the $3,000—leave it in the stock or sell the stock and use the $3,000 elsewhere.

1.5.8 Postponable Costs

A *postponable cost* is a cost that can be avoided or delayed for some period of time. As an example, the costs of certain types of maintenance or of personnel for certain planning functions may be postponable, while the cost of direct labor is unavoidable or not postponable if production is to continue.

1.5.9 Escapable Costs

When a reduction or elimination of business activity will result in certain costs being eliminated (with perhaps others increased), the net reduction in costs is considered the *escapable cost*. Escapable costs are related to declines in activity in a manner similar to the way variable costs are related to increases in business activity. The escapable cost when a business activity is decreased from X_2 to X_1 is frequently smaller than the variable cost that originally resulted when business level was expanded from X_1 to X_2. For example, it is usually a more difficult management task to reduce labor and other costs and commitments during a contraction than to increase them during an expansion. It is important in estimating net escapable costs that the amount of eliminated costs be reduced by the amount of any additional costs that would be incurred in related activities as a result of the change.

1.5.10 Replacement Costs

Replacement cost is, as the name implies, the cost of replacing an item. It is important to economic analyses because replacement cost rather than historical original cost is the relevant cost factor for most economic decisions. For example, if a storekeeper has been stocking an item costing $8, and selling that item for $12, and the price to the storekeeper for replacing the item is suddenly increased to $14, then the selling price should be raised to at least $14 before any additional units of that item are purchased.

1.5.11 Cash Costs Versus Book Costs

Costs that involve payments of cash or increases in liability are called *cash costs* to distinguish them from noncash (*book*) costs. Other common terms for cash costs are "out-of-pocket costs" or costs that are "cash flows." Book costs are costs that do not involve cash payments but rather represent the amortization of past expenditures for items of lengthy durability. The most common examples of book costs are depreciation and depletion charges for the use of assets such as plant and equipment. In economic analyses, the only costs that need to be considered are cash flows or potential cash flows. Depreciation, for example, is not a cash flow and is important only in the way it affects income taxes, which are cash flows.

1.6 Cost Factors

In an economic analysis, a listing of main factors that may be relevant for projects under consideration is as follows:

First cost, installed and ready to run (or net realizable value)

Insurance and property tax

The life period of the machine until displaced from the proposed job

The salvage value at the date of displacement

The degree and the pattern of utilization; that is, the percent of capacity at which the machine will operate on the intended job with allowances for possible future changes in utilization

Routine maintenance and repair costs

Major repair items or periodic overhauls

Direct operating costs, including operating labor, fuel or power, scrap material, and rework

Indirect costs: indirect labor, tooling, supplies, floor space, inventory

Fringe benefits

Hazards and losses relative to equipment, material, and labor time

Changes in sales volume or price resulting from the choice

Changes in unit cost of labor, power, supplies, etc., resulting in changes in operating costs

1.7 Objectives of Firm and Nonmonetary Factors

While the primary concern of this book is techniques for considering economic or monetary desirability, it should be recognized that the usual decision between alternatives involves many factors other than those that can be reasonably reduced to monetary terms. For example, a limited listing of objectives other than profit maximization or cost minimization that may be important to a firm are

minimization of risk of loss,

maximization of safety,

maximization of sales,

maximization of service quality,

minimization of cyclic fluctuation of the firm,

minimization of cyclic fluctuation of the economy,

maximization of well-being of employees,

creation or maintenance of a desired public image.

Economic analyses provide only for the consideration of those objectives or factors that can be reduced to monetary terms. The results of these analyses should be weighted together with other nonmonetary (irreducible) objectives or factors before a final decision can be made. Multiattribute techniques for weighting objectives and nonmonetary as well as monetary factors are given in Chapters 19 and 20.

1.8 The Role of the Engineer and Manager in Economic Decision Making

Economic analyses and decisions between alternatives can be made by the engineer considering alternatives in the design of equipment, facilities, or man–machine systems. However, the decisions are more commonly made by a manager acting on a number of investment opportunities and alternatives within each opportunity. Whenever the alternatives involve technical considerations, the engineer serves to provide estimates and judgment for the analyses upon which the final managerial decision can be made.

1.9 Scope and Importance

All analysis procedures covered in Chapters 2 through 9 are based on single estimates or amounts for each of the variable quantities considered. That is, if an analysis involves estimates of project investment, life, salvage value, operating expenses, etc., only single estimates for each are made even though it is recognized that each of the estimates may be subject to considerable variation or error. Analyses under these conditions are often called "assumed certainty" analyses. Parts II and III, beginning with Chapter 10, will show methods that explicitly consider the variation in estimated quantities and incorporate many refinements for rational economic analyses.

Regardless of who performs economic analyses or who makes the final investment decisions, the proper performance of these functions is critical to the economic progress of our country and of the world, as well as to the economic health or even survival of the individual firm. Business decisions frequently involve investments that must be planned and executed many years before the expected returns will be realized. Moreover, the scale of the investments in research and capital assets required for our expanding economy grows as new technologies develop. Hence, knowledge of the principles and techniques underlying economic analyses is extremely important.

PROBLEMS

1-1. A supplier purchased an Ajax charger five years ago for $5,000, intending to sell it at its usual markup for $5,800.

Before they were able to obtain delivery, a competitor brought out a radically new charger for the same type of service, better in every way, but selling at a retail price of only $3,000. As a result, the Ajax charger has been a white elephant in the supplier's hands—it is a large piece of obsolete equipment that has been occupying valuable floor space which is now vitally needed.

In discussing what to do, two members of the firm find themselves in disagreement. The president feels that the charger should be kept unless the $5,000 purchase price is realized on the sale. The accountant feels that the equipment should not be sold unless both the $5,000 cost and $500 cost of storage to date can be realized.

Which course of action would you recommend? Why?

1-2. Smith purchased his house several years ago for $75,000 and was just offered $125,000 cash for it. Smith and his family had not been planning to sell and move, even though they are willing to do so. A neighborhood economist has correctly computed that the pretax annual rate of profit on the cash Smith has invested in the house would be 35%, and on this basis he recommends that Smith sell the house. What additional information does Smith need to make a decision? What irrelevant information was given?

1-3. A merchant has been attempting to maintain his stock of goods at a constant physical volume even though prices have been rising. His stock of one item was originally purchased for $10 per unit, which is the cost for accounting purposes. He sold these goods at $16 per unit (applying his usual markup) and immediately replaced them by identical ones purchased at the new wholesale price of $18 per unit. What do you think of the profitableness of this transaction? What should he have sold them for to make $0.01 per unit?

1-4. Certain factory space cost $20.00 per square foot to build and is estimated to have an economic life of 25 years and $0 salvage value. The minimum attractive rate of return on invested capital is 10%. The annual out-of-pocket cost of property taxes, heat, lights, and maintenance is $1.00 per square foot whether or not the space is being used. What should be the cost per square foot considered in an economic analysis of a new project A that entails proposed use of that space under each of the following conditions?

a. The space is now being used for another project B, which will have to be moved to new quarters costing $4.00 per square foot per year.

b. The space is idle and there is no alternative use of it expected for the entire period in which the project under consideration would exist.

c. The space is part of a large area that is used normally; hence, it is thought reasonable to charge only long-run average costs.

1-5. A firm has a manufacturing division with a normal manufacturing capacity of 1,000,000 units that sell for $120 each. The price consists of variable labor and material costs, $60; fixed costs, $40; and profit, $20. During a severe recession, only 200,000 units can be sold annually and only if the price is reduced to $100 each. The total fixed cost can be reduced 15% below normal if the plant remains open and 30% below normal if it closes. The variable cost is directly proportional to output.

a. Disregard irreducible considerations. Should the plant remain open for the next year or two to produce 200,000 units a year, or should it shut down and reopen when business improves?

b. At the 200,000 unit production rate, to what level may the price be reduced during the recession before shutting down the plant becomes more economical than operating it?

1-6. Process A, designed to produce 10,000 units a year, has a fixed cost of $100,000 a year. Process B, with the same design capacity, has a fixed cost of $80,000 a year. Process A produces the initial 4,000 units at a variable cost of $10 and the next 6,000 units at a variable cost of $17. Process B produces the first 5,000 units at a variable cost of $9 each and produces the next 5,000 at $8 each. Show what load should be assigned to each plant if the demand for the product is varied from zero to 20,000 units. Assume that at no load the fixed costs will not be reduced.

1-7. A firm is considering whether to contract with vendors or use in-house crews for a type of maintenance work. The cost accounting system yields estimates per job as follows:

Direct labor	$5, 000
Materials	2, 200
Overhead	4, 000
Total	$11, 200

A vendor has offered to do the work for $10,000 per job. Discuss whether the firm should contract with the vendor under each of the following conditions:

a. The "in-house" crews would still be on the payroll and would be otherwise unoccupied if they were not doing this work.

b. The "in-house" crews on the payroll will always be productively occupied whether or not this maintenance work is done by them. That is, their labor time is worth $5,000 elsewhere if they do not work on this job. Of the standard overhead charge given, one-half is fixed and one-half is avoidable if the "in-house" crews do not do the job.

1-8. For the coming period, overhead costs for a firm are estimated to be $600,000 for production that is expected to sell for $2,800,000 and to require 200,000 direct labor hours costing $800,000 and direct materials costing $200.000. The firm allocates its overhead to various jobs on the basis of the following relation:

$$\text{Overhead cost} = \$100,000 + 0.5(\text{direct labor} + \text{direct materials}).$$

The firm is considering changing its method of manufacturing a particular job with estimates as follows:

	Old Method	New Method
Direct labor	$100,000	$90,000
Direct material	8,000	16,000

Which method is more economical if actual overhead costs for the two methods are thought to:

a. vary according to the allocation formula?

b. be the same for each method?

c. be $5,000 more for the new method than for the old method?

1-9. Explain by example the difference between book costs and sunk costs.

1-10. What is the difference between postponable costs and escapable costs?

1-11. Formulate a one-sentence rule for an engineering group indicating when they may eliminate consideration of overhead costs in the analysis of investment decisions.

1-12. List the costs associated with owning an automobile that you would classify as fixed costs and those that you would classify as variable costs according to miles driven during the next year. Assume that you plan to keep the car for many years.

1-13. Bad news—you have just wrecked your car! An automobile wholesaler offers you $2,000 for the car "as is." Also, your insurance company's claims adjuster estimates that there is $2,000 in damages to your car. Because you have collision insurance with a $1,000 deductibility provision, the insurance company mails you a check for $1,000. The odometer reading on the wrecked car is 58,000 miles.

What should you do? You need another car immediately. The best alternatives appear to be

(1) Sell the wrecked car tomorrow to the wholesaler and invest your $7,000 life savings in a used $10,000 car (odometer reading = 28,000 miles);

(2) Spend $2,000 immediately to restore the wrecked car to its previous condition;

(3) Give the car to a "shade tree" body shop where they will repair it for $1,100 but will take an extra month to do so. During this repair time, you will rent a car at a monthly rate of $400.

In alternatives (2) and (3), you also have the option of selling the repaired car for $4,500 and then investing $5,500 of your savings in the used $10,000 car. If you elect to sell your refurbished car for $4,500, any money remaining in your life savings will be able to earn 5% annually in a certificate of deposit.

State your assumptions, and provide an analysis of what course of action you should take. State any nonmonetary considerations that may influence your decision.

APPENDIX 1-A

Accounting Fundamentals

This section contains an extremely brief and simplified exposition of the elements of accounting in recording and summarizing transactions affecting the finances of the enterprise. The fundamentals apply to any entity (such as an individual, corporation, governmental unit, etc.), referred to here just as a "firm."

All accounting is based on the *fundamental accounting equation,*

$$\text{Assets} = \text{Liabilities} + \text{Ownership}, \tag{1-A-1}$$

where "Assets" are those things of monetary value that the firm *possesses,* "Liabilities" are those things of monetary value that the firm *owes,* and "Ownership" is the worth of what the firm *owns* (also referred to as "equity," "net worth," etc.).

The fundamental accounting equation defines the format of the *balance sheet,* one of the two most common accounting statements. It shows the financial position of the firm *at any given point in time.*

Another important, and rather obvious, accounting relationship is

$$\text{Revenue} - \text{Expenses} = \text{Profit (or Loss)}. \tag{1-A-2}$$

This relationship defines the format of the *income statement* (also commonly known as "profit-and-loss statement"), which summarizes the revenue and expense results of operations *over a period of time.*

It is useful to note that a revenue serves to increase the ownership amount for a firm, while an expense serves to decrease the ownership amount for a firm.

To illustrate the workings of accounts in reflecting the decisions and actions of a firm, suppose you decide to undertake an investment opportunity and that the following sequence of events occurs over a period of a year:

1. Organize a firm and invest $3,000 cash as capital.
2. Purchase equipment for a total cost of $2,000 by paying cash.
3. Borrow $1,500 through note to bank.
4. Manufacture year's supply of inventory through the following:
   a. Pay $1,200 cash for labor.
   b. Incur $400 account payable for material.
   c. Recognize the partial loss in value (depreciation) of the equipment amounting to $500.

5. Sell on credit all goods produced for year, 1,000 units at $3.00 each. Recognize that the accounting value of these goods is $2,100, resulting in an increase in equity (through profits) of $900.
6. Collect $2,200 of account receivable.
7. Pay $400 account payable and $1,000 of bank note.

A simplified version of the accounting entries recording the same information in a format that reflects the effects on the fundamental accounting equation (with a "+" denoting an increase and a "−" denoting a decrease) is shown in Table 1-A-1.

A balance sheet at the end of the year of enterprise operation would appear as follows:

Your Firm

Balance sheet as of end of year 19xx

Assets		*Liabilities and Ownership*	
Cash	$2,100	Bank note:	$ 500
Accounts receivable:	800		
Equipment:	1,500	Equity:	3,900
Total:	$4,400	Total:	$4,400

An income statement is not so directly determinable from the above simplified format as was the balance sheet. In this case, the statement for the year would appear as follows:

Your Firm

Income statement for year ending 19xx

Operating revenues (Sales):		$3,000
Operating costs (Inventory depleted):		
Labor:	$1,200	
Material:	400	
Depreciation:	500	
		$2,100
Net income (Profits):		$ 900

It should be noted that the profit for a period serves to increase the value of the ownership in the firm by that amount. Also, it is worth noting that the net cash flow of $1,400 (= $3,000 − $1,200 − $400) is not all profit. This was recognized in transaction 4c, in which a capital consumption for equipment of $500 was declared. Thus, the profit was $900, or $500 less than the net cash flow.

Figure 1-A-1 shows interrelationships of many asset and cost categories commonly used by accounting functions. Note that these are shown to result in "return on investment," which is merely profit expressed as a percent of capital investment or rate of return for a period and which is a common measure of financial performance.

TABLE 1-A-1 Accounting Effects of Transactions

Account		1	2	3	4	5	6	7	Balance at end of year
ASSETS	Cash	+$3,000	−$2,000	+$1,500	−$1,200		+$2,200	+$1,400	+$2,100
	Account receivable					+$3,000	−$2,200		+ $800
	Inventory		+$2,000		+$2,100	−$2,100			0
	Equipment				− $500				+$1,500
equals									
LIABILITIES	Account payable			+$1,500	+ $400			− $400	
	Bank note							−$1,000	+ $500
plus									
OWNERSHIP	Equity	+$3,000			+ $900				+$3,900

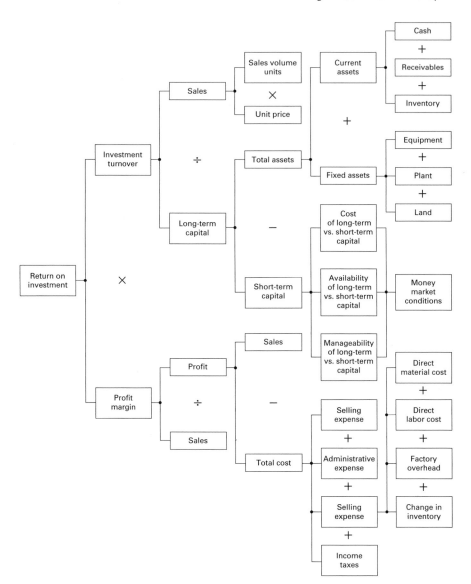

Figure 1-A-1   Interrelationships among many financial categories and terms commonly used in accounting and management practice.

2
Computations Involving Interest

The key to understanding the material in this and several subsequent chapters is recognizing that *money has time value!* To understand what it means for money to have time value, suppose you are offered the opportunity to receive $1,000 today versus receiving it one year from today. Except in unusual circumstances,[1] you would opt to receive the $1,000 today. Having money sooner is generally preferred to having it later. Why? Because you can invest the money, earn a return on it, and accumulate a greater sum than you would otherwise. By deferring receipt of money, you defer the opportunity for investment.

Just as the spacing of forces is a primary consideration in mechanics, the spacing of cash flows (receipts and disbursements) is important in economic analyses. Often, the timing of the cash flows makes *the difference* in the profitability of an investment.

The mechanism used to express the time value of money is the *interest rate.* Also called the *discount rate* and *opportunity cost rate,* the interest rate expresses the change in the value of money as it moves forward or backward in time. As an illustration, if your time value of money is 10%, then you would be indifferent to receiving the $1,000 today and receiving $1,100 one year later. Stated differently, $1,000 today is *equivalent* to $1,100 one year in the future if the time value of money is 10% or if money *is worth* 10%.

From this we see that people's time values of money can differ. Money might be worth 10% to you, but to another person it might be worth 8% or 12%. What determines the time value of money? Many factors contribute to the determination of the

---

[1] Some might be in a financial position such that deferring income is desirable due to income tax implications.

time value of money. For example, suppose you can receive either $1,000 now or $X one year from today. For what value of X would you be indifferent? Subtracting $1,000 from X and dividing the result by $1,000 provides a decimal estimate of your time value of money.[2]

We have been given estimates of X ranging from $1,100 and $2,000. In fact, some individuals would choose to receive $1,000 today rather than receive $3,000 or $4,000 a year from today. Why? If their survival or that of their family depends on having money for food or medicine today, it might not matter to them what is offered a year from today. Or perhaps they don't trust the offeror to deliver the money a year from today. Regardless of the reasons, people have different time values of money— and so do corporations!

Another factor that often arises in considering what value to assign to the time value of money is the rate of inflation. While it is true that the rate of inflation can affect the time value of money, it is also the case that *money has time value in the absence of inflation.* Some have difficulty understanding why the latter is true. If you have difficulty accepting the claim that money has a time value in the absence of inflation, then consider the following. If you needed a place to live while at college, you would think nothing of paying rent for an apartment; indeed, you would expect to pay a rate that more than covers the costs of ownership of the apartment. You expect the owner of the apartment to make a profit on the transaction. Otherwise, why tie money up in apartments and forego the opportunity to invest it elsewhere and earn a return on it?

Likewise, if you do not own a car and need to use one for the weekend, you know that you can rent cars from an auto rental agency. Again, you expect to pay a rate that exceeds the cost of ownership of the car being rented. The same is true for chain saws, computers, offices, land, and a host of other rental items. Furthermore, even in the absence of inflation, you expect to pay rental rates that result in profits for the owners of the rental items.

Now, let's add one more to the list of rental items, money! If you need money, you can rent it from someone or some institution that owns money. In doing so, they will charge you rent for the money. However, they generally call it interest, not rent. And, as with other rental items, even in the absence of inflation, you expect to pay interest rates that will result in a profit for the owner of the money. Likewise, if you own the money, you expect others (including banks) to pay you interest on your money if you loan it to them, rent it to them, or invest it in them, *even in the absence of inflation.* The expectation of inflation normally increases the interest rate one expects to receive (or pay) for the use of the money. Chapter 18 focuses on the consideration of inflation.

2.1 Interest Calculations

Interest calculations may be based on interest rates that are either *simple* or *compound.*

---

[2] The purpose of the exercise is to illustrate the factors contributing to an individual's time value of money. Undoubtedly, different estimates of X would be provided under different circumstances of need, risk, and alternative investment opportunities.

2.1.1 Simple Interest

Whenever the interest charge for any period is based on the principal amount only and not also on any accumulated interest charges, the interest is said to be *simple.* Calculations involving simple interest may be performed utilizing the following formula:

$$I = P \times s \times N,$$

where I = total amount of interest owed after N periods,

P = amount borrowed (invested),

s = simple interest rate,

N = number of periods before repayment (withdrawal).

Example 2-1
An individual borrows $1,000 at a simple interest rate of 8% per year and wishes to repay the principal and interest at the end of 4 years. How much must be repaid?

Solution

$$I = P \times s \times N$$
$$= (\$1,000)(0.08)(4)$$
$$= \$320$$

Therefore, $1,000 + $320 = $1,320 is repaid in 4 years.   ■

2.1.2 Compound Interest

Whenever the interest charge for any interest period is based on the remaining principal amount plus any accumulated interest charges up to the beginning of that period, the interest is said to be *compound.* To illustrate the effect of compounding, the following example is given.

Example 2-2
An individual borrows $1,000 at a compound interest rate of 8% per year and wishes to repay the principal and interest in 4 years. How much must be repaid?

Solution

Year	Amount owed at beginning of year	Interest charge for year	Amount owed at end of year
1	$1,000.00	$1,000.00 × 0.08 = $ 80.00	$1,080.00
2	1,080.00	1,080.00 × 0.08 = 86.40	1,166.40
3	1,166.40	1,166.40 × 0.08 = 93.31	1,259.71
4	1,259.71	1,259.71 × 0.08 = 100.78	1,360.49

Thus, $1,360.49 is repaid. The difference between this and the $1,320.00 answer in the previous example utilizing simple interest is due to the effect of compounding of interest over the 4 years.   ■

2.2 Equivalence

Before exploring various compound interest relationships, let's return to the notion of equivalence. Recall that we said that $1,000 is equivalent to $1,100 one year later if money is worth (has a time value of) 10%. Likewise, based on compound interest, $1,000 today is equivalent to $12,100 two years in the future if money is worth 10%. Generalizing this concept to a single sum or a series of money, it can have an infinite range of equivalent values over time, although it can have actual existence at only one point in time. Thus, to have precise meaning, an item of money must be identified in terms of timing as well as amount. For purposes of definition, two amounts of money or series of money at different points in time are said to be equivalent if they are equal to each other at some point in time at a given interest rate.

Since compound interest is encountered in practice much more often than simple interest, compound interest will be used throughout this book unless otherwise stated. The balance of the chapter deals with the use of compound interest formulas for equivalence conversions.

2.3 Compound Interest Formulas

A variety of compound interest formulas have been developed to facilitate the comparison of investment alternatives, the determination of loan payments, the determination of returns obtained from investments made, and the performance of other financial calculations. In general, the formulas were derived to facilitate calculations that would be laborious in the absence of computational hardware and software. However, given the widespread use of spreadsheets and sophisticated computers and hand-held calculators, there is less reliance today on compound interest formulas and their tabulated values than in the past.

Due to the existence of software to perform financial analyses, why have we chosen to include in this book compound interest formulas and tabulations of their values? The three principal reasons for doing so are as follows: not everyone has access to the technologies available; the assumptions underlying the software vary among products; and understanding how the formulas are obtained that underlie the software provides important insights into the implications of the values obtained from analysis. (The latter is the compelling reason, for even if everyone had access to the very same software, we would provide coverage of the compound interest formulas to ensure that readers understood what is behind the results obtained from the software.)

NOTATION AND CASH FLOW DIAGRAM

The following notation is used throughout this book for compound interest calculations:

i = effective interest rate per interest period,

N = number of compounding periods,

P = present sum of money (the equivalent worth of one or more cash flows at a relative point in time called the present),

F = future sum of money (the equivalent worth of one or more cash flows at a relative point in time called the future),

A = end-of-period cash flows (or equivalent end-of-period values) in a uniform series continuing for a specified number of periods,

and

G = uniform period-by-period increase or decrease in cash flows or amounts (the arithmetic gradient).

The use of time or cash flow diagrams is strongly recommended for most problems, at least whenever the analyst desires to visualize the cash flow situation. Whenever some distinction between types of cash flows seems desirable, it is recommended to use an upward arrow for a cash inflow and a downward arrow for a cash outflow.

2.3.1 Interest Formulas Relating Present and Future Sums

Figure 2-1 shows a time diagram involving a present single sum P and a future single sum F separated by N periods with interest at $i\%$ per period. Two formulas relative to those sums are presented below.

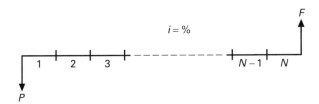

Figure 2-1   Time diagram for single sums.

Find F When Given P

If P dollars are deposited now in an account earning $i\%$ per period, the account will grow to $P(1 + i)$ by the end of one period; by the end of two periods, the account will be $P(1 + i)(1 + i) = P(1 + i)^2$; and by the end of N periods, the account will have grown to a future sum F, as given by

$$F = P(1 + i)^N \qquad (2\text{-}1)$$

where the quantity $(1 + i)^N$, designated (F/P), is tabled in Appendix A for numerous values of i and N. Symbolically, we shall use the notation

$$F = P(F/P, i\%, N), \qquad (2\text{-}2)$$

where the symbol in parentheses denotes the unknown and known, the interest rate, and the number of periods, respectively.

Find P When Given F

The reciprocal of the relationship between P and F, from above, is given mathematically as

$$P = F\left(\frac{1}{1 + i}\right)^N, \qquad (2\text{-}3)$$

where the quantity $1/(1 + i)^N$ is tabled in Appendix A. Symbolically,

$$P = F(P/F, i\%, N) \qquad (2\text{-}4)$$

2.3.2 Applying Interest Formulas to Cash Flow Series

Figure 2-2 depicts a time diagram involving a present single sum P and single sums $A_1, A_2, \ldots, A_N$ occurring at the end of periods $1, 2, \ldots, N$, respectively. With interest at $i\%$ per period, the *present worth equivalent* of the cash flow series $\{A_1, A_2, \ldots, A_N\}$ can be obtained by summing the present worth equivalents of each individual cash flow in the cash flow series. Hence,

$$P = A_1(P/F, i\%, 1) + A_2(P/F, i\%, 2) + \cdots + A_N(P/F, i\%, N). \tag{2-5}$$

Similarly, the *future worth equivalent* of the cash flow series is given by the sum of the future worth equivalents of the individual cash flows,

$$F = A_N + A_{N-1}(F/P, i\%, 1) + \cdots + A_1(F/P, i\%, N-1). \tag{2-6}$$

In a number of instances, the present worth and future worth equivalents of cash flow series can be obtained in a closed mathematical form by using summation of series relations. In particular, if each cash flow in the series has the same value, A, then the series is referred to as a *uniform series.* If the value of a given cash flow differs from the value of the previous cash flow by a constant amount, G, then the series is referred to as a *gradient series.* When the value of a given cash flow differs from the value of the previous cash flow by a constant percentage, $j\%$, then the series is referred to as a *geometric series.* Closed-form expressions are available for P and F for uniform, gradient, and geometric series and will be presented in subsequent sections. Additional series for which closed-form series can be developed are explored in the problems at the end of the chapter.

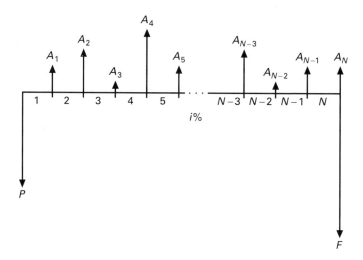

Figure 2-2   Time diagram for a series of cash flows.

2.3.3 Interest Formulas Relating Uniform Series of Payments to Their Present Worth and Future Worth

Figure 2-3 shows a time diagram involving a series of uniform cash flows of amount A occurring at the end of each period for N periods with interest at $i\%$ per period. As depicted in Fig. 2-3, the formulas and tables below are derived such that

1. P occurs one interest period before the first A; and
2. F occurs at the same point in time as the last A, and N periods after P.

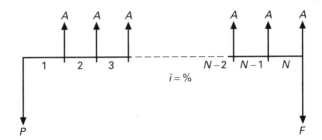

Figure 2-3   Standard time diagram for uniform series.

Four formulas relating A to F and P are given below.

Find F When Given A

If A dollars are deposited at the end of each period for N periods in an account earning $i\%$ per period, the future sum F accrued at the end of the Nth period is

$$F = A[1 + (1 + i) + (1 + i)^2 + \cdots + (1 + i)^{N-1}].$$

It can be shown that this reduces to

$$F = A\left[\frac{(1 + i)^N - 1}{i}\right], \tag{2-7}$$

where the quantity $\{[(1 + i)^N - 1]/i\}$ is tabled in Appendix A. Symbolically,

$$F = A(F/A, i\%, N). \tag{2-8}$$

Find A When Given F

The reciprocal of the relationship between A and F, from above, is given mathematically as

$$A = F\left[\frac{i}{(1 + i)^N - 1}\right], \tag{2-9}$$

where the quantity $\{i/[(1 + i)^N - 1]\}$ is tabled in Appendix A. Symbolically,

$$A = F(A/F, i\%, N). \tag{2-10}$$

Find P When Given A

If we take the relation

$$F = A\left[\frac{(1+i)^N - 1}{i}\right]$$

and substitute

$$F = P(1+i)^N,$$

then we find

$$P = A\left[\frac{(1+i)^N - 1}{i}\right]\left[\frac{1}{1+i}\right]^N$$

which simplifies to

$$P = A\left[\frac{(1+i)^N - 1}{i(1+i)^N}\right]. \qquad (2\text{-}11)$$

The factor in the brackets is tabled in Appendix A. Symbolically,

$$P = A(P/A, i\%, N). \qquad (2\text{-}12)$$

Find A When Given P

The reciprocal of the relationship between A and P, from above, is given mathematically as

$$A = P\left[\frac{i(1+i)^N}{(1+i)^N - 1}\right]. \qquad (2\text{-}13)$$

Again, the factor in brackets is tabled in Appendix A. Symbolically,

$$A = P(A/P, i\%, N). \qquad (2\text{-}14)$$

A summary of the formulas and their symbols, together with example problems, is given in Table 2-1. It should be noted that for all problems in this book involving uniform series, end-of-year payments are assumed unless stated otherwise.

2.3.4 Interest Factor Relationships

The following relationships exist among the six basic interest factors:

$$(P/F, i\%, N) = \frac{1}{(F/P, i\%, N)} \qquad (2\text{-}15)$$

$$(A/P, i\%, N) = \frac{1}{(P/A, i\%, N)} \qquad (2\text{-}16)$$

$$(A/F, i\%, N) = \frac{1}{(F/A, i\%, N)} \qquad (2\text{-}17)$$

$$(A/P, i\%, N) = i\% + (A/F, i\%, N) \qquad (2\text{-}18)$$

$$(F/A, i\%, N) = (P/A, i\%, N)(F/P, i\%, N) \qquad (2\text{-}19)$$

$$(P/A, i\%, N) = \sum_{j=1}^{N}(P/F, i\%, j) \qquad (2\text{-}20)$$

$$(F/A, i\%, N) = \sum_{j=0}^{N-1}(F/P, i\%, j) \qquad (2\text{-}21)$$

TABLE 2-1 Summarization of Discrete Compound Interest Factors and Symbols

To find	Given	Multiply "Given" by factor below	Factor name	Factor functional symbol	Example (answer for $i = 5\%$) (Note: All uniform series problems assume end-of-period payments.)
F	P	$(1+i)^N$	Single sum compound amount	$(F/P, i\%, N)$	A firm borrows \$1,000 for 5 years. How much must it repay in a lump sum at the end of the fifth year? *Ans.*: \$1,276
P	F	$\dfrac{1}{(1+i)^N}$	Single sum present worth	$(P/F, i\%, N)$	A company desires to have \$1,000 8 years from now. What amount is needed now to provide for it? *Ans.*: \$676.80
P	A	$\dfrac{(1+i)^N - 1}{i(1+i)^N}$	Uniform series present worth	$(P/A, i\%, N)$	How much should be deposited in a fund to provide for 5 annual withdrawals of \$100 each? First withdrawal 1 year after deposit. *Ans.*: \$432.95
A	P	$\dfrac{i(1+i)^N}{(1+i)^N - 1}$	Capital recovery	$(A/P, i\%, N)$	What is the size of 10 equal annual payments to repay a loan of \$1,000? First payment 1 year after receiving loan. *Ans.*: \$129.50
F	A	$\dfrac{(1+i)^N - 1}{i}$	Uniform series compound amount	$(F/A, i\%, N)$	If 4 annual deposits of \$2,000 each are placed in an account, how much money has accumulated immediately after the last deposit? *Ans.*: \$8,620
A	F	$\dfrac{i}{(1+i)^N - 1}$	Sinking fund	$(A/F, i\%, N)$	How much should be deposited each year in an account in order to accumulate \$10,000 at the time of the fifth annual deposit? *Ans.*: \$1,810

Key: i = Interest rate per interest period
N = Number of interest periods

A = Uniform series amount
F = Future worth

P = Present worth

2.3.5 Interest Formulas for Uniform Gradient Series

Some economic analysis problems involve receipts or disbursements that are projected to change by a constant amount each period. For example, maintenance and repair expenses on specific equipment may increase by a relatively constant amount of change, G, each period.

Figure 2-4 is a cash flow diagram of a series of end-of-period disbursements increasing at the constant amount of change, G dollars per period. For convenience in derivation of the formulas, it is assumed that a series of uniform payments of amount G is started at the end of the second period, another series of amount G is started at the end of the third period, and so on. Each of these series terminates at the same time, the end of the Nth period. The future sum (at the of the Nth period) equivalent to the gradient series shown in Fig. 2-4 is

$$F = G\left[(F/A, i\%, N - 1) + (F/A, i\%, N - 2) + \cdots + (F/A, i\%, 2) + (F/A, i\%, 1)\right]$$
$$= \frac{G}{i}\left[(1 + i)^{N-1} + (1 + i)^{N-2} + \cdots + (1 + i)^2 + (1 + i) - (N - 1)\right]$$
$$= \frac{G}{i}\left[(1 + i)^{N-1} + (1 + i)^{N-2} + \cdots + (1 + i)^2 + (1 + i) + 1\right] - \frac{NG}{i}.$$

The expression in the brackets reduces to

$$\frac{(1 + i)^N - 1}{i} = (F/A, i\%, N).$$

Hence,

$$F = \frac{G}{i}\left[\frac{(1 + i)^N - 1}{i} - N\right]. \tag{2-22}$$

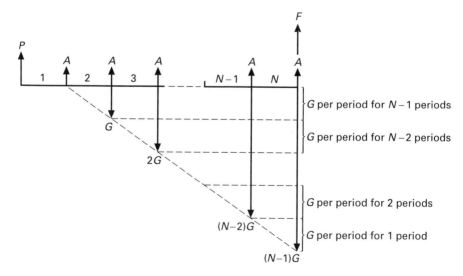

Figure 2-4 Cash flow diagram for uniform gradient of G dollars per period.

The equivalent uniform annual worth of the gradient series may be found by multiplying the above sum of compound amounts by $(A/F, i\%, N)$. Hence,

$$
\begin{aligned}
A &= F(A/F, i\%, N) \\
&= \frac{G}{i}\left[\frac{(1+i)^N - 1}{i} - N\right]\left[\frac{i}{(1+i)^N - 1}\right] \\
&= \frac{G}{i} - \frac{NG}{i}\left[\frac{i}{(1+i)^N - 1}\right] \qquad\qquad (2.23a) \\
&= G\left\{\frac{1}{i} - \left[\frac{N}{(1+i)^N - 1}\right]\right\}.
\end{aligned}
$$

The factor in the braces is given in Table A-21 for a wide range of i and N. Symbolically, the relationship to find the uniform series equivalent to the gradient series is

$$
A = G(A/G, i\%, N). \qquad\qquad (2.23b)
$$

To find the present worth of a uniform gradient series, by employing the relationship $P = A(P/A, i\%, N)$ and Eq. 2-23a, it can be shown that

$$
P = G\left\{\frac{1}{i}\left[\frac{(1+i)^N - 1}{i(1+i)^N} - \frac{N}{(1+i)^N}\right]\right\}. \qquad\qquad (2.24a)
$$

The factor in braces is given in Table A-20 for a wide range of i and N. Symbolically, the relationship is

$$
P = G(P/G, i\%, N). \qquad\qquad (2.24b)
$$

2.3.6 Interest Formulas for Geometric Series

Some economic analysis problems involve cash flows that are anticipated to increase over time by a constant percentage. Labor, energy, and material costs are examples of items that may increase by a constant $j\%$ each period.

Figure 2-5 gives a cash flow diagram of a series of end-of-period disbursements increasing at a constant rate of $j\%$ per period. If A_1 represents the size of the disbursement at the end of period 1, it can be seen that the size of the disbursement at the end of period k, A_k, is equal to $A_1(1+j)^{k-1}$.

The present sum equivalent to the geometric series shown in Fig. 2-5 is

$$
\begin{aligned}
P &= A_1(P/F, i\%, 1) + A_2(P/F, i\%, 2) + A_3(P/F, i\%, 3) \\
&\quad + \cdots + A_N(P/F, i\%, N) \\
&= A_1(1+i)^{-1} + A_2(1+i)^{-2} + A_3(1+i)^{-3} \\
&\quad + \cdots + A_N(1+i)^{-N} \qquad\qquad (2\text{-}25) \\
&= A_1(1+i)^{-1} + A_1(1+j)(1+i)^{-2} + A_1(1+j)^2(1+i)^{-3} \\
&\quad + \cdots + A_1(1+j)^{N-1}(1+i)^{-N} \\
&= A_1(1+i)^{-1}[1 + x + x^2 + \cdots + x^{N-1}],
\end{aligned}
$$

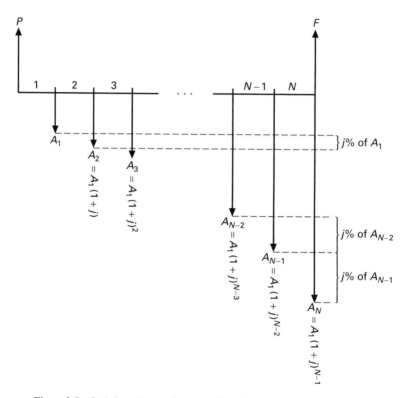

Figure 2-5 Cash flow diagram for geometric series increasing by $j\%$ per period.

where $x = (1 + j)/(1 + i)$. The expression in brackets in Eq. 2-25 reduces to $(1 - x^N)/(1 - x)$ when $x \neq 1$ or $j \neq i$. If $j = i$, then $x = 1$ and the expression in brackets reduces to N, the number of terms in the summation. Hence,

$$P = \begin{cases} A_1(1 + i)^{-1}(1 - x^N)/(1 - x) & j \neq i \\ A_1 N(1 + i)^{-1} & j = i, \end{cases} \tag{2-26}$$

which reduces to

$$P = \begin{cases} \dfrac{A_1[1 - (1 + i)^{-N}(1 + j)^N]}{i - j} & j \neq i \\ A_1 N(1 + i)^{-1} & j = i, \end{cases} \tag{2-27}$$

or

$$P = \begin{cases} \dfrac{A_1[1 - (P/F, i\%, N)(F/P, j\%, N)]}{i - j} & j \neq i \\ A_1 N(P/F, i\%, 1) & j = i. \end{cases} \tag{2-28}$$

By employing the appropriate interest formula from Table 2-1, we can convert the present sum equivalent of a geometric series into a future sum equivalent or a uniform series equivalent. In terms of the basic interest factors, the future sum equivalent is

$$F = \begin{cases} \dfrac{A_1[(F/P, i\%, N) - (F/P, j\%, N)]}{i - j} & j \neq i \\[2mm] A_1 N(F/P, i\%, N-1) & j = i. \end{cases} \qquad (2\text{-}29)$$

SOLVED PROBLEMS

1. Ms. Smith loans Mr. Brown $10,000 with interest compounded at a rate of 8% per year. How much will Mr. Brown owe Ms. Smith if he repays the loan at the end of 5 years?

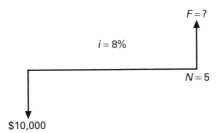

Since the problem is of the form "find F when given P," the formula to use is

$$\begin{aligned} F &= P(F/P, 8\%, 5) \\ &= \$10{,}000(1.4693) \\ &= \$14{,}693. \end{aligned}$$

2. Mr. Lee wishes to accumulate $10,000 in a savings account in 10 years. If the bank pays 5% compounded annually on deposits of this size, how much should Mr. Lee deposit in the account?

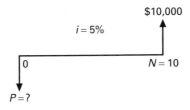

This problem is of the form "find P when given F," and the formula to use is

$$\begin{aligned} P &= F(P/F, 5\%, 10) \\ &= \$10{,}000(0.6139) = \$6{,}139. \end{aligned}$$

3. An individual has been making equal annual payments of $2,000 to repay a loan. The individual wishes to pay off the loan immediately after having made an annual payment. Four payments remain to be paid. With an interest rate of 8%, how much should be paid in the final payment?

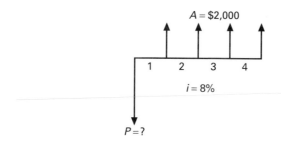

$$P = A(P/A, 8\%, 4)$$
$$= \$2,000(3.3121) = \$6,624$$

4. A person borrows $10,000 at 6% compounded annually. If the loan is repaid in ten equal annual payments, what will be the size of the payments if the first payment is made 1 year after borrowing the money?

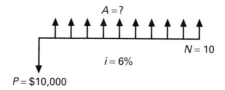

$$A = P(A/P, 6\%, 10)$$
$$= \$10,000(0.1359) = \$1,359$$

5. If $800 is deposited annually for 10 years in an account that pays 6% compounded annually, how much money will be in the fund immediately after the tenth deposit?

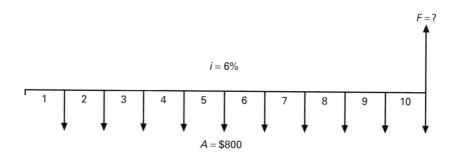

$$F = A(F/A, 6\%, 10)$$
$$= \$800(13.1808) = \$10,545$$

6. An individual wishes to accumulate $1,000,000 in 30 years. If 30 end-of-year deposits are made into an account that pays interest at a rate of 10% compounded annually, what size deposit is required to meet the stated objective?

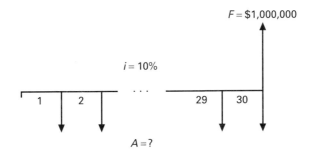

$$A = F(A/F, 10\%, 30)$$
$$= \$1{,}000{,}000(0.0061) = \$6{,}100$$

7. It is expected that a machine will incur operating costs of $4,000 the first year and that these costs will increase by $500 each year thereafter for the 10-year life of the machine. If money is worth 15% per year to the firm, what is the equivalent annual worth of the operating costs?

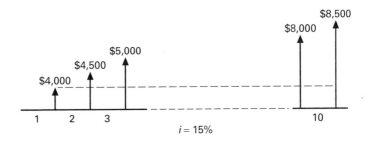

The problem reduces to a constant $4,000 per year plus the $500 gradient.

$$A = \$4{,}000 + G(A/G, i\%, N)$$
$$= \$4{,}000 + \$500(A/G, 15\%, 10)$$
$$= \$4{,}000 + \$500(3.3832) = \$5{,}692$$

8. Work Problem 7 if the timing of the costs is reversed as shown in the following diagram:

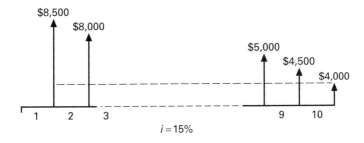

The problem reduces to a constant $8,500 per year minus a $500 gradient of payments.

$$A = \$8,500 - 500(A/G, 15\%, 10)$$
$$= \$8,500 - 500(3.3832) = \$6,808$$

9. In Problem 7 suppose the operating cost the first year is $4,000 and that each year there-after for the 10-year life of the machine operating costs increase by 6% per year. If money is worth 15% per year to the firm, what is the equivalent annual worth of the operating costs?

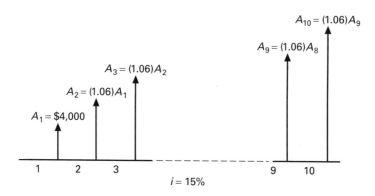

Since $j = 6\%$ and $i = 15\%$,

$$F = \frac{A_1[(F/P, 15\%, 10) - (F/P, 6\%, 10)]}{0.15 - 0.06}$$
$$= \frac{\$4,000[4.0456 - 1.7908]}{0.09} = \$100,213$$

and

$$A = F(A/F, 15\%, 10)$$
$$= \$100,213(0.0493) = \$4,941.$$

2.3.7 Deferred Uniform Payments

Frequently, uniform payments occur at points in time such that more than one interest formula must be applied in order to obtain the desired answer. For example, suppose a person borrows $50,000 to purchase a small business and does not wish to begin repaying the loan until the end of the third year after purchasing the business. With an interest rate of 5% compounded annually, it is desired to determine the amount of the equal annual payment if 15 payments are made.

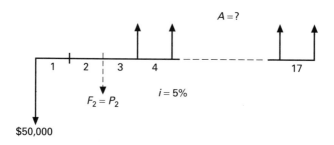

From the cash flow diagram it is apparent that direct use of the formula "find A when given P" is not possible, since P does not occur one period prior to the first A. However, the problem can be logically solved in two steps. First, the amount of the loan, plus interest, after 2 years is

$$F_2 = P(F/P, 5\%, 2)$$
$$= \$50,000(1.1025) = \$55,125.$$

Note the use of the subscript 2 to denote the point in time. The direct use of $A = P(A/P, i\%, N)$ is now possible, since the amount to be repaid (\$55,125) occurs one period prior to the first loan payment. Therefore, the loan payment size will be

$$A = P_2(A/P, 5\%, 15)$$
$$= \$55,125(0.0963) = \$5,309.$$

2.4 Spreadsheets

A variety of software packages is available to facilitate the use of spreadsheets in performing analyses, e.g., Excel®, Lotus 1-2-3®, and Quattro® Pro. Rather than endorsing a particular spreadsheet program, our purpose in this section is to demonstrate how a spreadsheet can be used to facilitate both financial calculations and the performance of sensitivity analyses.

If you want to construct your own spreadsheet, most of what we have presented in the chapter can be replicated by using the formula $V_2 = V_1(1 + i)$, where V_t denotes the value of a single sum of money at time t and i denotes the time value of money. Due to the computer's speed, there is no need to take advantage of the structure of series of cash flows; instead, each cash flow can be treated as a single sum of money. To move money forward in time by one time unit (or time period), multiply by $(1 + i)$; to move money backward in time by one time unit, divide by $(1 + i)$.

Consider the spreadsheet given in Figure 2-6. The columns are identified alphabetically and the rows are identified numerically. The interest rate is entered, using decimal format, in cell B1. The magnitudes of cash flows are entered in column B,

row/column	A	B	C	D
1	Discount Rate =	(enter interest rate)		
2				
3	Time	Cash Flow	Discount Factor	Present Value
4	0	A_1	1.0000	D4=B4*C4
5	1	A_2	C5=C4/(1.0+B1)	D5=B5*C5
6	2	A_3	C6=C5/(1.0+B1)	D6=B6*C6
7	3	A_4	C7=C6/(1.0+B1)	D7=B7*C7
8	4	A_5	C8=C7/(1.0+B1)	D8=B8*C8
9	5	A_6	C9=C8/(1.0+B1)	D9=B9*C9
10				D10=SUM(D4:D9)

Figure 2-6   Sample spreadsheet.

row/column	A	B	C	D
1	Discount Rate =	0.10		
2				
3	Time	Cash Flow	Discount Factor	Present Value
4	0	($10,000)	1.000000	($10,000.00)
5	1	$3,000	0.909091	$2,727.27
6	2	$3,000	0.826446	$2,479.34
7	3	$3,000	0.751315	$2,253.94
8	4	$3,000	0.683013	$2,049.04
9	5	$3,000	0.620921	$1,862.76
10				$1,372.36

Figure 2-7   Example of using a spreadsheet.

beginning with row 4. The value of the discount factor for a particular time period is obtained by dividing the discount factor for the previous period by the sum of one and the interest rate, $1 + i$. The present value of an individual cash flow is obtained by multiplying the value of the cash flow by the discount rate. Finally, the present value for the cash flow series is obtained by summing the individual present values, as shown in cell D10.

Entering data into the spreadsheet, as shown in Figure 2-7, and performing the indicated calculations yields a present value of $1,372.36 for this example.

When one uses Excel 5.0,[3] the net present value of a series of cash flows can be obtained easily. Specifically, by entering =NPV(0.10,B5:B9)+B4 in a cell, the net present value of $1,372.36 is obtained. It is important to note that the NPV function computes the present value of a series of cash flows at one period prior to the first value in the series. Using the NPV calculation shown, the net present value is determined for the cash flows entered in cells B5 through B9; hence, to obtain the present value for the entire cash flow series, it is necessary to add the entry in cell B4 to the value obtained by using the NPV function over the range of cells from B5 through B9.

Once the spreadsheet is set up, it is relatively easy to examine the impact of changes in the parameters of the investment. For example, if the annual receipts are reduced to $2,500, then as shown in Figure 2-8, the resulting NPV value will be equal to −$523.03. Likewise, as shown in Figure 2-9, if the interest rate in Figure 2-7 is changed to 16%, the net present value changes to −$177.12.

If the future value of the cash flow series is required, the division operation performed in column C of Figure 2-6 should be replaced by the multiplication operation. (Excel 5.0 does not include an equivalent NFV function. The FV function in Excel 5.0 is limited to the determination of the future worth of a uniform series of payments; similar limitations apply to the PV function and PMT, or annuity, function. Hence, care must be taken in using financial functions in software packages; even though functions may have the same names used in this book, they can take on very different

[3] Microsoft® Excel, Version 5.0, Microsoft Corporation, 1993–1994.

row/column	A	B	C	D
1	Discount Rate =	0.10		
2				
3	Time	Cash Flow	Discount Factor	Present Value
4	0	($10,000)	1.000000	($10,000.00)
5	1	$2,500	0.909091	$2,272.73
6	2	$2,500	0.826446	$2,066.12
7	3	$2,500	0.751315	$1,878.29
8	4	$2,500	0.683013	$1,707.53
9	5	$2,500	0.620921	$1,552.30
10				($523.03)

Figure 2-8   Using a spreadsheet to analyze changes in cash flows.

row/column	A	B	C	D
1	Discount Rate =	0.16		
2				
3	Time	Cash Flow	Discount Factor	Present Value
4	0	($10,000)	1.000000	($10,000.00)
5	1	$3,000	0.862069	$2,586.21
6	2	$3,000	0.743163	$2,229.49
7	3	$3,000	0.640658	$1,921.97
8	4	$3,000	0.552291	$1,656.87
9	5	$3,000	0.476113	$1,428.34
10				($177.12)

Figure 2-9   Using a spreadsheet to analyze changes in the interest rate.

meanings. Excel users will find it beneficial to become familiar with the various financial functions, e.g., FV, FVSCHEDULE, IPMT, IRR, MIRR, NPER, PMT, PPMT, PV, and RATE.)

2.5 Compounding Frequency; Nominal and Effective Rates

In most economic studies, interest is accounted for as if compounding occurs once a year. In practice, the interest accumulation may take place more frequently, so it is important to note the effects of compounding frequency and to treat properly those problems where the assumption of annual compounding is not appropriate.

As an example, an interest rate may be stated as 12% compounded quarterly. In this case, the 12% is understood to be an annual rate, and is called the *nominal interest rate*. The number of quarterly compounding periods in a year is four. Hence, the interest rate per interest period is 12% ÷ 4 = 3% per quarter. The *effective interest rate* is the

TABLE 2-2 Impact of Compounding Frequency Upon Effective Interest Rate.

		For a nominal rate of 12%	
Frequency of compounding	No. of compounding periods per year	Interest rate per period	Effective rate
Annual	1	12%	12.00%
Semiannual	2	6	12.36%
Quarterly	4	3	12.55%
Monthly	12	1	12.68%
Continuously	$\to \infty$	$\to 0$	12.75%

exact annual rate that takes into account the compounding that occurs within the year. The following formula may be used to calculate the effective interest rate:

$$\text{Effective rate} = (1 + r/M)^M - 1, \qquad (2\text{-}30)$$

where

$$M = \text{number of interest periods per year,}$$
$$r = \text{nominal interest rate.}$$

Recall that $(1 + r/M)^M$ is the single-sum compound amount factor; the effective interest rate may be determined directly from the interest tables using the relation

$$\text{Effective rate} = (F/P, r/M\%, M) - 1. \qquad (2\text{-}31)$$

Hence, for our example, the effective rate is $(F/P, 3\%, 4) - 1 = 12.55\%$.

The effective interest rate can be determined for the case of an infinite number of compounding periods from Eq. 2-30 by letting M, the number of interest periods per year, become infinitely large. Such a condition is termed *continuous compounding*. Thus, the effective rate of interest with continuous compounding is $e^r - 1$, where e is the base of the Naperian or natural logarithm and equals 2.7183. As an example, the effective rate of 12% compounded continuously is $e^{0.12} - 1 = 12.75\%$.

The effect of compounding frequency on the effective interest rate for a nominal rate of 12% is given in Table 2-2.

2.6 Continuous Compounding Interest Formulas

In some instances, particularly when payments occur rather frequently within periods rather than at the beginning or end of periods, the additional theoretical accuracy of continuous compounding may be significant. Continuous compounding means that the interest or profit growth is proportional to the amount of total principal and interest at each instant.

To find the future worth of a present single sum under continuous compounding, we let N denote the number of years involved and substitute for i the effective interest rate for continuous compounding. Thus,

$$
\begin{aligned}
F &= P(1 + i_{\text{eff}})^N \\
  &= P[1 + (e^r - 1)]^N \\
  &= Pe^{rN}.
\end{aligned}
\qquad (2\text{-}32)
$$

To find the present worth of a single sum under continuous compounding, the above equation can be transposed to

$$P = Fe^{-rN}. \tag{2-33}$$

As an example, the future worth of a present amount of $10,000 at 20% nominal interest compounded continuously for 5 years is $10,000e^{0.20(5)} = \$27,183$.

The effective rate of interest in this case can be calculated as $e^r - 1 = e^{0.20} - 1 = 22.14\%$.

We can find that the future worth of the same $10,000 at effective interest of 22.14% compounded at the end of each year for 5 years is

$$F = P(F/P, 22.14\%, 5) = \$10,000(1 + 0.2214)^5 = \$27,183.$$

This illustrates the general principle that continuous compounding at a given nominal interest rate is equivalent to discrete annual compounding at the corresponding effective rate for continuous compounding. Hence, by letting i equal $e^r - 1$ and N be expressed in years, the discrete interest factors can be used to deal directly with continuous compounding. The continuous compounding interest factors for discrete payments are given by the first six entries in Table 2-3. To distinguish between discrete and continuous compounding, the interest rate is underlined in the appropriate symbolic representation of the interest factor. Notice that the interest rate used in Table 2-3 is the nominal interest rate, r.

2.6.1 Continuous Payments Throughout the Year

In some cases, money is disbursed uniformly throughout the year. Consequently, rather than depicting a cash flow of, say, $1,000 occurring at the end of the year, it is assumed that the $1,000 is spread uniformly over the year, as depicted in Fig. 2-10. To distinguish the discrete cash flow from the continuous cash flow, a bar above the symbol is used. Thus, $\bar{A}_k$ will denote the continuous cash flow spread uniformly over period k.

Suppose a total of $\bar{A}_1$ dollars flows uniformly and continuously throughout period 1. Divide $\bar{A}_1$ into M equal amounts to be deposited at equally spaced points in time during a year. The interest rate per period is r/M, and the future worth of the series of M equal amounts is

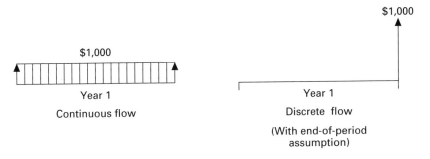

Figure 2-10   Comparison of continuous and discrete flows.

TABLE 2-3 Summary of Continuous Compounding Interest Factors and Symbols

To find	Given	Multiply "given" by factor below	Factor name	Factor functional symbol
	Discrete payments			
F	P	e^{rN}	Continuous compounding compound amount factor (Discrete, single sum)	$(F/P, r\%, N)$
P	F	e^{-rN}	Continuous compounding present worth factor (Discrete, single sum)	$(P/F, r\%, N)$
P	A	$\dfrac{e^{rN} - 1}{e^{rN}(e^r - 1)}$	Continuous compounding present worth factor (Discrete, uniform series)	$(P/A, r\%, N)$
A	P	$\dfrac{e^{rN}(e^r - 1)}{e^{rN} - 1}$	Continuous compounding capital recovery factor (Discrete, uniform series)	$(A/P, r\%, N)$
F	A	$\dfrac{e^{rN} - 1}{e^r - 1}$	Continuous compounding compound amount factor (Discrete, uniform series)	$(F/A, r\%, N)$
A	F	$\dfrac{e^r - 1}{e^{rN} - 1}$	Continuous compounding sinking fund factor (Discrete, uniform series)	$(A/F, r\%, N)$
	Continuous payments			
P	$\overline{A}$	$\dfrac{e^{rN} - 1}{re^{rN}}$	Continuous compounding present worth factor (Continuous, uniform flow)	$(P/\overline{A}, r\%, N)$
$\overline{A}$	P	$\dfrac{re^{rN}}{e^{rN} - 1}$	Continuous compounding capital recovery factor (Continuous, uniform flow)	$(\overline{A}/P, r\%, N)$
F	$\overline{A}$	$\dfrac{e^{rN} - 1}{r}$	Continuous compounding compound amount factor (Continuous, uniform flow)	$(F/\overline{A}, r\%, N)$
$\overline{A}$	F	$\dfrac{r}{e^{rN} - 1}$	Continuous compounding sinking fund factor (Continuous, uniform flow)	$(\overline{A}/F, r\%, N)$

$$F = \frac{\overline{A}_1}{M}(F/A, r/M\%, M),$$

which reduces to

$$F = \frac{\overline{A}_1}{r}\left[\left(1 + \frac{r}{M}\right)^M - 1\right].$$

Letting M approach infinity yields

$$F = \bar{A}_1 \frac{(e^r - 1)}{r}. \tag{2-34}$$

Hence, a continuous and uniform flow of $\bar{A}_k$ over year k is equivalent to a single discrete sum of $\bar{A}_k(e^r - 1)/r$ at the end of year k. Employing the previous convention of denoting by $\bar{A}_k$ the value of a single sum of money occurring at the end of year k, it is evident that a set of continuous cash flows $\{\bar{A}_1, \bar{A}_2, \ldots, \bar{A}_N\}$ spread uniformly over years 1 through N, respectively, is equivalent to a set of discrete cash flows $\{A_1, A_2, \ldots, A_N\}$ occurring at the end of years 1 through N, respectively, with $A_k = \bar{A}_k(e^r - 1)/r$.

In order to obtain the appropriate continuous compounding interest formulas for continuous flows of money, substitute $\bar{A}_k(e^r - 1)/r$ for A_k in the equation for continuous compounding of discretely spaced cash flows. The resulting interest formulas are summarized in Table 2-3. Tables in Appendix B provide factors for continuous compounding.

SOLVED PROBLEMS

1. A person makes six end-of-year deposits of \$1,000 in an account paying 5% compounded annually. If the accumulated fund is withdrawn 4 years after the last deposit, how much money will be withdrawn?

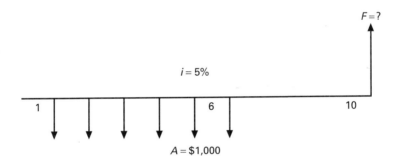

Since F does not occur at the time of the last A, it is necessary that the solution proceed in two steps. The amount of money in the account at the time of the last deposit may be computed as

$$F_6 = A(F/A, 5\%, 6)$$
$$= \$1,000(6.8019) = \$6,802.$$

The problem now is to find F_{10}, given $F_6 = P_6 = \$6,802$.

$$F_{10} = P_6(F/P, 5\%, 4)$$
$$= \$6,802(1.2155) = \$8,268$$

2. What is the effective interest rate for 4.75% compounded annually and 4.60% compounded quarterly?

$$\text{Effective rate} = (1 + r/M)^M - 1$$
$$= (1 + 0.0475)^1 - 1$$
$$= 4.75\%$$
$$\text{Effective rate} = (1 + r/M)^M - 1$$
$$= \left[1 + \frac{0.046}{4}\right]^4 - 1$$
$$= 4.68\%$$

3. A loan company advertises that it will loan $1,000 to be repaid in 30 monthly install-ments of $44.60. What is the effective interest rate?

$$A = P(A/P, i\%, 30)$$
$$\frac{\$44.60}{\$1,000.00} = (A/P, i\%, 30) = 0.0446$$

By inspection (with interpolation in tables), $i = 2\%$, and

$$\text{Effective rate} = (F/P, 2\%, 12) - 1 = 0.2682 = 26.82\%.$$

4. Annual deposits of $1,000 are made in an account that pays 4% compounded quarterly. How much money should be in the account immediately after the fifth deposit?

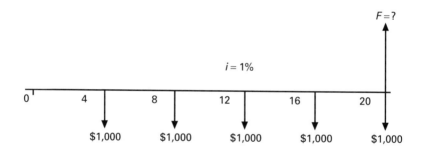

$$\text{Effective rate} = (F/P, 1\%, 4) - 1$$
$$= 4.06\%$$
$$F = A(F/A, 4.06\%, 5)$$
$$= \$1,000 \left[\frac{(1 + 0.0406)^5 - 1}{0.0406}\right] = \$5,423$$

or

$$A = \$1,000(A/F, 1\%, 4)$$
$$= \$1,000(0.2463) = \$246.30$$
$$F_{20} = \$246.30(F/A, 1\%, 20)$$
$$= 246.30(22.019) = \$5,423$$

An alternate solution method is to treat the five annual deposits as single sums of money. Therefore,

$$F = \$1{,}000[(F/P, 1\%, 16) + (F/P, 1\%, 12) + (F/P, 1\%, 8) + (F/P, 1\%, 4) + 1]$$
$$= \$1{,}000[1.1726 + 1.1268 + 1.0829 + 1.0406 + 1.000] = \$5{,}423.$$

5. Given the payments shown in the following cash flow diagram, what is the equivalent worth in 2005 with interest at 6%?

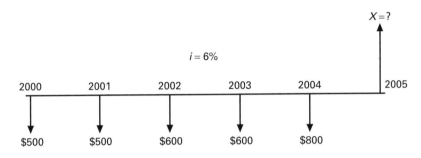

$$X = [\$500(F/A, 6\%, 5) + \$100(F/A, 6\%, 3) + \$200](F/P, 6\%, 1)$$
$$= [\$500(5.6371) + \$100(3.1836) + \$200](1.060) = \$3{,}537$$

6. With interest at 8% compounded annually, how long does it take for a certain amount to double in magnitude?

$$(F/P, 8\%, N) = 2.00$$

By inspection of 8% interest tables, $N = 9$ years.

7. An individual approaches the Loan Shark Agency for $1,000 to be repaid in 24 monthly installments. The agency advertises interest at $1\frac{1}{2}\%$ per month. They proceed to calculate the size of his payment in the following manner:

Amount requested:	$1,000
Credit investigation:	25
Credit risk insurance:	5
Total:	$1,030

Interest:	$(1030)(24)(0.015) = \$371$
Total owed:	$\$1{,}030 + \$371 = \$1{,}401$
Payment:	$\dfrac{\$1{,}401}{24} = \$58.50$

What effective interest rate is the individual paying?

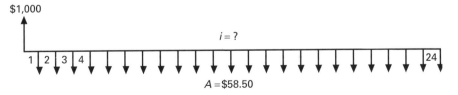

$$A = P(A/P, i\%, 24)$$
$$\$58.50 = \$1{,}000(A/P, i\%, 24)$$

By interpolation in tables, $i = 2.9\%$ per month, and

$$\text{Effective rate} = (F/P, 2.9\%, 12) - 1 = 1.41 - 1 = 41\%.$$

8. Money is to be invested for a child's college expenses. Annual deposits of $2,000 are made in a fund that pays 5% compounded annually. If the first deposit is made on the child's 5th birthday and the last on the child's 15th birthday, what is the size of 4 equal withdrawals on the child's 18th, 19th, 20th, and 21st birthdays that will just deplete the account?

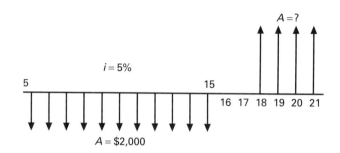

Amount in fund at $t = 15$:

$$F_{15} = A(F/A, 5\%, 11)$$
$$= \$2,000(14.2068) = \$28,414.$$

Amount in fund at $t = 17$:

$$F_{17} = P_{15}(F/P, 5\%, 2)$$
$$= \$28,414(1.1025) = \$31,326.$$

Amount of withdrawals:

$$A = P_{17}(A/P, 5\%, 4)$$
$$= \$31,326(0.2820) = \$8,834.$$

9. A college student borrows money in her senior year to buy a car. She defers payments for 6 months and makes 36 beginning-of-month payments thereafter. If the original note is for $12,000 and interest is $\frac{1}{2}\%$ per month on the unpaid balance, how much will her payments be?

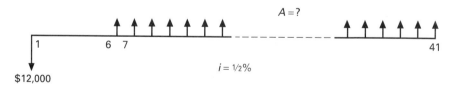

Amount owed at $t = 5$:

$$F_5 = P_0(F/P, \tfrac{1}{2}\%, 5)$$
$$= \$12,000(1.0253) = \$12,304.$$

Amount of monthly payments:

$$A = P_5(A/P, \tfrac{1}{2}\%, 36)$$

$$= \$12,304\left[\frac{i(1 + i)^N}{(1 + i)^N - 1}\right]$$

$$= \$12,304\left[\frac{(0.005)(1.005)^{36}}{(1.005)^{36} - 1}\right] = \$372.45.$$

10. What is the present worth of $100,000 ten years hence if interest is (a) 15% compounded annually? (b) 15% compounded continuously?

   a. $P = F(P/F, 15\%, 10) = \$100,000(0.2472) = \$24,720$

   b. $P = F(P/F, \underline{15\%}, 10) = \$100,000(0.2231) = \$22,310$

11. A firm spends $100,000 per year on materials, with the cost spread uniformly over each year. Annual rental payments for building and equipment total $300,000 per year, with the payments being made at the beginning of each year.

   Using an interest rate of 10% compounded continuously, what is the present worth equivalent for 10 years of activity?

$$P = \overline{A}(P/\overline{A}, \underline{10}\%, 10) + \left[A(P/A, \underline{10}\%, 9) + A\right]$$

$$= \$100,000(6.3212) + \$300,000(5.6425) + \$300,000 = \$2,624,870$$

or

$$P = \frac{\$100,000(e^{0.10} - 1)}{0.10}(P/A, \underline{10}\%, 10) + \left[\$300,000(5.6425) + \$300,000\right]$$

$$= \$105,170.92(6.0104) + \left[\$300,000(5.6425) + \$300,000\right] = \$2,624,869$$

PROBLEMS[4]

2-1. What is the smallest integer-valued annual compound interest rate that will result in an investment tripling in value in less than or equal to 12 years?

2-2. How much money today is equivalent to $50,000 in 10 years, with interest of 6% compounded annually?

2-3. How much money must be deposited today in a fund paying 6% compounded annually in order to accumulate $50,000 in 20 years?

2-4. How much money should be placed in a fund today and once every year thereafter to accumulate $50,000 in 5 years if money is worth 6% per year?

2-5. How many monthly payments are necessary to repay a loan of $5,000 with an interest rate of 1% per month and end-of-month payments of $225?

2-6. What annual interest rate makes $4,000 today equivalent to $9,000 six years from now?

2-7. If $125,000 is invested now at 8% compounded annually, how much will it be worth in 10 years?

2-8. If you want to accumulate $30,000 in 15 years, how much would you have to deposit today in an account paying 13.5% interest?

---

[4] Unless stated otherwise, assume discrete compounding and end-of-period payments.

2-9. If money is worth more than 0% to you, would you rather pay $10,000/year for 5 years or pay $5,000/year for 10 years?

2-10. Suppose you deposit $2,000 in an account at the end of 2002, $1,500 at the end of 2003, $1,000 at the end of 2004, $500 at the end of 2005, $1,000 at the end of 2006, $1,500 at the end of 2007, and $2,000 at the end of 2008. If the account pays interest of 12% compounded annually, how much will be in the fund at the end of 2009?

2-11. In 1970 first-class postage for a 1-ounce envelope was $0.06. In 1995 a first-class stamp for the same envelope cost $0.32. What compounded annual increase in the cost of first-class postage was experienced during this 25-year period?

2-12. What is the effective annual interest rate for 15% compounded (a) monthly? (b) quarterly? (c) semiannually? (d) continuously?

2-13. Suppose you borrow $15,000 at 10% annual interest and agree to pay it off with 10 equal annual payments, with the first payment occurring one year after receiving the $15,000. How much of the fifth payment will go toward reducing the principal?

2-14. Maria makes annual deposits of $2,000 in an investment fund. How much will be in the fund immediately after the fifteenth deposit if the fund pays 8% annual compound interest?

2-15. Roberta makes 10 equal annual deposits of $2,000 in an investment fund and then makes no additional deposits. If she withdraws all money in the fund 15 years after the final deposit, how much will she be able to withdraw if the fund pays 8% annual compound interest?

2-16. Paul makes annual deposits in a fund that pays 8% annual compound interest. If the first deposit totals $1,000 and he increases the size of his annual deposit by $100 each year thereafter, how much will be in the fund immediately after his twenty-fifth deposit?

2-17. Jaime makes annual deposits in a fund that pays 8% annual compound interest. If the first deposit totals $1,000 and each successive deposit is 5% larger than the previous deposit, how much will be in the fund immediately after his twenty-fifth deposit?

2-18. At 8% annual compound interest, what uniform series of deposits over a 10-year period would be equivalent to the 25 deposits in Problem 2-17? Assume the first cash flow in the 10-year series occurs at the same time at Jaime's $1,000 deposit.

2-19. At 8% annual interest, what uniform series of deposits over a 10-year period would be equivalent to the 25 deposits in Problem 2-16? Assume the first cash flow in the 10-year series occurs at the same time as Paul's $1,000 deposit.

2-20. Suppose you make a deposit of $1,000 in an account and each successive year increase the deposit by $100 over the previous year's deposit. If the account pays interest equal to 8% per year, how much will be in the account immediately after the twenty-fifth deposit?

2-21. Suppose you make a deposit of $1,000 in an account and each successive year you increase the deposit by 10% over the previous year's deposit. If the account pays 8% annual interest, how much will be in the account immediately after the twenty-fifth deposit?

2-22. Suppose you make a deposit of $10,000 in an account and each successive year the deposit is $250 less than the previous year's deposit. If the account pays 8% annual interest, how much will be in the account immediately after the twenty-fifth deposit?

2-23. Suppose you make a deposit of $10,000 in an account and each successive year the deposit is 8% more than the previous year's deposit. If the account pays 12% annual interest, how much will be in the account immediately after the twenty-fifth deposit?

2-24. If you make annual deposits of $2,500 in an account that pays annual compound interest of 4% the first 2 years, 5% the next 2 years, and 6% the following 2 years, how much will be in the account immediately after the sixth deposit?

2-25. On January 1, 1995, a person's retirement fund was worth $200,000. Every month thereafter this person made a cash contribution of $676 to the fund. If the fund is worth $400,000 on January 1, 2000, what annual rate of return was earned on this fund?

2-26. An individual purchased a refrigerator for $1,000. The store financed the refrigerator by charging 1% monthly interest on the unpaid balance. If the refrigerator is paid for with 30 equal end-of-month payments, what will the monthly payments be? If the first monthly payment is not made until 6 months after the purchase, what will the monthly payments be?

2-27. What interest rate makes $1,000 today equivalent to five annual amounts of $400, with the first $400 cash flow occurring one year in the future?

2-28. How much money today is equivalent to $100,000 in 20 years with interest of 8% compounded (a) monthly? (b) quarterly? (c) semiannually? (d) continuously?

2-29. What is the effective interest rate for 8% compounded (a) monthly? (b) quarterly? (c) semiannually? (d) continuously?

2-30. An individual borrows $10,000 at 10% compounded annually. Equal annual payments are to be made for 10 years. However, at the time of the fourth payment, the individual elects to pay off the loan. How much should be paid?

2-31. Over a 5-year period, a firm spends $20,000 per year for supplies with the cost occurring continuously and uniformly over each year. Using an interest rate of 8% compounded continuously, determine the present sum of money that is equivalent to the expenditures.

2-32. An individual receives an annual bonus and deposits it in a savings account paying interest at a rate of 6% compounded annually. The size of the annual bonus increases at a rate of 8% per year. How much money will be in the account immediately after the tenth deposit if the first deposit totals $500?

2-33. An individual borrows $10,000 and wishes to repay it with ten annual payments. Each payment is to be $250 greater than the previous payment. With an interest rate of 10% compounded annually, what will the last payment be?

2-34. A person deposits $1,000 in an account each year for 5 years; at the end of 5 years, one-half of the account balance is withdrawn; $2,000 is deposited annually for 5 more years, with the total balance withdrawn at the end of the fifteenth year. If the account earns interest at an annual rate of 7%, what is the amount of withdrawal?

2-35. Given the cash flow diagram shown below and an interest rate of 10% per period, solve for the value of an equivalent amount at (a) $t = 5$, (b) $t = 12$, and (c) $t = 15$. (*Note:* Upward arrows denote cash inflows and downward arrows denote cash outflows.)

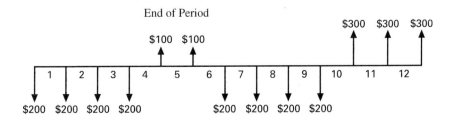

2-36. Consider the following cash flow profile:

EOY	Cash flow	EOY	Cash flow	EOY	Cash flow
0	−$25,000	3	+$ 8,000	6	+$11,000
1	+ 8,000	4	+ 8,000	7	+ 11,000
2	+ 8,000	5	+ 11,000	8	+ 11,000

With a 10% annual interest rate, what single sum of money at the end of the fifth year will be equivalent to the cash flow series?

2-37. Consider the cash flow profile in Problem 2-36. What uniform series over the interval [1,8] will be equivalent to the cash flow profile if money is worth 10%/year?

2-38. Revenues from a certain project are expected to be $50,000 at the end of the first year and decrease at the rate of $2,000 per year through the 15-year life of the project. If interest is 15% per year, what is the equivalent (a) present worth and (b) annual worth for the project?

2-39. What is the present worth of operating expenditures of $100,000 per year which are assumed to be incurred continuously throughout a 10-year period if the effective annual rate of interest is 15%?

2-40. What is the equivalent present worth of $1,000 obligations at the beginning of each year for 5 years with interest of 9.53% nominal, compounded continuously, i.e., an effective rate of 10%?

2-41. A person deposits $1,000 annually into an account that pays interest at an annual effective rate of 6% the first year after the initial deposit. The annual rate increases by 0.25% each year. If ten annual deposits are made, what will be the worth of the fund immediately after the last deposit?

2-42. An individual borrows $20,000 and repays the loan with three equal annual payments. The interest rate for the first year of the loan is 6% compounded annually, for the second year of the loan is 8% compounded annually, and for the third year of the loan is 10% compounded annually.

  a. Determine the size of the equal annual payments.

  b. Compare the result from (a) with that which results from reversing the interest rates for the first and third years.

2-43. An individual deposits $5,000 in a savings account each year for 5 years. The first year after the initial deposit the account pays interest at a rate of 5% compounded annually; the second year after the initial deposit the interest rate is 5% compounded continuously; the third year the interest rate is 6% compounded annually; an interest rate of 6% compounded semiannually exists the fourth year; and 6% compounded continuously is the interest rate the fifth year. How much money is accumulated in the account one year after the fifth deposit?

2-44. A firm purchases a unit of equipment for P, keeps it for N years, and sells it for F. With an interest rate of $i\%$, show that the equivalent uniform annual cost for the transaction is given by either

$$P(A/P, i\%, N) - F(A/F, i\%, N)$$

or

$$(P - F)(A/P, i\%, N) + Fi$$

or

$$(P - F)(A/F, i\%, N) + Pi.$$

2-45. Consider a cash flow series $\{A_1, \ldots, A_N\}$ consisting of N discrete cash flows. Suppose $A_k = pA_{k-1}$ when $0 < p < 1$. Develop a closed-form expression for the present sum equivalent based on an interest rate of $i\%$ per period.

2-46. Consider a cash flow series $\{A_1, \ldots, A_N\}$ consisting of N discrete cash flows. Suppose $A_k = kC$, where C is a positive constant. Develop a closed-form expression for the present sum equivalent based on an interest rate of $i\%$ per period.

2-47. Consider a cash flow series $\{A_1, \ldots, A_N\}$ consisting of N discrete cash flows. Suppose $A_k = A_1 e^{(k-1)C}$ where $0 < C < 1$. Develop a closed-form expression for the future sum equivalent based on an interest rate of $r\%$ compounded continuously.

2-48. In Problem 2-47 suppose $A_k = k(1 + j)^{k-1}A_1$. Develop a closed-form expression for the future sum equivalent based on an interest rate of $i\%$ per period.

2-49. Consider a cash flow series $\{A_1, A_2, \ldots\}$ consisting of an infinite number of discrete cash flows. Suppose $A_k = A_1(1 + j)^{k-1}/k!$, where $!$ denotes the factorial operation. Develop a closed-form expression for the present sum equivalent based on an interest rate of $i\%$ per period.

2-50. If $\$P$ is borrowed and repaid with N equal annual end-of-period payments of $\$A$, based on an interest rate of $i\%$, then show the amount of payment k that is an interest payment is given by

$$I_k = A[1 - (P/F, i\%, N - k + 1)],$$

and the amount of payment k that is a payment against the principal amount borrowed is given by

$$E_k = A(P/F, i\%, N - k + 1).$$

3
Equivalent Worth Methods for Comparing Alternatives

In comparing investment alternatives a number of different measures of economic effectiveness are often used. This and subsequent chapters consider various methods for studying the economic desirability of an individual project and for comparing the relative economic desirabilities of two or more projects. The methods are as follows:

1. present worth;
2. annual worth;
3. future worth;
4. rate of return;
5. benefit–cost ratio.

The first three methods are treated in this chapter, rate of return methods are presented in Chapter 4, and benefit–cost methods are included in Chapter 7. All of the methods are based on the concept of the time value of money, described in Chapter 2.

3.1 Measures of Economic Effectiveness

The various measures of economic effectiveness that will be used in this text are defined as follows:

Present worth: A determination of the present worth (PW) involves the conversion of each individual cash flow to its present worth equivalent and the summation of the individual present worths to obtain the net present worth.

Annual worth: The annual worth (AW) is determined by converting all cash flows to an equivalent uniform annual series of cash flows.

Future worth: The future worth (FW) is obtained by converting each individual cash flow to its future worth equivalent and determining the net future worth for the project.

Rate of return: Among the many definitions of rate of return, the most popular definition is the interest rate that yields a net present worth of zero; such a rate of return is referred to as the *internal rate of return* (IRR).

Benefit–cost ratio: There are several definitions of the benefit–cost ratio (B/C), but, in general, it can be defined as the ratio of the equivalent worth of benefits to the equivalent worth of costs.

3.2 Defining Investment Alternatives

In defining investment *alternatives,* it is important to ensure that the set of alternatives is mutually exclusive: the term "mutually exclusive" signifies that an "either–or, but not both" situation exists. That is, the choice of one excludes the choice of any other. Investment *opportunities* will denote projects, proposals, and other options available for investment. Distinct combinations of investment opportunities will be used to define the investment alternatives.

As an illustration, suppose two investment opportunities, A and B, are available. In this case, four mutually exclusive alternatives can be formed: neither A nor B, A only, B only, and both A and B. In general, if there are m investment opportunities, then 2^m investment alternatives can be formed. Of course, not all of the 2^m alternatives are necessarily feasible, because there are budget limitations, dependencies among investment opportunities, and other restrictions.

Example 3-1

Two industrial trucks, A and B, are being considered by the warehouse manager. In addition, two industrial truck attachments, C and D, are being considered. Opportunities A and B are mutually exclusive; opportunities C and D are mutually exclusive; and the purchase of an attachment is *contingent* upon the purchase of one of the trucks. The sixteen possible alternatives are

Do nothing	D only*	B and C	A, B, and D*
A only	A and B*	B and D	A, C, and D*
B only	A and C	C and D*	B, C, and D*
C only*	A and D	A, B, and C*	A, B, C, and D*

*Denotes an infeasible alternative.

Nine alternatives are infeasible because of the mutually exclusive and contingent dependencies among the investment opportunities. Thus, only seven alternatives (sometimes descriptively called *mutually exclusive combinations*) must be considered.   ■

3.3 A Systematic Procedure for Comparing Investment Alternatives

A systematic procedure for comparing investment alternatives can be stated as follows:

1. define the alternatives;
2. determine the study period;
3. provide estimates of the cash flows for each alternative;
4. specify the time value of money or interest rate;
5. select the measure(s) of effectiveness;
6. compare the alternatives;
7. perform sensitivity analyses;
8. select the preferred alternative.

As noted in the previous sections, one may wish to aggregate investment opportunities in various combinations to arrive at the alternatives to be compared. As an illustration, suppose exactly one tractor-trailer combination is to be purchased. There might exist two tractor alternatives (A and B) and two trailer alternatives (C and D). Hence, there can be formed four tractor-trailer alternatives (AC, AD, BC, BD). Additionally, the do-nothing alternative (maintain the status quo) should not be overlooked.

The study period defines the period of time over which the analysis is to be performed. Cash flows that occur prior to and after the study period are not considered, except as they might influence cash flows during the study period. The study period may or may not be the same as the useful lives for equipment involved. If the study period is less than the useful life of an asset, then an estimate should be provided of its salvage value at the end of the study period; if the study period is longer than the useful life of an asset, then estimates of cash flows should be provided for subsequent replacements for the asset.

Estimates of the cash flows should be provided for each alternative by using the approaches described in Chapter 5. Even though an asset may continue to be used for a period of time beyond the study period, an estimate of its salvage value should be provided assuming it will be disposed of at the end of the study period.

The interest rate or discount rate to be used is the minimum attractive rate for the firm. The *minimum attractive rate of return (MARR)*, treated in detail in Chapter 9, is defined to be the return that could be earned by investing elsewhere (opportunity cost concept).

As mentioned previously, a number of different measures of economic effectiveness can be selected. All of the methods we consider will yield the same choices. The recommended basis for selecting a particular measure of economic effectiveness is communication. Namely, the method that is best understood by management should be used so long as it yields recommendations consistent with those based on correct time value of money analyses. We will have more to say about this matter at the end of Chapter 4.

When the measure of economic effectiveness is used, the alternatives should be compared and the most economic alternative identified. In the case of present worth, annual worth, and future worth comparisons, one can rank the alternatives on the basis of their worths. When rate of return and benefit–cost ratio comparisons are used, an incremental approach is used.

Since the comparison of investment alternatives involves the use of estimates of future economic conditions, errors can occur. Consequently, it is worthwhile to consider the consequences of such errors on the decision to be made. Sensitivity analyses can be performed to determine the effect of estimation errors on the economic performance of each alternative. Methods for performing sensitivity analyses are presented in Chapters 11 and 12.

The final selection of the preferred alternative is complicated by the presence of nonmonetary factors, multiple objectives, and risk and uncertainties concerning future outcomes. Methods for treating such conditions are presented in Chapters 13, 14, 19, and 20.

3.4 Judging the Economic Worth of Investment Opportunities

The measures of economic effectiveness can be used to judge the economic worth of individual investment opportunities. In this section, it will be assumed that each investment opportunity is independent of other opportunities that may be under consideration. Furthermore, it is assumed that there are no restrictions on the number of such investment opportunities to be undertaken. The decision is not which of several opportunities is best, but rather, is an individual opportunity a worthwhile investment. We define an opportunity as being worthwhile if it has either a nonnegative present worth, annual worth, or future worth, a rate of return at least equal to the minimum attractive rate of return, or a benefit–cost ratio at least equal to one.

Example 3-2
Given the following investment opportunities, and using each of the five methods given in Section 3.1, determine which are worthwhile investment opportunities. A minimum attractive rate of return of 10% is to be used in the analysis.

Opportunity	Investment (P)	Study period (N)	Salvage value (S)	Net annual cash flow (A)
A	$10,000	5 yr	$10,000	+$2,000
B	12,000	5 yr	0	+ 3,000
C	15,000	5 yr	0	+ 4,167

Solution By using methods to be described subsequently, the following values are obtained:

Opportunity	Present worth	Annual worth	Future worth	Rate of return	Conventional benefit–cost ratio
A	+$3,790	+$1,000	+$6,106	20%	2.0000
B	− 630	− 170	− 1,015	8%	0.9475
C	+ 850	+ 217	+ 1,369	12%	1.0567

Thus, investments A and C (in that order) are deemed worthwhile.   ■

3.5 Present Worth (PW) Method

The term *present worth* (*PW*) means an amount at some beginning or base time that is equivalent to a particular schedule of receipts and/or disbursements under consideration. If disbursements only are considered, the term can be best expressed as *present worth-cost.*

3.5.1 Study Period in Comparisons of Alternative Projects

In comparing alternatives by the present worth (and future worth) method, it is essential that all alternatives be considered over the same length of time. If the alternatives all have the same expected life, there is no problem, for that life can be used. When the alternatives have different expected lives, it is common to use a study period equal to the lowest common multiple of the lives, or the length of time during which the services of the chosen alternative will be needed, whichever is less. For example, if two alternatives have expected lives of 3 and 4 years, respectively, then the lowest common multiple of the lives to use as a study period is 12 years. However, if the service for which the alternatives are being compared is expected to be needed for only 9 years, then 9 years should be the study period used.

3.5.2 Comparing Alternatives Using Present Worth Analysis When Receipts and Disbursements Are Known

When receipts (cash inflow) as well as disbursements (cash outflow) for more than one mutually exclusive project are known, the project with the highest net present worth should be chosen, as long as that present worth is greater than or equal to zero. As an example, consider two alternative lathes A and B, only one of which should be selected, if either.

Example 3-3

	Lathe	
	A	B
First cost:	$10,000	$15,000
Life:	5 yr	10 yr
Salvage value:	$2,000	$0
Annual receipts:	$5,000	$7,000
Annual disbursements:	$2,200	$4,000
Minimum attractive rate of return = 8%		
Study period = 10 yr		

Solution   The lowest common multiple of the lives is 10 years. Assuming that the service will be needed for at least that long and that what is estimated to happen in the first 5 years for project A will be repeated in the second 5 years, the solution is shown in the following table.

	Lathe	
	A	B
Annual receipts = $5,000(P/A, 8%, 10):	$33,551	
7,000(P/A, 8%, 10):		$46,970
Salvage value at year 10 = $2,000(P/F, 8%, 10):	926	
Total PW of cash inflow:	$34,477	$46,970
Annual disbursements = $2,200(P/A, 8%, 10):	−$14,762	
4,000(P/A, 8%, 10):		−$26,840
First cost:	− 10,000	− 15,000
Replacement = ($10,000 − $2,000)(P/F, 8%, 5):	− 5,445	
Total PW of cash outflow:	−$30,207	−$41,840
Net PW:	$4,270	$5,130

Thus, project B, having the higher net present worth greater than zero, is the better economic choice. ∎

3.5.3 Comparing Alternatives Using Present Worth Analysis When Receipts Are Constant or Not Known

When alternatives that perform essentially identical services involve only known cash outflows, it is possible to compare the alternatives on the basis of present worth-cost (PW-cost). The method of study is the same as illustrated for Example 3-3 except, of course, the alternative with the lowest present worth-cost is best. As an example, consider the following situation involving two compressors, each of which will do the desired job but differ as shown.

Example 3-4

	Compressor	
	I	II
First cost:	$6,000	$8,000
Life:	6 yr	9 yr
Salvage value:	$1,000	$0
Annual operating disbursements:	$4,000	$3,200
Minimum return on investment = 15%		
Study period = 18 yr		

Solution Again, assuming a study period equal to the lowest common multiple of lives, i.e., 18 years:

	I	II
First cost:	$6,000	$8,000
First replacement = ($6,000 − $1,000)(P/F, 15%, 6):	2,160	
$8,000(P/F, 15%, 9):		2,276
Second replacement = ($6,000 − $1,000)(P/F, 15%, 12):	936	
Operating disbursements = $4,000(P/A, 15%, 18):	24,450	
$3,200(P/A, 15%, 18):		19,600
Less Salvage value (yr. 18) = $1,000(P/F, 15%, 18):	− 82	
Net PW−cost:	$33,464	$29,876

Since compressor II has the lower net present worth-cost, it is the better economic choice. ■

3.6 Annual Worth (AW) Method

The term *annual worth* (*AW*) means a uniform annual series of money for a certain period of time that is equivalent in amount to a particular schedule of receipts and/or disbursements under consideration. If disbursements only are considered, the term is usually expressed as annual cost (AC) or equivalent uniform annual cost (EUAC).

3.6.1 Calculation of Capital Recovery Cost

The *capital recovery cost* (*CR*) for a project is the equivalent uniform annual cost of the capital invested. It is an annual amount that covers the following two items:

1. depreciation (loss in value of the asset);

2. interest (minimum attractive rate of return) on invested capital.

As an example, consider a machine or other asset having an investment cost (*P*) of $10,000, a life (*N*) of 5 years, and then have a salvage value (*S*) of $2,000. Further, the interest on invested capital, *i*, is 8%.

It can be shown that no matter which method of calculating depreciation is used, the equivalent annual cost of the capital recovery is the same. For example, if straight-line depreciation is used, the equivalent annual cost of interest is calculated to be $564, as shown in Table 3-1. The annual depreciation cost (*D*) by the straight-line method is $D = (P - S)/N = (\$10,000 - \$2,000)/5 = \$1,600$. The $564 added to $1,600 results in a calculated capital recovery cost of $2,164.

[margin handwritten notes: Annual Depreciation Cost — Straight Line Depreciation $D = (P-S)/N$]

TABLE 3-1   Calculation of Equivalent Annual Cost of Interest Assuming Straight-Line Depreciation

Year	Investment at beginning of year	Interest on beginning-of-year investment @ 8%	Present worth of interest @ 8%
1	$10,000	$800	$800(P/F, 8\%, 1) = \$ \ 741$
2	8,400	672	$672(P/F, 8\%, 2) = \ 576$
3	6,800	544	$544(P/F, 8\%, 3) = \ 434$
4	5,200	416	$416(P/F, 8\%, 4) = \ 306$
5	3,600	288	$288(P/F, 8\%, 5) = \underline{\ 196}$
			Total: $2,253

Annual equivalent cost of interest $= \$2,253(A/P, 8\%, 5) = \$564$

There are several convenient formulas by which capital recovery cost may be calculated in order to obtain the same answer as above. The most apparent formula is (using the same figures as for the machine above)

[margin handwritten note: CAPITAL RECOVERY COST]

$$CR = \overset{inv\ cost}{P}(A/P, i\%, N) - \overset{Salvage}{S}(A/F, i\%, N)$$
$$= \$10,000(A/P, 8\%, 5) - \$2,000(A/F, 8\%, 5)$$
$$= \$10,000(0.2505) - \$2,000(0.1705) = \$2,164.$$

Two other convenient formulas for calculating the capital recovery cost are

$$CR = P(i\%) + (P - S)(A/F, i\%, N)$$
$$= P(0.08) + (P - S)(A/F, 8\%, 5)$$
$$= \$10,000(0.08) + \$8,000(0.1705) = \$2,164$$

and

$$CR = (P - S)(A/P, i\%, N) + S(i\%)$$
$$= (P - S)(A/P, 8\%, 5) + S(0.08)$$
$$= \$8,000(0.2505) + \$2,000(0.08) = \$2,164.$$

The last of the above formulas will be used primarily for the calculation of captial recovery cost throughout the rest of this book.

3.6.2 Comparing Alternatives Using Annual Worth Analysis When Receipts and Disbursements Are Known

When receipts as well as disbursements figures are known for more than one mutually exclusive project, the project that has the highest net annual worth should be ✓ chosen, as long as that net annual worth is greater than or equal zero.

As an example, consider the following two alternative lathes treated previously.

Example 3-5

	Lathe	
	A	B
First cost:	$10,000	$15,000
Life:	5 yr	10 yr
Salvage value:	$2,000	$0
Annual receipts:	$5,000	$7,000
Annual disbursements:	$2,200	$4,000
Minimum attractive rate of return = 8%		
Study period = 10 yr		

Solution

	Lathe	
	A	B
Annual receipts:	$5,000	$7,000
Annual disbursements:	− 2,200	− 4,000
CR cost: = $8,000(A/P, 8%, 5)		
+ $2,000(8%):	− 2,164	$15,000(A/P, 8%, 10): − 2,235
Net AW:	$ 636	$ 765

Thus project B, having the higher net annual worth, is the better economic choice. ∎

3.6.3 Comparing Alternatives Using Annual Worth Analysis When Receipts Are Constant or Not Known

Very often alternative projects are expected to perform almost identical functions so that each results in the same receipts, savings, or benefits. Sometimes the savings or benefits are intangible or cannot be estimated; hence, the alternatives are judged on the basis of negative net annual worth, or annual cost. As an example, consider the same alternative compressors as were compared earlier using the PW method:

Example 3-6

	Compressor	
	I	II
First cost:	$6,000	$8,000
Life:	6 yr	9 yr
Salvage value:	$1,000	$0
Annual operating disbursements:	$4,000	$3,200
Minimum return on investment = 15%		
Study period = 18 yr		

Solution

	Compressor	
	I	II
CR cost = ($6,000 − $1,000)(A/P, 15%, 6) + $1,000(15%):	− $1,470	
CR cost = $8,000(A/P, 15%, 9):		− $1,676
Annual operating disbursements:	− 4,000	− 3,200
Net AW:	− $5,470	− $4,876

Thus, compressor II, having the lower annual cost (least negative annual worth), is apparently the more economical choice. It should be noted that this analysis of competing alternatives with different lives makes certain assumptions that will be discussed subsequently. ∎

3.7 Future Worth (FW) Method

The term *future worth* (*FW*) means an amount at some ending or termination time that is equivalent to a particular schedule of receipts and/or disbursements under consideration. If disbursements only are considered, the term can best be expressed as *future worth-cost* (FW − cost) or *future cost.* The future worth is also referred to as the *terminal worth.*

3.7.1 Comparing Alternatives Using Future Worth Analysis When Receipts and Disbursements Are Known

When receipt as well as disbursement figures are known for more than one ✓ mutually exclusive project, the project that has the highest net future worth should be chosen, as long as that net future worth is greater than or equal zero.

As an example, consider the same two alternative lathes treated previously:

Example 3-7

	Lathe	
	A	B
First cost:	$10,000	$15,000
Life:	5 yr	10 yr
Salvage value:	$2,000	$0
Annual receipts:	$5,000	$7,000
Annual disbursements:	$2,200	$4,000
Minimum attractive rate of return = 8%		
Study period = 10 yr		

Solution

	Lathe	
	A	B
Annual receipts = $5,000($F/A$, 8%, 10):	$72,433	
$7,000($F/A$, 8%, 10):		$101,406
Salvage value at year 10:	2,000	
Total FW of cash inflow:	$74,433	$101,406
Annual disbursements = $2,200($F/A$, 8%, 10):	− 31,871	
= $4,000($F/A$, 8%, 10):		− 57,946
First cost = $10,000($F/P$, 8%, 10):	− 21,589	
= $15,000($F/P$, 8%, 10):		− 32,384
Replacement = ($10,000 − $2,000)($F/P$, 8%, 5):	− 11,754	
Total FW of cash inflow:	−$65,214	−$90,330
Net FW:	$9,219	$11,076

Thus project B, having the higher net future worth, is the better economic choice. ■

3.7.2 Comparing Alternatives Using Future Worth Analysis When Receipts Are Constant or Not Known

As noted for present worth and annual worth analyses, alternative projects can be expected to perform identical functions such that only known cash outflows are given. It is possible to compare the alternatives on the basis of future worth-cost. The method of study is the same as illustrated for the preceding example; however, the alternative with the lowest future worth–cost is best. As an example, consider the same two compressors treated earlier.

Example 3-8

	Compressor	
	I	II
First cost:	$6,000	$8,000
Life:	6 yr	9 yr
Salvage value:	$1,000	$0
Annual operating disbursements:	$4,000	$3,200
Minimum attractive rate of return = 15%		
Study period = 18 yr		

Solution   It is assumed that replacements have identical cash flows.

	Future worth-cost compressor	
	I	II
First cost = $6,000(F/P, 15%, 18):	$ 74,254	
$8,000(F/P, 15%, 18):		$ 99,004
First replacement = ($6,000 − $1,000)(F/P, 15%, 12):	26,750	
$8,000(F/P, 15%, 9):		28,144
Second replacement = ($6,000 − $1,000)(F/P, 15%, 6):	11,566	
Operating disbursements = $4,000(F/A, 15%, 18):	303,346	
$3,200(F/A, 15%, 18):		242,676
Less salvage value @ yr 18:	−$ 1,000	
Net FW-cost:	$414,916	$369,824

Since compressor II has the lower net future worth–cost, it is the better economic choice. ■

3.8 Assumptions in Comparisons of Alternatives with Different Lives

In the comparison of projects A and B and also the comparison of compressors I and II, the alternatives compared had different expected lives. The solutions as shown are fully valid only if the following assumptions are reasonable:

1. The period of needed service for which the alternatives are being compared (study period) is either indefinitely long or a length of time equal to a common multiple of the lives of the alternatives.

   Note: The student should recognize that any point in time equal to a common multiple of lives would be a point at which each alternative would have just exhausted a life cycle.

2. What is estimated to happen in the first life cycle will happen in all succeeding life cycles, if any, for each alternative.

These assumptions are commonly made in economic analyses by default; i.e., they are made because there is no good basis for estimates to the contrary. The assumptions are sometimes referred to as the *repeatability* assumptions. They are implicitly contained in all examples and problems illustrating all methods of economic evaluations herein unless there is a statement to the contrary.

Applied to the earlier examples in which the compressors were compared, the assumptions mean the following:

1. The services of a compressor will be needed either 18 years, 36 years, 54 years, etc., or indefinitely.
2. When either compressor I or II is replaced at the end of a life cycle, it will be replaced with a compressor having characteristics affecting cost (i.e., first cost, life, salvage value, and annual operating disbursements) identical to the estimates used for the first life cycle.

Whenever alternatives to be compared have different lives and one or both of the conditions are not appropriate, then it is necessary to enumerate what receipts and what expenses are expected to occur at what points in time for each alternative for as long as service will be needed or the irregularity is expected to exist (i.e., the study period). This is sometimes referred to as the *coterminated* assumption. Various assumptions can be used to adjust the terminal value of each alternative at the cotermination time. For instance, if an alternative has a useful life shorter than the study period, the estimated annual cost of contracting for the services involved might be utilized during the remaining years. Similarly, if the useful life is longer than the study period, an estimated salvage value is customarily employed to serve as a terminal cash flow at the project's coterminated life. The enumerated information can then be converted into an equivalent PW, AW, FW, or other measure of merit by ordinary time value of money computations.

Example 3.9

Suppose that for the earlier compressor illustration it is expected that the standard assumptions are not met as follows: (a) a compressor is needed for only 12 years; (b) the replacement for compressor II is expected to cost $14,000 rather than $8,000, and its salvage value after 3 years' service (end of the twelfth year of study) is expected to be $400. Compare the two compressors by the annual cost method.

Solution   The annual cost (AC) for compressor I remains the same at $5,470. For compressor II, the cash flow diagram and solution are

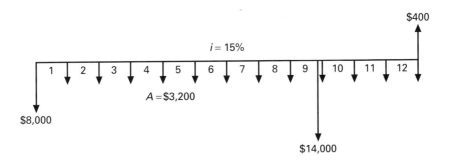

$$AC = [\$8,000 + \$14,000(P/F, 15\%, 9) - \$400(P/F, 15\%, 12)](A/P, 15\%, 12)$$
$$+ \$3,200$$
$$= [\$8,000 + \$14,000(0.2843) - \$400(0.1869)](0.1845) + \$3,200$$
$$= \$5,396$$

Thus, under the changed conditions, compressor II is still less costly, but by much less margin.

The easiest way to calculate the present worths (at time 0) and the future worths (at time 12) for comparison of the alternatives, given the above annual costs, is as follows:

Alternative	PW-cost = AC$(P/A,15\%,12)$
I	$\$5,470(5.4206) = \$29,650$
II	$\$5,396(5.4206) = \$29,250$

Alternative	FW-cost = AC$(F/A,15\%,12)$
I	$\$5,470(29.0017) = \$158,640$
II	$\$5,396(29.0017) = \$156,494$

3.9 Relationship of Various Analysis Methods

All methods of economic evaluation considered to this point have the reassuring property of providing consistent results regarding the economic desirability or relative ranking of projects compared. In fact, it can be shown that the annual worths, present worths, and future worths for any projects under consideration are linearly proportional to each other.

Example 3-10

Show the consistency of economic comparison results for compressors I and II by the various methods given in this chapter.

    Solution

$$\frac{PW_I}{PW_{II}} = \frac{AW_I}{AW_{II}} = \frac{FW_I}{FW_{II}}; \quad \frac{\$33,464}{\$29,876} = \frac{\$5,470}{\$4,876} = \frac{\$414,916}{\$369,824} = 1.12$$

PROBLEMS

3-1. Based on a MARR of 15%, recommend which of the five independent investment opportunities to undertake. Use the (a) PW method, (b) AW method, (c) FW method.

	Investment Opportunity				
	A	B	C	D	E
Initial investment:	$8,000	$10,000	$12,000	$15,000	$16,000
Annual net receipts:	2,000	3,000	3,200	3,500	4,000
Salvage value:	− 1,000	2,000	1,000	2,000	0
Study period = 5 yr					

3-2. In Problem 3-1, suppose A through E constitute a set of feasible, mutually exclusive investment alternatives. Which would be chosen using (a) the PW method? (b) the AW method? (c) the FW method?

3-3. Using a MARR of 12% and a study period of 10 years, determine which of the following independent (nonrepeating) investment opportunities are economically worthwhile. Make your determination using (a) the PW method, (b) the AW method, (c) the FW method.

			Cash Flows (CF) for Investment Opportunity			
EOY	A	B	C	D	E	F
0	−$5,000	−$5,000	−$8,000	−$12,000	0	−$15,000
1	− 5,000	500	1,000	1,500	0	400
2	1,000	1,000	1,000	1,500	0	800
3	1,000	1,500	1,000	1,500	−$6,000	1,200
4	2,000	2,000	1,000	1,500	2,000	1,600
5	2,000	2,500	1,000	1,500	2,000	2,000
6	3,000	3,000	1,000	1,500	2,000	2,400
7	3,000	0	1,000	1,500	2,000	2,800
8	4,000	0	1,000	1,500	2,000	3,200
9	4,000	0	1,000	1,500	0	3,600
10	− 5,000	0	− 2,000	3,000	0	4,000

3-4. In Problem 3-3, suppose A through F constitute a set of feasible, mutually exclusive investment alternatives. Which would be chosen using (a) the PW method? (b) the AW method? (c) the FW method?

3-5. If money is worth 8% to you, which (if any) of the following investment opportunities would you favor? Why? Use the PW method in making your recommendation.

EOY	CF(A)	CF(B)	CF(C)
1	$1,000	$5,000	$3,000
2	2,000	4,000	3,000
3	3,000	3,000	3,000
4	4,000	2,000	3,000
5	5,000	1,000	3,000

3-6. In Problem 3-5, suppose the three cash flow profiles are for feasible, mutually exclusive investment alternatives and one must be chosen. Further, suppose your time value of money is 15%. Which would you choose?

3-7. Three investment proposals are being considered: A, B, and C. C is contingent on either A or B. How many feasible, mutually exclusive investment alternatives can be formed for this situation?

3-8. Four investment proposals are being considered: A, B, C, and D. A and C are mutually exclusive proposals; C is contingent on B; and D is contingent on A. How many feasible, mutually exclusive investment alternatives can be formed for this situation?

3-9. Five investment proposals are being considered: A, B, C, D, and E. A and B are mutually exclusive proposals; C is contingent on A; D is contingent on either B or C; E is contingent on either D or A; and the "do nothing" alternative is not feasible. How many feasible, mutually exclusive investment alternatives can be formed for this situation?

3-10. Five investment proposals are being considered: A, B, C, D, and E. A and B are mutually exclusive proposals. C is contingent on A; D is contingent on B; and E is contingent on either A or B. How many feasible, mutually exclusive investment alternatives can be formed?

3-11. A choice between two machines must be made. It is estimated that Machine Q will have an initial cost of $1,500,000; first-year annual operating and maintenance costs of $150,000, increasing by 15% annually over the 5-year planning horizon; and a negligible salvage value at the end of the planning horizon. Machine R is estimated to have an initial cost of $1,750,000; $75,000 first-year operating and maintenance costs, increasing by 15% per year over the 5-year planning horizon; and a negligible salvage value at the end of the planning horizon. With a MARR of 10%, compare the two machines using (a) the PW method, (b) the AW method, (c) the FW method.

3-12. Three mutually exclusive investment alternatives are being considered, including the "do nothing" alternative. The cash flows (CF) are shown for a 5-year planning horizon. The MARR is 25%. Compare the alternatives using (a) the PW method, (b) the AW method, (c) the FW method.

EOY	CF(A)	CF(B)	CF(C)
0	$0	−$100,000	−$150,000
1	0	+ 25,000	+ 50,000
2	0	+ 50,000	+ 75,000
3	0	+ 75,000	+ 100,000
4	0	+ 100,000	+ 125,000
5	0	+ 125,000	+ 150,000

3-13. Three mutually exclusive investment alternatives are being considered; the "do nothing" alternative is not feasible. The cash flows (CF) are shown for a 10-year planning horizon. The MARR is 25%. Determine which is best using (a) the PW method, (b) the AW method, (c) the FW method.

EOY	CF(A)	CF(B)	CF(C)
0	−$ 50,000	−$100,000	−$200,000
1	− 100,000	− 100,000	+ 50,000
2–5	+ 50,000	+ 70,000	+ 50,000
6–9	+ 50,000	+ 70,000	+ 75,000
10	+ 75,000	+ 70,000	+ 75,000

3-14. Three mutually exclusive investment alternatives are being considered, including the "do nothing" alternative. The cash flows (CF) are shown below over a 7-year planning horizon. The MARR is 12%. Determine which is best using (a) the PW method, (b) the AW method, (c) the FW method.

EOY	CF(A)	CF(B)	CF(C)
0	$0	−$100,000	−$180,000
1	0	− 100,000	− 20,000
2	0	+ 0	+ 100,000
3	0	+ 25,000	+ 75,000
4	0	+ 50,000	+ 50,000
5	0	+ 75,000	+ 25,000
6	0	+ 100,000	+ 0
7	0	+ 125,000	+ 100,000

3-15. Three mutually exclusive investment alternatives are being considered; the "do nothing" alternative is not feasible. The cash flows (CF) are shown for a 7-year planning horizon. Based on a MARR of 20%, determine which is best using (a) the PW method, (b) the AW method, (c) the FW method.

EOY	(CF)A	CF(B)	CF(C)
0	−$ 50,000	−$100,000	−$200,000
1	− 100,000	− 100,000	+ 75,000
2–7	+ 50,000	+ 70,000	+ 75,000

3-16. Three order-picking systems are being considered for use in a catalog distribution center. Alternative A involves picking items manually from bin shelving. With alternative B, items are picked manually from a horizontal carousel conveyor; hence, essentially no walking is required by the order picker. With alternative C, items are picked automatically by an automatic item retrieval machine. The alternatives have different space and labor requirements. Likewise, they have different acquisition, maintenance, and operating costs.

For the study period of 10 years, the estimates for the alternatives are given below.

	A	B	C
First cost:	$1,150,000	$1,250,000	$2,000,000
Salvage value:	750,000	750,000	875,000
Annual disbursements:	425,000	400,000	275,000

Based on a MARR of 15%, determine which alternative is preferred using (a) the PW method, (b) the AW method, (c) the FW method.

3-17. In Problem 3-16, suppose there is uncertainty regarding the annual disbursements required for alternative C; what is the upper limit on annual disbursements for it to be the preferred alternative?

3-18. In Problem 3-16, how sensitive is the selection to the MARR being (a) 10%? (b) 20%? (c) 30%?

3-19. A machine can be repaired today for $7,500. If repairs are not made, operating expenses are expected to increase by $1,200 each year for the next 5 years. The MARR is 10%. Assuming the machine will have a negligible salvage value at the end of the 5-year period and using the PW method, should the machine be repaired?

3-20. Three alternatives are available to fill a given need that is expected to last for at least 12 years. Each is expected to have a negligible salvage value at the end of the life cycle.

	Plan A	Plan B	Plan C
First cost:	$4,000	$12,000	$24,000
Life cycle:	6 yr	3 yr	4 yr
Annual disbursements:	$7,000	$2,000	$500

Using a MARR of 15%, compare the alternatives using (a) the PW method, (b) the AW method, (c) the FW method.

3-21. Two manufacturing methods have been proposed for a new production requirement. One method involves two general-purpose machines that cost $15,000 each, installed. Each will produce 10 pieces per hr and will require an operator costing $10.00 per hr during operation. The other method requires a special-purpose machine costing $45,000 that will produce 20 pieces per hr and will require an operator costing $8.00 per hr during operation. Both types of machines are expected to last 10 years and have negligible salvage values. Other relevant data are as follows:

	General-purpose machine (each)	Special-purpose machine
Power cost per hr:	$ 0.50	$ 0.65
Fixed maintenance per yr:	750.00	900.00
Variable maintenance per hr:	0.25	0.15
Insurance and floor space per yr:	4,200.00	5,500.00

a. If the expected output is 20,000 pieces per yr and the minimum before-tax rate of return is 20%, which method has the lower total annual cost?

b. At what annual output rate would one be indifferent between the two methods?

3-22. It is desired to determine the most economical thickness of insulation for a large cold-storage room. Insulation is expected to cost $150 per 1,000 sq ft of wall area per in. of thickness installed and to require annual property taxes and insurance of 5% of first cost. It is expected to have negligible net salvage value after a 20-year life. The following are estimates of the heat loss per 1,000 sq ft of wall area for several thicknesses:

Insulation, in.	Heat loss, Btu per hr
3	4,400
4	3,400
5	2,800
6	2,400
7	2,000
8	1,800

The cost of heat removal is estimated at $0.01 per 1,000 Btu per hr. The minimum required yield on investment is 20%. Assuming continuous operation throughout the year, which thickness is the most economical?

3-23. Alternative methods I and II are proposed for a plant operation. The following is comparative information:

	Method I	Method II
Initial investment:	$10,000	$40,000
Life:	5 yr	10 yr
Salvage value:	$1,000	$5,0000
Annual disbursements:		
Labor:	$12,000	$4,000
Power:	$250	$300
Rent:	$1,000	$500
Maintenance:	$500	$200
Property taxes and insurance:	$400	$2,000

All other expenses are equal for the two methods, and the income from the operation is not affected by the choice. If the minimum attractive rate of return is 15% and the study period is 10 years, which is the better choice using the annual worth (cost) method?

3-24. Compare the net future worths of two temporary structures that will be retired at the end of 10 years. Assume the minimum attractive rate of return = 10% and that estimates are as follows:

	Structure A	Structure B
First cost:	$20,000	$25,000
Net salvage value:	5,000	− 1,000
Annual maintenance and	5,000	1,500
property taxes:		

3-25. A proposed material for covering the roof of a building will have an estimated life of 8 years and will cost $5,000. A heavier grade of this roofing material will cost $800 more but will have an estimated life of 12 years. Installation costs for either material will be $1,300. Compare the annual costs using a minimum attractive return of 10% and the "repeatability" assumptions.

3-26. A small tractor is required for snow removal. It can be purchased for $3,500 and is expected to have a $500 salvage value at the end of its economic life of 5 years. Its annual operating cost is $1,000 and maintenance will be $300 the first year and increase by $100 per year. If the minimum attractive rate of return is 10%, and if a contractor will provide this service for $2,200 per year, which alternative has lower total present worth-cost?

3-27. Compare the annual costs of pumps A and B for a 15-year service life using an interest rate of 15%.

	Pump A	Pump B
First cost:	$3,500	$5,000
Estimated salvage value:	0	2,000
Annual pumping cost:	450	300
Annual repair cost:	150	80

3-28. It is desired to determine the optimal height for a proposed building that is expected to last 30 years and then be demolished at zero salvage. The following are pertinent data:

	Number of floors			
	2	3	4	5
Building first cost:	$200,000	$250,000	$320,000	$400,000
Annual revenue:	40,000	60,000	85,000	100,000
Annual cash disbursements:	15,000	25,000	25,000	45,000

In addition to the building first cost, the land requires an investment of $100,000 and is expected to retain that value throughout the life period. If the minimum required rate of return is 20%, show which height, if any, should be built based on annual worth comparisons.

3-29. A manufacturing process can be designed for varying degrees of automation. The minimum required rate of return is 20%. Which degree should be selected if the economic life is 10 years and the salvage values are negligible? Use a present worth analysis.

	Machine A	Machine B	Machine C
First cost:	$ 20,000	$ 25,000	$ 35,000
Annual receipts:	150,000	180,000	200,000
Annual disbursements:	138,000	170,000	184,000

3-30. An individual is faced with two mutually exclusive investment alternatives. By investing $10,000 a single sum of $15,000 will be received 4 years after the investment; alternatively, by investing $15,000 a single sum of $18,500 will be received 2 years after the investment. For a MARR of 15%, determine the preferred alternative using the following method(s): (a) PW, (b) AW, (c) FW.

4
Rate of Return Methods
for Comparing Alternatives

In the previous chapter, present worth, annual worth, and future worth analyses were considered in the comparison of investment alternatives. In this chapter rate of return methods are presented. The use of rate of return methods in measuring the acceptability of individual investment *opportunities* was mentioned at the beginning of Chapter 3. It was noted that an individual investment opportunity is worthwhile if its rate of return is not less than the minimum attractive rate of return (MARR).

Any *rate of return* (*RR*) method of economic comparison involves the calculation of a rate or rates of return and comparison against a minimum standard of desirability (i.e., the MARR).

Two common techniques for calculating rates of return can be said to be theoretically sound because they directly take into account the effects of any particular timing of cash flows throughout the study period considered. These methods, which can lead to slightly different calculated results, will be referred to as follows:

1. the internal rate of return (IRR) method,
2. the external rate of return (ERR) method.

The most common method of calculation of the internal rate of return for a single project involves merely finding the interest rate at which the present worth of the cash inflow (receipts or cash savings) equals the present worth of the cash outflow (disbursements or cash savings foregone). That is, one finds the interest rate at which the PW of cash inflow equals the PW of cash outflow; or, at which the PW of cash inflow minus the PW of cash outflow equals zero; or, at which the PW of net cash flows equals zero. Because the PW, AW, and FW differ only by a constant, the IRR can also be calculated by finding the interest rate that equates to zero the AW and/or FW of net cash flows.

To solve for the IRR, one can use trial-and-error methods and manually compute the PW, AW, or FW—attempting to find an interest rate that yields a value of zero for the particular measure of worth. Generally, the process concludes by interpolating between interest rates that yield positive and negative values for the measure of worth. Alternately, one can use one of the financial calculators or computer software packages available.

To demonstrate the errors inherent in using linear interpolation to determine the IRR, several situations are given ahead. For our purposes, cash inflows are denoted by a positive sign and cash outflows are denoted by a negative sign.

4.1 Computation of Internal Rate of Return (IRR) for a Single Investment Opportunity

Example 4-1

First cost:	$10,000
Project life:	5 yr
Salvage value:	$2,000
Annual receipts:	$5,000
Annual disbursements:	$2,200

Solution Expressing PW of net cash flow:

$$-\$10,000 + (\$5,000 - \$2,000)(P/A, i\%, 5) + \$2,000(P/F, i\%, 5) = 0,$$

$$@ \ i = 15\%: -\$10,000 + \$2,800(P/A, 15\%, 5) + \$2,000(P/F, 15\%, 5) \overset{?}{=} 0$$

$$\$365 \neq 0,$$

$$@ \ i = 20\%: -\$10,000 + \$2,800(P/A, 20\%, 5) + \$2,000(P/F, 20\%, 5) \overset{?}{=} 0$$

$$-\$598 \neq 0.$$

Since we have both a positive and a negative PW of net cash flow, the answer is bracketed. Linear interpolation for the answer can be set up as follows:

i	PW of net cash flow
15%	$365
x%	0
20%	−598

The answer x% can be found by solving either

$$\frac{15\% - x\%}{15\% - 20\%} = \frac{\$365 - \$0}{\$365 - (-\$598)}$$

or

$$x\% = 15\% + \frac{\$365}{\$365 + \$598}(20\% - 15\%).$$

Solving, $x\% = 16.9\%$.   ∎

Using Excel 5.0™ and a Hewlett-Packard 92005LX palm computer, about the same IRR value was obtained, 16.4763%. For linear interpolation to provide a very accurate estimate of IRR, a very small interval is needed. As examples, using interest rates of 16% and 17% yield present worths of $120.25 and −$129.61, respectively. Using linear interpolation, a value of 16.4813% (rounded to 16.5%) is obtained—this is much closer to the value obtained using the Excel and Hewlett-Packard financial software than that obtained by interpolating between 15% and 20%.

4.1.1 Principles in Comparing Alternatives by a Rate of Return Method

When comparing alternatives by a RR method when at most one alternative will be chosen, there are two main principles to keep in mind:

1. Each increment of investment capital must justify itself (by sufficient RR on that increment).
2. Compare a higher investment alternative against a lower investment alternative only if that lower investment alternative is justified.

The usual criterion for choice when using a RR method is, "Choose the alternative requiring the highest investment for which each increment of investment capital is justified."

This choice criterion assumes that the firm wants to invest any capital needed as long as the capital is justified by earning a sufficient RR on each increment of capital. In general, a sufficient RR is any RR greater than or equal to the MARR.

4.1.2 Alternative Ways to Find the Internal Rate of Return on Incremental Investment

The internal rate of return on the incremental investment for any two alternatives can also be found by

1. finding the rate at which the PW (or AW or FW) of the net cash flow for the difference between the two alternatives is equal to zero, or
2. finding the rate at which the PWs (or AWs or FWs) of the two alternatives are equal.

4.1.3 Comparing Alternatives When Receipts and Disbursements Are Known

Consider the same two alternative lathes A and B compared in the last chapter, and determine which is better by the internal rate of return method, using the first way outlined above.

Example 4-2

	Lathe	
	A	B
First cost:	$10,000	$15,000
Life:	5 yr	10 yr
Salvage value:	$2,000	$0
Annual receipts:	$5,000	$7,000
Annual disbursements:	$2,200	$4,000
Minimum attractive rate of return = 10%		
Study period = 10 yr		

Solution   The first increment of investment to be studied is the $10,000 for lathe A. This project is the same as illustrated in the "single project" solution in Example 4-1. The IRR for the lathe, and hence the first increment of investment, was shown to be approximately 16.5%. Since 16.5% is greater than the minimum required rate of return of 10%, the increment of investment in lathe A is justified.

The next step is to determine if the second increment of investment (i.e., increasing the investment from $10,000 in lathe A to $15,000 in lathe B) is justified. An easy way to obtain the solution is to calculate the year-by-year difference in net cash flow for the two projects and then to find the IRR on the difference. In order for this year-by-year difference in net cash flow to be computed, the cash flows for each project must be shown for the same number of years (length of study period). The study period should be a common multiple of the lives of the projects under consideration, or the length of time during which the services of the chosen alternatives will be needed, whichever is less. For the example lathes, a study period of 10 years will be used.

Year	Lathe A	Lathe B	Difference Lathe B − Lathe A
1	−$10,000	−$15,000	−$5,000
2	+ 2,800	+ 3,000	+ 200
3	+ 2,800	+ 3,000	+ 200
4			
5	− $8,000		+ $8,000
6			
7			
8			
9			
10	+ 2,800 + $2,000	+ 3,000	+ 200 − $2,000

The equation expressing the present worth of the net cash flow for the difference between the two lathes is

$$-\$5,000 + \$200(P/A, i\%, 10) + \$8,000(P/F, i\%, 5) - \$2,000(P/F, i\%, 10) = 0.$$

Using financial calculators or computer software, a value of 12.1337% (~12.1%) yields a present worth of zero for the increment of investment. Thus, the IRR on the incremental investment is approximately 12.1%. Since the MARR is 10% and the IRR > the MARR, the increment of investment is justified. Hence, lathe B is preferred to lathe A.

The IRR for the individual alternatives can be obtained by finding the interest rate that equates their individual present worths to zero. For lathe A (the same "single investment opportunity" given in Example 4-1) the IRR is approximately 16.5%. For lathe B, a value of approximately 15.1% is obtained for the IRR by finding the interest rate that satisfies the following equality:

$$-\$15,000 + \$3,000(P/A, i\%, 10) = 0.$$

Based on the results obtained, it is clear that B would be chosen so long as the MARR is not greater than approximately 12.1%. Even though the IRR for B is slightly greater than 15%, for a MARR of 15% lathe A should be chosen, not lathe B. In effect, lathe B earns 16.5% on the first $10,000 of investment and 12.1% on the additional $5,000 increment; overall, its IRR is 15.1%. If the MARR is 15%, then we would be better off financially to invest $10,000 in lathe A and earn 16.5%; the balance of $5,000 should remain in the "opportunity fund" where it is expected to earn at least the MARR of 15%.   ∎

4.1.4 Comparing Numerous Alternatives

The following example is given to further illustrate the principle that the return on each increment of investment capital should be justified. To make the computations easier, each alternative in this example has a salvage value equal to the investment. In such cases, the IRR can be calculated directly by dividing the annual net cash inflow or savings by the investment amount. In the tabulated solution shown, the symbol Δ is used to mean "incremental" or "change in." The letters on each end of the arrows designate the projects for which the increment is considered.

Example 4-3

	Alternative project					
	A	B	C	D	E	F
Investment:	$1,000	$1,500	$2,500	$4,000	$5,000	$7,000
Annual savings in cash disbursements:	150	375	500	925	1,125	1,425
Salvage value:	1,000	1,500	2,500	4,000	5,000	7,000

If the company is willing to invest any capital that will earn at least 18%, find which alternative, if any, should be chosen using the IRR method.

Solution   It should be noted that the alternatives are arranged in order of increasing investment amount and that calculations regarding an increment must be completed before one knows which increment to consider next. The symbol ΔIRR means internal rate of return (IRR) on incremental investment.

Increment considered	A	B	B → C	B → D	D → E	E → F
ΔInvestment:	$1,000	$1,500	$1,000	$2,500	$1,000	$2,000
ΔAnnual savings:	$150	$375	$125	$550	$200	$300
ΔIRR:	15%	25%	12.5%	22%	20%	15%
Is increment justified?	No	Yes	No	Yes	Yes	No

By the above analysis, alternative E would be chosen because it is the alternative requiring the highest investment for which each increment of investment capital is justified. Note that the analysis was performed without even considering the IRR on the total investment for each of the alternatives.

In choosing alternative E, several increments of investment were justified, as shown below:

Increment	Investment	Internal rate of return on increment (ΔIRR)
B	$1,500	25%
B → D	2,500	22
D → E	1,000	20
Total:	$5,000	

As a side note, the IRR on the total investment for each alternative is as follows:

	Alternative					
	A	B	C	D	E	F
IRR:	15%	25%	20%	23%	22.5%	20.4%

Note that alternative B has the highest overall IRR and that alternative F has an overall IRR that is greater than the minimum of 18%. Nevertheless, alternative E would be chosen on the rationale that the company wants to invest any increment of capital when and only when that increment will earn at least the minimum return.   ■

✓ 4.1.5 Comparing Alternatives When Disbursements Only Are Known

When disbursements only are known, IRRs can be calculated for incremental investments only and not for the investment in any one alternative. Thus, the lowest investment has to be assumed to be justified (or necessary) without being able to calculate the IRR on that alternative. As an example, consider the same alternative compressors compared in the last chapter, and determine which is the better alternative.

Example 4-4

	Compressor	
	I	II
First cost:	$6,000	$8,000
Life:	6 yr	9 yr
Salvage value:	$1,000	$0
Annual operating disbursements:	$4,000	$3,200
MARR = 15%		
Study period = 18 yr		

Solution   Listing the cash flows for the lowest common multiple of lives yields the values given below. Note that for compressor I the net cash flow in year 6 is obtained by summing the annual disbursement in year 6 (−$4,000), the salvage value at the end of year 6 ($1,000), and the first cost for the replacement compressor (−$6,000), yielding a value of −$9,000.

Year	Compressor I	Compressor II	Difference Comp. II − Comp. I
0	($6,000)	($8,000)	($2,000)
1	(4,000)	(3,200)	800
2	(4,000)	(3,200)	800
3	(4,000)	(3,200)	800
4	(4,000)	(3,200)	800
5	(4,000)	(3,200)	800
6	(9,000)	(3,200)	5,800
7	(4,000)	(3,200)	800
8	(4,000)	(3,200)	800
9	(4,000)	(11,200)	(7,200)
10	(4,000)	(3,200)	800
11	(4,000)	(3,200)	800
12	(9,000)	(3,200)	5,800
13	(4,000)	(3,200)	800
14	(4,000)	(3,200)	800
15	(4,000)	(3,200)	800
16	(4,000)	(3,200)	800
17	(4,000)	(3,200)	800
18	(3,000)	(3,200)	(200)

To use the IRR method, rank the alternatives in increasing order of investment (compressor I first and then compressor II). Next, determine the incremental investment required to invest in the alternative with the greatest initial investment (compressor II) by subtracting the cash flows for compressor I from those required for compressor II. The IRR for the difference between the compressors (i.e., on the incremental investment) is obtained by solving the following equation for the difference in the net cash flows:

$$-\$2,000 - \$8,000(P/F, i\%, 9) - \$1,000(P/F, i\%, 18) + \$800(P/A, i\%, 18)$$
$$+ \$5,000(P/F, i\%, 6) + \$5,000(P/F, i\%, 12) = 0.$$

By trial and error, the IRR can be found to be approximately 47%. Since this return on the increment of investment is greater than the MARR of 15%, compressor II is the economical choice. (Using Excel 5.0, the IRR obtained for the incremental investment is 46.8%.) It should be noted that the IRR method can be used even though we are unable to compute the IRR for either compressor I or compressor II.

The use of AW calculations in determining the IRR on an incremental investment provides a convenient shortcut for problems where the two "repeatability" assumptions stated in Chapter 3 (regarding the period of needed service and the repeatability of cash flows) hold true. For the compressor problem above, the IRR may be found by solving the following:

$$(\$6,000 - \$1,000)(A/P, i\%, 6) + \$1,000(i\%) + \$4,000$$
$$= \$8,000(A/P, i\%, 9) + \$3,200.$$

As with the PW approach, the equation is satisfied for an interest rate of approximately 47%. ■

4.1.6 Differences in Ranking of Investment Opportunities

It was pointed out previously that the IRR method will always give results that are consistent (regarding project acceptance or rejection) with results using the PW, AW, or FW method. However, the IRR method can give a different *ranking* of the order of desirability of individual investment opportunities than the PW, AW, or FW method. As an example, consider Fig. 4-1 depicting the relation of IRR to net present worth for two investment opportunities, projects X and Y.

✓The IRR for each project is the rate at which the net present worth for that project is zero. The net present worth for each project is shown for a typical interest rate. For the hypothetical but quite feasible relationship shown in Fig. 4-1, project Y has the higher IRR while project X has the higher net present worth for all IRRs less than the rate at which the net present worths are equal. This illustrates a case in which the IRR method does result in a different ranking of alternatives as compared to the PW (AW or FW) method. However, since both projects had a net present worth greater than zero, and the IRR for both projects is greater than the minimum attractive rate of return, the determination of acceptance of both projects is consistently shown by either method.

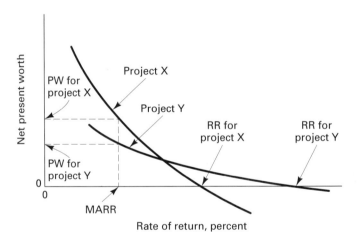

Figure 4-1   Relation of IRR to PW for projects X and Y.

4.1.7 Problems in Which Either No Solution or Several Solutions for Rate of Return Exist

It is possible, but not commonly experienced in practice, to have situations in which there is no single IRR solution by the discounted cash flow method. Descartes' rule of signs indicates that multiple solutions can occur whenever the cash flow series reverses sign (from net outflow to net inflow or the opposite) more than once over the period of study. As an example, consider the following project for which the IRR is desired.

Example 4-5

Find the IRR(s) for these data:

Year	Net cash flow
−1	+$ 500
0	− 1,000
1	0
2	+ 250
3	+ 250
4	+ 250

Solution

Year	Net cash flow	PW @ 35% Factor	PW @ 35% Amount	PW @ 63% Factor	PW @ 63% Amount
−1	+$ 500	1.35	+$ 676	1.63	+$ 813
0	− 1,000	1.00	− 1,000	1.00	− 1,000
1	0				
2	+ 250	0.55	+ 137	0.38	95
3	+ 250	0.41	+ 102	0.23	57
4	+ 250	0.30	+ 75	0.14	35
		Net PW:	$\Sigma = 0$		$\Sigma = 0$

Thus, the present worth of the net cash flows equals 0 for interest rates of 35% and 63%. Whenever multiple answers such as these exist, it is likely that neither is correct. ∎

An effective way to overcome this difficulty and obtain a "correct" answer is to ✓ manipulate cash flows as little as necessary so that there is only one reversal of the cumulative net cash flow. This can be done by using the minimum attractive rate of return to manipulate the funds, and then solving by the discounted cash flow method. For Example 4-5, if the minimum attractive rate of return is 10%, the + $500 at year − 1 can be compounded to year 0 to be $500(F/P,10%,1) = + $550. This, added to the − $1,000 at year 0, equals −$450. The − $450, together with the remaining cash flows, which are all positive, now fits the condition of only one reversal in the cumulative net cash flow. The interest rate at which the present worth of the net cash flows equals 0 can now be shown to be 19% per the following table:

Timing	Net cash flow	PW @ 19% Factor	PW @ 19% Amount
0	−$450	1.00	−$450
2	+ 250	0.70	+ 177
3	+ 250	0.59	+ 150
4	+ 250	0.48	+ 123
	Net PW:		$\Sigma = 0$

It should be noted that whenever a manipulation of net cash flows is performed, the calculated IRR will vary depending on what cash flows are manipulated and at what interest rate. The less manipulation and the closer the minimum rate of return to the calculated rate of return, the less the variation in the final calculated rate of return.

Appendix 4-A gives another example of the above type of problem and shows how the ERR method can be used to obtain a solution.

4.2 External Rate of Return (ERR) Method

Implicit in the internal rate of return method is the assumption that recovered funds are reinvested at a rate equal to the internal rate of return. Since it is often the case that opportunities do not exist for investing recovered funds and earning such a rate, the notion of an explicit reinvestment rate of return method has appeal. Additionally, the possibility of multiple solutions using the internal rate of return method can result in misinterpretations and misunderstandings of the rate of return figure obtained.

Because of the desire to incorporate explicitly the reinvestment rate in rate of return calculations and the desire to use a method that will yield a unique solution, an external rate of return method was developed. If complicated manipulations are to be avoided, the explicit reinvestment rate of return method is limited to a single investment (negative cash flow) and a uniform series of returns (positive cash flows). A more general approach is to define the external rate of return as the interest rate that equates the future worth of investments to the accumulation of reinvested returns. Recovered funds are assumed to be reinvested at the minimum attractive rate of return, based on the opportunity cost concept. To illustrate, consider the same single project for which rates of return were calculated at the beginning of the chapter.

Example 4-6

Determine the ERR of these cash flows:

First cost:	$10,000
Project life:	5 yr
Salvage value:	$2,000
Annual receipts:	$5,000
Annual disbursements:	$2,200

Solution   The following solution by the external rate of return (ERR) method uses a reinvestment rate or minimum attractive rate of return (MARR) of 10%:

Year	Net annual cash flow
0	−$10,000
1–5	2,800
5	2,000

Future accumulation of recovered monies $= \$2,800(F/A, 10\%, 5) + \$2,000 = \$19,094$. Future worth of investments $= \$10,000(F/P, i\%, 5)$. Thus,

$$\$10,000(F/P, i\%, 5) = \$19,094$$

or

$$(F/P, i\%, 5) = 1.9094.$$

From the interest tables it is seen that

i	$(F/P, i\%, 5)$
12%	1.7623
15%	2.0114

By using linear interpolation, a value of 1.9094 is obtained for the $(F/P, i\%, 5)$ factor when $i\%$ equals, approximately, 13.77%. The exact value of $i\%$ can be obtained by using logarithms and recalling the mathematical relation represented by the $(F/P, i\%, N)$ factor. Namely,

$$(1 + i)^N = (F/P, i\%, N).$$

Hence, for the example,

$$(1 + i)^5 = 1.9094,$$
$$5 \ln(1 + i) = \ln 1.9094,$$
$$\ln(1 + i) = 0.1293,$$
$$1 + i = \ln^{-1}(0.1293),$$
$$i = 0.1381.$$

Thus, an external rate of return of 13.81% will result from the project. Since 13.81% is greater than the MARR of 10%, the project is justified economically.

In general, a project justified economically using the internal rate of return method will also be justified using the external rate of return method. A unique solution always occurs using the ERR method.   ∎

4.2.1 Comparing Alternatives Using the External Rate of Return Method

An incremental approach is required when comparing investment alternatives using any RR method. Specifically, the external rate of return on incremental investment is defined to be the interest rate that equates the future worth of the incremental investment and the future accumulation of incremental positive-valued cash flows. To illustrate, consider the lathe example presented previously.

Example 4-7

Use the ERR method to select either lathe A or lathe B.

	Lathe	
	A	B
First cost:	$10,000	$15,000
Life:	5 yr	10 yr
Salvage value:	$2,000	$0
Annual receipts:	$5,000	$7,000
Annual disbursements:	$2,200	$4,000
Minimum attractive rate of return = 10%		
Study period = 10 yr		

Solution   It was shown previously that lathe A has an ERR of 13.81% and is economically justified. It remains to determine if the incremental investment required to purchase lathe B is justified. Over the 10-year study period the incremental cash flows have the following profile:

End of year	CF(B) – CF(A)
0	–$5,000
1–4	+ 200
5	+ 8,200
6–9	+ 200
10	– 1,800

Future accumulation of positive-valued cash flows equals

$$\$8,000(F/P, 10\%, 5) + \$200(F/A, 10\%, 9)(F/P, 10\%, 1) = \$15,871.$$

Future worth of negative-valued cash flows equals

$$\$5,000(F/P, i\%, 10) + \$1,800.$$

Hence,

$$\$5,000(F/P, i\%, 10) + \$1,800 = \$15,871$$

or

$$(F/P, i\%, 10) = 2.8142.$$

Thus,

$$\log(1 + i\%)^{10} = \log 2.8142.$$

Solving for the ERR gives a value of 10.9%, which is greater than the MARR; hence, the incremental investment required for B is justified.   ■

4.2.2 Comparing Alternatives When Disbursements Only Are Known

When disbursements only are known, the external rate of return method is applied to incremental cash flows in the same manner as other rate of return methods. Thus, the alternative having the lowest investment is assumed to be justified (or necessary). As an illustration, consider again the compressor example.

Example 4-8

	Compressor	
	I	II
First cost:	$6,000	$8,000
Life:	6 yr	9 yr
Salvage value:	$1,000	$0
Annual operating disbursements:	$4,000	$3,200
Minimum attractive rate of return = 15%		
Study period = 18 yr		

Solution

End of year	II − I
0	−$2,000
1	800
2	800
3	800
4	800
5	800
6	800 + $5,000
7	800
8	800
9	800 − $8,000
10	800
11	800
12	800 + $5,000
13	800
14	800
15	800
16	800
17	800
18	800 − $1,000

Future accumulation of positive-valued cash flows equals

$$\$800(F/A, 15\%, 18) + \$5,000(F/P, 15\%, 12) + \$5,000(F/P, 15\%, 6) = \$98,986.$$

Future worth of negative-valued cash flows equals

$$\$2,000(F/P, i\%, 18) + \$8,000(F/P, i\%, 9) + \$1,000.$$

$$@\, i = 20\% \qquad \$95,526 < \$98,986$$
$$@\, i = 25\% \qquad \$171,628 > \$98,986$$

Interpolation gives ERR = 20.2%, which is greater than the MARR of 15%; hence, compressor II is recommended.   ■

4.3 Using Spreadsheets to Perform IRR and ERR Analyses

Recall that in Chapter 2 we discussed the use of spreadsheets in performing economic analyses. The presentation focused on computing present worths. We also demonstrated the use of Excel 5.0 in computing the net present worth (PW) for a cash flow series. While the use of spreadsheets is advantageous in computing present worths, their advantages are very evident in solving for rates of return. The tedious trial-and-error approach is essentially eliminated by using spreadsheets. In this section, we will illustrate the use of spreadsheets in computing the IRR and ERR for a cash flow series.

4.3.1 Internal Rate of Return Calculation

Consider the spreadsheet given in Fig. 4-2. Note that we have modified the spreadsheet by placing the *P/F* column between the EOY column and the first cash flow (CF) column. Since the discount rate (MARR) applies to all of the alternatives, it is more convenient to place it as shown in cell B1.

row/column	A	B	C	D	E	F	G	H
1		B1=DR						
2								
3	Time	(P/F,i%,N)	CF(1)	PW	CF(2)	PW(2)	CF(2-1)	PW(2-1)
4	0	B4=1.0000	C4	D4=B4*C4	E4	F4=B4*E4	G4=E4-D4	H4=B4*G4
5	1	B5=B4/(1.0+B1)	C5	D5=B5*C5	E5	F5=B5*E5	G5=E5-D5	H5=B5*G5
6	2	B6=B5/(1.0+B1)	C6	D6=B6*C6	E6	F6=B6*E6	G6=E6-D6	H6=B6*G6
7	3	B7=B6/(1.0+B1)	C7	D7=B7*C7	E7	F7=B7*E7	G7=E7-D7	H7=B7*G7
8	4	B8=B7/(1.0+B1)	C8	D8=B8*C8	E8	F8=B8*E8	G8=E8-D8	H8=B8*G8
9	5	B9=B8/(1.0+B1)	C9	D9=B9*C9	E9	F9=B9*E9	G9=E9-D9	H9=B9*G9
10	PW			=SUM(D4:D9)		=SUM(F4:P9)		=SUM(H4:H9)

Figure 4-2 Illustration of the structure of a spreadsheet that can be used to facilitate IRR analysis.

Note: DR denotes the discount rate to obtain the PW values; when performing a PW analysis, the DR = the MARR.

To determine the IRR for the cash flow series, CF(1), vary the value in cell B1 until the PW value obtained in cell D10 equals zero. The same procedure can be applied to determine the IRR for cash flow series CF(2) and CF(2 − 1). (Using Excel 5.0, the IRR for cash flow series CF(1) can be obtained by entering in a cell the following: =IRR(C4:C9).)

Example 4-9

Use spreadsheets to perform an IRR analysis of the investments involving the two alternative lathes.

    Solution   The spreadsheet based on an 8% discount rate is given in Fig. 4-3. Notice that the PW values obtained are the same as those obtained in Chapter 3. Increasing the discount rate to 10% yields the spreadsheet given in Fig. 4-4; since neither PW is negative, the process of increasing the discount rate was continued. Figure 4-5 depicts the spreadsheet for a 15% discount rate; note that the present worth for alternative B is negative and that the present worth for the increment of investment is also negative. Figure 4-6 depicts the spreadsheet for a 12.1337% discount rate, which is the IRR for the increment of investment; also shown are the IRR values for alternatives A and B and the increment of investment required to go from A to B.

    The IRR values shown in Fig. 4-6 were obtained by using Excel 5.0, rather than continuing to search for a discount rate that would yield a value of zero for the present worths for A, B, and the difference between B and A. (We could have determined immediately the IRR value by using the Excel software, rather than using trial-and-error methods to converge on the value. However, we wanted to illustrate the behavior of the various PW values as the discount rate was changed.)   ■

4.3.2 External Rate of Return Calculation

To determine the external rate of return, Fig. 4-2 is modified as shown in Fig. 4-7. Two columns (B and C) are provided, each containing single-sum, compound amount factors. Column B provides the compound amount factors for the trial discount rate provided in cell B1; column C provides the compound amount factors for the MARR provided in cell C1. Depending on whether a cash flow entry is negative or positive, it is multiplied by the entry in column B or C. The ERR is that value of the discount rate (B1 entry) that results in the sum of the FW values for the cash flow series (columns E, G, and I) being zero.

Example 4-10

Using spreadsheets, perform an ERR analysis of the investments involving the two alternative lathes.

    Solution   From Fig. 4-8, it is seen that by varying the entry in cell B1, the ERR for alternative A is found to be 12.5819%. From Fig. 4-9, the ERR for alternative B is found to be 12.2910%. From Fig. 4-10, the ERR for the difference in cash flows between B and A is found to be 10.9015%. Since ERR(B − A) > MARR, the increment of investment required to purchase lathe B is justified.

    Notice that the ERR for A, 12.5819%, is not the same value obtained when a single investment cycle was considered. Previously, we computed a value of 13% for the ERR. The reason for the difference in values is easy to understand when you compare the cash flow series for a single investment cycle with that for a repetitive investment. Recalling the rationale of the

R/C	A	B	C	D	E	F	G	H
1		0.0800						
2								
3	Time	(P/F,i%,N)	CF(A)	PW(A)	CF(B)	PW(B)	CF(B-A)	PW(B-A)
4	0	1.0000	($10,000)	($10,000)	($15,000)	($15,000)	($5,000)	($5,000)
5	1	0.9259	$2,800	$2,593	$3,000	$2,778	$200	$185
6	2	0.8573	$2,800	$2,401	$3,000	$2,572	$200	$171
7	3	0.7938	$2,800	$2,223	$3,000	$2,381	$200	$159
8	4	0.7350	$2,800	$2,058	$3,000	$2,205	$200	$147
9	5	0.6806	($5,200)	($3,539)	$3,000	$2,042	$8,200	$5,581
10	6	0.6302	$2,800	$1,764	$3,000	$1,891	$200	$126
11	7	0.5835	$2,800	$1,634	$3,000	$1,750	$200	$117
12	8	0.5403	$2,800	$1,513	$3,000	$1,621	$200	$108
13	9	0.5002	$2,800	$1,401	$3,000	$1,501	$200	$100
14	10	0.4632	$4,800	$2,223	$3,000	$1,390	($1,800)	($834)
15	PW			$4,270		$5,130		$860

Figure 4-3   Spreadsheet for the lathe example, with an 8% discount rate.

R/C	A	B	C	D	E	F	G	H
1		0.1000						
2								
3	Time	(P/F,i%,N)	CF(A)	PW(A)	CF(B)	PW(B)	CF(B-A)	PW(B-A)
4	0	1.0000	($10,000)	($10,000)	($15,000)	($15,000)	($5,000)	($5,000)
5	1	0.9091	$2,800	$2,545	$3,000	$2,727	$200	$182
6	2	0.8264	$2,800	$2,314	$3,000	$2,479	$200	$165
7	3	0.7513	$2,800	$2,104	$3,000	$2,254	$200	$150
8	4	0.6830	$2,800	$1,912	$3,000	$2,049	$200	$137
9	5	0.6209	($5,200)	($3,229)	$3,000	$1,863	$8,200	$5,092
10	6	0.5645	$2,800	$1,581	$3,000	$1,693	$200	$113
11	7	0.5132	$2,800	$1,437	$3,000	$1,539	$200	$103
12	8	0.4665	$2,800	$1,306	$3,000	$1,400	$200	$93
13	9	0.4241	$2,800	$1,187	$3,000	$1,272	$200	$85
14	10	0.3855	$4,800	$1,851	$3,000	$1,157	($1,800)	($694)
15	PW			$3,009		$3,434		$425

Figure 4-4 Spreadsheet for the lathe example, with a 10% discount rate.

R/C	A	B	C	D	E	F	G	H
1		0.1500						
2								
3	Time	(P/F,i%,N)	CF(A)	PW(A)	CF(B)	PW(B)	CF(B-A)	PW(B-A)
4	0	1.0000	($10,000)	($10,000)	($15,000)	($15,000)	($5,000)	($5,000)
5	1	0.8696	$2,800	$2,435	$3,000	$2,609	$200	$174
6	2	0.7561	$2,800	$2,117	$3,000	$2,268	$200	$151
7	3	0.6575	$2,800	$1,841	$3,000	$1,973	$200	$132
8	4	0.5718	$2,800	$1,601	$3,000	$1,715	$200	$114
9	5	0.4972	($5,200)	($2,585)	$3,000	$1,492	$8,200	$4,077
10	6	0.4323	$2,800	$1,211	$3,000	$1,297	$200	$86
11	7	0.3759	$2,800	$1,053	$3,000	$1,128	$200	$75
12	8	0.3269	$2,800	$915	$3,000	$981	$200	$65
13	9	0.2843	$2,800	$796	$3,000	$853	$200	$57
14	10	0.2472	$4,800	$1,186	$3,000	$742	($1,800)	($445)
15	PW			$570		$56		($513)

Figure 4-5 Spreadsheet for the lathe example, with a 15% discount rate.

R/C	A	B	C	D	E	F	G	H
1		0.1213						
2								
3	Time	(P/F,i%,N)	CF(A)	PW(A)	CF(B)	PW(B)	CF(B-A)	PW(B-A)
4	0	1.0000	($10,000)	($10,000)	($15,000)	($15,000)	($5,000)	($5,000)
5	1	0.8918	$2,800	$2,497	$3,000	$2,675	$200	$178
6	2	0.7953	$2,800	$2,227	$3,000	$2,386	$200	$159
7	3	0.7092	$2,800	$1,986	$3,000	$2,128	$200	$142
8	4	0.6325	$2,800	$1,771	$3,000	$1,897	$200	$126
9	5	0.5641	($5,200)	($2,933)	$3,000	$1,692	$8,200	$4,625
10	6	0.5030	$2,800	$1,408	$3,000	$1,509	$200	$101
11	7	0.4486	$2,800	$1,256	$3,000	$1,346	$200	$90
12	8	0.4000	$2,800	$1,120	$3,000	$1,200	$200	$80
13	9	0.3568	$2,800	$999	$3,000	$1,070	$200	$71
14	10	0.3182	$4,800	$1,527	$3,000	$954	($1,800)	($573)
15	PW			$1,858		$1,858		$0
16	IRR		16.4763%		15.0984%		12.1337%	

Figure 4-6 Spreadsheet for the lathe example, with a 12.1337% discount rate.

R/C	A	B	C	D	E	F	G	H	I
1		B1=DR	C1=MARR						
2	Time	B2=N							
3	(t)	(F/P,B1%,N-t)	(F/P,C1%,N-t)	CF(1)	FW(1)	CF(2)	FW(2)	CF(2-1)	FW(2-1)
4	0	B4=B5*(1+B1)	C4=C5*(1+C1)	D4	E4=B4*D4	F4	G4=B4*F4	H4=F4-D4	I4=B4*H4
5	1	B5=B6*(1+B2)	C5=C6*(1+C1)	D5	E5=C5*D5	F5	G5=C5*F5	H5=F5-D5	I5=C5*H5
6	2	B6=B7*(1+B2)	C6=C7*(1+C1)	D6	E6=C6*D6	F6	G6=C6*F6	H6=F6-D6	I6=C6*H6
7	3	B7=B8*(1+B2)	C7=C8*(1+C1)	D7	E7=C7*D7	F7	G7=C7*F7	H7=F7-D7	I7=C7*H7
8	4	B8=B9*(1+B2)	C8=C9*(1+C1)	D8	E8=C8*D8	F8	G8=C8*F8	H8=F8-D8	I8=C8*H8
9	5	B9=1.0	C9=1.0	D9	E9=C9*D9	F9	G9=C9*F9	H9=F9-D9	I9=C9*H9
10					=SUM(E4:E9)		=SUM(G4:G9)		=SUM(I4:I9)
11	assumes only D4, F4, and H4 are negative;								
12	negative CFs are multiplied by values in column B;								
13	positive CFs are multiplied by values in column C								

Figure 4-7 Spreadsheet for performing an ERR analysis.

R/C	A	B	C	D	E	F	G	H	I
1		12.5819%	10.0000%						
2	Time	N=10							
3	(t)	(F/P,B1%,N-t)	(F/P,B2%,N-t)	CF(A)	FW(A)	CF(B)	FW(B)	CF(B-A)	PW(B-A)
4	0	3.2710	2.5937	($10,000)	($32,710)	($15,000)	($49,066)	($5,000)	($16,355)
5	1	2.9055	2.3579	$2,800	$6,602	$3,000	$7,074	$200	$472
6	2	2.5808	2.1436	$2,800	$6,002	$3,000	$6,431	$200	$429
7	3	2.2923	1.9487	$2,800	$5,456	$3,000	$5,846	$200	$390
8	4	2.0362	1.7716	$2,800	$4,960	$3,000	$5,315	$200	$354
9	5	1.8086	1.6105	($5,200)	($9,405)	$3,000	$4,832	$8,200	$13,206
10	6	1.6065	1.4641	$2,800	$4,099	$3,000	$4,392	$200	$293
11	7	1.4269	1.3310	$2,800	$3,727	$3,000	$3,993	$200	$266
12	8	1.2675	1.2100	$2,800	$3,388	$3,000	$3,630	$200	$242
13	9	1.1258	1.1000	$2,800	$3,080	$3,000	$3,300	$200	$220
14	10	1.0000	1.0000	$4,800	$4,800	$3,000	$3,000	($1,800)	($1,800)
15	FW				$0		($1,253)		($2,284)

Figure 4-8 Spreadsheet for determining ERR for A.

R/C	A	B	C	D	E	F	G	H	I
1		12.2910%	10.0000%						
2	Time	N=10							
3	(t)	(F/P,B1%,N-t)	(F/P,B2%,N-t)	CF(A)	FW(A)	CF(B)	FW(B)	CF(B-A)	PW(B-A)
4	0	3.1875	2.5937	($10,000)	($31,875)	($15,000)	($47,812)	($5,000)	($15,937)
5	1	2.8386	2.3579	$2,800	$6,602	$3,000	$7,074	$200	$472
6	2	2.5279	2.1436	$2,800	$6,002	$3,000	$6,431	$200	$429
7	3	2.2512	1.9487	$2,800	$5,456	$3,000	$5,846	$200	$390
8	4	2.0048	1.7716	$2,800	$4,960	$3,000	$5,315	$200	$354
9	5	1.7854	1.6105	($5,200)	($9,284)	$3,000	$4,832	$8,200	$13,206
10	6	1.5899	1.4641	$2,800	$4,099	$3,000	$4,392	$200	$293
11	7	1.4159	1.3310	$2,800	$3,727	$3,000	$3,993	$200	$266
12	8	1.2609	1.2100	$2,800	$3,388	$3,000	$3,630	$200	$242
13	9	1.1229	1.1000	$2,800	$3,080	$3,000	$3,300	$200	$220
14	10	1.0000	1.0000	$4,800	$4,800	$3,000	$3,000	($1,800)	($1,800)
15	FW				$957		$0		($1,866)

Figure 4-9 Spreadsheet for determining ERR for B.

R/C	A	B	C	D	E	F	G	H	I
1		10.9015%	10.0000%						
2	Time	N=10							
3	(t)	(F/P,B1%,N-t)	(F/P,B2%,N-t)	CF(A)	FW(A)	CF(B)	FW(B)	CF(B-A)	FW(B-A)
4	0	2.8143	2.5937	($10,000)	($28,143)	($15,000)	($42,215)	($5,000)	($14,072)
5	1	2.5377	2.3579	$2,800	$6,602	$3,000	$7,074	$200	$472
6	2	2.2882	2.1436	$2,800	$6,002	$3,000	$6,431	$200	$429
7	3	2.0633	1.9487	$2,800	$5,456	$3,000	$5,846	$200	$390
8	4	1.8605	1.7716	$2,800	$4,960	$3,000	$5,315	$200	$354
9	5	1.6776	1.6105	($5,200)	($8,723)	$3,000	$4,832	$8,200	$13,206
10	6	1.5127	1.4641	$2,800	$4,099	$3,000	$4,392	$200	$293
11	7	1.3640	1.3310	$2,800	$3,727	$3,000	$3,993	$200	$266
12	8	1.2299	1.2100	$2,800	$3,388	$3,000	$3,630	$200	$242
13	9	1.1090	1.1000	$2,800	$3,080	$3,000	$3,300	$200	$220
14	10	1.0000	1.0000	$4,800	$4,800	$3,000	$3,000	($1,800)	($1,800)
15	FW				$5,249		$5,597		$0

Figure 4-10 Spreadsheet for determining ERR for the incremental investment between B and A.

ERR—determining the return that would have to be earned on invested capital to equate to the future value of the reinvestment of recovered capital at the MARR—with one life cycle only one investment occurs ($10,000 @ $t = 0$), whereas with consecutive life cycles there are two investments ($10,000 @ $t = 0$ and $8,000 @ $t = 5$). The time value of money causes different ERR values because of the differences in timing of the net cash flows.

When using the ERR method, it is useful to remember its relationship with the IRR and MARR. Namely, the following ordering will occur: either IRR > ERR > MARR, or IRR < ERR < MARR, or IRR = ERR = MARR.

An Excel 5.0 financial function closely related to the ERR is the MIRR function. Used to compute a modified internal rate of return based on reinvestment of recovered capital, it allows the specification of both the reinvestment rate, *rrate,* and an interest rate used to finance the investment, *frate.* The MIRR function is defined by MIRR(series, *frate, rrate*). Since it is designed for a series of periodic cash flows, the MIRR function can be used for a single life cycle for lathe A or lathe B but cannot be used for the cash flow series for the incremental invest-ment. (Likewise, it cannot be used for lathe A for a 10-year period, due to the intermediate net negative cash flow.) By letting *frate* = 0 and letting *rrate* = MARR, the ERR can be computed when the cash flow profile satisfies the periodicity requirements of the MIRR function.

For the cash flow series given in the spreadsheet in Fig. 4-11, the MIRR function can be applied to determine the ERR for alternative A by entering the following in cell B13: = MIRR(B2:B7,0,0.10); likewise, for lathe B enter in cell E13 the following: = MIRR (E2:E12,0,0.10). For lathe A, ERR = 13.8101%; for lathe B, ERR = 12.2910%.   ∎

In addition to the two methods of calculating rates of return that have been dis-cussed above, there are numerous other so-called investment recovery measures used in practice. Appendix 4-B briefly explains three of these measures.

R/C	A	B	C	D
	MARR=	10%		
1	EOY	CF(A)	EOY	CF(B)
2	0	($10,000)	0	($15,000)
3	1	$2,800	1	$3,000
4	2	$2,800	2	$3,000
5	3	$2,800	3	$3,000
6	4	$2,800	4	$3,000
7	5	$4,800	5	$3,000
8			6	$3,000
9			7	$3,000
10			8	$3,000
11			9	$3,000
12			10	$3,000
13	ERR=	13.8101%		12.2910%

Figure 4-11 Spreadsheet used to determine the ERR for individual life cycles of lathes A and B using the modified internal rate of return (MIRR) function in Microsoft Excel 5.0.

PROBLEMS

4-1. Determine which of the following independent investment opportunities should be undertaken. Use a MARR of 15% and (a) the IRR method, (b) the ERR method.

	Investment opportunity				
	A	B	C	D	E
Initial investment:	$8,000	$10,000	$12,000	$15,000	$16,000
Annual net receipts:	2,000	3,000	3,200	3,500	4,000
Salvage value:	– 1,000	2,000	1,000	2,000	0
Study period = 5 yr					

4-2. In Problem 4-1, suppose A through E constitute a set of feasible, mutually exclusive investment alternatives. Determine which should be chosen using (a) the IRR method, (b) the ERR method.

4-3. Determine which of the following independent (nonrepeating) investment opportunities are economically worthwhile. Use a MARR of 12%, a study period of 10 years, and (a) the IRR method, (b) the ERR method.

	Cash flows for investment opportunity					
EOY	A	B	C	D	E	F
0	–$5,000	–$5,000	–$8,000	–$12,000	$ 0	–$15,000
1	– 5,000	500	1,000	1,500	0	400
2	1,000	1,000	1,000	1,500	0	800
3	1,000	1,500	1,000	1,500	– 6,000	1,200
4	2,000	2,000	1,000	1,500	2,000	1,600
5	2,000	2,500	1,000	1,500	2,000	2,000
6	3,000	3,000	1,000	1,500	2,000	2,400
7	3,000	0	1,000	1,500	2,000	2,800
8	4,000	0	1,000	1,500	2,000	3,200
9	4,000	0	1,000	1,500	0	3,600
10	– 5,000	0	– 2,000	3,000	0	4,000

4-4. In Problem 4-3, suppose A through F constitute a set of feasible, mutually exclusive investment alternatives. Which should be chosen based on (a) the IRR method? (b) the ERR method?

4-5. Determine which of the following independent (nonrepeating) investment opportunities are economically worthwhile. Use a MARR of 15%, a study period of 5 years, and (a) the IRR method, (b) the ERR method.

	Cash flows for investment opportunity					
EOY	A	B	C	D	E	F
0	–$10,000	–$5,000	–$8,000	–$12,000	–$6,000	–$15,000
1	2,000	1,000	3,000	5,000	1,000	4,500
2	2,000	1,500	3,000	5,000	1,500	4,500
3	2,000	2,000	5,000	6,000	2,000	4,500
4	3,000	2,500	0	0	2,500	4,500
5	3,000	0	0	0	3,000	0

4-6. In Problem 4-5, suppose A through F constitute a set of feasible, mutually exclusive investment alternatives. Determine the economic choice using (a) the IRR method, (b) the ERR method.

4-7. If money is worth 8% to you, use the IRR method to determine which (if any) of the following investment opportunities you would favor.

EOY	CF(A)	CF(B)	CF(C)
1	$1,000	$5,000	$3,000
2	2,000	4,000	3,000
3	3,000	3,000	3,000
4	4,000	2,000	3,000
5	5,000	1,000	3,000

4-8. In Problem 4-7, suppose the three cash flow profiles are for feasible, mutually exclusive investment alternatives and one must be chosen. Further, suppose your time value of money is 15%. Using the ERR method, which would you choose?

4-9. Suppose you compute the internal rate of return for an investment alternative and find that it equals 18%. Will the external rate of return be equal to, less than, or greater than 18% if the MARR is equal to (a) 0%, (b) 15%, (c) 18%, (d) 20%?

4-10. Three mutually exclusive investment alternatives are being considered; the "do nothing" alternative is not feasible. The cash flows are shown for a seven-year planning horizon. The MARR is 20%. Determine the preferred alternative using (a) the IRR method, (b) the ERR method.

EOY	CF(A)	CF(B)	CF(C)
0	-$ 50,000	-$100,000	-$200,000
1	- 100,000	- 100,000	+ 75,000
2-7	+ 50,000	+ 70,000	+ 75,000

4-11. Three mutually exclusive investment alternatives are being considered; the "do nothing" alternative is not feasible. The cash flows are shown for a ten-year planning horizon. The MARR is 25%. Determine the preferred alternative using (a) the IRR method, (b) the ERR method.

EOY	CF(A)	CF(B)	CF(C)
0	-$ 50,000	-$100,000	-$200,000
1	- 100,000	- 100,000	+ 50,000
2-5	+ 50,000	+ 70,000	+ 50,000
6-9	+ 50,000	+ 70,000	+ 75,000
10	+ 75,000	+ 70,000	+ 75,000

4-12. A construction firm is considering leasing a crane needed on a project for 4 years for $300,000 payable now. The alternative is to buy a crane for $350,000 and sell it at the end of 4 years for $100,000. Annual maintenance costs for ownership only are expected to be $10,000 per year for the first 2 years and $15,000 per year for the last 2 years. At what interest rate (IRR) are the two alternatives equivalent?

4-13. An industrial machine costing $5,000 will produce net savings of $2,000 per year. The machine has a 5-year economic life but must be returned to the factory for major repairs after 3 years of operation. These repairs will cost $2,500. The company's MARR is approximately 15%. What IRR will be earned on the purchase of this machine? Would you recommend it?

4-14. An improved facility costing $250,000 has been proposed. Construction time will be 2 years with expenditures of $100,000 the first year and $150,000 the second year. Savings beginning the first year after construction completion are as follows:

Year	Savings
1	$ 50,000
2	65,000
3	80,000
4	95,000
5	110,000

The facility will not be required after 5 years and will have a salvage value of $25,000. Determine the (a) IRR and (b) ERR using a reinvestment rate of 10%.

4-15. A distillery is considering the erection of a bottle-making plant. The number of bottles needed annually is estimated at 600,000. The initial cost of the facility would be $150,000 with an estimated life of 20 years. Annual operation and maintenance costs are expected to be $40,000, and annual taxes and insurance $15,000. Should the distillery erect the bottle-producing facility or buy the bottles from another company at $0.15 each? Use the two rate of return methods with a minimum attractive rate of 12%.

4-16. What is the internal rate of return on an investment of $1,000 that will yield $325 per year for 5 years?

4-17. There are five alternative machines to do a given job. Each is expected to have a salvage value of 100% of the investment amount at the end of its life of 4 years. If the firm's minimum attractive rate of return is 12%, which machine is the best choice based on the following data? Use (a) an internal rate of return comparison, and (b) an external rate of return comparison.

	Machine				
	A	B	C	D	E
Investment:	$1,000	$1,400	$2,100	$2,700	$3,400
Net cash flow per yr:	110	180	280	340	445

4-18. Consider two investment alternatives having cash flow profiles as given below. Use both an internal and an external rate of return comparison to determine the recommended alternative. Base your analysis on a study period of 5 years and a minimum attractive rate of return of (a) 8%, (b) 15%, (c) 25%.

End of year	A	B
0	$ 0	−$10,000
1	− 20,000	5,000
2	30,000	3,500
3	250	40,000
4	20,000	2,576
5	3,864	30,000

4-19. Two alternative machines will produce the same product, but one machine will produce higher-quality items that can be expected to return greater revenue. Given the following data, determine which machine is better using the IRR and ERR methods, a MARR of 15%, and a study period of 10 years.

	Machine A	Machine B
First cost:	$ 20,000	$ 30,000
Salvage value:	2,000	0
Annual receipts:	150,000	180,000
Annual disbursements:	138,000	163,000

4-20. An individual is faced with two mutually exclusive investment alternatives. By investing $15,000 a single sum of $18,500 will be received 2 years after the investment; alternately, by investing $20,000 a single sum of $30,000 will be received 4 years after the investment. Using IRR and ERR methods, determine the preferred alternative based on a MARR of (a) 5%, (b) 10%, (c) 20%.

4-21. Using a 10-yr study period, determine the preferred incinerator alternative using the following method(s): (a) IRR, (b) ERR.

	Incinerator			
	A	B	C	D
First cost:	$3,000	$3,800	$4,500	$5,000
Salvage value:	0	0	0	0
Annual operating disbursements:	1,800	1,770	1,470	1,320
MARR = 10%				

4-22. Three mutually exclusive investment alternatives are being considered, including the "do nothing" alternative. The cash flows are shown below over a seven-year planning horizon. The MARR is 12%. Using the ERR method, determine the preferred alternative.

EOY	CF(A)	CF(B)	CF(C)
0	$0	-$100,000	-$180,000
1	0	- 100,000	- 20,000
2	0	0	+ 100,000
3	0	+ 25,000	+ 75,000
4	0	+ 50,000	+ 50,000
5	0	+ 75,000	+ 25,000
6	0	+ 100,000	0
7	0	+ 125,000	+ 100,000

4-23. Three order-picking systems are being considered for use in a catalog distribution center. Alternative A involves picking items manually from bin shelving. With alternative B, items are picked manually from a horizontal carousel conveyor; hence, essentially no walking is required by the order picker. With alternative C, items are picked automatically by an automatic item retrieval machine. The alternatives have different space and labor requirements. Likewise, they have different acquisition, maintenance, and operating costs.

A study period of 10 years and a MARR of 15% are to be used. For the study period, the cash flow estimates for the alternatives are given below.

	A	B	C
First cost:	$1,150,000	$1,250,000	$2,000,000
Salvage value:	750,000	750,000	875,000
Annual disbursements:	425,000	400,000	275,000

    a. Using the IRR method, which alternative is preferred?

    b. Would your preference change if the MARR were (i) 10%? (ii) 20%? (iii) 30%?

4-24. Given the following mutually exclusive projects and assuming $N = \infty$, which investment should be selected if the company's MARR is 25%? Use the IRR method.

	A	B	C	D	E	F
Investment:	$30,000	$60,000	$75,000	$50,000	$55,000	$70,000
Annual savings:	10,000	18,000	21,500	14,000	16,000	20,500

4-25. For the case of a zero salvage value, single negative cash flow at $t = 0$, and a uniform series of cash flows for $t = 1, \ldots, N$, show that if IRR > MARR, then IRR > ERR > MARR, and if IRR < MARR (where IRR ≥ 0), then IRR < ERR < MARR.

4-26. Let $\{C_t\}$ denote the set of investments and $\{R_t\}$ denote the set of returns for an investment alternative. Given below are various plots of future worth versus the discount rate. What would be the recommendation in each case using the IRR and ERR methods?

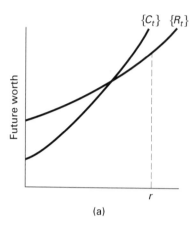

(a)

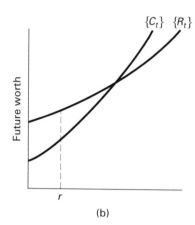

(b)

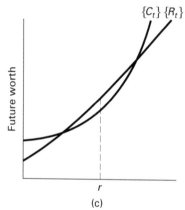

(c)

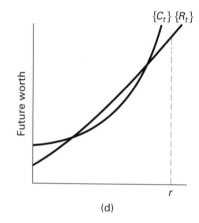

(d)

Use of the ERR Method to Overcome Multiple-Solutions Difficulty with the IRR Method

In order to illustrate the differences in approach using the IRR and ERR methods, consider Solomon's classic pump problem.[1]

Example

A firm is considering installing a new pump that will move oil faster than the present pump. The new pump will finish the job in one year; the present pump will require two years. The total value of the oil to be pumped is $20,000. The new pump costs $1,600, and neither pump will have any salvage value.

The cash flow profiles for each pump and the incremental cash flow of the new pump over the old are presented in Table 4-A-1. If the firm's MARR is 25%, determine which pump is more economical using the IRR and ERR methods.

TABLE 4-A-1   Cash Flow Profiles

End of year	Old pump	New pump	Increment New pump – Old pump
0	–$ 0	–$ 1,600	–$ 1,600
1	10,000	20,000	10,000
2	10,000	0	–10,000

Solution   Using FW calculations to solve for the internal rate of return on the incremental net cash flows gives

$$-1,600(1 + i)^2 + 10,000(1 + i) - 10,000 = 0.$$

This equation has two roots yielding internal rates of return of $i = 0.25$ (25%) and $i = 4.0$ (400%). With two (or more) roots, there may be some confusion as to which should be compared to the MARR in order to select the correct alternative. Actually, neither may be correct.

Using the same incremental net cash flows to compute the external rate of return results in

$$1,600(1 + i)^2 + 10,000 = 10,000(1 + r).$$

[1] Solomon, Ezra, "The Arithmetic of Capital Budgeting Decisions," *The Journal of Business* Vol. 29, No. 2(1956):124–129.

Solving for i as a function of the reinvestment rate r,

$$i = 2.5\sqrt{r} - 1.$$

Table 4-A-2 gives different values of r, the corresponding external rate of return i, and the recommended alternative to select.

Table 4-A-2 Values of r, i, and Recommended Alternative

r Reinvestment rate	$i = $ ERR External rate of return	Recommended alternative
0.10	−0.209	Old pump
0.15	−0.032	Old pump
0.20	0.118	Old pump
0.25	0.250	Either pump
0.30	0.369	New pump
0.50	0.768	New pump
1.00	1.500	New pump
2.00	2.536	New pump
3.00	3.330	New pump
4.00	4.000	Either pump
5.00	4.590	Old pump

Note that for those ERR values in excess of the reinvestment rate, the decision is to purchase the new pump. Also, note the correspondence between the internal rates of return (25% and 400%) and the values of the external rate of return at which one is indifferent about the alternatives. Naturally, for ERR values less than r, it is preferable to invest elsewhere and earn the MARR, r, rather than invest in the new pump. ■

Other Rate of Return and Investment Recovery Measures of Merit for Financial and Management Purposes

Chapter 4 demonstrated two theoretically sound methods for calculating rates of return over a project life or study period. Several additional methods that often have significant weaknesses, but are sometimes used for computational simplicity or to relate to commonly understood accounting figures, are briefly explained below.

Consider the project (in Examples 4-1 and 4-6) with the following given data:

First cost:	$10,000
Project life:	5 yr
Salvage value:	$2,000
Annual receipts:	$5,000
Annual disbursements:	$2,200

The calculated IRR for the project was 16.9%, and the ERR for the project was 13.81% (for an assumed reinvestment rate of 10%). Both of these methods are theoretically correct (keeping in mind that the IRR method assumes reinvestment at the computed rate, 16.9% in this case).

A. Other Rates of Return (Based on Accounting Averages over Life)

We will now demonstrate calculation of other (nontheoretically correct) rates of return, with minimal explanations and assuming straight-line depreciation:

$$\text{Depreciation/yr} = \frac{\text{First cost} - \text{Salvage value}}{\text{Project life}}$$

$$= \frac{\$10,000 - \$2,000}{5} = \$1,600/\text{yr}.$$

1. RR: Average Return/Yr on Original Investment

$$= \frac{\text{Net profit/yr}}{\text{Original investment}} = \frac{\text{Net cash flow/yr} - \text{Depreciation/yr}}{\text{Original investment}}$$

$$= \frac{\$5,000 - \$2,200 - \$1,600}{\$10,000} = \frac{\$1,200}{\$10,000} = 12\%$$

2. RR: Average Return/Yr on Average Investment

$$= \frac{\text{Net profit/yr}}{\dfrac{\text{Original investment} + \text{Salvage value}}{2}}$$

$$= \frac{\$5,000 - \$2,200 - \$1,200}{\dfrac{\$10,000 + \$2,000}{2}} = \frac{\$1,200}{\$6,000} = 20\%$$

B. Rate of Return for a Given Period (Year)

This measure is commonly and correctly applied using accounting statement results for a business or firm, where:

$$RR = \frac{\text{Book profit for period (year)}}{\text{Book value (i.e., Undepreciated investment)}}.$$

However, when applied to a particular project, such as given above, its results are very dependent on the period for which the measure is made and do not reflect the RR over the life of the project. For example, if the RR were calculated at the end of the fifth year when the book value had been depreciated down to the salvage value of $2,000, the RR would be $1,200/$2,000 = 60%.

C. Measures of Investment Recovery

Analysts and managers often use quick screening measures of merit based on how quickly the investment is to be recovered, with or without interest included. Two common such measures are shown below.

1. Payback (Payout) Period

$$= \frac{\text{Investment}}{\text{Net cash flow/yr}} = \frac{\$10,000}{\$5,000 - \$2,200} = 3.57 \text{ yr}$$

2. Payback (Capital Recovery) Period [with Interest at, Say 10%]

$$= N \text{ at which } [-\$10,000 + (\$5,000 - \$2,000)(P/A, 10\%, N) = 0]$$

$$(P/A, 10\%, N) = \frac{\$10,000}{\$2,800} = 3.57$$

By interpolation, $N = 4.7$ yr.

The above measures have the significant weaknesses of not considering what is expected to happen after the payback period is reached, but they are often used as indicators of riskiness.

5
Estimating for Economic Analyses

5.1 Introduction

Probably the most difficult and expensive part of any economic analysis is to determine the estimates needed to complete the analysis. This chapter will attempt to provide a perspective and approaches to estimating, with emphasis on making single-valued estimates for the traditional assumed-certain analyses that have been the topic of Part I of this book. Chapter 10 will cover estimating in terms of probabilities and other measures of variability to reflect the risk inherent in predicting future outcomes. Chapter 17 discusses how activity-based cost data are used in the estimating process.

5.2 Estimating: Difficulty and Perspective

The basic difficulty in estimating for economic evaluations is that *forecasting* of critical elements associated with the manufacture of a product or the delivery of a service is unavoidable. In this regard, recall that in Chapter 1 we described various cost concepts that are important to economic analyses. The key point was that the costs important to economic comparisons are the marginal (variable, or incremental) costs for the *future*.

Another difficulty in estimating for economic analyses is that most prospective projects for which estimations are to be made are unique; that is, substantially similar projects have not been undertaken in the past under conditions that are the same as expected for the future. Hence, outcome data that can be used in estimating directly and without modification often do not exist. It may be possible, however, to gather data on certain past outcomes that are related to the outcomes being estimated and to adjust and project that data based on expected future conditions. Techniques for collecting and projecting estimation data and also for making probabilistic estimates are founded in the field of statistics. Chapter 10 contains some very useful techniques for making and using probabilistic estimates. Also, Chapter 15 covers statistical decision analysis that emphasizes Bayesian revision of probabilities.

Whenever an economic analysis is for a major new product or process, the estimating for that analysis should be an integral part of comprehensive planning procedure. Such comprehensive planning would require the active participation of at least the marketing, design engineering, manufacturing, finance, and top management functions. It would generally include the following features:

1. a realistic master plan for product development, testing phase into production, and operation;
2. provision for working capital and facilities requirements;
3. integration with other company plans;
4. evaluation against company objectives for market position, sales volume, profit, and investment;
5. provision of a sound basis for operating controls if the project is adopted.

Obviously, such comprehensive planning is costly in time and effort, but when a new product or process has major implications for the future of a firm, it is generally a sound rule to devote a greater rather than a lesser amount of effort to complete planning, including estimates for the economic analysis that is a partial result of the planning. The application of this rule, of course, is bounded by constraints of limited time and talent; however, following the rule will tend to minimize the chance of poor decisions or lack of preparedness to implement projects once the decision to invest has been made.

5.3 Estimation Accuracy

Estimates or forecasts, by their nature, are evaluations of incomplete evidence indicating what the future may hold. They may be based on empirical observations of only somewhat similar or analogous situations, adjusted on the basis of the kind of personal hunch that grows out of the accumulation of the experiences. Or they may be inferences drawn from various kinds of available objective data, such as trade statistics, results experienced in analogous situations, or personal observations.

Regardless of the estimate source, the estimate user should have specific recognition that the estimate will be in error to some extent. Even the use of formalized estimation techniques will not, in itself, eliminate error, although it will hopefully reduce error somewhat, or will at least provide specific recognition of the anticipated degree of error.

The level of detail and accuracy of an estimate should depend on the following:

1. the estimability of that which is to be estimated;
2. methods or techniques employed;

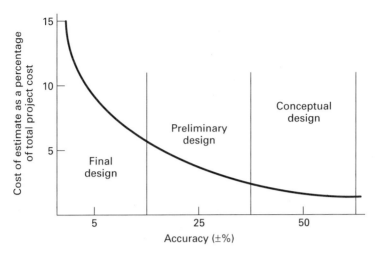

Figure 5-1   Estimation accuracy versus cost of the estimate.

   3. qualifications of estimator(s);

   4. time and effort available and justified by the importance of the study;

   5. sensitivity of study results to the particular estimate.

As estimates differ from conceptual to more detailed, the cost of preparing the estimate increases. As one might expect, the more expensive detailed estimates, which are usually associated with the final design of a system, are also more accurate. The general relationship between phase of design (conceptual, preliminary and final) and its typical estimation accuracy versus cost of the estimate as a percentage of total project cost is shown in Fig. 5-1.

5.4 Sources of Data

The variety of sources from which estimating information can be obtained is too great for complete enumeration. The following four major sources, which are ordered roughly according to decreasing importance, are described in subsequent sections:

   1. accounting records;

   2. other sources within the firm;

   3. sources outside the firm;

   4. research and development.

5.4.1 Accounting Records

It should be emphasized that although data available from the records of the accounting function are a prime source of information for economic analyses, these data are very often not suitable for direct, unadjusted use.

A very brief and oversimplified description of the accounting process is contained in Appendix 1-A. In its most basic sense, accounting consists of a series of procedures for keeping a detailed record of monetary transactions between established categories of assets, each of which has an accepted interpretation useful for its own purposes. The data generated by the accounting function are often inherently misleading for economic analyses, not only because they are based on past results, but also because of the following limitations.

First, the accounting system is rigidly categorized. These categories for a given firm may be perfectly appropriate for operating decisions and financial summaries, but rarely are they fully appropriate to the needs of economic analyses for longer-term decisions.

Another limitation of accounting data for obtaining estimates is the misstatements imbedded by convention into accounting practice. These underestimates are based philosophically on the idea that management should avoid overstating the value of its assets and should therefore assess them very conservatively. This leads to such practices as not changing the stated value of one's resources as they appreciate due to rising market prices and depreciating assets over a much shorter life than actually expected. As a result of such accounting customs, the analyst should always be careful about treating such resources as cheaply (or, sometimes, as expensively!) as they might be represented.

The final limitations of accounting data are their illusory precision and implied authoritativeness. Although it is customary to present data to the nearest dollar or the nearest cent, the records are not nearly that accurate in general.

In summary, accounting records are a good source of historical data, but they have severe limitations when used in making estimates for economic analyses. Further, accounting records rarely directly contain the variable costs, especially opportunity costs, appropriate for economic analyses.

5.4.2 Other Sources Within the Firm

The usual firm has a large number of people and records that may be excellent sources of estimates or information from which estimates can be made. Colleagues, supervisors, and workers can provide insights or suggest sources that can be obtained readily.

Examples of records that exist in most firms are sales, production, inventory, quality, purchasing, industrial engineering, and personnel. Table 5-1 provides a list of the types of data that might be needed for cost-estimating purposes together with typical sources (mostly intrafirm) for the data.

5.4.3 Sources Outside the Firm

There are innumerable sources outside the firm that can provide information helpful for estimating. The main problem is to determine which sources are potentially most fruitful for particular needs.

Published information such as technical directories, trade journals, U.S. government publications, and comprehensive reference books offer a wealth of information to the knowledgeable or persistent searcher.

TABLE 5-1   Types and Sources of Cost Estimating Data

Description of data	Sources
General design specifications	Product engineering and/or sales department
Quantity and rate of production	Request for estimate or sales department
Assembly or layout drawings	Product engineering or sales department or customer's contact person
General tooling plans and list of proposed subassemblies of product	Product engineering or manufacturing engineering
Detail drawings and bill of material	Product engineering or sales department
Test and inspection procedures and equipment	Quality control or product engineering or sales department
Machine tool and equipment requirements	Manufacturing engineering or vendors of materials
Packaging and/or transportation requirements	Sales department or shipping department or product engineering (government specifications)
Manufacturing routings and operation sheets	Manufacturing engineering or methods engineering
Detail tool, gauge, machine, and equipment requirements	Manufacturing engineering or material vendors
Operation analysis and workplace studies	Methods engineering
Standard time data	Special charts, tables, time studies, and technical books and magazines
Material release data	Manufacturing engineering and/or purchasing department or materials vendors
Subcontractor cost and delivery data	Manufacturing engineering and/or purchasing department or customer
Area and building requirements	Manufacturing engineering or plant layout or plant engineer
Historical records of previous cost estimates (for comparison purposes, etc.)	Manufacturing engineering or cost department or sales department
Current costs of items presently in production	Cost department or treasurer or comptroller

Personal contacts are excellent potential sources. Vendors, salespersons, professional acquaintances, customers, banks, government agencies, chambers of commerce, and even competitors are often willing to furnish needed information if the request is serious and tactful.

Probably the most valuable estimating sources outside the firm, which are available and updated continuously, are cost indexes. Cost indexes provide a means for converting past costs to present costs through the use of dimensionless numbers, called *indexes,* to reflect relative costs for two or more points in time.

There are many cost indexes, and they cover almost every area of interest. Some are based on national averages; others are very specialized. Indicative values of indexes from the *Engineering News-Record* are shown in Table 5-2.

The Bureau of Labor Statistics of the U.S. Department of Labor publishes much data on price changes of many types of products and earnings of workers in almost every industry. Some of these data are components of many cost indexes, and others are useful in constructing highly specialized indexes.

TABLE 5-2   Typical Engineering Indexes*

Year	Materials price index	Skilled labor index	Building cost index	Construction cost index
1975	862	1921	1378	2128
1976	971	2061	1504	2322
1977	1077	2208	1620	2513
1978	1177	2350	1750	2693
1979	1303	2487		2886
1980	1449	2670	1915	3159
1981	1480	2902	2014	3384
1982	1547	3244	2192	3721
1983	1641	3507	2352	4006
1984	1632	3691	2412	4118
1985	1604	3765	2406	4151
1986	1612	3808	2447	4231
1987	1648	3937	2518	4359
1988	1693	4061	2586	4484
1989	1677	4153	2612	4574
1990	1708	4283	2673	4691
1991	1693	4387	2715	4772
1992	1738	4536	2799	4927
1993	1846	4665	2915	5106
1994	2109	4764	3116	5381

*These index values are from the March 28, 1994 issue of *Engineering News-Record*, published by the McGraw-Hill Publishing Company. The base year is 1913, with an index value of 100.

Cost indexes are limited in their accuracy and, like all statistical devices, must be used with caution. Most indexes are based on data combined in more or less arbitrary fashion. A cost index, like cost data themselves, normally will reflect only average changes, and an average often has little meaning when applied to a specific case. Under favorable conditions a ±10% accuracy is the most that can be expected in projecting a cost index over a 4- or 5-year period.

5.4.4 Research and Development

If the information is not published and cannot be obtained by consulting someone who knows, the only alternative may be to undertake research to generate it. Classic examples are developing a pilot plant and undertaking a test market program. These activities are usually expensive and may not always be successful; thus, this final step is taken only when there are very important decisions to be made and when the sources mentioned above are known to be inadequate.

5.5 Quantitative Estimating Techniques

An estimate, or forecast, is useful if it reduces the uncertainty surrounding a revenue or cost element. In doing this, a decision should result that creates increased value relative to the cost of making the estimate. This section describes three groups of estimating techniques that have proven to be very useful in preparing estimates for economic analysis. They are (1) time-series techniques, (2) subjective techniques, and (3) cost engineering techniques.

When revenue and/or cost elements are a function of time, such as unit sales per quarter, they are often referred to as a time series. Time-series data should be collected for the element under study and then carefully examined for underlying patterns. For example, a sudden increase in sales may be explained by increased government spending or a vendor filling its distribution pipeline. We will examine the use of *regression* for estimating causal relationships within time-series data and *exponential smoothing* for estimating future extensions to historical data patterns.

Frequently, the next stage of estimating is to apply expert judgment to the results of time-series techniques. Examples of subjective estimating approaches to be examined are the Delphi technique and technology forecasting. A highly effective estimating strategy is to couple a time-series technique, based on past data, with a subjective technique that introduces human judgment in attempting to discover how future revenue and cost elements will differ from those of the past.

Cost engineering techniques identify and utilize various revenue/cost drivers to compute estimates. They include models for estimating capital, material, labor and many other factors of production. These models may utilize correlation and regression analysis or they may be as simple as extensions of ratios of relevant cost indexes. For example, capital costs can often be accurately estimated by knowing the weight of a particular structure. Similarly, operating costs such as that for fuel can be computed from forecasts of kilowatt-hours generated by an electric power station.

5.5.1 Time-Series Techniques[1]

Two relatively simple, yet extremely useful, techniques for obtaining initial time-series forecasts of elements being estimated are described in this section: (1) linear regression analysis and (2) exponential smoothing.

5.5.1.1 Correlation and Regression Analysis.   Sometimes it is possible to correlate an element, such as revenue for a product line, with one or more economic indices, such as construction contracts awarded, disposable personal income, etc. Correlation concerns the explainable association between variables. When an index can be found to which an element to be estimated is highly correlated, but with a time lag, formal correlation analysis may be highly useful. In cases where the lag is insufficient for longer-term forecasts, correlation of an element to be estimated with the available index still leaves the estimator with the need to predict the future value(s) of the index itself.

Regression is a statistical method of fitting a line through data to minimize squared error. It is exact; however, graphing might be used to provide a satisfactory approximation. With linear regression, approximated model coefficients can be used to obtain an estimate of a revenue/cost element.

---

[1] Section 5.5.1 is adapted from J. R. Canada and W. G. Sullivan, *Economic and Multiattribute Evaluation of Advanced Manufacturing Systems* (Englewood Cliffs, NJ: Prentice Hall, 1989). Reprinted by permission of the publisher.

In linear regression involving one independent variable, x, and one dependent variable, y, the relationship that is used to fit n data points $(1 \leq i \leq n)$ is

$$y = a + bx. \tag{5-1}$$

A mathematical statement of expressions used to estimate a and b in the simple linear regression equation 5-1 is as follows:

$$b = \frac{\sum\limits_{i=1}^{n} x_i y_i - \bar{x} \sum\limits_{i=1}^{n} y_i}{\sum\limits_{i=1}^{n} x_i^2 - \bar{x} \sum\limits_{i=1}^{n} x_i}, \tag{5-2}$$

$$a = \bar{y} - b\bar{x}. \tag{5-3}$$

Here $\bar{x}$ and $\bar{y}$ are averages of the independent variable and dependent variable, respectively, for the n data points.

Example 5-1

A durable goods manufacturer has found personal disposable income in its market region in a given quarter to be strongly related to sales in the following quarter. These data are listed and summarized in Table 5-3. Because a plot of these data indicates an approximately linear relationship between the dependent variable (on the y-axis) and the independent variable (on the x-axis), linear regression is used to fit an equation to the data. The data and calculations summarized in Table 5-3 are utilized below to determine the linear regression equation:

$$b = \frac{\sum\limits_{i=1}^{n} x_i y_i - \bar{x} \sum\limits_{i=1}^{n} y_i}{\sum\limits_{i=1}^{n} x_i^2 - \bar{x} \sum\limits_{i=1}^{n} x_i} = \frac{2{,}626{,}817 - 250.6(9{,}788)}{1{,}416{,}926 - 250.6(5{,}012)}$$

$$= \frac{173{,}944.2}{160{,}918.8} = 1.081,$$

$$a = \bar{y} - b\bar{x} = 489.4 - 1.081(250.6) = 218.5.$$

Thus,

$$y = 218.5 + 1.081x.$$

Figure 5-2 shows the plotted data and the calculated regression line. As an example of how the regression equation is used, suppose that disposable income for the previous quarter is 310 (or $\$310 \times 10^6$). Then our forecast or estimate of sales (in thousands of dollars) for the current quarter, $\hat{y}$, is

$$\hat{y} = 218.5 + 1.081(310) = 553.6 \text{ (or } \$553.6 \times 10^3 \text{)}.$$

The correlation coefficient is a measure of the strength of the relationship between two variables only if the variables are linearly related. In Example 5-1 the correlation coefficient, r, which measures the degree of strength, can be determined as follows:

$$r = \frac{S_{xy}}{\sqrt{S_{xx} \cdot S_{yy}}} \qquad (-1 \leq r \leq 1), \tag{5-4}$$

TABLE 5-3   Calculations for Simple Linear Regression

Data point, i (period)	y_i	x_i	$x_i y_i$	x_i^2
1	360	121	43,560	14,641
2	260	118	30,680	13,924
3	440	271	119,240	73,441
4	400	190	76,000	36,100
5	360	75	27,000	5,625
6	500	263	131,500	69,169
7	580	334	193,720	111,556
8	560	368	206,080	135,424
9	505	305	154,025	93,025
10	480	210	100,800	44,100
11	602	387	232,974	149,769
12	540	270	145,800	72,900
13	415	218	90,470	47,524
14	590	342	201,780	116,964
15	492	173	85,116	29,929
16	660	370	244,200	136,900
17	360	170	61,200	28,900
18	410	205	84,050	42,025
19	680	339	230,520	114,921
20	594	283	168,102	80,089
Totals	9,788	5,012	2,626,817	1,416,926

$$\sum_{i=1}^{n} x_i = 5,012 \qquad\qquad \sum_{i=1}^{n} y_i = 9,788$$

$$\bar{x} = \frac{\sum_{i=1}^{n} x_i}{n} = \frac{5,012}{20} = 250.6 \qquad \sum_{i=1}^{n} x_i^2 = 1,416,926$$

$$\bar{y} = \frac{\sum_{i=1}^{n} y_i}{n} = \frac{9,788}{20} = 489.4 \qquad \sum_{i=1}^{n} x_i y_i = 2,626,817$$

Key: y_i = actual quarterly sales ($\$10^3$) for period i
x_i = disposable income in *preceding* period ($\$10^6$)

where

$$S_{xy} = \sum_{i=1}^{n} x_i y_i - \left(\sum_{i=1}^{n} x_i \right) \left(\sum_{i=1}^{n} y_i \right) / n \qquad (5\text{-}5)$$

$$S_{xx} = \sum_{i=1}^{n} x_i^2 - \left(\sum_{i=1}^{n} x_i \right)^2 / n \qquad (5\text{-}6)$$

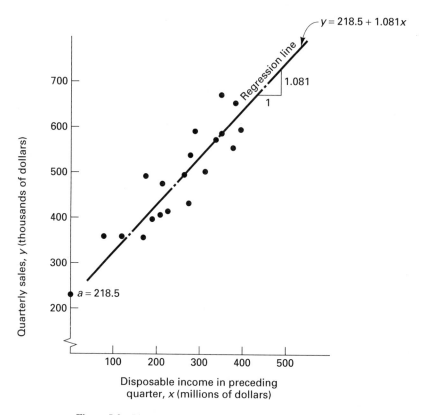

Figure 5-2 Plotted points and regression line for Example 5-1.

$$S_{yy} = \sum_{i=1}^{n} y_i^2 - \left(\sum_{i=1}^{n} y_i \right)^2 / n. \qquad (5\text{-}7)$$

If there is no relationship at all (a shotgun effect) between the dependent and independent variables, r will be zero, or nearly zero. A negative value of r indicates that one variable decreases as the other increases. When r is positive, the dependent and the independent variable both increase at the same time. The closer r is to -1 or $+1$, the more "perfect" is the correlation.

Using Eqs. 5-4 through 5-7, the correlation coefficient for Example 5-1 is

$$S_{xy} = 2,626,817 - \frac{(5,012)(9,788)}{20} = 173,944$$

$$S_{xx} = 1,416,926 - \frac{(5,012)^2}{20} = 160,919$$

$$S_{yy} = 5,030,754 - \frac{(9,788)^2}{20} = 240,507$$

$$r = \frac{173,944}{\sqrt{(160,919)(240,507)}} = 0.88.$$

The positive value of r indicates that as the independent variable (disposable income in the previous period) increases, the dependent variable (quarterly sales) will also tend to increase. This value of r indicates a good (but not great) relationship between the independent and dependent variables. One measure of the goodness of fit between x and y is called the coefficient of determination, which equals r^2. In our example, the value of r^2 is 0.77. The coefficient of determination measures the proportion of total variation that is explained by the regression line. Thus in Example 5-1 the regression line, $y = 218.5 + 1.081x$, accounts for 77% of the variation in quarterly sales activity ($\$10^3$) in period i for the 20 observations of disposable income in the preceding time period ($\$10^6$). ∎

5.5.1.2 Exponential Smoothing.

An advantage of the exponential smoothing method compared to simple linear regression for time-series estimates is that it permits the estimator to place relatively more weight on current data rather than treating all prior data points with equal importance. Forecasting equations can quickly be revised with a relatively small number of calculations as each new data point is collected. Also, exponential smoothing does not assume linearity.

The main disadvantage of exponential smoothing is the basic assumption that trends and patterns of the past will continue into the future. However, it is more sensitive to changes than is linear regression. Because time-series analysis cannot predict turning points in the future, expert judgment and/or analysis of suspected causal factors should be used in interpreting results.

The basic exponential smoothing model that we shall discuss and illustrate is as follows:

$$S_t = \alpha'x_t + (1 - \alpha')S_{t-1} \qquad (0 \le \alpha' \le 1)$$

or

$$\begin{pmatrix} \text{Forecast for period } t + 1, \\ \text{made in period } t \end{pmatrix} = \alpha'\begin{pmatrix} \text{Actual data point} \\ \text{in period } t \end{pmatrix}$$
$$+ (1 - \alpha')\begin{pmatrix} \text{Forecast for period } t, \\ \text{made in period } t - 1 \end{pmatrix}. \qquad (5\text{-}8)$$

This term, α', the *smoothing constant,* merely provides a relative weighting for the new datum point compared to previous estimates. In general, α' should lie between 0.01 and 0.30, but the analyst should not hesitate to use a value outside this range if it gives better results with representative historical data.

An advantage of this technique of forecasting is its flexibility of weighting. If the weighting constant α' is 1, the mathematical model reduces to using the most recent period's outcome as the forecast. If α' is very close to 0, this is essentially equivalent to using an arithmetic average of actual outcomes over a large number of previous periods as the best estimate of the future outcome. Intermediate choices for α' between 0 and 1 provide forecasts that have more or less emphasis on long-run average outcomes versus current outcomes.

In estimating revenues, sales demand is an essential element for an economic analysis. Single exponential smoothing is illustrated with sales data listed in Table 5-4 that are graphed in Fig. 5-3. Based on Eq. 5-8, the following are example calculations for S_t, which is termed a "smoothed statistic" (i.e., forecast) for period $t + 1$ (but made in period t):

$$S_t = \alpha'x_t + (1 - \alpha')S_{t-1},$$
$$S_1 = 0.3(50) + 0.7(50) = 50,$$
$$S_2 = 0.3(52) + 0.7(50) = 50.6,$$
$$S_3 = 0.3(47) + 0.7(50.6) = 49.52,$$
$$S_4 = 0.3(51) + 0.7(49.52) = 49.96.$$

In determining S_1 above, a value of S_0 must be estimated. Because no trend is assumed in the data for this model, an estimate based on the average of the first few data points is adequate. Here S_0 was chosen to be 50.

To better understand the meaning of exponential smoothing, the following expression shows how demand data $x_t, x_{t-1}, x_{t-2}, \ldots$ are included in the forecast:

$$S_t = \alpha'x_t + (1 - \alpha')S_{t-1}, \tag{5-9}$$

where

$$S_{t-1} = \alpha'x_{t-1} + (1 - \alpha')S_{t-2} \tag{5-10}$$

and then

$$S_t = \alpha'x_t + \alpha'(1 - \alpha')x_{t-1} + (1 - \alpha')^2 S_{t-2}. \tag{5-11}$$

TABLE 5-4   Exponential Smoothing Example

Period number, t	Demand, x_t (1,000 units)	S_t ($\alpha' = 0.30$)
0	—	50.00
1	50	50.00
2	52	50.60
3	47	49.52
4	51	49.96
5	49	49.67
6	48	49.17
7	51	49.72
8	40	46.80
9	48	47.16
10	52	48.61
11	51	49.33
12	59	52.23
13	57	53.66
14	64	56.76
15	68	60.13
16	67	62.19
17	69	64.23
18	76	67.76
19	75	69.93
20	80	72.95

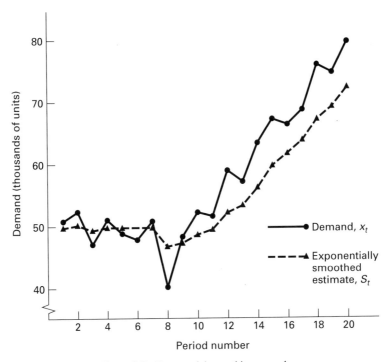

Figure 5-3   Exponential smoothing example.

It is possible to continue substituting smoothed values in the same fashion until we get the following:

$$S_t = \alpha'x_t + \alpha'(1 - \alpha')x_{t-1} + \alpha'(1 - \alpha')^2 x_{t-2}$$
$$+ \alpha'(1 - \alpha')^3 x_{t-3} + \cdots + (1 - \alpha')^t S_0. \qquad (5\text{-}12)$$

A forecast T periods into the future is simply S_t because simple exponential smoothing assumes a constant pattern of data over time (i.e., no trend).

As one can see, every previous value of x is included in S_t. The x values are weighted so that the values more distant in time have successively smaller weighting factors. A large α' will place very little weight on remote data. The following calculations illustrate the weighting of the different data points included in S_4 above as described by Eq. 5-12:

$$S_4 = 0.3(51) + 0.3(0.7)47 + 0.3(0.7)^2 52 + 0.3(0.7)^3 50 + (0.7)^4 50$$
$$= 0.30(51) + 0.21(47) + 0.147(52) + 0.1029(50) + 0.2401(50)$$
$$= 49.96.$$

This agrees with our calculation of S_4 earlier.

Table 5-4 shows a continuation of the preceding example for hypothetical demand data over twenty periods. It should be noted that the higher the value of α', the closer the new estimate will be to the most recent datum point.

5.5.2 Subjective Techniques

The time-series techniques discussed earlier are underpinned by the premise that the future is an extension of the past. However, the future will contain events that today are poorly understood or completely unanticipated. Hence the aim of this section is to explain and illustrate two techniques for developing subjective information for purposes of making an estimate: (1) the Delphi method and (2) technology forecasting.

5.5.2.1 The Delphi Method.

Most decision makers draw on the advice of experts as they form their judgments. Often the decision situation is highly complex and poorly understood, so that no single person can be expected to make an informed decision. The traditional approach to decision making in such cases is to obtain expert opinion through open discussions and to attempt to determine a consensus among the experts. However, results of panel discussions are sometimes unsatisfactory because group opinion is highly influenced by dominant individuals and/or because a majority opinion may be used to create the "bandwagon effect."

The Delphi method attempts to overcome these difficulties by forcing persons involved in the forecasting exercise to voice their opinions anonymously and through an intermediary. The intermediary acts as a control center in analyzing responses to each round of opinion gathering and in feeding back opinion to participants in subsequent rounds. By following such a procedure, it is hoped that the responses will converge on a consensus forecast that turns out to be a good estimator of the true outcome.

Two premises underlie the Delphi method. The first is that persons who are highly knowledgeable in a particular field make the most plausible forecasts. Second, it is believed that the combined knowledge of several persons is at least as good as that of one person.

Typically, the technique is initiated by writing an unambiguous description of the forecasting problem and sending this, along with relevant background information, to each participant in the study. Often the participants are invited to list major areas of concern in their particular specialty as they may relate to the problem being addressed by the study. The first questionnaire sent out might request the opinion of each expert regarding likely dates for the occurrence of an event identified in the problem statement. Because responses to this type of question will normally reveal a spread of opinions, *interquartile* ranges are customarily computed and presented to the experts at the beginning of the second round. Interquartile ranges identify upper and lower quartile values in the continuum of responses such that 50% of the responses fall within that range.

In the second round of the Delphi technique, the participants are asked to review their response in the first round relative to interquartile ranges from that round. They then have the opportunity to revise their estimates in light of the group response. At this point, participants can request that additional information relevant to the forecasting problem be gathered and sent to them.

If an estimate departs appreciably from the group median, the respondent who furnished it is asked to give reasons for his or her position. Frequently, all panelists are urged to conceive statements that challenge or support estimates falling outside the interquartile or some other range of responses. These reasons, along with routine second-round estimates for the entire group, are again analyzed and statistically summarized (usually as interquartile ranges, although other measures capable of showing group convergence or divergence could be used).

In those cases where a third-round questionnaire is felt necessary, participants receive a summary of second-round responses plus a request to reconsider and/or explain their estimate in view of group responses in the second round. They are again asked to reassess their earlier responses and possibly to explain why their estimates do not conform to the majority of group opinion.

An example of quantitative results of the Delphi technique is summarized in Fig. 5-4. The problem of concern to a manufacturer of large earth-moving equipment was to develop a forecast of total company sales during fiscal year 19XX. Six marketing and sales experts (A through F) were asked to consider historical company and industry data and anonymously prepare a forecast of bookings for a particular product. First-round questionnaire results are shown in the top part of Fig. 5-4. The median response in round 1 was 229, and the interquartile range of the responses was 85. Results of round 1 were fed back to each participant along with additional information pertaining to the forecast that each person requested. The second and third rounds were completed in a similar manner. Notice how forecasts in the three iterations of questioning tend to converge, with a final median forecast value of 260 units and an interquartile range of 47.

5.5.2.2 Technological Forecasting.

Technological forecasting is a name given to a myriad of specialized forecasting techniques. It provides procedures for data collection and analysis to predict future technological developments and the impacts such developments will have on the environment and lifestyles of people. These techniques seek to make potential technological developments explicit, but more important, they force decision makers to try to anticipate future developments.

Technology forecasting is a method that can be used to estimate the growth and direction of a technology. A typical question that technology forecasting attempts to answer is: What will be the machining tolerances of numerically controlled machine tools in the future? Or, what will be the operating characteristics of the next generation of integrated circuits for engineering workstations?

Trend extrapolation is often used to make technological forecasts. This technique is based on a historic time series for a selected technological parameter. It often is assumed that the factors influencing historical data are likely to remain constant rather than to change in the future. Usually, a single-function parameter such as speed, horsepower, or weight is extrapolated.[2] A good trend extrapolation depends on selection and prediction of key parameters of performance. The trend under study should be capable of quantification in order that it can be portrayed numerically, and an adequate database should exist on which to base a reliable trend line. An example of trend extrapolation is presented in Fig. 5-5. Notice that the y-axis is a logarithmic scale.

---

[2] Linear regression may be used for extrapolating data into the future or, more simply, a straight line can be "eye-balled."

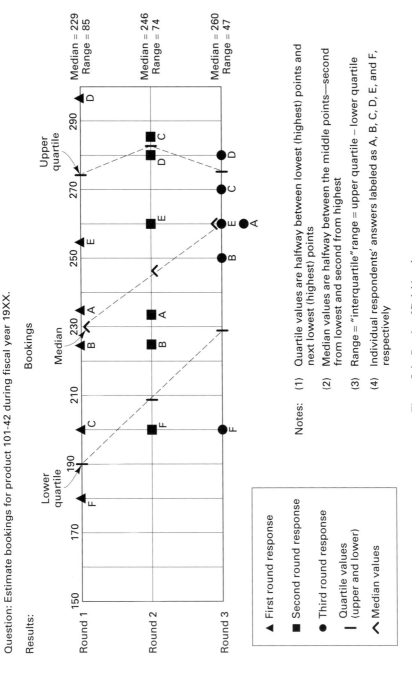

Figure 5-4 Results of Delphi study.

An advantage of trend extrapolation is that historical data are often readily obtainable. A straight-line or fitted-curve projection of the future is easily understood and used. A drawback to extrapolation stems from the assumption that factors that shaped the past will continue to hold basically unchanged in the future. Trend extrapolation techniques

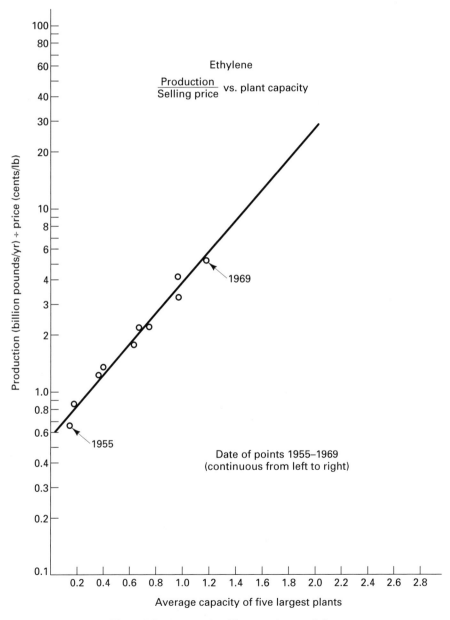

Figure 5-5   An example of linear trend extrapolation.

cannot predict unforeseen technology interactions such as unprecedented changes or inexplicable discoveries.

The *substitution curve* is based on the belief that a product or technology that exhibits a relative increase in performance over an older (i.e., established or conventional) product or technology will eventually substitute for the one having lesser performance. The relative increase in performance is the important factor in the substitution of one technology for another. A basic assumption with this method is that once the substitution of one technology for another has begun it will irreversibly continue to completion. Listed below are some common examples of the substitution effect:

Old technology	New technology
Petroleum lamps	Electric lamps
Horse-drawn carriages	Automobiles
Steam locomotive	Diesel locomotive
Cotton	Synthetic fibers
Leather	Vinyl
Soap	Detergent
Reciprocating engines	Turbojet engines
Hardwood floor	Plastic flooring

The forecast starts with the observation that a new technology is starting to displace an older technology. A measurement term that best defines the fraction of total usage of each technology must be selected, and time-series data are gathered for both technologies. These data are used to establish the initial takeover rate and to predict the year in which takeover will reach 50%. A typical substitution effect for two technologies is shown in Fig. 5-6.

Forecasting by analysis of *precursor events* uses the correlation of performance trends between two innovative technologies. Because technological advance usually follows a pattern of continuous increase, situations frequently occur in which an indicator of technical progress lags another by a given period of time. It is thus possible to utilize the leading technology to predict the status of the lagging technology over a time period equal to the lag time. The frequently cited example of precursor events shown in Fig. 5-7 concerns the historical relationship between maximum speed of military aircraft to the maximum speed of commercial transport aircraft.

In this example, it was found that the speed of commercial aircraft followed the speed of military aircraft by six years in the 1920s and eleven years in the 1950s. As a result, it was predicted that commercial transport aircraft with speeds of Mach 2 would be expected no later than 1970, or if such aircraft were not introduced at this time, aircraft with speeds of Mach 3 would be introduced near 1976.

5.5.3 Cost Engineering Techniques

A wide array of techniques exists for estimating investment and working capital requirements associated with products, processes, and materials.[3] In this section, we

---

[3] An excellent reference is P. F. Ostwald, *Engineering Cost Estimating,* 3rd ed. (Englewood Cliffs, NJ: Prentice Hall, 1992).

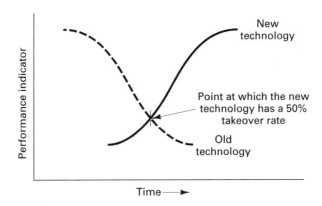

Typical technology life cycle

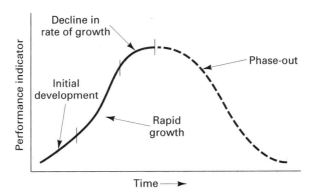

Figure 5-6   An example of substitution curves.

consider a small collection of useful techniques that permit these types of estimates to be easily made. Many of them utilize various kinds of revenue/cost indexes in the preparation of an estimate.

5.5.3.1 Unit Method.   A popular cost engineering technique is the *unit method,* which involves using an assumed or estimated "per unit" factor. Some examples are

- capital cost of plant per kilowatt of capacity,
- fuel cost per kilowatt-hour generated,
- capital cost per installed phone,
- revenue per long-distance call,
- operating cost per mile,
- maintenance cost per day of use.

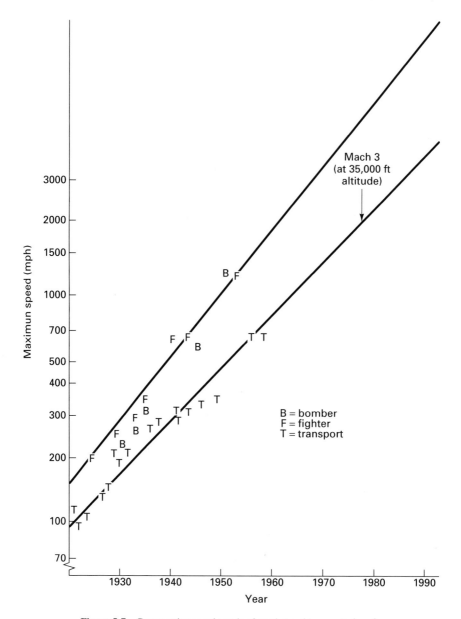

Figure 5-7   Comparative speed trends of combat and transport aircraft.

These factors may be multiplied by the appropriate unit to provide the total estimate. The following examples may be used for breaking quantities to be estimated
into units that can be estimated readily:

   1. in different units (*example:* dollars per week, to convert to dollars per
   year);

2. a proportion instead of a number (*example:* percent defective, to convert to number of defects);

3. a number instead of a proportion (*example:* number defective and number produced, to convert to percent defective);

4. a rate instead of a number (*example:* miles per gallon, to convert to gallons consumed);

5. a number instead of a rate (*example:* miles and hours traveled, to convert to average speed);

6. using an adjustment factor to increase or decrease a known or estimated number (*example:* defectives reported, to convert to total defectives).

Although the unit method is useful for preliminary estimating, the values can be very misleading because there is no consideration of the principle of economies of *scale* or economies of *scope*.

5.5.3.2 Factor Technique.   The *factor technique* is an extension of the unit method in which one sums the product of one or more quantities or components involving unit factors and adds these to any components estimated directly. That is,

$$C = \Sigma\, C_d + \Sigma\, f_i \times U_i, \tag{5-13}$$

where

$$
\begin{aligned}
C &= \text{value (cost, price, etc.) being estimated,}\\
C_d &= \text{cost of selected components estimated directly,}\\
f_i &= \text{cost per unit of component } i,\\
U_i &= \text{number of units of component } i.
\end{aligned}
$$

Example 5-2
Suppose that we need a slightly refined estimate of the cost of a house consisting of 1,500 ft^2, two porches, and a garage. Using unit factors of \$40/ft^2, \$2,000/porch, and \$3,000/garage, we can calculate the estimate as

$$\$40 \times 1,500 + \$2,000 \times 2 + \$3,000 = \$67,000. \qquad \blacksquare$$

5.5.3.3 Exponential Costing.   *Exponential costing* may be used when the proposed plant has a different production capacity than the existing plant. According to de la Mare, "the principle of exponential costing states that for many real-world production systems proportionate increases in production capacity can be achieved by less than proportionate increases in capital cost. This principle is a special manifestation of the law of increasing returns to scale, and is known as the law of increasing technical returns to scale."[4]

The following general equation represents most types of equipment:

$$\frac{C_a}{C_b} = \left(\frac{Q_a}{Q_b}\right)^{\beta}, \tag{5-14}$$

---

[4] R. F. de la Mare, *Manufacturing Systems Economics* (New York: Holt, Rinehart and Winston, 1982), p. 151.

where

C_a = capital cost of the proposed facility,
C_b = known capital cost of an existing facility,
Q_a = production capacity of the proposed facility,
Q_b = production capacity of the existing plant,
β = cost-exponent factor, which can range from 0.4 to greater
 than 1.00, but is usually in the range from 0.5 to 0.8.

Table 5-5 provides typical cost-exponent factors for selected types of industrial equipment. Equation 5-15 permits one to obtain an estimate of the capital cost of a proposed project by including factors to adjust for the effects of price increases as follows:

$$C_a = C_b \left(\frac{Q_a}{Q_b} \right)^\beta I_t, \qquad (5\text{-}15)$$

where $I_t = I_n/I_k$ = index of price change (inflation) from period k to period n. A fairly detailed treatment of inflation, as it affects cost estimating and comparisons of alternatives, is provided in Chapter 18.

The accuracy of the exponential costing method depends largely on the similarity between the two projects, and on the accuracy of the cost-exponent factor β. Generally, error ranges from ±10 to ±30% of the actual final cost.

Example 5-3

A certain steam-generating boiler in the utility plant of a manufacturing complex produces 50,000 lb/hr of saturated steam. This boiler was purchased for $250,000 eight years ago. If the price index for this type of boiler has increased at an average rate of 12% per year for the past 8 yearstion, how much would a 150,000 lb/hr boiler cost now? The cost-exponent factor for this boiler is 0.50.

Solution   Utilizing Eq. 5-15,

$$C_a = C_b \left(\frac{Q_a}{Q_b} \right)^\beta I_t,$$

with $I_t = (1 + 0.12)^8$, the cost of the proposed boiler can be estimated as follows:[5]

$$C_a = \$250,000 \left[\left(\frac{150,000}{50,000} \right)^{0.5} (1.12)^8 \right]$$

$$= \$250,000[(3.00)^{0.5}(2.476)]$$

$$= \$1,072,140. \qquad \blacksquare$$

Example 5-4

Assume that 6 years ago an 80 kW diesel electric set cost $160,000. The plant engineering staff is considering a 120 kW unit of the same general design to power a small isolated plant. The cost-exponent factor is 0.6. The price index for this class of equipment 6 years ago was 187 and is now 194. Assume we want to add a precompressor, which, when isolated and estimated separately, now costs $18,000. Determine the total cost now of the 120 kW unit.

---

[5] See Section 18.5 of Chapter 18. The index of price change, $I_t = I_n/I_k$, equals $(F/P, 12\%, 8)$.

TABLE 5-5 Typical Cost-Exponent Factors, β*

Process industrial equipment		Material-handling equipment		General industrial equipment	
Item	β	Item	β	Item	β
Agitators	0.3–0.5	Bagging machines	0.8	Air compressors	0.4
Centrifuges	0.7–1.3	Conveyors	0.7	Air driers	0.6
Evaporators	0.5–0.7	Conveyors (bucket)	0.6–0.8	Cranes	0.6
Heat exchangers	0.7–0.9	Conveyors (roller)	0.9	Driers (product)	0.4–0.5
Piping	0.7–0.9	Elevators	0.4	Electric motors	0.8
Pumps	0.5–0.9	Hoppers	0.7–0.9	Steam boilers	0.5
Tanks (rectangular)	0.5			Building	
				Single story	0.8
				Two story	0.7–0.8

*For other factors, see D. S. Remer and L. H. Chai, "Design Cost Factors for Scaling-up Engineering Equipment," *Chemical Engineering Progress* (August 1990):77–82.

Solution   The required calculations are shown below.

The index of price change, $I_t = 194 / 187 = 1.0374$.

$$C_{now}(120 \text{ kW}) = \$160,000\left[\left(\frac{120}{80}\right)^{0.6}(1.0374)\right] = \$18,000.$$ ■

Total cost $= \$211,700 + \$18,000 = \$229,700$.

5.5.3.4 Learning Curves.

In repetitive operations involving direct labor, the average time to produce an item or provide a service is typically found to decrease over time because of learning that occurs. This phenomenon is observed for purchasing activities, assembly operations, food preparation, etc. As a result, cumulative average and unit times required to complete a task will drop dramatically as output increases. This can substantially reduce costs for large production runs, and failure to include this phenomenon can create large errors in cost estimates.

As a general rule, when cumulative production/service doubles, the total time required per output unit is reduced by $x\%$. For example if $x = 4\%$ the time per unit required to go from, say, 4 to 8 units of output declines by 4% and we have a 96% ($= 100\% - 4\%$) learning curve. To illustrate this concept, suppose it takes 1 hour to produce the first unit of output. Assuming a 96% learning curve, it would take 0.96 hours to produce the second unit, $(0.96)^2 = 0.9216$ hours to produce the fourth unit, $(0.96)^3 = 0.8847$ hours to produce the eighth unit and so on.

A simple equation to estimate the time requirements of repetitive labor activities is given below.[6]

$$Y_i = Y_1 i^b, \tag{5-16}$$

where

$Y_i =$ direct labor hours (or cost) for the ith production unit,
$Y_1 =$ direct labor hours (or cost) for the initial (first) unit,
$i =$ cumulative count of units of output,
$b =$ the learning curve exponent.

For instance, Y_{10} represents the time (or cost) required for the tenth unit of output. Furthermore, b equals the ratio of log(learning curve expressed as decimal) $\div$ log 2. If we have a 96% learning curve,

$$b = \log(0.96)/\log 2 = -0.017728/0.30103 = -0.0589,$$

and from Eq. 5-16 we have

$$Y_i = Y_1 i^{-0.0589}.$$

If $Y_1 = 1.0$ hour, the estimate of Y_{10} becomes

$$Y_{10} = 1.0(10)^{-0.0589} = 0.873 \text{ hrs.}$$

---

[6] T. P. Wright, "Factors Affecting the Cost of Airplanes," *Journal of Aeronautical Sciences,* vol. 3, No. 4 (Feb. 1936).

The cumulative average time for i production units ($i \geq 1,000$) can be approximated by:

$$C_i = \frac{Y_1}{i(1 + b)}\left[(i + 0.5)^{1+b}\right], \text{ where } b < 0. \tag{5-17}$$

Hence, the value of C_{10} from Eq. 5-17 is

$$C_{10} = \frac{1.0}{10(0.9411)}\left[(10.5)^{0.9411}\right] = 0.97 \text{ hrs.}$$

Example 5-5

As the number of manually assembled wire harnesses doubles from 20 to 40, there is a 20% reduction in labor hours per harness. If the first unit required 24 hours to assemble, how much time will it take for the fortieth harness?

  Solution   The required calculations are as follows:
  The learning curve is $100\% - 20\% = 80\% (= 0.80)$

$$b = \log(0.8)/\log 2 = -0.322.$$

Using Eq. 5-16,

$$Y_{40} = Y_1(40)^{-0.322} = 24(0.305) = 7.32 \text{ hrs.}$$

  For convenience in making calculations, a range of learning curve exponents is provided in Table 5-6.
  The following example further illustrates the use of a learning curve for estimating labor costs.

Example 5-6

One thousand units of a new product are to be produced. Direct labor has been estimated to average 8 hours per unit. Six months after production has begun, management has requested the following information:

  1. average labor hours per unit, to date;

  2. labor hours per unit, for the latest month's units;

  3. the % learning curve that production has followed. The cumulative labor hours and units produced are listed in Table 5-7. Dividing the cumulative labor hours by the cumulative production shows that the average labor hours per unit to date is 15.75 labor hours. For the latest month's production, an average of $410/42 = 9.76$ labor hours per unit was reached.

TABLE 5-6 Typical Learning Curve Exponents, b

Percent learning	b	Percent learning	b
100	0 (no learning)	70	−0.515
95	−0.074	65	−0.621
90	−0.152	60	−0.737
85	−0.234	55	−0.861
80	−0.322	50	−1.000
75	−0.415		

TABLE 5-7 Production Data

Month, i	Month's production	Cumulative production	Month's labor hours	Cumulative labor hours	Month's avg. labor hours/unit	Cumulative avg. labor hours/unit $(= Y_i)$
1	11	11	301	301	27.36	27.36
2	17	28	420	721	24.71	25.75
3	16	44	307	1028	19.19	23.36
4	22	66	356	1384	16.18	20.97
5	24	90	285	1669	11.87	18.54
6	42	132	410	2079	9.76	15.75

In examining the data, it can be seen that the cumulative production at the end of month 6 (132 units) is double the cumulative production at the end of month 4 (66 units). The learning curve can be approximated by dividing the cumulative average labor hours per unit during month 6 by the cumulative average labor hours per unit during month 4. The learning curve in effect is approximately 75%, as shown below.

$$\frac{Y_{132}}{Y_{66}} = \frac{15.75 \text{ labor hours/unit}}{20.97 \text{ labor hours/unit}} = 0.75$$

5.6 Additional Examples of Cost Engineering

Three examples are provided that further illustrate the techniques discussed in Section 5.5. Example 5-7 involves exponential costing in the process industry, and Example 5-8 demonstrates the use of linear regression in a highly automated drilling operation. Finally, Example 5-9 uses a learning curve and the factor technique for estimating the selling price of a manufactured item.

Example 5-7

The R Square Corporation is considering two plans, A and B, for expanding the capacity of its urea manufacturing plant. The relevant data and analysis for both plans are given below.

Plan A

R Square Corporation's office of agricultural and chemical development operates a urea plant that was built in 1978 at a cost of $4.3 million and capacity of 300,000 lb/yr. Due to increasing use of urea-based fertilizers and increasing maintenance costs on the existing unit, the feasibility of constructing a new plant with a 750,000 lb/yr capacity is being studied. Time does not permit an in-depth cost estimation for the new plant, so engineers decide to obtain the cost of the new plant by scaling up the cost of the old plant. The cost-exponent factor is known to be 0.65, and the construction cost index has increased an average of 10.5% for the past 15 years. What cost should the engineers report to the project review committee?

Urea plant estimates:

C_b = $4.3 million: cost of plant in 1978,

Q_b = 300,000 lb/yr: capacity of existing plant,

C_a = ?: cost of 750,000 lb/yr plant (but based on 1978 pricing),

Q_a = 750,000 lb/yr: capacity of plant,

C_a' = cost reported to committee in 1993 dollars,

β = cost-exponent factor = 0.65.

Solution   From Eq. 5-14,

$$C_a = C_b \left(\frac{Q_a}{Q_b} \right)^\beta,$$

we find

$$C_a = \$4.3 \text{ million} \left(\frac{750,000 \text{ lb/yr}}{300,000 \text{ lb/yr}} \right)^{0.65} = \$7.8 \text{ million (in 1978)},$$

$$C_a' = C_a (F/P, 10.5\%, 15) = \$7.8 \text{ million}(1.105)^{15},$$

or

$$C_a' = \$34.88 \text{ million (estimated cost of new urea unit in 1993 dollars)}.$$

Plan B

As an alternative plan, it is learned that a manufacturer of these units can prefabricate and install a unit on site for a total cost of $22 million. The R Square Corporation's constructed unit has an estimated life of 15 years, and the prefabricated unit has a life of 12 years. The operating costs for the company unit are $40,000 per year for years 1–10 and $30,000 per year for years 10–15. The operating costs for the prefabricated unit are $50,000 per year. The interest rate is assumed to be 10%. Which alternative should they select? What assumptions are involved?

Solution   *Note:* Assume negligible salvage value for both alternatives, and compare the alternatives using the annual cost (AC) method.

	Prefabricated unit	Company unit
Initial cost:	$22,000,000	$34,880,000
Annual costs:	$50,000/yr	$40,000/yr, years 1–10
		$30,000/yr, years 10–15
Life:	12 yr	15 yr

Prefabricated unit:

$$AC = \$50,000 + \$22,000,000(A/P, 10\%, 12)$$
$$= \$3,279,600.$$

Company unit:

$$AC = \$30,000 + \$10,000(P/A, 10\%, 10) (A/P, 10\%, 15)$$
$$+ \$34,880,000 (A/P, 10\%, 15)$$
$$= \$4,893,884.$$

By a wide margin, the R Square Corporation should let the outside manufacturer construct and install the urea unit. However, numerous nonmonetary considerations could shift the decision to plan A. The monetary risks associated with expanding in an uncertain and highly competitive market may well cause neither plan to be acceptable. An indefinitely long study period has been assumed in the analysis above. ∎

Example 5-8

For a certain drilling operation within a flexible machining cell, data regarding the time (in hours) to drill 1, 2, 3, and 4 holes in a $\frac{1}{4}$-inch sheet of carbon steel have been obtained. These data are listed below.

(a) Develop a linear regression equation for estimating the time (dependent variable) required for the number of holes drilled (independent variable).

(b) What is the correlation coefficient, r, for these data?

(c) Make an estimate for the time required to drill six holes, and discuss the danger in using the regression equation for this purpose.

Data point	x (No. of holes)	y (Hours)
1	2	0.0381
2	3	0.0720
3	4	0.1078
4	3	0.0815
5	1	0.0360
6	2	0.0605
7	1	0.0382
8	4	0.1318
9	3	0.0985
10	1	0.0468
11	2	0.0721
12	4	0.0950

Solution

(a) For the data given, various values must be calculated and substituted into Eqs. 5-2 and Eq. 5-3: $\Sigma x = 30$, $\Sigma y = 0.8783$, $\Sigma x^2 = 90.0$, $\Sigma y^2 = 0.0748$, $\Sigma xy = 2.5568$, and $\bar{x} = 2.5$. From these values, we find that

$$b = \frac{2.5568 - 2.5(0.8783)}{90.0 - 2.5(30.0)}$$

$$= 0.0241,$$

$$a = \frac{0.8783}{12} - 0.0241(2.5)$$

$$= 0.0129.$$

Thus, the linear regression equation is $y = 0.0129 + 0.0241x$. Estimates of time required to drill 1, 2, 3, and 4 holes are listed below.

No. of holes, x	Estimated time (hours), y
1	0.0370
2	0.0611
3	0.0852
4	0.1093

(b)   The correlation coefficient for these data can be determined by using Eq. 5-4. It is $r = 0.909$, so the relationship between x and y is reasonably well estimated by a linear function.

(c)   Finally, an estimate for drilling six holes is

$$y = 0.0129 + 0.0241(6) = 0.1575 \text{ hrs.}$$

Caution must be exercised in using linear regression to make estimates when the value of the independent variable lies outside the range of values (for holes drilled) that was utilized to develop coefficients (a and b) of the regression equation. In practice, it is not uncommon to observe this use of regression equations, but the practice cannot be recommended as a sound one.   ■

Example 5-9

The ABC manufacturing company is trying to determine the unit selling price for a high-density, double-sided disk with 2 MB storage capacity. The disks are produced by installing a magnetic film into a plastic cartridge.

A total of three machining operations need to be performed:

- Cut out disks from magnetic film.
- Apply disk control centerpiece.
- Insert into plastic cartridge.

The film, centerpiece, and cartridges are purchased from an outside manufacturer. A total of 10,000 disks are to be produced. Relevant information is listed below.

- The magnetic film is bought in rolls that cost $90 each. From each roll, 2,000 circular disks can be cut out.
- One person is needed to operate and supervise the cut-out machine. Installing a new roll takes 8 minutes, and cutting out 2,000 circular disks takes 25 minutes.
- No learning curve is applicable for the cut-out operation.
- The disk control centerpieces cost $0.12 per unit.
- One person is required to apply the centerpieces to the magnetic disks. Applying the first centerpiece takes 30 seconds, and for the remaining centerpieces a 80% learning curve is applicable.
- The plastic cartridges cost $0.15 per unit.
- One person is needed to supervise the disk-insertion operation. This operation is done automatically by a machine that can insert 1,500 disks per hour.
- No learning curve is applicable for inserting disks.
- The direct labor rate is $15.00/hour.
- Planning and liaison are 15% of factory labor.
- Quality control is 30% of factory labor.
- Factory overhead is 800% of total labor.
- General and administrative expense is 50% of total labor.
- Packing costs are 100% of total labor.
- The profit margin is 15% of total manufacturing cost.

Solution   The various components of total cost and required selling price are computed and summarized as follows.

Production Material Cost. To produce 10,000 disks, 10,000/2,000 = 5 magnetic rolls are required. This costs 5 × $90 = $450.

10,000 centerpieces and cartridges cost 10,000($0.12 + $0.15) = $2,700. Thus, total production material cost = $450 + $2,700 = $3,150.

Direct Labor Hours. Cutting out 2,000 disks from magnetic film takes: 8 + 25 = 33 minutes. Hence, producing 10,000 disks takes

$$5 \times 33 = 165 \text{ minutes.}$$

By summing Eq. 5-16, the cumulative time for applying 10,000 centerpieces (in minutes) with an 80% learning curve is

$$0.5 \text{ minutes} \times \sum_{i=1}^{10,000} i^{\log 0.8/\log 2} = 379.71 \text{ minutes.}$$

The disk-insertion operation requires

$$10,000/1,500 = 6.67 \text{ labor hours.}$$

Thus, a total of

$$(165 + 379.71)/60 + 6.67 = 15.75 \text{ labor hours}$$

is needed.

A popular cost-estimating template appears below.[7]

Customer:	Apex		
Model:	HDDS012X	Estimator:	Chas. Everyperson
Part Name:	HD/DS Disk		
Part No:	012	Date:	February 20, 1994
No. Parts Required:	10,000	Page:	1 of 1

MANUFACTURING COST	HOURS	CHARGE RATE	DOLLARS
A. Factory Labor	15.75	$15.00	$ 236.25
B. Planning & Liaison Labor		15% of A	$ 35.44
C. Quality Control		30% of A	$ 70.88
D. TOTAL LABOR			$ 342.57
E. Factory Overhead		800% of D	$2,740.50
F. General & Admin. Expense		50% of D	$ 171.28
G. Production Material			$3,150.00
H. Outside Manufacture			$ 0.00
I. SUBTOTAL			$6,404.34
J. Packing Costs		100% of D	$ 342.57
K. TOTAL DIRECT CHARGE			$6,746.91
L. OTHER DIRECT CHARGE			$ 0.00
M. Facility Rental			$ 0.00
N. TOTAL MANUFACTURING			$6,746.91
O. Profit/Fee		15% of N	$1,012.04
TOTAL SELLING PRICE			$7,758.94
QUANTITY			10,000
UNIT SELLING PRICE			$ 0.78

[7] T. F. McNeill and D. S. Clark, *Cost Estimating and Contract Pricing* (New York: American Elsevier Publishing Co., 1966), p. 71.

Hence, this estimating template gives a unit selling price of $0.78.   ■

5.7 Summary

In this chapter techniques for obtaining and/or developing data required in economic analyses have been discussed and illustrated. Specifically, we have concentrated on making single-valued estimates for the engineering economy topics treated in Part I of this book. Three groups of quantitative estimating techniques were covered: (1) time-series techniques, (2) subjective techniques, and (3) cost engineering techniques. The first two groups are useful in forecasting elements of revenue and cost, whereas the third group is widely employed for developing more detailed estimates of costs.

PROBLEMS

5-1. Your firm is considering replacing its conventional trucks with turbine-powered vehicles. What information would you like to have in studying the decision? Where would you expect to get the information?

5-2. How would you obtain information needed to study the economics of the following?

   a. Leasing versus purchasing a computer system.

   b. Maintaining equipment with in-house personnel or purchased services.

   c. Keeping versus replacing an old machine system.

   d. Coal-fired steam versus gas turbine generation plant.

5-3. Use the Delphi method to estimate the following:

   a. The ticket price for a 5,000-mile SST one-way trip.

   b. The cost of a year's college education in a state-supported university 5 years from now.

   c. The time to dig a trench $2' \times 4' \times 10'$ in soft clay using hand tools.

   d. The average annual growth rate for cellular telephone service over the next 10 years.

   e. The price of premium gasoline in 2002.

   f. The average length of time required to change an automobile tire.

   g. The percentage of football (or other) games to be won next year by the team of your choice.

5-4. Suppose that you own a small company that manufactures metal castings for several large automotive companies. Over the past several years you have found that quarterly new-car sales tend to lag behind the prime interest rate by 3 months. You would like to make a forecast of next quarter's car sales so that the size of your work force can be anticipated. There is a direct relationship between car sales and demand for castings that your company produces. The data on the following page are gathered:

   a. Calculate a linear regression equation for these data, assuming that the interest rate is the independent variable.

   b. Calculate the correlation coefficient.

   c. Make a forecast of sales for the next quarter based on this quarter's prime interest rate of 7.50%.

Year	Quarter	Interest rate (%)	Next quarter sales ($M)	Year	Quarter	Interest rate (%)	Next quarter sales ($M)
1	1	8.00	$23	4	1	7.00	$25
	2	8.25	17		2	7.50	26
	3	8.50	18		3	7.50	17
	4	8.25	20		4	8.25	20
2	1	7.75	21	5	1	8.75	15
	2	7.25	25		2	8.50	18
	3	7.70	24		3	7.50	22
	4	7.25	29		4	7.00	23
3	1	7.50	24	6	1	7.50	?
	2	7.75	23				
	3	7.25	26				
	4	7.00	30				

5-5. Total operating costs and the corresponding production volumes for a particular process have been found to be as follows:

Operating costs ($M)	Production volume (hundreds of units)
800	10.0
1,000	11.0
700	9.0
600	8.5

a. Calculate the least-squares linear regression line to relate total operating costs as a function of production volume.

b. Estimate the operating costs for a production volume of 950 units.

c. Calculate the coefficient of correlation, and comment on whether this indicates a relatively good or poor fit of the regression line to the data.

5-6. In the packaging department of a large automotive parts distributor, a fairly reliable estimate of packaging and processing costs can be determined by knowing the weight of an order. Thus, weight is a cost driver that accounts for a sizable fraction of the packaging and processing costs at this company. Data for the past ten orders are given on the following page.

a. Estimate the a and b coefficients, and write the linear regression equation to fit these data.

b. What is the correlation coefficient (r)?

c. If an order weighs 250 pounds, how much should it cost to package and process it?

Packaging and processing costs ($), y	Weight (lbs.), x
97	230
109	280
88	210
86	190
123	320
114	300
112	280
102	260
107	270
86	190

5-7. Suppose that comparison of Norcar Company sales with many economic indicators shows that sales correlate best with, say, the state's construction volume committed. The nature of the correlation is shown in the table below.

Norcar Company Sales and Construction
Volume Committed

Year	Sales ($ million)	Construction volume committed ($ million)
1995	3	40
1996	2	25
1997	5	50
1998	4	45

a. Determine the least-squares regression line that expresses the correlation between annual sales and construction volume.

b. Determine the correlation coefficient, r, between the two variables.

c. Obtain a linear extrapolation of sales for a committed construction index of $70 million.

d. What is the danger in using a construction volume outside the original range of $25 – $50 million to forecast future sales?

5-8. The following is an exercise to illustrate the use of various weighting constants for exponential smoothing forecasting. Suppose the actual sales of a firm were 500 units for year 1 and 600 for year 2. You forecasted it would be 550 units for year 2, and now you wish to forecast for year 3 and beyond.

a. What would be your forecast for year 3 if your smoothing constant, α', was, respectively, 0.1, 0.5, and 0.97?

b. Suppose actual sales turn out to be as follows:

Year	Actual sales (units)
3	700
4	800
5	700
6	600
7	600

What would have been the forecast for each year (4, 5, 6, and 7) using each of the three smoothing constants?

c. Distinguish between the actual results and the forecast for each year using each of the three smoothing constants. What are your conclusions on the desirability and nondesirability of using a low value of α'?

5-9. Consider the following time-series data of demand for a certain company's product.

Month	Demand (booked orders)	Month	Demand (booked orders)
1	3,009	11	3,387
2	2,641	12	3,138
3	2,934	13	2,908
4	3,239	14	3,512
5	3,490	15	3,291
6	2,569	16	2,804
7	3,205	17	3,096
8	2,561	18	3,106
9	3,047	19	3,195
10	2,607	20	3,605

a. Plot these data on a piece of graph paper.

✓**b.** Apply single exponential smoothing to the data when $\alpha' = 0.20$, and make a forecast for $T = 1$ month into the future.

✓**c.** Repeat part (b) when $\alpha' = 0.05$ and $T = 3$ months.

5-10. Discuss the principal advantages and disadvantages of the Delphi method of forecasting.

5-11. In your class, attempt to run a Delphi study to determine the price of a compact disc player 3 years from now. Was group consensus affected by conducting two or three rounds of the procedure?

5-12. Try to think of some products that are presently in the early stages of a substitution curve effect. List them and try to estimate when the newer product will take more than half the market.

5-13. How could trend extrapolation be used to forecast future innovations in the aerospace industry? What performance characteristics do you believe are important here?

5-14. Use the factor technique to estimate the cost of installing a local area network in a factory environment having the following characteristics. One large building on a single level will require a total of 3,000 ft of coaxial (broadband) cable to network its six departments. Six network interface units (NIUs) will be required, and a total of 50 taps will have to be made to connect all the anticipated workstations and programmable devices. Two modems are needed in addition to one network manager/analyzer that costs $30,000. The information necessary to make the estimate may be obtained from the worksheet shown on the following page. How accurate do you think such an estimate would be?

Component	Cost-estimating relationship		
1. Interbuilding connections	$100–$150 per foot	× ____	= ____
2. Intrabuilding connections	$20–$50 per foot	× ____	= ____
3. Cable installation	$20 per foot	× ____	= ____
4. Equipment			
a. Broadband			
CATV amplifier	$500–$1500	× ____	= ____
Taps	$17–$20 each	× ____	= ____
Splitters	$5–$15	× ____	= ____
NIUs	$500–$1,000 per port	× ____	= ____
Modems	$1,000 each	× ____	= ____
b. Baseband			
NIUs	$600 per port	× ____	= ____
Repeaters	$1,200–$1,500 each	× ____	= ____
Taps/transceivers	$200–$300 each	× ____	= ____
c. Network manager	$10,000–$30,000		____
Network analyzer	$30,000		____

5-15. A residential builder just finished constructing a 3,000 ft^2 home for $240,000. This cost did not include the lot or utility access fees. A detailed breakout of costs for this job is as follows:

Item	Fraction of finished cost
Lumber and carpentry	0.20
Electrical wiring	0.10
Plumbing	0.14
Concrete and masonry	0.09
Wallboard	0.04
Flooring	0.06
Foundation preparation	0.02
Accessories and appliances	0.05
Heating and air conditioning	0.09
Roofing	0.10
Painting	0.07
Miscellaneous	0.04
	1.00

    a. What is the unit cost for the just-finished home?

    b. If a 4,000-ft^2 home is to be built, estimate the total cost from the answer to part (a) and compare it with the total of estimated item costs based on the breakout given above.

5-16. A 100 kW diesel generator cost $140,000 seven years ago when a certain equipment cost index was arbitrarily set at 100. A similarly designed generator rated at 150 kW is now being proposed and the cost index is 140. The cost-exponent factor, β, is 0.7 for this type of equipment.

    a. Determine the estimated cost of the proposed generator by using the appropriate cost-estimating relationship.

    b. Repeat part (a) when $\beta = 0.4$.

5-17. The Neptune Manufacturing Company is considering abandoning its old plant, built 23 years ago, and constructing a new plant that has 50% more square footage. The original cost of the old plant was $300,000 and its capacity, in terms of standardized production units, is 250,000 units per year. Capacity of the proposed plant is to be 500,000 units per year. During the past 23 years, costs of plant construction have risen by an average of 5% per year. If the cost exponent factor is 0.8, what is the estimated cost of the new plant?

5-18. A 60 kW diesel electric set, without a precompressor, cost $32,000 in 1974. A similar design, but using 140 kilowatts, is planned for an isolated installation. The cost-exponent factor $\beta = 0.7$, and the cost index in 1974 was 230. Now the cost index is 350. A precompressor is estimated separately at $1,900 now. Using the exponential costing model, find the estimated total equipment cost now.

5-19. A small plant has been constructed and the costs are known. A new plant is to be estimated using the exponential costing model. Major equipment, costs, and factors are as follows (note $mW = 10^6$ watts):

Equipment	Reference size	Unit reference cost	Cost-exponent factor	New design size
Two boilers	6 mW	$300,000	0.80	10 mW
Two generators	6 mW	400,000	0.60	9 mW
Tank	80,000 gal	106,000	0.66	91,500 gal

    If ancillary equipment will cost an additional $200,000, find the cost for the proposed plant.

5-20. Your company is now making a product that has a raw material cost of exactly $0.53 per unit out of a total cost of $1.63 per unit. You are responsible for an analysis of the economics of tripling the present capacity. Should you automatically assume that the raw-materials cost for the added capacity will be $0.53 per unit? Why or why not?

5-21. A rule of thumb sometimes used is that when a unit is being operated at 50% capacity the maintenance costs will be approximately 75% of the maintenance costs at 100% capacity.

    a. Why would the maintenance costs not be 50% of the maintenance costs at 100% capacity?

    b. Is it reasonable that maintenance costs at 120% of capacity could be less than 120% of the maintenance costs at 100% capacity? More than 120%? What might cause the difference?

✓**5-22.** The mechanical engineering department has a student team that is designing a formula car for national competition this coming spring.

    The time required for the team to assemble the first car is 100 hours. Their improvement (or learning rate) is 0.8, which means that as output is doubled their time to assemble a car is reduced by 20%. For instance, $K = 100$ hours for unit 1, and so unit 2 will take 80 hours. Unit 4 will require 80 hours $(0.8) = 64$ hours, and so on.

    a. How much time will it take the team to assemble the tenth car?

    b. How much total time will be required to assemble the first ten cars?

    c. What is the estimated cumulative average assembly time for the first ten cars?

5-23. The structural engineering design section within the engineering department of a regional electrical utility corporation has developed several standard designs for a group of similar transmission line towers. The detailed design for each tower is based on one of the standard designs. A transmission line project involving 50 towers has been approved. The estimated number of engineering hours to accomplish the first detailed tower design is 126. Assuming a 95% learning curve, what is your estimate of the number of engineering hours needed to design the eighth tower and to design the last tower in the project?

5-24. You have been asked to estimate the *per unit selling price* of a new line of widgets. Pertinent data are as follows:

Direct labor rate:	$15.00 per hour
Production material:	$375.00 per 100 widgets
Factory overhead:	125% of direct labor
Packing costs:	75% of direct labor
Desired profit:	20% of total manufacturing cost

Past experience has shown that an 80% learning curve applies to the labor required for producing widgets. The time to complete the first widget has been estimated to be 1.76 hours. Use the estimated time to complete the fiftieth widget as your standard time for the purpose of estimating the unit selling price.

5-25. If 846.2 labor hours are required for the third production unit, and 783.0 labor hours are required for the fifth production unit, determine the learning curve parameter (b).

✓**5-26.** You have been asked to estimate the *per unit selling price* of a new model of widgets. Pertinent data are as follows:

Direct labor rate:	$15.00 per hour
Production material:	$625.00 per 100 widgets
Factory overhead:	75% of factory labor cost
Packing costs:	40% of production material cost
Desired profit:	12% of total manufacturing cost

Initial studies have shown that an 85% learning curve applies to the labor required for producing the widgets. The time to complete the first widget has been estimated to be 1.5 hours. Use the estimated time to complete the fiftieth widget as your standard time for the purpose of estimating the unit selling price.

6
Consideration of Depreciation and Income Taxes

This chapter provides a brief overview of the principal tax and depreciation consid-erations and a general technique for including the effect of income taxes in economic studies.

Only cash flows need be considered in determining the economic desirability of an alternative in an economic analysis. Income taxes are relevant cash flows and should be considered whenever their omission may cause the selection of an uneco-nomical alternative.[1] Even though depreciation write-offs are not, in themselves, cash flows, they do affect income taxes, and hence affect cash flows.

6.1 Introduction to Depreciation

The primary purpose of depreciation accounting is to provide for the recovery of cap-ital invested in property that is expected to decline in value as a result of time and/or use. This is done through the mechanism of *depreciation charges,* or write-offs, which are allocations made periodically for the purpose of distributing the cost of capital assets over their useful lives. Thus, depreciation is the decrease in value of physical properties with the passage of time. More specifically, depreciation is an accounting concept that establishes an annual deduction against before-tax income such that the effect of use and time on an asset's value can be reflected in a firm's financial state-ments. Although depreciation does occur and is easily recognized, the determination of its magnitude in advance is not easy. The actual amount of depreciation can never be established until the asset is retired from service. Because depreciation is a *noncash*

---

[1] This is demonstrated in Example 6-25.

cost that affects income taxes, we must learn to consider it properly when making after-tax engineering economy studies.

Depreciable property is property used in a business to produce income. The cost of depreciable property can be deducted from business income for income tax purposes over a future period of time. Property is depreciable if it meets these requirements:

1. It must be used in business or held for the production of income.
2. It must have determinable life, and the life must be longer than one year.
3. It must be something that wears out, decays, gets used up, becomes obsolete, or loses value from natural causes.

In general, if property does not meet all three of these conditions, it is not depreciable.

Depreciable property may be classified as *tangible* or *intangible*. Tangible property is any property that can be seen or touched, and intangible property is property, such as a copyright or franchise, that is not tangible. Additionally, depreciable property may be classified as *real* or *personal*. Personal property is property that can be transported, such as machinery or equipment, which is not real estate. Real property is land and generally anything that is erected on, growing on, or attached to land. However, land itself is never depreciable.

The purpose of the following sections is to acquaint the student with several depreciation methods that have been employed for many decades. These methods are of interest to us because they form the foundation for modern depreciation models in widespread use today.

6.1.1 Straight-Line Method

Straight-line depreciation is the simplest depreciation method. It assumes that a constant amount is depreciated each year over the life of the asset. The following definitions are used in the equations below. If we define

N = depreciable life (write-off period) of the asset in years,

B = cost basis,

d_k = annual depreciation deduction in year k ($1 \le k \le N$),

BV_k = book value at end of year k,

BV_N = estimated book (salvage) value in year N,

d_k^* = cumulative depreciation through year k,

then

$$d_k = (B - \mathrm{BV}_N)/N, \tag{6-1}$$

$$d_k^* = k d_k \quad \text{for } 1 \le k \le N, \tag{6-2}$$

$$\mathrm{BV}_k = B - d_k^*. \tag{6-3}$$

Notice that for this method you must have an estimate of the final BV, which will also be the final book value at the end of year N.

Example 6-1

A machine costs $15,000 installed. The allowable write-off period is 12 years, at which time the salvage value is estimated to be $1,500. What will be the annual depreciation charge and what will be the book value at the end of the third year?

Solution

$$d = \frac{(B - BV_N)}{N} = \frac{\$15,000 - \$1,500}{12} = \$1,125$$

$$BV_3 = B - 3d = \$15,000 - 3\left[\frac{\$15,000 - \$1,500}{12}\right]$$

$$= \$11,625 \qquad \blacksquare$$

6.1.2 Declining Balance Method

In the declining balance method, sometimes called the *constant percentage method,* it is assumed that the cost of depreciation for any year is a fixed percentage (designated R) of the book value at the *beginning* of that year. In this method, $R = 2/N$ when a 200% declining balance is being used (i.e., twice the straight-line rate of $1/N$), and N equals the allowable life of an asset. If the 150% declining balance method is specified, then $R = 1.5/N$. The following relationships hold true for the declining balance method:

$$d_1 = B(R), \qquad (6\text{-}4)$$

$$\text{Depreuation} \rightarrow d_k = B(1 - R)^{k-1}(R), \qquad (6\text{-}5)$$
$$\text{charge}$$

$$d_k^* = B[1 - (1 - R)^k], \qquad (6\text{-}6)$$

$$\text{Book Value} \rightarrow BV_k = B(1 - R)^k, \qquad (6\text{-}7)$$

$$BV_N = B(1 - R)^N. \qquad (6\text{-}8)$$

Notice that Eqs. 6-4 through 6-8 do not contain a term for BV_N.

In order for the book value to equal the estimated salvage value at the end of the write-off period, N years, R should be calculated as

$$R = 1 - \sqrt[N]{\frac{BV_N}{B}}. \qquad (6\text{-}9)$$

Example 6-2

For the previous example using straight-line depreciation, determine the book value at the end of the third year and the depreciation charge for the fourth year using the 200% declining balance method of depreciation.

Solution

$$R = 2/12 = 0.1667$$

$$\text{Book Value} \rightarrow BV_3 = B(1 - R)^3 = \$15,000(0.8333)^3 = \$8,679.51$$

$$d_4 = B(1 - R)^3(R) = \$1,446.88 \qquad \blacksquare$$

Example 6-3
Repeat Example 6-2 when it is desired for the terminal book value in year 12 to equal exactly $1,500.

Solution

$$R = 1 - \sqrt[N]{BV_N/B} = 1 - \sqrt[12]{1,500/15,000}$$

$$= 1 - 0.826 = 0.174$$

$$BV_3 = \$15,000(1 - 0.174)^3 = \$8,460$$

$$d_4 = \$8,460(0.174) = \$1,470$$ ∎

6.1.3 Sum-of-the-Years'-Digits (SYD) Method

To compute the depreciation deduction by the SYD method, the digits corresponding to the number for each permissible year of life are first listed in reverse order. The sum of these digits is then determined. The depreciation factor for any year is the number from the reverse-ordered listing for that year divided by the sum of the digits. For example, for a property having a life of five years, SYD depreciation factors are as follows:

Year	Number of the year in reverse order (digits)	SYD depreciation factor
1	5	5/15
2	4	4/15
3	3	3/15
4	2	2/15
5	1	1/15
Sum of the digits =	15	

The depreciation for any year is the product of the SYD depreciation factor for that year and the difference between the cost basis (B) and the BV_N. The general expression for the annual cost of depreciation for any year k, when N equals the depreciable life of an asset, is

$$d_k = (B - BV_N) \cdot \left[\frac{2(N - k + 1)}{N(N + 1)} \right]. \tag{6-10}$$

The book value at the end of year k is

$$BV_k = B - \left[\frac{2(B - BV_N)}{N} \right] k + \left[\frac{(B - BV_N)}{N(N + 1)} \right] k(k + 1), \tag{6-11}$$

and the cumulative depreciation through the kth year is simply

$$d_k^* = B - BV_k. \tag{6-12}$$

Example 6-4
Work Example 6-1 using the SYD method.

Solution

$$SYD = 1 + 2 + \cdots + 12 = \frac{12(13)}{2} = 78$$

$$BV_3 = \$15,000 - \left[\frac{2(\$15,000 - \$1,500)}{12}\right] \cdot 3 + \left[\frac{\$15,000 - \$1,500}{12 \cdot 13}\right] \cdot 3 \cdot 4$$

$$= \$9,288$$

$$\left(\text{or } BV_3 = \$15,000 - \left[\frac{12}{78}(\$13,500) + \frac{11}{78}(\$13,500) + \frac{10}{78}(\$13,500)\right]\right)$$

$$d_4 = (\$15,000 - \$1,500)\left[\frac{2(12 - 4 + 1)}{12(13)}\right] = \$1,558 \qquad \blacksquare$$

6.1.4 Sinking Fund Depreciation

The *sinking fund depreciation method* is of more interest in some special types of engineering economy studies than for its use for normal accounting and income tax purposes. Its historical origin stems from times when it was fairly common for public utilities and other regulated firms to set aside depreciation charges in an interest-bearing account so that the accumulated charges and interest would just equal the estimated depreciable part of the asset investment as of the end of the investment life.

The annual sinking fund deposit, s, which is assumed to earn interest at $i\%$ per year, can be calculated as

$$s = (B - BV_N)(A/F, i\%, N). \qquad (6\text{-}13)$$

The total depreciation charge for any year is the sinking fund deposit plus accumulated interest for that year. It is also the difference in book value for that year and the previous year. Thus,

$$d_k = BV_k - BV_{k-1}.$$

The book value for any year is the first cost minus the accumulated sinking fund deposits and interest. Thus,

$$BV_k = B - s(F/A, i\%, k). \qquad (6\text{-}14)$$

Example 6-5

Work Example 6-1 except use the sinking fund depreciation method with an interest (fund or reinvestment) rate of 6%.

    Solution

$$s = (\$15,000 - \$1,500)(A/F, 6\%, 12) = \$801$$

$$BV_3 = B - s(F/A, 6\%, 3)$$

$$= \$15,000 - \$801(F/A, 6\%, 3) = \$12,450$$

Similarly, $BV_2 = \$13,350$, so $d_3 = \$13,350 - \$12,450 = \$900$. $\blacksquare$

6.1.5 Comparison of Depreciation Methods

To provide a common basis for comparing the four methods of depreciation examined here so far, Table 6-1 shows year-by-year depreciation charges for a typical asset that costs $16,000, is expected to last 5 years, and then has a terminal book value of $1,000.

TABLE 6-1   Comparison of Depreciation Charges Using Four Methods for a Machine Having a $16,000 Investment, Five-Year Life, and $1,000 Terminal Book Value

End of year	Straight line	Double declining balance	Sum-of-years' digits	Sinking fund @ 10%
1	$ 3,000	$ 6,400	$ 5,000	$ 2,457
2	3,000	3,840	4,000	2,703
3	3,000	2,304	3,000	2,973
4	3,000	1,382	2,000	3,270
5	3,000	1,074*	1,000	3,597
Totals	$15,000	$15,000	$15,000	$15,000

*Year 5 depreciation set equal to $15,000 less cumulative depreciation through the fourth year ($13,926) so that $BV_5 = \$1,000$.

6.1.6 Units-of-Production Depreciation

All the depreciation methods discussed to this point are based on elapsed time on the theory that the decrease in value of property is mainly a function of time. When the decrease in value is mostly a function of use, depreciation may be based on the ✓ *units-of-production method.*

This method results in the total depreciable investment being allocated equally over the units produced and requires an estimate of the total lifetime of productive use. The depreciation rate is calculated as

$$\text{Depreciation per unit of production} = \frac{B - BV_N}{\text{Estimated lifetime production}} \qquad (6\text{-}15)$$

Example 6-6
A vehicle used in a business has a first cost of $50,000 and is expected to have a $10,000 salvage value when traded after 100,000 miles of use. It is desired to find its depreciation rate based on functional use and to find its book value after 20,000 miles of use.

Solution

$$\text{Depreciation per unit of production} = \frac{\$50,000 - \$10,000}{100,000 \text{ miles}} = \$0.40 \text{ per mile}$$

$$\text{After 20,000 miles, } BV = \$50,000 - \frac{\$0.40}{\text{mile}}(20,000 \text{ miles}), \text{ or } BV = \$42,000$$

6.2 The Tax Reform Act of 1986 and Its Depreciation Provisions[2]

The Modified Accelerated Cost Recovery System (MACRS) method was created by the Tax Reform Act of 1986 (TRA 86) and is now the principal means for computing depreciation expenses (termed *recovery allowances* under TRA 86). MACRS is mandatory for most tangible depreciable assets *placed in service* after July 31, 1986. Under MACRS, $BV_N > 0$, and useful life estimates are not directly utilized in calculating depreciation.

[2] The remainder of this chapter has been excerpted from P. E. DeGarmo, W. G. Sullivan, and J. A. Bontadelli, *Engineering Economy*, 9th ed. (New York: Macmillan Publishing Company, 1993). Adapted by permission of the publisher.

MACRS consists of two methods for depreciating property. The main method is called the General Depreciation System (GDS), while the second is called the Alternate Depreciation System (ADS). Unless required by law or specifically elected, GDS is normally used to determine the appropriate depreciation deduction.

MACRS allows a business to recoup the *cost basis* of recovery property over a *recovery period.* The cost basis is normally the cash purchase price of a property plus the cost of making the asset serviceable, thus including shipping and handling, insurance, installation, and training expenses.

Example 6-7

In 1995, your firm purchased a used machine for $10,500 for use in producing income. An additional $1,000 was spent to recondition and install the machine. What is the cost basis?

   Solution   The cost basis of the machine is $11,500 (total costs of $10,500 plus $1,000).

The procedure for computing MACRS depreciation deductions in any given year of an asset's useful life is given in Fig. 6-1. Step 1 is to obtain the property's Asset Depreciation Range (ADR) value from Table 6-2. The ADR value, which depends on the type of property and industry involved, is used in Step 2 to establish the asset's MACRS class life in Table 6-3.

Based on the MACRS class life, a set of GDS recovery rates is located in Table 6-4 (Step 3) and then used in Step 4 to determine MACRS depreciation deductions in year k ($1 \leq k \leq N + 1$):

$$d_k(p) = r_k(p) \cdot B, \qquad (6\text{-}16)$$

where $d_k(p)$ = depreciation deduction in year k for recovery property class p,

$r_k(p)$ = MACRS rate (a decimal) for year k in recovery property class p,

B = cost basis of the recovery property.

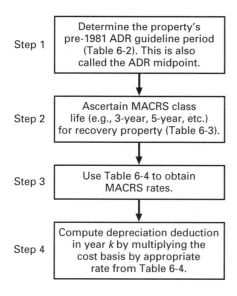

Figure 6-1   Flow diagram for computing depreciation GDS deductions under MACRS per TRA 86.

TABLE 6-2   Selected Asset Depreciation Range
Guideline Periods

Description of depreciable assets	Guideline period (years)
Transportation	
Automobile, taxis	3
Buses	9
General-purpose trucks:	
Light	4
Heavy	6
Air transport (commercial)	12
Petroleum	
Exploration and drilling assets	14
Refining and marketing assets	16
Manufacturing	
Sugar and sugar products	18
Tobacco and tobacco products	15
Carpets and apparel	9
Lumber, wood products, and furniture	10
Chemical and allied products	9.5
Cement	20
Fabricated metal products	12
Electronic components	6
Rubber products	14
Communication	
Telephone	
Central office buildings	45
Distribution poles, cables, etc.	34
Radio and television broadcasting	6
Electric utility	
Hydraulic plant	50
Nuclear	20
Transmission and distribution	30
Services	
Office furniture and equipment	10
Computers and peripheral equipment	6
Recreation—bowling alleys, theater, etc.	10

SOURCE: *Depreciation*, IRS Publication 534, Washington, D.C.:
U.S. Government Printing Office, Dec. 1987 (rev.).

MACRS does not allow deductions for the cost of property, such as land, which
has no determinable life.

Notice that in Table 6-4 there are $N + 1$ rates shown for a MACRS recovery
period of N years. If a depreciable asset is disposed of *after* year $N + 1$, the final BV
of the item will be zero. Also, from Eq. 6-16 it is apparent that MACRS rates in Table
6-4 are applied to the cost basis (B) only, regardless of an asset's expected market
value when disposal occurs.

Example 6-8

In May 1994, your company traded in a computer, used in its business, that had a book value at
that time of $25,000. A new, faster computer system having a fair selling price of $400,000 was

Table 6-3   MACRS Class Lives and Permissible Methods for Calculating Depreciation Rates

MACRS class life and depreciation method	ADR guideline period	Special rules
3-year, 200% declining balance	4 years or less	Includes some racehorses. Excludes cars and light trucks.
5-year, 200% declining balance	More than 4 years to less than 10 years	Includes cars and light trucks, semiconductor manufacturing equipment, qualified technological equipment, computer-based central office switching equipment, some renewable and biomass power facilities, and research and development property.
7-year, 200% declining balance	10 years to less than 16 years	Includes single-purpose agricultural and horticultural structures and railroad track. Includes property with no ADR midpoint.
10-year, 200% declining balance	16 years to less than 20 years	None.
15-year, 150% declining balance	20 years to less than 25 years	Includes sewage treatment plants, telephone distribution plants, and equipment for two-way voice and data communication.
20-year, 150% declining balance	25 years or more	Excludes real property with ADR midpoint of 27.5 years or more. Includes municipal sewers.
27.5-year, straight line	N/A	Residential rental property.
31.5-year, straight line	N/A	Nonresidential real property.

SOURCE: *Depreciation*, IRS Publication 534, Washington, D.C.: U.S. Government Printing Office, Dec. 1987 (rev.).

acquired. Because the vendor accepted the older computer as a trade-in, a deal was agreed to whereby your company would pay $325,000 cash for the new computer system.

    (a)   What is the MACRS class life of the new computer?

    (b)   How much depreciation can be deducted each year based on this class life? (Refer to Fig. 6-1).

Solution

    (a)   The new computer has an ADR guideline period of 6 years (see Table 6-2). Hence, its MACRS recovery period is 5 years (see Table 6-3).

    (b)   The cost basis for this property is $350,000, which is the sum of the $325,000 cash price of the computer and the $25,000 book value remaining on the trade-in (in this case the trade-in was treated as a nontaxable transaction).   ■

    MACRS rates that apply to the $350,000 cost basis are found in Table 6-4. An allowance (half-year) is built into the year 1 rate, so it does not matter that the com-

TABLE 6-4   Modified ACRS Rates per TRA 86

Year	Class life (i.e., recovery period)					
	3-year[a]	5-year[a]	7-year[a]	10-year[a]	15-year[b]	20-year[b]
1	0.3333	0.2000	0.1429	0.1000	0.0500	0.0375
2	0.4445	0.3200	0.2449	0.1800	0.0950	0.0722
3	0.1481	0.1920	0.1749	0.1440	0.0855	0.0668
4	0.0741	0.1152	0.1249	0.1152	0.0770	0.0618
5		0.1152	0.0893	0.0922	0.0693	0.0571
6		0.0576	0.0892	0.0737	0.0623	0.0528
7			0.0893	0.0655	0.0590	0.0489
8			0.0446	0.0655	0.0590	0.0452
9				0.0656	0.0591	0.0447
10				0.0655	0.0590	0.0447
11				0.0328	0.0591	0.0446
12					0.0590	0.0446
13					0.0591	0.0446
14					0.0590	0.0446
15					0.0591	0.0446
16					0.0295	0.0446
17						0.0446
18						0.0446
19						0.0446
20						0.0446
21						0.0223

[a] These rates are determined by applying the 200% declining balance method to the appropriate class life with the half-year convention applied to the first and last years. Rates for each class life must sum to 1.0000.

[b] These rates are determined with the 150% declining balance method instead of the 200% declining balance method and are rounded off to four significant digits.

SOURCE: *Depreciation*, IRS Publication 534, Washington, D.C.: U.S. Government Printing Office, Dec. 1987 (rev.).

puter was purchased in May 1994 instead of, say, November 1994. The GDS depreciation deductions for 1994 through 1999 can be computed with Eq. 6-16 and the 5-year MACRS rates as follows:

Property	Date placed in service	Cost basis	ADR guideline period	MACRS recovery period
Computer system	May 1994	$350,000	6 years	5 years

Year	Depreciation deductions
1994	$0.20 \times \$350,000 = \$\ 70,000$
1995	$0.32 \times\ 350,000 =\ 112,000$
1996	$0.192 \times\ 350,000 =\ \ 67,200$
1997	$0.1152\times\ 350,000 =\ \ 40,320$
1998	$0.1152\times\ 350,000 =\ \ 40,320$
1999	$0.0576\times\ 350,000 =\ \ \ 20,160$
	Total $350,000

■

From Example 6-8 we can conclude that the following relationship is true:

Cost basis = BV of the trade-in (if any)

+ cash price of new equipment after trade-in. (6-17)

6.2.1 Half-Year Convention

We must be careful to observe that MACRS uses a half-year convention. The IRS assumes that assets are purchased halfway through the year, no matter when the asset is actually purchased. This means the rate for year 1 in Table 6-4 is for only a half-year of depreciation. This is built into the calculations used to determine the rates, so there is no need to adjust the rate. Thus, the cost of three-year property is recovered by using the half-year convention for the first year; then the full allowance for the second and third years is deducted, and the remaining balance is deducted for the fourth year. The "recovery period" is still three years, but the deductions are spread over four years if the asset is kept in service for four years or more.

When an asset is disposed of, the half-year convention is also used. If disposal occurs at the end of year $N + 1$ of the recovery period or later, nothing is changed. *If the asset is disposed of before this period, then only half of the normal deduction can be taken for that year. This means that the rate in Table 6-4 should be divided by 2 to compute depreciation in the year of disposal if the asset is sold before year $N + 1$.* In the case of early disposal, the BV in the year of disposal is

$$BV_k = B - d_k^* \text{ through } k \text{ years of depreciation.}$$

Example 6-9

A firm purchased a new piece of semiconductor manufacturing equipment in July 1992. The cost basis for the equipment is $100,000. Determine (a) the depreciation charge permissible in the fourth year (1995), (b) the BV at the end of 1995, (c) the cumulative depreciation through 1994, and (d) the BV at the end of 1996 if the equipment is disposed of (sold) in 1996.

Solution From Table 6-2, it may be seen that the semiconductor (electronic) manufacturing equipment has an ADR guideline period of 6 years, and from Table 6-3 it has a 5-year MACRS class life. The rates that apply are given in Table 6-4.

(a) The depreciation deduction, or cost recovery allowance, that is allowable in 1995 (d_4) is 0.1152($100,000) = $11,520.

(b) The BV at the end of 1995 (BV_4) is the cost basis less depreciation charges in years 1–4:

$$BV_4 = \$100,000 - \$100,000(0.20 + 0.32 + 0.192 + 0.1152)$$
$$= \$17,280.$$

(c) Accumulated depreciation through 1994, d_3^*, is the sum of depreciation in 1992, 1993, and 1994, or

$$d_3^* = d_1 + d_2 + d_3$$
$$= \$100,000(0.20 + 0.32 + 0.192)$$
$$= \$71,200.$$

(d) The depreciation deduction in 1996 (year 5) can only be $(0.5)(0.1152)(\$100,000)$ $= \$5,760$ when the equipment is disposed of (sold) prior to 1997 (year 6). Thus, the BV at the end of year 5 is $BV_4 - \$5,760 = \$11,520.$ ■

6.2.2 The Alternate MACRS Method

— Allows you to slow down depreciation

TRA 86 provides for the use of an alternate method (the ADS). Election to adopt the Alternate Depreciation System for a class of property applies to all property in that class placed in service during the tax year. The decision to use the alternate MACRS method, once made, is irrevocable for the assets involved. Under this method, depreciation is calculated using (1) the straight-line method of depreciation with $BV_N = 0$, and (2) a half-year convention for the depreciable (ADR) life of the property (see Table 6-2).

Example 6-10
A large manufacturer of sheet metal products purchased a new, modern, computer-controlled flexible manufacturing system in October 1995 for $3.0 million. Because this company would not be profitable until the new technology had been in place for several years, it elected to utilize the ADS in computing its depreciation deductions. Thus, the company could slow down its depreciation allowances in hopes of postponing income tax advantages until it became a profitable concern. What depreciation allowances can be claimed for the new system?

Solution   From Table 6-2, the ADR guideline period for a manufacturer of fabricated metal products is 12 years. This flexible manufacturing system would normally be depreciated using 7-year MACRS rates (Table 6-3). However, under the ADS, the straight-line method with no terminal book value is applied to the 12-year depreciation period using half-year convention. Consequently, depreciation in year 1 (1995) would be

$$\frac{1}{2}\left(\frac{\$3,000,000}{12}\right) = \$125,000.$$

Depreciation deductions in years 2–12 (1996–2006) would be $250,000 each year, and depreciation in year 13 (2007) would be $125,000. Notice how the half-year convention extends depreciation deductions over 13 years. ■

6.3 A Comprehensive Example

We now consider an asset for which depreciation is computed by most methods discussed to this point. Be careful to observe the differences in the mechanics of each method, as well as the differences in the annual depreciation amounts themselves.

Example 6-11
The La Salle Bus Company has decided to purchase a new bus for $85,000, which includes a trade-in of its old bus. The old bus has a book value of $10,000 at the time of the trade-in. The new bus will be kept for 10 years before being sold. Its estimated book (salvage) value at that time is expected to be $5,000.

First, we must calculate the cost basis. The basis is the original purchase price of the bus plus the book value of the old bus that was traded in. Thus, the cost basis is $85,000 + $10,000, or $95,000. We will also need to determine the ADR guideline period. Hence, we need to look at Table 6-2 and find "buses" under the Transportation category. We find that buses have a 9-year

ADR guideline period. This will be used as the number of years over which we depreciate the bus with historical methods discussed earlier.

Solution: Straight-Line Method

For the straight-line method, we use the ADR guideline period of 9 years even through the bus will be kept for 10 years. By using Eqs. 6-1 and 6-3, we obtain the following information:

$$d_k = \frac{\$95,000 - \$5,000}{9 \text{ years}} = \$10,000 \quad \text{for } k = 1 \text{ to } 9.$$

Straight-Line Method

EOY k	d_k	BV_k
0	—	$95,000
1	$10,000	85,000
2	10,000	75,000
3	10,000	65,000
4	10,000	55,000
5	10,000	45,000
6	10,000	35,000
7	10,000	25,000
8	10,000	15,000
9	10,000	5,000

Notice that no depreciation was taken after year 9 because the ADR life was only 9 years. Also, the BV will remain at $5,000 from the last ADR year until the bus is sold.

Solution: Declining Balance Method

To demonstrate this method, we will use the 200% declining balance equations. By using Eqs. 6-4, 6-5, and 6-7, we derive these example calculations:

$$R = 2/9 = 0.2222$$
$$d_1 = \$95,000(0.2222) = \$21,111$$
$$d_5 = \$95,000(1 - 0.2222)^{5-1}(0.2222) = \$7,726$$
$$BV_5 = \$95,000(1 - 0.2222)^5 = \$27,040$$

200% Declining Balance Method

EOY k	d_k	BV_k
0	—	$95,000
1	$21,111	73,889
2	16,420	57,469
3	12,771	44,698
4	9,932	34,765
5	7,726	27,040
6	6,009	21,031
7	4,674	16,357
8	3,635	12,722
9	2,827	9,895

Solution: Sum-of-the-Years'-Digits Method

Once again, we use the ADR guideline period of 9 years. The SYD depreciation amounts are as follows:

Year	Number of the year in reverse order	SYD deprec. factor	$d_k =$ $(B - \mathrm{BV}_N) \cdot$ factor	BV_k
0	—	—	—	$95,000
1	9	9/45	$18,000.00	77,000
2	8	8/45	16,000.00	61,000
3	7	7/45	14,000.00	47,000
4	6	6/45	12,000.00	35,000
5	5	5/45	10,000.00	25,000
6	4	4/45	8,000.00	17,000
7	3	3/45	6,000.00	11,000
8	2	2/45	4,000.00	7,000
9	1	1/45	2,000.00	5,000
	Sum = 45			

Using Eqs. 6-10 and 6-11, we compute the following:

$$d_5 = (\$95,000 - \$5,000)\left[\frac{2(9 - 5 + 1)}{9(9 + 1)}\right] = \$10,000,$$

$$\mathrm{BV}_5 = \$95,000 - \frac{2(\$95,000 - \$5,000)}{9}5 + \frac{(\$95,000 - \$5,000)5(5 + 1)}{9(9 + 1)} = \$25,000.$$

Solution: MACRS Method

For this method, we will follow the steps given in Fig. 6-1.

1. Determine the ADR guideline period from Table 6-2. We have already found it to be 9 years.
2. Ascertain the MACRS class life for recovery property. By Table 6-3, an ADR guideline period of 9 years corresponds to a 5-year MACRS class life.
3. Now use Table 6-4 to determine d_k.

EOY	MACRS rate	d_k	BV_k
0	—	—	$95,000
1	0.2000	$19,000	76,000
2	0.3200	30,400	45,600
3	0.1920	18,240	27,360
4	0.1152	10,944	16,416
5	0.1152	10,944	5,472
6	0.0576	5,472	0
7	—	—	0
8	—	—	0
9	—	—	0
10	—	—	0

Notice that the depreciation life is much shorter than for the previous methods, and the BV at the end of year $N + 1$ is zero. The estimated salvage value of $5,000 does not affect MACRS depreciation amounts.

Solution: MACRS Half-Year Convention

To demonstrate the half-year convention, we will change the La Salle bus problem so that the bus is now sold in year 5 in part (a) and in year 6 for part (b).

(a) Selling bus in year 5:

EOY	Factor	d_k	BV_k
0	—	—	$95,000
1	0.2000	$19,000	76,000
2	0.3200	30,400	45,600
3	0.1920	18,240	27,360
4	0.1152	10,944	16,416
5	0.0576	5,472	10,944

(b) Selling bus in year 6:

EOY	Factor	d_k	BV_k
0	—	—	$95,000
1	0.2000	$19,000	76,000
2	0.3200	30,400	45,600
3	0.1920	18,240	27,360
4	0.1152	10,944	16,416
5	0.1152	10,944	5,472
6	0.0576	5,472	0

Notice that when we sold the bus in year 5 before the recovery period had ended, we claimed only half of the normal depreciation (MACRS rate = 0.1152/2). The other years (years 1–4) were not changed. When the bus was sold in year 6 (at the end of the recovery period), we did not divide the last year amount by 2.

Selected methods of depreciation, illustrated in Example 6-11, are compared in Fig. 6-2. ■

6.4 Introduction to Income Taxes

The main types of taxes important to economic analyses are as follows:

1. *Property taxes* are based on the valuation of property owned, such as land, equipment, buildings, inventory, etc., and the established tax rates. They do not vary with income and are usually much lower in amount than income taxes.

2. *Sales taxes* are taxes imposed on product sales, usually at the retail level. The are relevant in engineering economy studies only to the extent that they add to the cost of items purchased.

3. *Excise taxes* are taxes imposed upon the manufacture of certain products, such as alcohol and tobacco.

4. *Income taxes* are taxes on pretax income of an organization in the course of regular business. Income taxes are also levied on gains on the disposal of

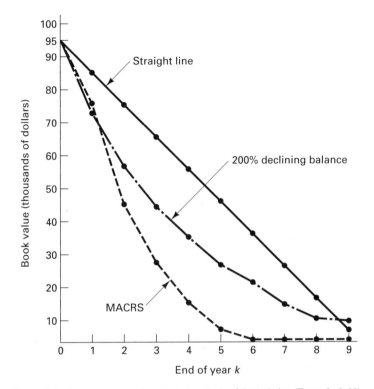

Figure 6-2   BV comparisons for selected methods of depreciation (Example 6–11).

capital property. They are usually the most significant type of tax to consider in economic analyses and are the subject of the remainder of this chapter.

6.4.1 When Income Taxes Should Be Considered

In the preceding chapters we have treated income taxes as if they are either not applicable, or taken into account by using a before-tax rate of return that is larger than the after-tax rate of return. The intention in this second case is to adjust resulting cash flows so they will be sufficient to provide for both the after-tax rate of return and the income tax.

An approximation of the before-tax rate of return (RR) requirement to include the effect of income taxes in studies using before-tax cash flows can be determined from the following relationship:

[Before-tax RR][1 − {Effective income tax rate}] = [After-tax RR].

Thus,

$$[\text{Before-tax RR}] = \frac{[\text{After-tax RR}]}{[1 - \{\text{Effective income tax rate}\}]}. \qquad (6\text{-}18)$$

If the property is nondepreciable and there is no capital gain or loss of investment tax credit, then the above relationship is exact, not an approximation. If the salvage value is less than 100% of first cost and if the life of the property is finite, then the depreciation method selected for income tax purposes affects the timing of income tax payments. Therefore, error can be introduced by using the relationship in Eq. 6-18.

In practice, it is important to make after-tax analyses in any income-tax–paying organization. After-tax analyses can be performed by exactly the same methods (PW, IRR, etc.) as before-tax analyses. *The only difference is that after-tax cash flows must be used in place of before-tax cash flows, and the calculation of a measure of merit such as PW is based on an after-tax MARR.*

✓ The mystery behind the sometimes complex computation of income taxes is reduced when one recognizes that income taxes paid are just another type of expense, while income taxes saved (through business deductions, expenses, or direct tax credits) are identical to other kinds of reduced expenses (e.g., savings).

The basic concepts underlying federal and state income tax regulations that apply to most economic analyses of capital investments generally can be understood and applied without difficulty. This chapter is not intended to be a comprehensive treatment of federal tax law. Rather, we utilize some of the more important provisions of the Tax Reform Act of 1986 (TRA 86), followed by illustrations of a general procedure for computing after-tax cash flows (ATCFs) and conducting after-tax economic analyses. Where appropriate, important changes to income tax provisions enacted by the Omnibus Reconciliation Act of 1993 (OBRA 93) are also included in this chapter.

6.4.2 Taxable Income of Business Firms

At the end of each tax year, a corporation must calculate its net (i.e., taxable) before-tax income or loss. Several steps are involved in this process, beginning with the calculation of *gross income.* Gross income represents the gross profits from operations (revenues from sales minus the cost of goods sold) plus income from dividends, interest, rent, royalties, and gains (or losses) on the exchange of capital assets. The corporation may deduct from gross income all ordinary and necessary operating expenses, including interest, to conduct the business. Deductions for depreciation are permitted each tax period as a means of consistently and systematically recovering capital investment. Consequently, allowable expenses and depreciation deductions may be used to determine *taxable income,* as shown in Eq. 6-19:

$$\begin{aligned}
\text{Taxable income} = {} & \text{gross income} \\
& - \text{ all expenses (except capital expenditures)} \qquad (6\text{-}19) \\
& - \text{ depreciation deductions.}
\end{aligned}$$

Taxable income is often referred to as *net income before taxes,* and when income taxed are subtracted from it, the remainder is called *net income after taxes.* There are two types of income for tax computation purposes: ordinary income (and losses) and capital gains (and losses). These types of income are explained in subsequent sections.

Example 6-12

In 199X a company generates $1,500,000 of gross income and incurs operating expenses of $800,000. Interest payments on borrowed capital amount to $48,000. The total depreciation deductions in 199X equal $114,000. (a) What is the taxable income of this firm? (b) If interest expenses had been $590,000, what would the net operating loss (NOL) have been?

> **Solution**
>
> **(a)**   Based on Eq. 6-19, this company's taxable income in 199X would be
>
> $1,500,000 - $800,000 - $48,000 - $114,000 = $538,000.
>
> **(b)**   NOL = |$1,500,000 - $800,000 - $590,000 - $114,000| = $52,000   ■

6.4.3 Taxable Income of Individuals

An individual's total income is essentially what is called *adjusted gross income* (AGI) for federal income tax purposes. Adjusted gross income consists of wages, salaries, and/or tips—plus other income such as interest, corporate dividends, and pensions—less adjustments to income such as penalties on early withdrawal from savings accounts and alimony payments. From this amount individuals may subtract personal exemptions and allowable deductions to determine their *taxable income.* In 1993 personal exemptions were provided at the rate of $2,350 (for AGI under $162,700 for married/joint returns) for the taxpayer and each dependent, and this exemption is indexed to inflation. Allowable deductions can also be claimed, within some limits, for such items as large medical costs, state and local taxes, interest on borrowed money, charitable contributions, and casualty losses. A standard deduction may be taken by people who do not itemize their allowable deductions.

Taxpayers who *do not* itemize deductions are permitted to reduce their adjusted gross income by a specified standard amount. These standard deductions are indexed for inflation based on the following 1993 standard deductions categorized by filing status:

Filing status	1993 standard deduction
Married/joint return	$6,200
Head of household	5,450
Single	3,700
Married/separate return	3,100

Thus, taxable income of individual taxpayers is determined by using Eq. 6-20:

Taxable income = adjusted gross income − personal exemption deductions

  − other allowable deductions (such as standard deductions

  discussed above).   (6-20)

Example 6-13

Jayne Doe has an adjusted gross income of $60,000 in 1993, resulting from her salary and interest on savings accounts. She files her federal tax return as a *single* taxpayer and elects to take the standard deduction, which has been indexed for inflation. What is Jayne's taxable income in 1993?

Solution   In 1993 the personal exemption deduction that Jayne can claim is $2,350. Furthermore, she can take the standard deduction, which amounts to $3,700. Therefore, from Eq. 6-20, Jayne's taxable income is $60,000 − $2,350 − $3,700 = $53,950.   ∎

6.4.4 Calculation of Effective Income Tax Rates

Although the regulations of most of the states with income taxes have the same basic features as the federal regulations, there is great variation in income tax rates. State income taxes are in most cases much lower than federal taxes and can often be closely approximated as a constant percentage of federal taxes, typically ranging from 6% to 12% of taxable income. No attempt will be made here to discuss the details of state income taxes. An understanding of the applicable federal income tax regulations usually will enable the analyst to apply the proper procedures if state income taxes must also be considered.

To illustrate the calculation of an effective (combined federal and state) income tax rate for a very large corporation, suppose that the federal income tax rate is 34% and the state income tax rate is 9%. Further assume the common case in which taxable income is computed the same way for both types of taxes, except that state taxes are deductible from taxable income for federal tax purposes but federal taxes are *not* deductible from taxable income for state tax purposes. The general expression for the effective income tax rate (t) is

$$t = \text{federal rate} + \text{state rate} - (\text{federal rate})(\text{state rate}). \qquad (6\text{-}21)$$

In this example the effective income tax rate would be

$$t = 0.34 + 0.09 - 0.34(0.09) = 0.3994, \text{ or about 40\%.}$$

It is the effective income tax rate on increments of taxable income that is of importance in engineering economy studies. In this chapter, we will use an effective income tax rate of 40% in most of the examples that follow.

6.4.5 Income Taxes on Ordinary Income (and Losses)

Ordinary (taxable) income is the net income before taxes that results from the routine business operations (such as the sale of products or services) performed by a corporation or individual. For federal income tax purposes, virtually all ordinary income adds to taxable income and is subject to a graduated rate scale (higher rates for higher taxable income).

The federal income tax rates as of 1993 are given in Table 6-5. For example, suppose that a firm in 1993 has a gross income of $5,270,000, expenses (excluding capital) of $2,927,500, and depreciation of $1,874,300. Its taxable income and federal income tax would be determined with Eq. 6-19 and Table 6-5 as follows.

$$\begin{aligned}
\text{Taxable income} &= \text{gross income} - \text{expenses} - \text{depreciation} \\
&= \$5,270,000 - \$2,927,500 - \$1,874,300 \\
&= \$468,200
\end{aligned}$$

	Taxable income	Income tax
Income tax = 15% of first	$ 50,000	$ 7,500
+ 25% of next	25,000	6,250
+ 34% of next	25,000	8,500
+ 39% of next	235,000	91,650
+ 34% of remaining	133,200	45,288
Total	$468,200	Total $159,188

TABLE 6-5   Corporate Federal Income Tax

If taxable income is:		The tax is:	of the amount over
Over	but not over		
$ 0	$ 50,000	15%	$ 0
50,000	75,000	$ 7,500 + 25%	50,000
75,000	100,000	13,750 + 34%	75,000
100,000	335,000	22,250 + 39%	100,000
335,000	10,000,000	113,900 + 34%	335,000
10,000,000	15,000,000	3,400,000 + 35%	10,000,000
15,000,000	18,333,333	5,150,000 + 38%	15,000,000
18,333,333	....	6,416,667 + 35%	18,333,333

The total income tax liability in this case is $159,188. Because engineering economy studies are concerned with incremental differences among alternatives, we shall use a 34% *incremental* federal income tax rate, as shown in Fig. 6-3, for most

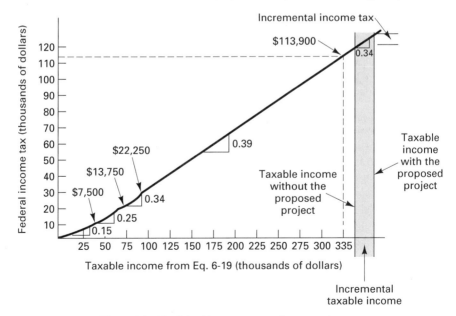

Figure 6-3   The federal income tax rates for corporations (assuming taxable income is $10,000,000 or less).

studies by assuming that corporations have taxable incomes between $335,000 and $10,000,000. Thus, an approximate *effective* income tax rate of 40% is appropriate in several examples to follow.

Example 6-14

A corporation is expecting an annual taxable income of $45,000. It is considering investing an additional $100,000, which is expected to create an added annual net cash flow (receipts minus expenses) of $35,000 and an added annual depreciation charge of $20,000. What is the corporation's federal income tax liability based on rates in effect in 1993 (a) without the added investment, and (b) with the added investment?

Solution

(a)

Income taxes	Rate	Amount
On first $45,000	15%	$ 6,750
	Total	$ 6,750

(b)

Taxable income		
Before added investment		$45,000
+ added net cash flow		35,000
− depreciation		(20,000)
	Total	$60,000

Income taxes on $60,000	Rate	Amount
On first $50,000	15%	$ 7,500
On next $10,000	25%	2,500
	Total	$10,000

The increased income tax liability from the investment is $3,250.

As an added note, the change in tax liability can usually be determined more readily by an incremental approach. For instance, this example involved changing the taxable income from $45,000 to $60,000 as a result of the new investment. Thus, the change in income taxes per year (for 1993) could be calculated as follows:

$$\text{First } \$50,000 - \$45,000 = \$5,000 \text{ at } 15\% = \$ \ 750$$
$$\text{Next } \$60,000 - \$50,000 = \$10,000 \text{ at } 25\% = \ \underline{2,500}$$
$$\text{Total} \quad \$3,250 \quad \blacksquare$$

A simple example of computing income taxes for an individual is now provided to complete Section 6.4.5.

Example 6-15

Jayne Doe, whose taxable income was determined to be $53,950 in Example 6-13, would now like to know how much money she owes in federal income taxes in 1993. According to 1993 tax tables, as a head of a household she must pay 15% of the first $29,600 in taxable income and 28% on the remainder up to a threshold of $76,400. Determine her tax liability in 1993.

Solution   She owes 0.15($29,600) + 0.28($53,950 − $29,600) = $11,258 in 1993 federal income taxes.

6.4.6 Income Taxes on Capital Gains (and Losses)

When a capital asset[3] is disposed of for more (or less) than its book value, the resulting capital gain (or capital loss) historically had been taxed (or saved taxes) at a rate different from that for ordinary income. However, TRA 86 eliminated the differential between income taxes on capital gains and ordinary income. Thus, for firms with federal taxable income between $335,000 and $10,000,000, capital gains are taxed at a rate of 34% and capital losses create tax savings amounting to 34% of the loss. Furthermore, there is no longer a distinction between long-term and short-term capital gains and losses.

In equation form, the determination of capital gains and losses is straightforward:

$$\text{Capital gain (loss)} = \text{net selling price (i.e., market value)} - \text{book value} \tag{6-22}$$
$$= MV - BV$$

or

$$\text{Capital gain (loss)} = \text{net selling price (i.e., MV)} - \text{original first cost}$$
$$\text{(i.e., cost basis)} + \text{accumulated depreciation deductions}$$
$$= MV - B + d_k^*. \tag{6-23}$$

In many cases, personal property used in a trade or business is sold for an amount that is greater than its BV but less than its cost basis at the time of disposal. Such gains are technically *not* capital gains, but instead are referred to as *depreciation recapture*. TRA 86 specifies that this difference between the disposal price (i.e., MV) of a depreciable asset and its BV at the time of disposal is to be taxed as ordinary income. In this chapter capital gains are taxed at an effective income tax rate of 40%. When an asset's selling price is less than its BV, the resulting loss on disposal creates an income tax credit that is 40% of the loss.

Example 6-16

A large, profitable company has just made two transactions in which a gain and a loss have been recorded. The first transaction involved the sale of long-term bonds having a face value of $100,000. Because interest rates in the economy are substantially higher than those in force when the bond was purchased five years ago, the company was able to obtain only $95,800 from the sale of these bonds. The second transaction was the sale of a piece of earth-moving equipment for $23,100. This depreciable asset has a current BV of $18,000.

(a) Classify each transaction and determine its federal income tax consequence.

(b) What is the net income tax liability (or credit)?

---

[3] Capital assets include all property owned by a taxpayer except (1) property held mainly for sale to customers, (2) most accounts or notes receivable, (3) depreciable property utilized to carry out the production process, (4) real property used by the business, and (5) copyrights and certain types of short-term, non-interest-bearing state/federal notes. Stock owned in another company is a familiar example of a capital asset.

Solution

(a) The first transaction involves a capital asset, and the *capital loss* is $100,000 − $95,800 = $4,200. The resultant tax *credit* is ($4,200)(0.34) = $1,428. The second transaction is an illustration of depreciation recapture because the selling price is higher than the BV of the equipment. Here a *gain on disposal* is $23,100 − $18,000 = $5,100, and the income tax *liability* is $5,100(.34) = $1,734.

(b) The net consequence of these transactions is a tax liability of $1,734 − $1,428 = $306. This figure could also be obtained by offsetting the capital loss against the gain from depreciation recapture and taxing the difference:

$$\text{Taxes owed} = 0.34[\$5,100 \text{ (gain)} − \$4,200 \text{ (loss)}] = \$306. \qquad \blacksquare$$

6.4.7 Investment Tax Credit

A special provision of the federal income tax law, *when in force,* is the investment tax credit (ITC). Originally enacted in 1962, it permits businesses to subtract from their overall tax liability a stated percentage of the cost basis in qualifying property purchased during a particular tax year; thus, it encourages capital investment. If the ITC is 10% of the value of a qualifying property, for instance, the net impact is to reduce the after-tax cost of the asset to 90% of its cost to a company. Qualifying property is depreciable, tangible property used in a trade or business.

TRA 86 *repealed the ITC* in the interest of broadening the corporate tax base and lowering income tax rates on ordinary income. Because the ITC has been repealed and reinstated several times since 1962, it is highly likely that Congress will restore the credit in some form in the future. A simple example is provided here to illustrate the mechanics of the investment tax credit.

Example 6-17

A firm purchased a computerized machine tool for $250,000. Suppose that Congress has restored a 10% investment tax credit that applies to domestically manufactured machine tools. (a) What is the after-tax cost of acquiring this equipment? (b) What is the MACRS depreciation deduction in the first year of ownership if this property's class life is 7 years?

Solution

(a) The 10% investment tax credit is $25,000, assuming that the entire investment is qualifying property. This financial incentive applies dollar for dollar against tax liabilities that the firm incurs, and it effectively reduces the machine tool's first cost (after taxes) to $225,000.

(b) The basis in the machine tool is often unaffected by the 10% ITC. This represents another financial incentive of ITCs. Hence, the first-year MACRS depreciation deduction is 0.1429($250,000) = $35,725. $\qquad \blacksquare$

6.4.8 Section 179 Deduction

In general, Section 179 of the Internal Revenue Code as of 1993 permits taxpayers to deduct as an expense up to $17,500 of the cost of qualifying tangible property used in a trade or business. This provision allows businesses and corporations essentially to claim an extra depreciation deduction in the initial tax year, subject to certain limitations. OBRA 93 limits corporations to an extra $17,500 deduction if *more than*

$200,000 of property is acquired during the year. Thus, Section 179 is primarily intended to benefit small to medium-sized businesses. Where applicable, we shall assume that companies qualify for the full $17,500 rather than a prorated share of it.

Example 6-18

In 1994 a medium-sized firm having taxable income of $750,000 purchases a new coordinate measuring machine (CMM) for $250,000 and takes a Section 179 deduction. How much *federal* income tax would this firm pay in 1994?

Solution   From Table 6-5, we determine the pre-Section 179 income tax liability to be $0.15(\$50,000) + 0.25(\$25,000) + 0.34(\$25,000) + 0.39(\$235,000) + 0.34(\$415,000) = \$255,000$. By taking the $17,500 Section 179 deduction (because the CMM costs more than $200,000), the firm's income tax liability becomes $255,000 - \$17,500 = \$237,500$. (In this case, the depreciable cost basis of the CMM would be reduced to $250,000 - \$17,500 = \$232,500$.) ∎

Only minimal coverage of some of the main provisions of TRA 86 and OBRA 93 that are important to engineering economy studies has been provided here. It is by no means complete, but it is intended to establish a basis for illustrating after-tax (i.e., after income tax) economic analyses. In general, the analyst should either search out specific provisions of the federal and/or state income tax law affecting projects being studied or seek information from persons qualified in income tax law.[4]

The remainder of the chapter illustrates various types of after-tax problems by using a tabular form for computing after-tax cash flows (ATCFs).

6.5 General Procedure for Making After-Tax Economic Analyses

After-tax economic analyses can be performed by using exactly the same methods as before-tax analyses. The only difference is that ATCFs are used in place of before-tax cash flows (BTCFs) by including expenses (or savings) due to income taxes and then making equivalent worth calculations using an after-tax MARR. The income tax rates and governing regulations may be complex and subject to changes, but once those rates and regulations have been translated into their effect on ATCFs, the remainder of the after-tax analysis is relatively straightforward.

To formalize the procedure described in previous sections for determining net income before taxes (NIBT), net income after taxes (NIAT), and ATCF of the incremental project shown in Fig. 6-3, the following notation and equations are provided. For any given year k of the study period, $k = 0, 1, 2, \ldots, N$, let

R_k = revenues from the project; this is the positive cash flow from the project during period k,

E_k = cash outflows during year k for deductible expenses and interest,

---

[4] Some applicable publications are as follows:

a. *Tax Guide for Small Business,* IRS Publication 334, Washington, D.C.: U.S. Government Printing Office, published annually.

b. *Your Federal Income Tax,* IRS Publication 17, Washington, D.C.: U.S. Government Printing Office, published annually.

c. J. K. Lasser, *Your Income Tax* (New York: Simon and Schuster, published annually).

d_k = sum of all noncash, or book, costs during year k, such as depreciation and depletion,

t = effective income tax rate on *ordinary* income (federal, state, and other); t is assumed to remain constant during the study period,

T_k = income taxes paid during year k,

ATCF_k = ATCF from the project during year k.

Because the NIBT (i.e., taxable income) is $(R_k - E_k - d_k)$, the *ordinary income tax liability* when $R_k > (E_k + d_k)$ is computed with Eq. 6-24:

$$T_k = -t(R_k - E_k - d_k). \tag{6-24}$$

The NIAT is then simply taxable income (i.e., net income before taxes) minus the tax liability amount determined by Eq. 6-24:

$$\text{NIAT}_k = \underbrace{(R_k - E_k - d_k)}_{\text{taxable income}} - \underbrace{t(R_k - E_k - d_k)}_{\text{income taxes}}$$

or

$$\text{NIAT}_k = (R_k - E_k - d_k)(1 - t). \tag{6-25}$$

The ATCF associated with a project equals the NIAT plus noncash items such as depreciation:

$$\begin{aligned}\text{ATCF}_k &= \text{NIAT}_k + d_k \\ &= (R_k - E_k - d_k)(1 - t) + d_k \end{aligned} \tag{6-26}$$

or

$$\text{ATCF}_k = (R_k - E_k)(1 - t) + td_k. \tag{6-27}$$

In many economic analyses of engineering and business projects, ATCFs in year k are computed in terms of BTCF_k (i.e., year k before-tax cash flows):

$$\text{BTCF}_k = R_k - E_k. \tag{6-28}$$

Thus,[5]

$$\text{ATCF}_k = \text{BTCF}_k + T_k \tag{6-29}$$

$$\begin{aligned} &= (R_k - E_k) - t(R_k - E_k - d_k) \\ &= (1 - t)(R_k - E_k) + td_k. \end{aligned} \tag{6-30}$$

Tabular headings to facilitate the computation of after-tax cash flows with Eqs. 6-24 and 6-30 are as follows:

Year	(A) Before-tax cash flow	(B) Depreciation	(C) = (A) − (B) Taxable income	(D) = −t(C) Cash flow for income taxes	(E) = (A) + (D) After-tax cash flow
k	$R_k - E_k$	d_k	$R_k - E_k - d_k$	$-t(R_k - E_k - d_k)$	$(1 - t)(R_k - E_k) + td_k$

[5] In Eq. 6-24 and Fig. 6-3 we use $-t$ in column D, so subtraction of income taxes in Eq. 6-29 is accomplished.

Column A consists of the same information used in before-tax analyses, namely, the cash revenues (or savings) less the deductible expenses. Column B contains depreciation that can be claimed for tax purposes. Column C is the taxable income, or amount subject to income taxes. Column D contains the income taxes paid (or saved). Finally, column E shows the ATCFs to be used directly in after-tax economic analyses.

A summary of the process of determining the NIAT and ATCF during each year of an N-year study period is provided in Fig. 6-4. NIAT is well understood in many companies, and it can be easily obtained from Fig. 6-4 for making presentations to upper-level management. The format of Fig. 6-4 is used extensively throughout the remainder of Chapter 6, and it provides a convenient way to organize data in after-tax studies.

The column headings of Fig. 6-4 indicate the arithmetic operation for computing columns C, D, and E when year $k = 1, 2, \ldots , N$. When $k = 0$ and $k = N$, capital expenditures are usually involved and their tax treatment (if any) is illustrated in the examples that follow. The table should be used with the conventions of + for cash inflow or savings and − for cash outflow or opportunity foregone.

Example 6-19

If the revenue from a project is $10,000 in 1994, out-of-pocket expenses are $4,000, and depreciation claimed for income tax purposes is $2,000, what is the ATCF when $t = 0.40$? What is the NIAT?

Solution   From Eq. 6-26, we have

$$\text{ATCF}_{1994} = (\$10,000 - \$4,000 - \$2,000)(1 - 0.4) + \$2,000 = \$4,400.$$

The same result can be obtained with Eq. 6-27 (or 6-30):

$$\text{ATCF}_{1994} = (\$10,000 - \$4,000)(1 - 0.4) + 0.4(\$2,000) = \$4,400.$$

Equation 6-27 clearly shows that depreciation contributes a credit of td_k to the after-tax cash flow in operating year k. The NIAT from Eq. 6-26 is $4,400 − $2,000 = $2,400.   ∎

> The ATCF attributable to depreciation (a tax savings) is td_k in year k. After income taxes, an expense becomes $(1 - t)E_k$ and a revenue becomes $(1 - t)R_k$.

Example 6-20

Suppose that an asset, with a cost basis of $100,000 and a depreciable life of five years, is being depreciated with *alternate MACRS* as follows:

Year	1994	1995	1996	1997	1998	1999
Depreciation deduction	$10,000	$20,000	$20,000	$20,000	$20,000	$10,000

If the firm's effective income tax rate remains constant at 40% during 1994–1999, what is the PW of after-tax savings resulting from depreciation when MARR = 10% (after income taxes)?

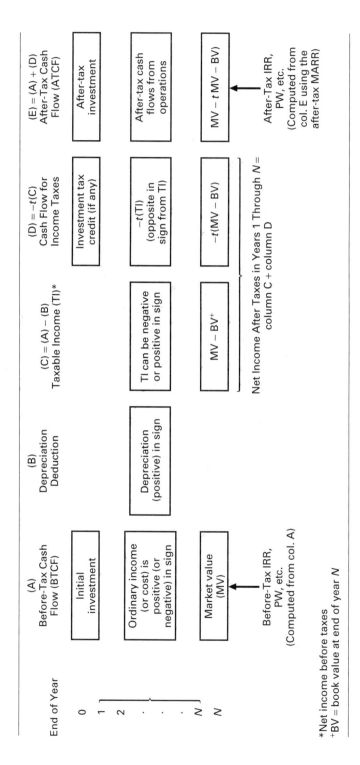

Figure 6-4 General format for determining ATCF.

Solution   The PW of tax credits (savings) due to this depreciation schedule is

$$\text{PW}(10\%) = \sum_{k=1994}^{1999} 0.4d_k(1.10)^{-k} = \$4,000(0.9091) + \$8,000(0.8264) + \cdots$$
$$+ \$4,000(0.5645) = \underline{\$28,948}.\qquad\blacksquare$$

Example 6-21
The asset in Example 6-20 is expected to produce net cash inflows (net revenues) of $30,000 per year during the 1994–1999 period, and its terminal market value is negligible. If the effective income tax rate is 40%, how much can a firm afford to spend for this asset and still earn the MARR? What is the meaning of any excess in affordable amount over the $100,000 cost basis given in Example 6-19?

Solution   After income taxes, the PW of net revenues is $(1 - 0.4)(\$30,000) \cdot (P/A, 10\%, 6)$ = $18,000(4.3553) = $78,395. After adding to this the PW of tax savings computed in Example 6-20, the affordable amount is $107,343. Because the cost of the investment is $100,000, the net PW equals $7,343 (which is the excess in affordable amount over $100,000). This same result can be obtained by using the general format of Fig. 6-4:

EOY	(A) BTCF	(B) Depreciation deduction	(C) = (A) – (B) Taxable income	(D) = –0.4(C) Income taxes $\times PW\%$	(E) = (A) + (D) ATCF
0	–$100,000	—	—	—	–$100,000
1 (1994)	30,000	$ 10,000	$20,000	–$8,000 $\times ?$	22,000
2 (1995)	30,000	20,000	10,000	– 4,000	26,000
3 (1996)	30,000	20,000	10,000	– 4,000	26,000
4 (1997)	30,000	20,000	10,000	– 4,000	26,000
5 (1998)	30,000	20,000	10,000	– 4,000	26,000
6 (1999)	30,000	10,000	20,000	– 8,000	22,000
Totals	$ 80,000	$100,000	$80,000		$ 48,000
					PW(10%) = $ 7,343

$\blacksquare$

6.5.1 Illustration of Computations of ATCFs

The following problems (Examples 6-22 and 6-23) illustrate the computation of ATCFs, as well as many common situations that affect income taxes. All problems include the assumption that income tax expenses (or savings) occur at the same time (year) as the income or expense that gives rise to the taxes. For purposes of comparing the effects of various situations, the after-tax IRR and PW are computed for each example. One can observe from the results of Example 6-22 that the faster (i.e., earlier) the depreciation deduction, the more favorable the after-tax IRR and PW will become.

Example 6-22
Certain new machinery placed in service in 1994 is estimated to cost $180,000. It is expected to *reduce* net annual operating expenses by $36,000 per year for 10 years and to have a $30,000 market value (MV) at the end of the tenth year. (a) Develop the before-tax and after-tax cash flows, and (b) calculate the before-tax and after-tax IRR using MACRS depreciation with a federal income tax rate of 34% plus a state income tax rate of 6%. State income taxes are deductible from federal taxable income. This machinery has a MACRS class life of 5 years. (c) Calculate the after-tax PW when the after-tax MARR = 10%. In this example the study period is 10 years but the tax life is 6 years (which includes the carryover effect of the half-year convention).

Solution

(a) Table 6-6 applies the format illustrated in Fig. 6-4 to calculate the BTCF and ATCF for this example. In column D the effective income tax rate is very close to 0.38 (from Eq. 6-21) based on the information just provided.

(b) The before-tax IRR is computed from column A:

$$0 = -\$180,000 + \$36,000(P/A, i'\%, 10) + \$30,000(P/F, i'\%, 10).$$

By trial and error, $i' = 16.1\%$.

The entry in the last year is shown to be \$30,000 since the machinery will have this estimated MV. However, the asset was depreciated to zero with the MACRS method. Therefore, when the machine is sold at the end of year 10, there will be \$30,000 of *recaptured depreciation,* which is taxed at the effective income tax rate of 38%. This tax entry is shown in column D (EOY 10).

By trial and error, the after-tax IRR for Example 6-22 is found to be 12.4%.

(c) When MARR = 10% is inserted into the PW equation at the bottom of Table 6-6, it can be determined that the after-tax PW of this investment is \$17,209.   ∎

If the machinery in Example 6-22 has been classified as 10-year MACRS property instead of 5-year property, depreciation deductions would be slowed down in the early years of the study period and shifted into later years, as shown in Table 6-7. Compared to Table 6-6, entries in columns C, D, and E of Table 6-7 are less favorable in the sense that a fair amount of ATCF is deferred until later years, producing a lower after-tax IRR and PW. For instance, the PW is reduced from \$17,209 in Table 6-6 to \$9,061 in Table 6-7. The only difference between Table 6-6 and Table 6-7 is the *timing of the ATCF,* which is a function of the timing and magnitude of the depreciation deductions. In fact, the curious reader can confirm that the sums of entries in columns A through E of Tables 6-6 and 6-7 are identical (except for round-off differences) in both tables—timing of cash flows does, of course, make a difference!

It is also of interest to note the impact that the ITC would have on the after-tax IRR in Example 6-22. If a 10% ITC could have been claimed for the investment, it would have reduced other income taxes payable by the firm, thereby reducing the out-of-pocket investment by \$18,000 to \$162,000. As a result, the after-tax PW is increased by \$18,000 and the after-tax IRR is increased to 15.3% (the depreciation schedule is assumed not to change). The affected portion of the cash flow table would be as follows:

Year, k	BTCF	Depreciation deduction	Taxable income	Cash flow for income taxes	ATCF
0	-\$180,000			+\$18,000	-\$162,000
1	36,000	\$36,000	\$ 0	0	36,000
2	36,000	\$57,000	- 21,600	- 8,202	44,208
·	·	·	·	·	·
·	·	·	·	·	·
·	·	·	·	·	·

TABLE 6-6 ATCF Analysis of Example 6-22 (5-year MACRS Life)

End of year, k	(A) BTCF	(B) Basis	×	MACRS rate	=	Deduction	(C) = (A) − (B) Taxable income	(D) = − 0.38(C) Cash flow for income taxes	(E) = (A) + (D) ATCF
0	−$180,000	—		—		—	—	—	−$180,000
1	36,000	$180,000	×	0.2000	=	$36,000	$ 0	$ 0	36,000
2	36,000	180,000	×	0.3200	=	57,600	− 21,600	+ 8,208	44,208
3	36,000	180,000	×	0.1920	=	34,560	1,440	− 547	35,453
4	36,000	180,000	×	0.1152	=	20,736	15,264	− 5,800	30,200
5	36,000	180,000	×	0.1152	=	20,736	15,264	− 5,800	30,200
6	36,000	180,000	×	0.0576		10,368	25,632	− 9,740	26,260
7–10	36,000	0		—		0	36,000	− 13,680	22,320
10	30,000	—		—		—	30,000[a]	− 11,400[b]	18,600
Total	$210,000								$130,201
								PW(10%) =	$ 17,209

[a] Depreciation recapture = MV − BV = $30,000 − 0 = $30,000.

[b] Tax on depreciation recapture = $30,000(0.38) = $11,400.

After-tax IRR: Set PW of Column E = 0 and solve for i' in the following equation.

$0 = -\$180{,}000 + 36{,}000(P/F, i', 1) + 44{,}208(P/F, i', 2) + 35{,}453(P/F, i', 3) + 30{,}200(P/F, i', 4) + 30{,}200(P/F, i', 5) + 26{,}260(P/F, i', 6) + 22{,}320(P/A, i', 4)(P/F, i', 6)$
$+ 18{,}600(P/F, i', 10);$ IRR = 12.4%

TABLE 6-7 Reworked Example 6-22 with a 10-year MACRS Class Life

End of year, k	(A) BTCF	(B) Basis	×	MACRS rate	=	Deduction	(C) = (A) = (B) Taxable income	(D) = −0.38(C) Cash flow for income taxes	(E) = (A) + (D) ATCF
0	−$180,000	—		—		—			−$180,000
1	36,000	$180,000	×	0.1000	=	$18,000	$18,000	−$ 6,840	29,160
2	36,000	180,000	×	0.1800	=	32,400	3,600	− 1,368	34,632
3	36,000	180,000	×	0.1440	=	25,920	10,080	− 3,830	32,170
4	36,000	180,000	×	0.1152	=	20,736	15,264	− 5,800	30,200
5	36,000	180,000	×	0.0922	=	16,596	19,404	− 7,374	28,626
6	36,000	180,000	×	0.0737	=	13,266	22,734	− 8,639	27,361
7	36,000	180,000	×	0.0655	=	11,790	24,210	− 9,200	26,800
8	36,000	180,000	×	0.0655	=	11,790	24,210	− 9,200	26,800
9	36,000	180,000	×	0.0656	=	11,808	24,192	− 9,193	26,807
10	36,000	180,000	×	0.0655/2	=	5,895	30,105	− 11,440	24,560
10	30,000						18,210*	− 6,920	23,080

Total $130,196
PW(10%) = $ 9,061
IRR = 11.2%

*MV − BV = $30,000 − 0.0655($180,000)
= $18,210

The investment tax credit is normally treated as a "year 0" cash flow because of the quarterly income tax payments required of corporations. The end of the first quarter is as long as a firm would wait to claim an ITC, and three months is obviously closer to the beginning of the first year than it is to the end of the year.

A minor complication is introduced in ATCF analyses when the study period is shorter than an asset's MACRS recovery period (e.g., for a 5-year recovery period, the study period is 5 years or less). In such cases, we shall assume that the asset is sold for its MV in the last year of the study period. Due to the half-year convention, only one-half of the normal MACRS depreciation can be claimed in the year of disposal or end of the study period, so there will usually be a difference between an asset's BV and its MV. Resulting income tax adjustments will be made at the time of sale (*note:* see last row in Fig. 6-4), unless the situation clearly specifies that the asset in question is to be relegated to standby service for an indefinite period of time. In such a case, depreciation deductions usually continue through the end of the asset's MACRS recovery period. Our assumption of project termination at the end of the study period makes good economic sense, as illustrated in Example 6-23.

Example 6-23

A highly specialized piece of optical character recognition equipment has a first cost of $50,000. If this equipment is purchased in 1996, it will be used to produce income (through rental) of $20,000 per year for only 4 years. At the end of year 4, the equipment will be sold for a negligible amount. Estimated annual expenses for upkeep are $3,000 during each of the 4 years. The MACRS recovery period for the equipment is 7 years, and the firm's effective income tax rate is 40%.

(a)   If the after-tax MARR is 7%, should the equipment be purchased?

(b)   Rework the problem, assuming that the equipment is placed on standby status such that depreciation is taken over the full MACRS recovery period.

Solution

(a)

EOY	(A) BTCF	(B) Depreciation	(C) = (A) − (B) Taxable income	(D) = −0.4(C) Income taxes	(E) = (A) + (D) ATCF
0	−$50,000				−$50,000
1	17,000	$ 7,145	$ 9,855	−$3,942	13,058
2	17,000	12,245	4,755	− 1,902	15,098
3	17,000	8,745	8,255	− 3,302	13,698
4	17,000	3,123[a]	13,877	− 5,551	11,449
4	0	0	− 18,742[b]	7,497	7,497
					Total = $10,800
					PW(7%) = $ 1,026

[a] Under MACRS only one half-year of depreciation can be claimed in the year an asset is sold, when disposal occurs before year $N + 1$.
[b] Remaining BV.

Because the PW > 0, the equipment should be purchased.

(b)

EOY	(A) BTCF	(B) Depreciation	(C) Taxable income	(D) Income taxes	(E) ATCF
0	−$50,000	—	—	—	−$50,000
1	17,000	$ 7,145	$ 9,855	−$3,942	13,058
2	17,000	12,245	4,755	− 1,902	15,098
3	17,000	8,745	8,255	− 3,302	13,698
4	17,000	6,245	10,755	− 4,302	12,698
5	0	4,465	− 4,465	1,786	1,786
6	0	4,460	− 4,460	1,784	1,784
7	0	4,465	− 4,465	1,786	1,786
8	0	2,230	− 2,230	892	892
8	0	—	—	—	0
					Total $10,800
					PW(7%) = $ 353

The present worth is $673 higher in part (a), which equals the PW of deferred depreciation credits in part (b). A firm would opt for the situation in part (a) if it had a choice.   ■

6.5.2 Computer Spreadsheets for Calculation of ATCFs

This section includes two illustrations of using computer spreadsheets for solving engineering economy problems on an after-tax basis. In the first, a spreadsheet shows the cell formulas for MACRS depreciation over several different recovery periods. A 5-year recovery period is then utilized to work Example 6-22 to obtain the printout beneath the cell formulas. As expected, the PW(10%) shown in the printout equals the $17,209 obtained in Table 6-6.

The second illustration reworks Example 6-22 with straight-line depreciation of $30,000 per year for 5 years and with $BV_5 = $30,000$. The cell formulas are again provided, followed by the printout. In the second illustration we notice that PW(10%) = $11,928, which is less than the PW with the MACRS depreciation. The difference is due to the fact that straight-line depreciation is "slower" than MACRS, thus shifting income taxes forward in time and lowering the after-tax PW.

R/C	A	B	C	D	E	F	G	H	I	J
1	MACRS Depreciation (Table)									
2										
3	Class Life	5				MACRS Depreciation Percentages				
4										
5	Year	Rates		Year			MACRS Class Life			
6					3	5	7	10	15	20
7	1	0.2000		1	0.3333	0.2000	0.1429	0.1000	0.0500	0.0375
8	2	0.3200		2	0.4445	0.3200	0.2449	0.1800	0.0950	0.0722
9	3	0.1920		3	0.1481	0.1920	0.1749	0.1440	0.0855	0.0668
10	4	0.1152		4	0.0741	0.1152	0.1249	0.1152	0.0770	0.0618
11	5	0.1152		5		0.1152	0.0893	0.0922	0.0693	0.0571
12	6	0.0576		6		0.0576	0.0892	0.0737	0.0623	0.0528
13	7			7			0.0893	0.0655	0.0590	0.0489
14	8			8			0.0446	0.0655	0.0590	0.0452
15	9			9				0.0656	0.0591	0.0447
16	10			10				0.0655	0.0590	0.0447
17	11			11				0.0328	0.0591	0.0446
18	12			12					0.0590	0.0446
19	13			13					0.0591	0.0446
20	14			14					0.0590	0.0446
21	15			15					0.0591	0.0446
22	16			16					0.0295	0.0446
23	17			17						0.0446
24	18			18						0.0446
25	19			19						0.0446
26	20			20						0.0446
27										
	The rate shown in B7 is obtained using the following formula:									
		=HLOOKUP(LIFE,MACRS,A7+1)								

Spreadsheet Illustration #1 (formulas)—Example 6-22

R/C	A	B	C	D	E	F	G
1	Effective Tax Rate		TR				
2	After-tax MARR		DR				
3	MACRS Class Life		LIFE				
4							
5	Cost Basis		CB				
6	Investment Tax Credit		ITC				
7							
8						Cash Flow	
9			MACRS	MACRS	Taxable	for Income	
10	Year	BTCF	Percent	Depr'n	Income	Taxes	ATCF
11	0	(B11)	- -	- -	- -	CB*ITC	B11+F11
12	1	B12	HLOOKUP(LIFE,MACRS,A12+1)	CB*C12	B12-D12	-TR*E12	B12+F12
13	2	B13	HLOOKUP(LIFE,MACRS,A13+1)	CB*C13	B13-D13	-TR*E13	B13+F13
14	3	B14	HLOOKUP(LIFE,MACRS,A14+1)	CB*C14	B14-D14	-TR*E14	B14+F14
15	4	B15	HLOOKUP(LIFE,MACRS,A15+1)	CB*C15	B15-D15	-TR*E15	B15+F15
16	5	B16	HLOOKUP(LIFE,MACRS,A16+1)	CB*C16	B16-D16	-TR*E16	B16+F16
17	6	B17	HLOOKUP(LIFE,MACRS,A17+1)	CB*C17	B17-D17	-TR*E17	B17+F17
18	7	B18	HLOOKUP(LIFE,MACRS,A18+1)	CB*C18	B18-D18	-TR*E18	B18+F18
19	8	B19	HLOOKUP(LIFE,MACRS,A19+1)	CB*C19	B19-D19	-TR*E19	B19+F19
20	9	B20	HLOOKUP(LIFE,MACRS,A20+1)	CB*C20	B20-D20	-TR*E20	B20+F20
21	10	B21	HLOOKUP(LIFE,MACRS,A21+1)	CB*C21	B21-D21	-TR*E21	B21+F21
22	10	B22	- -	- -	B22-(CB-SUM(D12:D21))	-TR*E22	B22+F22
23							
24				PW=	NPV(+DR,G12:G21)+G22/(1+DR)^10+G11		
25				AW=	-PMT(DR,10,+G24)		
26							

Spreadsheet Illustration #1 (printout)

R/C	A	B	C	D	E	F	G
1	Effective Tax Rate		0.38				
2	After-tax MARR		0.1				
3	MACRS Class Life		5				
4							
5	Cost Basis		180,000				
6	Investment Tax Credit		0				
7							
8						Cash Flow	
9			MACRS	MACRS	Taxable	for Income	
10	Year	BTCF	Percent	Depr'n	Income	Taxes	ATCF
11	0	($180,000)	- -	- -	- -		($180,000)
12	1	$36,000	0.2	$36,000			$36,000
13	2	$36,000	0.32	$57,600	($21,600)	$8,208	$44,208
14	3	$36,000	0.192	$34,560	$1,440	($547)	$35,453
15	4	$36,000	0.1152	$20,736	$15,264	($5,800)	$30,200
16	5	$36,000	0.1152	$20,736	$15,264	($5,800)	$30,200
17	6	$36,000	0.0576	$10,368	$25,632	($9,740)	$26,260
18	7	$36,000			$36,000	($13,680)	$22,320
19	8	$36,000			$36,000	($13,680)	$22,320
20	9	$36,000			$36,000	($13,680)	$22,320
21	10	$36,000			$36,000	($13,680)	$22,320
22	10	$30,000	- -	- -	$30,000	($11,400)	$18,600
23							
24						PW=	$17,209
25						AW=	$2,800.66
26							

Spreadsheet Illustration #2 (formulas)—Example 6-22

R/C	A	B	C	D	E	F
1	Effective Tax Rate		TR			
2	After-tax MARR		DR			
3	ADR Life		LIFE			
4						
5	Cost Basis		CB			
6	Salvage Value		SV			
7	Investment Tax Credit		ITC			
8						
9					Cash Flow	
10			Straight Line	Taxable	for Income	
11	Year	BTCF	Depreciation	Income	Taxes	ATCF
12	0	(B12)	- -	- -	ITC*CB	B12+E12
13	1	B13	IF(A13<=LIFE,(CB-SV)/LIFE,0)	B13-C13	-TR*D13	B13+E13
14	2	B14	IF(A14<=LIFE,(CB-SV)/LIFE,0)	B14-C14	-TR*D14	B14+E14
15	3	B15	IF(A15<=LIFE,(CB-SV)/LIFE,0)	B15-C15	-TR*D15	B15+E15
16	4	B16	IF(A16<=LIFE,(CB-SV)/LIFE,0)	B16-C16	-TR*D16	B16+E16
17	5	B17	IF(A17<=LIFE,(CB-SV)/LIFE,0)	B17-C17	-TR*D17	B17+E17
18	6	B18	IF(A18<=LIFE,(CB-SV)/LIFE,0)	B18-C18	-TR*D18	B18+E18
19	7	B19	IF(A19<=LIFE,(CB-SV)/LIFE,0)	B19-C19	-TR*D19	B19+E19
20	8	B20	IF(A20<=LIFE,(CB-SV)/LIFE,0)	B20-C20	-TR*D20	B20+E20
21	9	B21	IF(A21<=LIFE,(CB-SV)/LIFE,0)	B21-C21	-TR*D21	B21+E21
22	10	B22	IF(A22<=LIFE,(CB-SV)/LIFE,0)	B22-C22	-TR*D22	B22+E22
23	10	SV	- -	B23-(CB-SUM(C13:C22))	-TR*D23	B23+E23
24						
25			PW=	NPV(+DR,F13:F22)+F23/(1+DR)^10+F12		

Spreadsheet Illustration #2 (printout)

R/C	A	B	C	D	E	F
1	Effective Tax Rate		0.38			
2	After-tax MARR		0.1			
3	ADR Life		5			
4						
5	Cost Basis		180,000			
6	Salvage Value		30000			
7	Investment Tax Credit		0			
8					Cash Flow	
9			Straight Line	Taxable	for Income	
10	Year	BTCF	Depreciation	Income	Taxes	ATCF
11	0	($180,000)	- -	- -		($180,000)
12	1	$36,000	$30,000	$6,000	($2,280)	$33,720
13	2	$36,000	$30,000	$6,000	($2,280)	$33,720
14	3	$36,000	$30,000	$6,000	($2,280)	$33,720
15	4	$36,000	$30,000	$6,000	($2,280)	$33,720
16	5	$36,000	$30,000	$6,000	($2,280)	$33,720
17	6	$36,000		$36,000	($13,680)	$22,320
18	7	$36,000		$36,000	($13,680)	$22,320
19	8	$36,000		$36,000	($13,680)	$22,320
20	9	$36,000		$36,000	($13,680)	$22,320
21	10	$36,000		$36,000	($13,680)	$22,320
22	10	$30,000	- -			$30,000
23						
24						
25					PW =	$11,928

6.5.3 Illustration of After-Tax Economic Comparisons of Cost-Only Alternatives

In this section the after-tax comparison of mutually exclusive alternatives involving costs only is demonstrated and discussed. Example 6-24 presents a break-even type of analysis for purchase versus lease evaluation, and Example 6-25 demonstrates that after-tax analyses may produce different recommendations than do before-tax analyses.

Example 6-24

An engineering consulting firm can purchase a fully configured computer-aided design (CAD) workstation for $20,000. It is estimated that the useful life of the workstation is 7 years, and its MV in 7 years should be $2,000. Operating expenses are estimated to be $40 per 8-hour work day, and maintenance will be performed under contract for $8,000 per year. The MACRS class life is 5 years, and the effective income tax rate is 40%.

As an alternative, sufficient computer time can be leased from a company at an end-of-year cost of $20,000. If the after-tax MARR is 10%, how many work days per year must the workstation be needed in order to justify *leasing* it?

Solution   This example involves an after-tax evaluation of purchasing depreciable property versus leasing it. We are to determine how heavily the workstation must be utilized so that the lease option is a good economic choice. A *key* assumption is that the cost of engineering design time (i.e., operator time) is unaffected by whether the workstation is purchased or leased. Variable operations expenses associated with ownership result from the purchase of supplies, utilities, and so on. Hardware and software maintenance cost is contractually fixed at $8,000 per year. It is further assumed that the maximum number of working days per year is 250.

Lease fees are treated as an expense, and the consulting firm (the lessee) may *not* claim depreciation of the equipment to be an additional expense. (The leasing company presumably has included the cost of depreciation in its fee.) Determination of ATCF for the lease option is relatively straightforward and is not affected by how much the workstation is utilized:

$$\text{After-tax annual cost of the lease } = \$20,000(1 - 0.40) = \$12,000.$$

ATCFs for the purchase option involve expenses that are fixed (not a function of equipment utilization) in addition to expenses that vary with equipment usage. If we let X equal the number of working days per year that the equipment is utilized, the variable cost per year of operating the workstation is $40X$. The after-tax analysis of the purchase alternative is shown in Table 6-8.

The after-tax annual cost of purchasing the workstation is

$$\$20,000(A/P, 10\%, 7) + 24X + [\$3,200(P/F, 10\%, 1 + \cdots + \$4,800(P/F, 10\%, 7)] \cdot$$
$$(A/P, 10\%, 7) - \$1,200(A/F, 10\%, 7) = \$24X + \$7,511.$$

To solve for X, we equate the after-tax annual costs of both alternatives:

$$\$12,000 = \$24X + \$7,511.$$

Thus, $X = 187$ days per year. Therefore, if the firm expects to utilize the CAD workstation in its business *more than* 187 days per year, the equipment should be leased. The graphic summary of Example 6-24 shown in Fig. 6-5 provides the rationale for this recommendation. The

TABLE 6-8   After-Tax Analysis of Purchase Alternative

End of year, k	(A) BTCF	(B) Depreciation deduction[a]	(C) = (A) − (B) Taxable income	(D) = −t(C) Cash flow for income taxes	(E) = (A) + (D) ATCF
0	−$20,000	—	—	—	−$20,000
1	−40X − $8,000	$4,000	−40X − $12,000	−16X + $4,800	−24X − $3,200
2	−40X − 8,000	6,400	−40X − 14,400	−16X + 5,760	−24X − 2,240
3	−40X − 8,000	3,840	−40X − 11,840	−16X + 4,736	−24X − 3,264
4	−40X − 8,000	2,304	−40X − 10,304	−16X + 4,122	−24X − 3,878
5	−40X − 8,000	2,304	−40X − 10,304	−16X + 4,122	−24X − 3,878
6	−40X − 8,000	1,152	−40X − 9,152	−16X + 3,661	−24X − 4,339
7	−40X − 8,000	0	−40X − 8,000	−16X + 3,200	−24X − 4,800
7	2,000	—	2,000	− 800	1,200

[a] Depreciation deduction$_k$ = $20,000 × (MACRS rate)$_k$. Refer to Table 6-4.

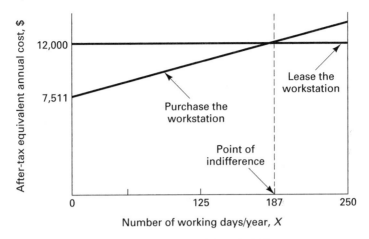

Figure 6-5   Summary of Example 6-24.

importance of the workstation's estimated utilization, in work days per year, is now quite apparent.   ■

Example 6-25a (Before-Tax Analysis)

A company is going to install a new plastic-molding press. Four different presses are available. The essential differences in initial investment and operating costs are shown in the following table.

	Press			
	A	B	C	D
Investment (installed):	$6,000	$7,600	$12,400	$13,000
Useful life:	5 yr	5 yr	5 yr	5 yr
Annual operation and maintenance costs:				
Power:	$680	$680	$1,200	$1,260
Labor:	$6,600	$6,000	$4,400	$3,200
Maintenance:	$400	$450	$650	$500
Property taxes and insurance:	$120	$152	$248	$260
Total annual costs:	$7,800	$7,282	$6,298	$5,220

Each press will produce the same number of units. However, because of different degrees of mechanization, each requires different amounts and grades of labor and has different operation and maintenance costs. None is expected to have a market value, and the study period is 5 years. Any capital invested is expected to earn at least 25% before taxes. Which press should be chosen?

Solution by the PW Method

When alternatives for which revenues are nonexistent or considered equal are compared with the present worth method, the alternative that has the minimum total PW is judged to be the most desirable. Table 6-9 shows the analysis of the four presses by the PW method. The economic criterion is to choose the alternative that has the minimum PW of costs, which is press A. The order of preference among the alternatives in *decreasing order* is press A, press B, press D, and press C. This rank ordering is identical in a before-tax analysis for all equivalent worth methods, when correctly applied.   ■

TABLE 6-9 Comparison of Four Molding Presses Using PW Method
(Example 6-25)

	Press			
	A	B	C	D
Present worth of				
Investment:	$ 6,000	$ 7,600	$12,400	$13,000
Costs:				
(Total annual costs) × (*P/A*, 25%, 5)	19,594	18,293	15,821	13,867
Total PW (costs)	$25,594	$25,893	$28,221	$26,867

Example 6-25b (After-Tax Analysis)

It is now desired to compare the four presses in Example 6-25 on an after-tax basis by using the present worth (PW) method. Let's assume an effective income tax rate (t) of 40%. Further, we utilize MACRS to compute depreciation with a 3-year recovery period. None of the presses will have a market value at the end of the five-year study period. Finally, the after-tax MARR is set equal to $(1 - t)(25\%) = 15\%$ for purposes of discounting after-tax cash flows. The data are given in Table 6-10, which also calculates after-tax cash flows associated with each press.

From Table 6-10, it is seen that the after-tax PW(15%) rank orderings are press D (best), press A, press B, and press C (worst). This is a different ordering of presses than we obtained in the before-tax analysis (see Table 6-9). Because project desirability in an after-tax analysis can differ from what is obtained in a before-tax analysis, we strongly recommend that after-tax evaluations be performed in engineering economy studies. ∎

6.6 Summary

This chapter has presented important aspects of TRA 86 and the Omnibus Reconciliation Act of 1993 relating to depreciation and income taxes. It is essential to understand these topics so that correct after-tax engineering economy evaluations of proposed projects and ventures may be conducted. Depreciation and income taxes are also integral parts of subsequent chapters in this book.

Many concepts regarding current federal income tax laws were described—such as taxable income, effective income tax rates, and taxation of ordinary income and capital gains. A general format for pulling together and organizing all these apparently diverse subjects is presented in Fig. 6-4. This format offers the student or practicing engineer a means of collecting on one worksheet information that is required for determining ATCFs for properly evaluating the after-tax financial results of a proposed capital investment.

REFERENCES

American Telephone and Telegraph Company, Engineering Department. *Engineering Economy*, 3d ed. New York: American Telephone and Telegraph Co., 1977.

Commerce Clearing House, Inc. *Explanation of Tax Reform Act of 1986.* Chicago, 1987.

TABLE 6-10 After-Tax Annual Cash Flow Computations for Comparison of the Four Molding Presses

Press	EOY	(A) BTCF	(B) Depreciation	(C) = (A) − (B) Taxable income	(D) = −0.4(C) Income taxes at 40%	(E) = (A) + (D) ATCF	PW(15%)
A	0	−$6,000	—	—	—	−$6,000	
	1	−7,800	$2,000	−$9,800	$3,920	−3,880	
	2	−7,800	2,667	−10,467	4,187	−3,613	
	3	−7,800	889	−8,689	3,476	−4,324	−$19,850
	4	−7,800	444	−8,244	3,298	−4,502	
	5	−7,800	0	0	3,120	−4,680	
B	0	−$7,600	—	—	—	−$7,600	
	1	−7,282	$2,533	−$9,815	$3,926	−3,356	
	2	−7,282	3,378	−10,660	4,264	−3,018	
	3	−7,282	1,126	−8,408	3,363	−3,919	−$19,919
	4	−7,282	563	−7,845	3,138	−4,144	
	5	−7,282	0	−7,282	2,913	−4,369	
C	0	−$12,400	—	—	—	−$12,400	
	1	−6,298	$4,133	−$10,431	$4,172	−2,126	
	2	−6,298	5,512	−11,810	4,724	−1,574	
	3	−6,298	1,836	−8,134	3,254	−3,044	−$21,269
	4	−6,298	919	−7,217	2,887	−3,411	
	5	−6,298	0	−6,298	2,519	−3,779	
D	0	−$13,000	—	—	—	−$13,000	
	1	−5,220	$4,333	−$9,553	$3,821	−1,399	
	2	−5,220	5,779	−10,999	4,400	−820	
	3	−5,220	1,925	−7,145	2,858	−2,362	−$19,517
	4	−5,220	963	−6,183	2,473	−2,747	
	5	−5,220	0	−5,220	2,088	−3,132	

Engineering Economist, The. A quarterly journal jointly published by the Engineering Economy Division of the American Society for Engineering Education and the Institute of Industrial Engineers, published by IIE, 25 Technology Park, Atlanta, Georgia 30092.

Lasser, J. K. *Your Income Tax.* New York: Simon & Schuster (see latest edition).

Smith, G. W. *Engineering Economy: The Analysis of Capital Expenditures,* 4th ed. Ames, IA: Iowa State University Press, 1987.

U. S. Department of the Treasury. *Your Federal Income Tax,* IRS Publication 17. Washington, D.C.: U.S. Government Printing Office, revised annually.

U. S. Department of the Treasury. *Tax Guide for Small Business,* IRS Publication 334. Washington, D.C.: U.S. Government Printing Office, revised annually.

U. S. Department of the Treasury. *Depreciation,* IRS Publication 534. Washington, D.C.: U.S. Government Printing Office (see latest edition).

U. S. Department of the Treasury. *Investment Credit,* IRS Publication 572. Washington, D.C.: U.S. Government Printing Office (see latest edition).

PROBLEMS

Note: Terminal book (salvage) value (BV) and market value (MV) are given in each problem, as appropriate. Unless otherwise stated, use an effective incremental income tax rate of 40%.

6-1. A machine costs $40,000. Its life for depreciation purposes is estimated at 10 years, and its terminal book value is assumed to be $4,000. Determine (1) the depreciation charge for the fifth year and (2) the BV at the end of the fifth year using each of the following methods:

  a. Straight line

  b. Sum-of-the-years'-digits

  c. Double declining balance

6-2. A new machine has just been purchased by a manufacturer for $25,000. Freight and trucking charges were $500, and the installation cost was $300. The machine has an estimated useful life of 8 years, at which time it is expected that $1,000 dismantling costs will have to be paid in order to sell it for $5,000. Compute (1) the depreciation charge for the first year and (2) the BV at the end of the first year using each of the following methods:

  a. Straight line

  b. Sum-of-the-years'-digits

  c. Declining balance

  d. Sinking fund with reinvestment rate of 12%

6-3. An asset costs $10,000 and is expected to have $1,000 salvage value at the end of 5 years. Graph its BV as a function of year using each of the following methods.

  a. Straight line

  b. Sum-of-the-years'-digits

  c. Double declining balance

6-4. A special-purpose machine is to be depreciated as a linear function of use. It costs $25,000 and is expected to produce 100,000 units and then be sold for $5,000. Up to the end of the third year it had produced 60,000 units, and during the fourth year it produced 10,000 units. What are the depreciation charge for the fourth year and the BV at the end of the fourth year?

6-5. A company purchased a machine for $15,000. It paid shipping costs of $1,000 and non-recurring installation costs amounting to $1,200. At the end of 3 years, the company had no further use for the machine, so it spent $500 to have the machine dismantled and was able to sell the machine for $1,500.

    a. What is the total investment cost for this machine?

    b. The company had depreciated the machine on a straight-line basis, using an estimated life of 5 years and a terminal BV of $1,000. By what amount did the recovered depreciation fail to cover the actual depreciation?

6-6. An asset for drilling that was purchased by a petroleum company in 1996 had a first cost of $60,000 and an estimated salvage value of $12,000. The ADR guideline period for useful life is taken from IRS Publication 534 (Table 6-2), and the MACRS recovery period is 7 years. Compute the depreciation amount in the third year and the BV at the end of the fifth year of life by each of these methods:

    a. the straight-line method over useful (ADR) life;

    b. the SYD method over useful (ADR) life;

    c. the 200% declining balance method (with $R = 2/N$) over useful (ADR) life;

    d. the MACRS method, using rates from Table 6-4.

6-7. By each of the following methods, calculate the BV of a highpost binding machine at the end of 4 years if the item originally cost $1,800 and had an estimated salvage value of $400. The ADR guideline period is 8 years, and the MACRS recovery period is 5 years.

    a. The machine was purchased in 1995, and MACRS rates in Table 6-4 are applicable.

    b. Alternate MACRS is to be utilized over the ADR period of 8 years with half-year convention.

6-8. An optical scanning machine was purchased in 1992 for $150,000. It is to be used for reproducing blueprints of engineering drawings, and its ADR guideline period is 9 years. The estimated MV of this machine at the end of 9 years is $30,000.

    a. What is the MACRS recovery period of the machine?

    b. Based on your answer to (a), what is the depreciation deduction in 1995?

    c. What is the BV at the beginning of 1996?

6-9. An asset was purchased 6 years ago for $6,400. At the same time its MACRS recovery period and SV were estimated to be 5 years and $1,000, respectively. The ADR guideline period is 9 years. If the asset is sold now for $1,500, what is the difference between its MV of $1,500 and its present BV if depreciation has been by

    a. the straight-line method (based on ADR guideline period)?

    b. the SYD method (based on ADR guideline period)?

    c. the MACRS method?

6-10. During the first quarter of 1995, a pharmaceutical company purchased a mixing tank that had a retail (fair market) price of $120,000. It replaced an older, smaller mixing tank that had a BV of $15,000 in the second quarter of 1995. Because a special promotion was underway, the old tank was used as a trade-in for the new one, and the cash price (including delivery and installation) was set at $99,500. The ADR guideline period for the new mixing tank was 9.5 years.

    a. Under MACRS, what is the depreciation deduction in 1997?

    b. Under MACRS, what is the BV at the end of 1998?

    c. If 200% declining balance depreciation had been applied to this problem, what would be the cumulative depreciation through the end of 1998?

6-11. Determine the more economical means of acquiring a business machine if you may either (1) purchase the machine for $5,000 with a probable resale value of $2,000 at the end of 5 years or (2) rent the machine at an annual rate of $900 per year for 5 years with an initial deposit of $500 refundable upon returning the machine in good condition. If you own the machine, you will depreciate it for tax purposes at the annual rate of $600. Of course, all leasing rental charges are deductible from taxable income. As either owner or lessee you will assume liability for all expenses associated with the operation of the machine. Compare the alternatives using the annual cost method. The after-tax minimum attractive rate of return is 10%, and the income tax rate is 40%.

6-12. A corporation in year 199X expects a gross income of $500,000, total operating (cash) expenditures of $400,000, and capital expenditures of $20,000. In addition, the corporation is able to declare $60,000 depreciation charges for the year. What is the expected taxable income and total federal income taxes owed for the year?

6-13. BIG Corporation is considering making an investment of $100,000 that will increase its annual taxable income from $40,000 to $60,000. What annual federal income taxes will be owed as a result of this investment? What is the average rate to be paid on the taxable income?

6-14. Suppose the investment by BIG Corporation (in Problem 6-13) has expected results as follows:

Annual receipts:	$75,000
Annual disbursements:	$45,000
Life:	8 yr
Terminal book value:	$20,000

If straight-line depreciation is used and if the after-tax minimum attractive rate of return is 15%, determine if the investment is attractive using the PW method.

6-15. Two alternative machines will produce the same product, but one is capable of higher-quality work, which can be expected to return greater revenue. The following are relevant data:

	Machine A	Machine B
First cost:	$20,000	$30,000
Life:	12 yr	8 yr
Terminal BV (and MV):	$4,000	$0
Annual receipts:	$150,000	$188,000
Annual disbursements:	$138,000	$170,000

Determine which is the better alternative assuming "repeatability" and based on using straight-line depreciation, an income tax rate of 40%, and an after-tax minimum attractive rate of return of 10% using the following methods:

a. Annual worth

b. Present worth

c. Internal rate of return

6-16. Alternative methods I and II are proposed for a plant operation. The following is comparative information:

	Method I	Method II
Initial investment:	$10,000	$40,000
Useful (ADR) life:	5 yr	10 yr
Terminal market value:	$ 1,000	$ 5,000
Annual disbursements:		
Labor:	$12,000	$ 4,000
Power:	$ 250	$ 300
Rent:	$ 1,000	$ 500
Maintenance:	$ 500	$ 200
Property taxes & insurance:	$ 400	$ 2,000
Total annual disbursements:	$14,150	$ 7,000

Determine which is the better alternative based on an after-tax annual cost analysis with an effective income tax rate of 40% and an after-tax MARR of 12% assuming the following methods of depreciation:

a. Straight line

b. MACRS

6-17. A manufacturing process can be designed for varying degrees of automation. The following is relevant cost information:

Degree	First cost	Annual labor cost	Annual power and maintenance cost
A	$10,000	$9,000	$ 500
B	14,000	7,500	800
C	20,000	5,000	1,000
D	30,000	3,000	1,500

Determine which is best by after-tax analysis using an income tax rate of 40%, an after-tax MARR of 15%, and straight-line depreciation. Assume each has a life of five years and no book or market value. Use each of the following methods:

a. Annual worth

b. Present worth

c. Internal rate of return

6-18. A centerless grinder can be purchased new for $18,000. It will have an eight-year useful life and no MV. Reductions in operating costs (savings) from the machine will be $8,000 for the first 4 years and $3,000 for the last 4 years. Depreciation will be by the MACRS method using a recovery period of 5 years. A used grinder can be bought for $8,000 and will have no MV in 8 years. It will save a constant $3,000 per year over the 8-year period and will be depreciated at $1,000 per year for 8 years. The effective income tax rate is 40%. Determine the after-tax PW on the *incremental* investment required by the new centerless grinder. Let the MARR be 12% after taxes.

6-19. Currently, a firm has annual operating revenues of $190,000, cost of sales of $50,000, and depreciation charges of $40,000 annually. A new project is proposed that will raise annual revenues by $30,000 and increase the cost of sales by $10,000 per year. If this new project necessitates a total capital cost of $50,000, which can be depreciated to zero BV and has no MV at the end of its 6-year life, what is the IRR of the project after federal income taxes are paid? Assume that MACRS depreciation is used with a class life of five years.

6-20. A certain piece of real estate has a first cost of $50,000. If this building is purchased in 199X, it is believed that the property will be held for 10 years and then sold for an estimated $30,000. The estimated receipts from rental are $10,000 a year throughout the 10-year life. Estimated annual upkeep costs are $3,000 the first year and will increase by $300 each year to $5,700 in the tenth year. In addition, it is estimated that there will be a single outlay of $2,000 for maintenance overhaul at the end of the fifth year. Assume MACRS depreciation with a 20-year class life and an effective income tax rate of 40%. Also assume that the $2,000 overhaul cost can be treated for tax purposes as a current expense in the fifth year. If an investor has a 10% after-tax MARR, should he purchase the rental property? Use the PW method.

6-21. A firm must decide between two systems, A and B, shown below. Their effective income tax rate is 40%, and MACRS depreciation is used. If the after-tax desired return on investment is 10%, which system should be chosen? Use the AW method and state your assumptions.

	System A	System B
Initial cost:	$100,000	$200,000
Class life:	5 yr	5 yr
Useful ADR life:	7 yr	6 yr
Market value at end of useful life:	$30,000	$50,000
Annual revenues less expenses over useful life:	$20,000	$40,000

6-22. The owners of a small TV repair shop are planning to invest in some new circuit testing equipment. The details of the proposed investment are as follows:

> First cost: $5,000
> Terminal market value: $0
> Extra revenue: $2,000/yr
> Extra expenses: $800/yr
> Expected life: 5 yr (also equal to the MACRS recovery period)
> Effective income tax rate: 15%

a. If MACRS depreciation is used, calculate the PW of ATCFs when the MARR (after taxes) is 12%. Should the equipment be purchased?

b. Use the alternate MACRS depreciation method over the ADR recovery period to calculate ATCFs, and recommend whether the new equipment should be purchased if the after-tax MARR is 12%. Include the half-year convention when determining depreciation.

6-23. Your firm can purchase a machine for $12,000 to replace a rented machine. The rented machine costs $4,000 per year. The machine that you are considering would have a life of 8 years and a $5,000 terminal BV at the end of its life. By how much could annual operating expenses increase and still provide a return of 10% after taxes? The firm is in the 40% income tax bracket, and revenues produced with either machine are identical. Assume that alternate MACRS (straight line with half-year convention) is utilized to recover the investment in the machine and that the ADR period is eight years.

6-24. Your company has purchased equipment (for $50,000) that will reduce materials and labor costs by $14,000 each year for N years. After N years there will be no further need for the machine, and since the machine is specially designed, it will have no market value at any time. However, the IRS has ruled that you must depreciate the equipment on a straight-line basis with a tax life of five years. If the effective income tax rate is 40%, what is the minimum number of years your firm needs to operate the equipment to earn 10% after taxes on its investment?

6-25. Entropy Enterprise, Ltd., is considering a $100,000 heat recovery incinerator (7-year MACRS class life) that is expected to cause a net reduction in out-of-pocket costs of $30,000/year for 6 years (the study period). The incinerator will be depreciated using MACRS recovery allowances from Table 6-4, and no market value is expected. The effective income tax rate is 40%. Determine the PW of the ATCFs when the after-tax MARR is 8%.

6-26. Your company has to obtain some new production equipment to be used for the next six years, and leasing is being considered. You have been directed to perform an after-tax study of the leasing approach. The pertinent information for the study is as follows:

Lease costs: First year, $80,000; second year, $60,000; third through sixth years, $50,000 per year. Assume that a 6-year contract has been offered by the lessor that fixes these costs over the 6-year period. Other costs (not covered in the contract) are $4,000 per year, and the effective income tax rate is 40%.

a. Develop the annual ATCFs for the leasing alternative.

b. If the MARR after taxes is 8%, what is the equivalent annual cost for the leasing alternative?

6-27. Your company is contemplating the purchase of a large stamping machine. The machine will cost $180,000. With additional transportation and installation costs of $5,000 and $10,000, respectively, the cost basis for depreciation purposes is $195,000. Its market value at the end of 5 years is estimated to be $40,000. The IRS has assured you that this machine will fall under a 3-year MACRS recovery period. The justifications for this machine include $40,000 savings per year in labor, $20,000 per year in improved quality, and $10,000 per year in reduced materials. The before-tax MARR is 20% per year, and the effective income tax rate is 40%.

a. The simple payback period (before taxes) is most nearly
   (i) 0.36 years
   (ii) 2.0 years
   (iii) 2.2 years
   (iv) 2.8 years
   (v) 3.5 years

b. The net income after taxes at the end of year 1 is most nearly
   (i) −$10,000
   (ii) $3,000
   (iii) $10,000
   (iv) $15,000
   (v) $40,000

c. The MACRS depreciation in year 4 is most nearly
  (i) $0
  (ii) $13,000
  (iii) $14,000
  (iv) $31,000
  (v) $45,000

d. The book value at the end of year 2 is most nearly
  (i) $33,000
  (ii) $36,000
  (iii) $42,000
  (iv) $43,000
  (v) $157,000

e. The before-tax cash flow (total) in year 5 is most nearly
  (i) $9,000
  (ii) $40,000
  (iii) $70,000
  (iv) $80,000
  (v) $110,000

f. The taxable income for year 3 is most nearly
  (i) $9,000
  (ii) $18,000
  (iii) $39,000
  (iv) $41,000
  (v) $70,000

g. The present worth of the after-tax savings from the machine, in labor, quality and materials (neglecting the first cost, depreciation, and the salvage value), is most nearly
  (i) $12,000
  (ii) $95,000
  (iii) $151,000
  (iv) $184,000
  (v) $193,000

6-28. A firm can purchase a centrifugal separator (5-year MACRS property) for $20,000. Its estimated market value is $3,000 after a useful life of 6 years. Operating and maintenance costs for the first year are expected to be $2,000. These costs are projected to increase by $1,000 per year each year thereafter. The income tax rate is 40%, and the MARR is 15% after taxes.

  What must the annual benefits be for the purchase of the centrifugal separator to be economical on an after-tax basis? Be sure to show your calculations for MACRS depreciation, taxable income, and after-tax cash flows for each year.

6-29. Individual industries will use energy as efficiently as economically possible, and there are several incentives to improve the efficiency of energy consumption. One incentive for the purchase of more energy-efficient equipment is to reduce the time allowed to write off the initial cost. Another "incentive" might be to raise the price of energy in the form of an energy tax.

  To illustrate these two incentives, consider the selection of a new motor-driven centrifugal pump for a refinery. The pump is to operate 8,000 hours per year. Pump A costs $1,600, consumes 10 hp, and has an overall efficiency of 65% (it delivers 6.5 hp). The other available alternative, pump B, costs $1,000, consumes 13 hp, and has an over-all efficiency of 50% (delivers 6.5 hp). *Note:* 1 hp = 0.746 kW.

Compute the after-tax rate of return on extra investment in pump A, assuming an effective income tax rate of 40%, an ADR useful life of 10 years [parts (a) and (c) only], zero market values, and straight-line depreciation for each of these situations:

a. The cost of electricity is $0.04/kWh.

b. A 5-year depreciation write-off period is allowed, the expected life of both pumps is still 10 years, and the cost of electricity is $0.04/kWh.

c. Repeat part (a) when electricity costs $0.07/kWh.

6-30. Consider the justification of an advanced engineering workstation for the ABC Manu-facturing Company. The following are known data:

- The combined hardware and software cost is $80,000.
- Contingency costs have been set at $15,000.
- A service contract on the hardware costs $500/month.
- The company is located in Virginia, where a 6% sales tax applies to all purchases.
- The company is in a 40% (effective) income tax bracket.
- Company management has established a 15% (after-tax) hurdle rate.

In addition, the following assumptions/projections have been made:

- In order to support the system on an ongoing basis, a programmer/analyst will be required: starting salary is $24,000 per year.
- The system is expected to yield a labor savings of 3 personnel (to be reduced through normal attrition) at an average salary of $16,200/year/person (base plus fringes).
- A 3% reduction in purchased material costs is expected; current year purchases are $1,000,000.
- Project life is expected to be six years, and capital expenditures will be fully depreci-ated in that time period using the MACRS method (recovery period = 5 years).
- Terminal market value of the total investment is negligible.

a. Based on the above data, construct an after-tax cash flow analysis over the worksta-tion's six-year life.

b. Is this project acceptable based on economic factors alone?

7
Analyses for Government Agencies and Public Utilities

Public organizations such as government agencies and privately owned public utilities are confronted with selections between alternatives that involve differing patterns of cash flows and intangible considerations much the same as within private competitive enterprises. Some differences in methods and emphases for economic studies applicable to these organizations will be described herein, first for government agencies and then for public utilities.

7.1 Investments by Government Agencies

The government (public sector) analyst has a somewhat more difficult problem of measuring benefits than does the private sector analyst because the government sector analyst knows he or she should include the social benefits (and costs) of the project. The private sector manager may choose to take refuge in profit maximization or cost minimization for the firm and not include the social costs and benefits.

On the other hand, economic studies for government agencies are made easier because such agencies are not subject to income taxes. While funds for many such agencies are obtained through income taxes, none of the agencies pays income taxes to others, and thus the relative economic merits of alternatives considered by the agencies are not affected by income taxes.

The bulk of government expenditures is not allocated on the basis of economic criteria. However, economic analysis is used to a considerable extent for such expenditures as water resource development, highway location and sizing, and weapons systems selection.

When capital expenditures are involved, government analyses notably include the following:

1. measuring benefits and costs, often under conditions in which no market yardstick is available, and sometimes weighting the incidence of benefits on different individuals and groups;
2. determining criteria for judging acceptability;
3. selecting the appropriate interest rate to convert benefits and costs into equivalent terms.

The latter two are discussed in the following sections.

7.1.1 Criteria for Judging Acceptability of Government Projects

Capital expenditures by various agencies of government are applied to a vast array of purposes, such as for the control and preservation of natural resources, providing economic services, protection, and cultural development. For the most part, these highly diverse purposes are independent of one another. Analyses of capital expenditures for *independent* projects can take many forms, such as:

1. maximization of the ratio of benefits to costs on an annual or present equivalent basis;
2. minimization of combined annual cost to the public user and the agency supplier of the facility when benefits are fixed;
3. minimization of the present value sacrificed if the program were cut.

The very difficult, subjective part of most decisions involving competing classes of projects revolves around divergent views on what is the "general welfare" of the public. What is perceived to be in the best interests of the public may vary sharply over time, depending on changes in such conditions as the state of the economy, foreign relations, public health, environmental degradation, and public service needs. Furthermore, intangibles or irreducibles are generally much more important in public works than they are in private enterprises.

Judging the importance of irreducibles for public projects in many respects presents the same kinds of problems as occur in private economy studies. The best that an individual can do in dealing with personal problems of economy may be to note the satisfactions that will come from particular alternatives and to consider them in the light of their long-run costs and in the light of the individual's capacity to pay.

Analysts who perform economic studies for a government agency sometimes fall into the trap of making the study based on just the disbursements and receipts, if any, for that particular agency or government body it represents. Thus, for example, a state or local government might rationalize that a particular project is justified if the federal government pays most of the cost, even though the project is grossly unjustified based on total costs and benefits. In principle, if the ideal of democratic government to "promote the general welfare" is to be followed, one should consider the effects of governmental alternatives on all of the people, not merely on the economic considerations of a particular unit of the government.

Example 7-1

A county is considering installing a water treatment system that is expected to cause environmental and direct benefits of $1,000,000 per year for its inhabitants. The system would require an investment of $9,000,000 and have operating and maintenance costs of $300,000 per year for an expected life of 20 years, after which it would have no value. If money for this type of project costs the county 6%, is the project justified on an economic basis? Suppose the state government is willing to pay $4,000,000 of the investment. Now is it justified?

Solution   Using the net present worth method based on total benefits and costs, we get

Net PW $= (\$1,000,000 - \$300,000)(P/A, 6\%, 20) - \$9,000,000 = -\$971,000$

$\qquad = -\$971,000 < \$0.$

Thus, it is not justified in total.

Based only on benefits and costs to the county:

Net PW $= (\$1,000,000 - \$300,000)(P/A, 6\%, 20)$

$\qquad\qquad - (\$9,000,000 - \$4,000,000) = \$3,029,000$

$\qquad = \$3,029,000 > \$0.$

Thus, it is apparently justified to the county.

It would be common for county decision makers to consider only the net PW = $3,029,000 on the county investment and conclude that the project must be worthwhile, never seriously considering that the net PW on the total investment is well below zero. This is the equivalent of considering the state's money as free, which is sometimes rationalized on the basis that ". . . we ought to get it, because if we don't some other agency or organization will." Nevertheless, in principle, one should consider the total benefits and costs to determine if the investment of state, as well as county, funds is justified.   ■

7.1.2 Interest Rate for Government Projects

The interest rate plays the same formal role in the evaluation of public sector investment projects that it does in the private sector. The rationale for its use is somewhat different. It is used in the private sector, because it leads directly to the private sector goal of profit maximization or cost minimization. Its basic function in the public sector is similar, in that it should lead to an optimization of economic and social net benefits, providing these have been appropriately measured. It will lead to the determination of how available funds may be allocated best among competing projects, particularly considering the scale and capital intensity of those projects.

Three main choices for the interest rate to use in government economic studies are as follows:

1. borrowing rate;
2. the opportunity cost to the governmental agency;
3. the opportunity cost to the taxpayer.

In general, it is appropriate to use the borrowing rate only for cases in which money is borrowed specifically for the project(s) under analysis and where use of that money will not cause other worthy projects to be foregone.

Opportunity cost is the interest rate on the best investment opportunity foregone. If projects are chosen so that the return rate on all accepted projects is higher than the return rate on any of the rejected projects, then the interest rate for use in the economic analysis is equal to the opportunity cost.

If this is done for all projects and investment capital available within a government agency, then the result is a *government opportunity cost.* If, on the other hand, one considers the best opportunities available to the taxpayers if the money were not obtained through taxes for use by the government agency, the result is a *taxpayers' opportunity cost.*

Theory suggests that in government economic analyses the interest rate should be the largest of the three listed above. Generally, the taxpayers' opportunity cost is substantially the highest of the three. As an indicative example, a federal government directive in 1972 specified that an interest rate of 10% should be used in economic studies for a wide range of federal projects. This 10%, it can be argued, was at least a rough approximation of the average return taxpayers could be obtaining from the use of that money. In any case, it was substantially greater than the 6% to 8% the federal government was paying for the use of borrowed money at that time.

7.2 Comparison of Alternatives: The Benefit–Cost Ratio Method

A method of comparing investment alternatives that has experienced considerable usage in the public sector is the benefit–cost ratio method. Many federal government agencies and departments, as well as the United States Postal Service and a number of public utilities, use benefit–cost ratio methods in performing economic analyses.

The *benefit–cost ratio (B/C)* can be defined as the ratio of the equivalent worth of benefits to the equivalent worth of costs. The equivalent worths can be PWs, AWs, or FWs. The B/C method is often used by a government agency to measure the economic effectiveness of an investment that will benefit some segment of the general public: *hence, the B/C is sometimes defined as the ratio of the present or annual worth of benefits for the user public to the present or annual worth of the total costs of supplying the benefits.* The B/C is also referred to as the *savings–investment ratio* (SIR) by some government agencies and departments.

Two commonly used formulations of the B/C ratio are as follows:

1. Conventional B/C:

$$\text{B/C} = \frac{\text{PW(benefits to user)}}{\text{PW(total costs to supplier)}} = \frac{\text{PW}(B)}{\text{PW}[CR + (O + M)]} \quad (7\text{-}1)$$

or

$$\text{B/C} = \frac{\text{AW(benefits to user)}}{\text{AW(total costs to supplier)}} = \frac{B}{CR + (O + M)}, \quad (7\text{-}2)$$

where

B = annual worth of benefits to user,

CR = capital recovery cost or the equivalent annual cost of the
initial investment, considering any salvage value,

O = uniform annual operating cost,

M = uniform annual maintenance cost.

2. Modified B/C:

$$B/C = \frac{PW[B - (O + M)]}{PW(CR)} \qquad (7\text{-}3)$$

or

$$B/C = \frac{B - (O + M)}{CR}. \qquad (7\text{-}4)$$

The numerator of the modified B/C expresses the present or annual worth of the net of benefits and operating and maintenance costs; the denominator includes only the investment costs, expressed on a present or annual basis.

7.2.1 Comparing Alternatives Using Benefit–Cost Analysis When Receipts and Disbursements Are Known

Example 7-2

To illustrate the use of the B/C method in comparing alternatives when both receipts and disbursements are known, an example involving two alternative lathes will be considered. Annual receipts are treated as annual benefits in the analysis; annual disbursements are treated as annual operating and maintenance costs.

	Lathe	
	A	B
First cost:	$10,000	$15,000
Life:	5 yr	10 yr
Salvage value:	$2,000	$0
Annual receipts:	$5,000	$7,000
Annual disbursements:	$2,200	$4,300
Minimum attractive rate of return = 8%		
Study period = 10 yr		

Solution As previously, it is assumed that after 5 years lathe A will be replaced with another lathe having an identical cash flow profile. Hence, annual worths for individual life cycles can be used in computing the benefit–cost ratio.

The first increment of investment (the $10,000 for lathe A) was shown, has a B/C of 1.294 > 1.0. Hence, the increment of investment in lathe A is justified. The next step is to determine if the second increment of investment is justified. The simplest approach is to divide the difference in the annual worths of net annual benefits by the differences in the capital recovery costs (see Eq. 7-4).

For lathe B,

$$CR = \$15,000(A/P, 8\%, 10) = \$2,235.$$

For the second increment of investment,

$$\Delta B/C = \frac{(\$7,000 - \$4,300) - (\$5,000 - \$2,200)}{\$2,235 - \$2,163}$$

$$= \frac{\$2,700 - \$2,800}{\$2,235 - \$2,163}$$

$$= \frac{-\$100}{\$72} = -1.39 < 1.0.$$

Since the modified B/C ratio for the increment of investment from lathe A to lathe B has a value less than 1.0, the increment cannot be justified and thus lathe A is recommended. ∎

7.2.2 Comparing Numerous Alternatives

The following example is given to further illustrate the principle that each increment of investment capital must be justified when using the B/C ratio method. To simplify the computations involved, each alternative has a salvage value equal to the investment. Hence, there is no depreciation cost, and the incremental capital recovery cost equals the product of the MARR and the incremental investment.

Example 7-3

	Alternatives					
	A	B	C	D	E	F
Investments:	$1,000	$1,500	$2,500	$4,000	$5,000	$7,000
Annual savings in						
cash disbursements:	150	375	500	925	1,125	1,425
Salvage value:	1,000	1,500	2,500	4,000	5,000	7,000
MARR = 18%						

Solution   The alternatives are listed in increasing order of investment. The two primary principles employed are the same as for rate of return methods, i.e., each increment of capital investment must justify itself and an alternative should be compared with an alternative requiring a lower investment only if the latter investment is justified. The symbol ΔB/C denotes the benefit–cost ratio on the incremental cost.

Increment considered	A	B	B → C	B → D	D → E	E → F
ΔInvestment	$1,000	$1,500	$1,000	$2,500	$1,000	$2,000
ΔCR (at 18%)	$180	$270	$180	$450	$180	$360
ΔAnnual savings (ΔB)	$150	$375	$125	$550	$200	$300
ΔB/C	0.833	1.389	0.694	1.222	1.111	0.833
Is increment justified?	No	Yes	No	Yes	Yes	No

Since the last justified increment of cost is that required to obtain alternative E, it is recommended. It is interesting to note that alternative E does not have the greatest individual B/C, just as it did not have the greatest individual rate of return.

	Alternatives					
	A	B	C	D	E	F
B/C ratio	0.833	1.389	1.111	1.285	1.250	1.131

Note that alternative B has the highest B/C and that alternative F has a B/C greater than 1.0. Nevertheless, alternative E would be recommended on the basis that the firm wishes to incur incremental cost if and only if its ΔB/C $\geq$ 1.0. ∎

7.2.3 Comparing Alternatives When Disbursements Only Are Known

When disbursements only are known, benefit–cost ratios are to be calculated for incremental investments only and not for the investment in any one alternative. The alternative requiring the smallest initial investment is assumed to be justified (or necessary). As an illustration, consider the compressor example presented in Chapters 3 and 4.

Example 7-4

Data for Compressors I and II are shown below.

	Compressor	
	I	II
First cost:	$3,000	$4,000
Life:	6 yr	9 yr
Salvage value:	$500	$0
Annual operating disbursements:	$2,000	$1,600
MARR = 15%		
Study period = 18 yr		

Solution

$$CR\ (I) = (\$3{,}000 - \$500)(A/P, 15\%, 6) + \$500(0.15) = \$735$$
$$CR\ (II) = \$4{,}000(A/P, 15\%, 9) = \$840$$
$$\Delta B/C = \frac{(\$0 - \$1{,}600) - (\$0 - \$2{,}000)}{\$840 - \$735}$$
$$= \frac{-\$1{,}600 + \$2{,}000}{\$840 - \$735}$$
$$= \frac{\$400}{\$105} = 3.81 > 1.0$$

Since ΔB/C > 1.0, then the incremental cost is justified and compressor II is recommended. Notice that the repeatability assumption was again used. Furthermore, negative net annual benefits exist for each individual alternative; however, the incremental annual benefit (−$1,600 + $2,000) is positive. ∎

7.3 Investments by Privately Owned Utilities

Privately owned, regulated utilities are an important part of the U.S. economy. These include many suppliers of electricity, gas, water, telephone service, cable TV, and various types of transportation services. Regulation is conducted by various local, state, and federal commissions and includes such matters as rate structures, standards of service, and financing allowable.

Because a utility typically requires large amounts of capital to be invested in fixed plant and equipment, economy of operation for the company and low rates to the public are possible only if there are high use factors for such assets. This is generally achieved through government bodies that limit competition and in many cases grant the utility company a virtual monopoly for providing the service in a given locale. Public regulation of utility companies and the prices they charge is a substitute for regulation by competition. Limiting competition not only results in reducing uneconomical duplication of facilities, but it can also reduce public nuisances such as the duplication of utility service lines and transportation facilities.

Utility firm earnings, after income taxes, usually are not permitted to exceed 10% to 15% on owner's equity capital. It should be noted that, although there is a maximum limit put on utility earnings by regulatory bodies, there is no guarantee of any such profits, and there is no assurance against loss. However, if the utility can show that it is operating efficiently, it usually can obtain permission to earn whatever is necessary to attract needed capital.

Utilities have much greater stability of income than normal competitive enterprises. Because of their relative stability and great capital needs, utilities commonly use a much higher proportion of debt capital to total capital than do normal competitive businesses. Whereas nonutility firms seldom use more than 30% debt capital, it is very common for utilities to borrow over 50% of their total capital.

7.3.1 Differences Between Economy Studies in Regulated Public Utilities and in Competitive Industry

Economic studies for public utilities can be performed by virtually the same methods as for nonutility firms. The following are several main differences in circumstances that can affect the numbers used and the simplicity with which economic studies for utilities can be made.

First, funds for investment in new fixed assets are generally much more available for utilities than they are for competitive firms. Public utilities commonly raise new capital for expansion at frequent intervals, and consideration of capital rationing due to limitation of funds does not enter into the determination of the minimum attractive rate of return. The MARR for such expansion projects is likely to be at or only slightly higher than the weighted after-tax cost of capital (to be discussed later). On the other hand, if the utility has numerous cost reduction proposals and limited capital, such proposals are likely to be subject to capital rationing concepts, with the

MARRs for such proposals usually determined by the opportunity costs. These opportunity costs of capital are normally not much higher than the weighted after-tax cost of capital for a public utility, but for a competitive firm they can be very high.

Another distinction is that the utility is supposed to select alternatives that will minimize the revenue requirements from customers paying for the services of the utility, whereas the competitive firm is expected to select alternatives that will be most beneficial to the owners of the firm. The interests of customers and stockholders of a utility firm are generally the same in the long run, but this may not be true in the short run.

Utilities do not normally undertake engineering economy studies to determine whether or not a given service should be provided as competitive firms commonly do. Instead, utility economy studies are normally focused on which alternative for providing the service is preferable, that is, which alternative minimizes the revenue required to pay for the service.

7.4 Comparison of Alternatives: The Revenue Requirement Method[1]

The economic evaluation method most widely used by privately owned, regulated utilities is called the *minimum revenue requirement method*. This method provides a basis for comparing mutually exclusive alternatives. It can be applied to a wide spectrum of regulated businesses having various characteristics and concepts discussed in the previous sections.

> In essence, this method calculates revenues that a given project must provide to just meet all the costs associated with it, including a fair return to investors.

The relationship between a project's revenue requirements and its costs is shown in Fig. 7-1. Because regulatory commissions act on behalf of the consumers of a utility's services, investment project selection should be made in such a way that revenue requirements are minimized.

The payback method is another evaluation method used by utility companies. However, this method is often considered a screening technique, permitting the reduction of the number of candidate alternatives to a manageable size, rather than a final selection method. In electric power utilities the payback method is particularly used to evaluate small discretionary investments such as spare parts or retrofit activities.

The following sections focus on the development and illustration of the revenue requirement method. Examples will be used to illustrate various facets of this method as it applies to privately owned, regulated utilities.

[1] This section is reprinted from E. P. DeGarmo, W. G. Sullivan, and J. A. Bontadelli, *Engineering Economy*, 9th ed. (New York: Macmillan Publishing Co., 1993) pp. 483–495 (with permission of the publisher).

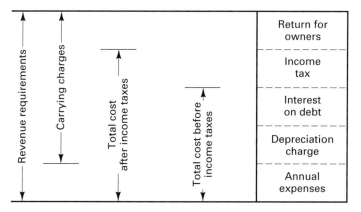

Figure 7-1   Relationship of revenue requirements and costs for an investor-owned utility.

As shown in Fig. 7-1, the minimum revenue requirement consists of carrying charges resulting from capital investments that must be recovered, plus all associated expenses that occur periodically (i.e., fuel, O&M expenses, property taxes, and insurance). Carrying charges are also called *total fixed charges.* They include the following:

- interest on bonds used to partially finance the project;
- equity return requirements for the stockholders;
- income taxes to be paid to state and local governments; and
- depreciation charges on the investment.

The concept of a fixed-charge rate is widely used in the utility industry. The fixed-charge rate is defined as the annual owning cost of an investment (carrying charges) expressed as a percentage of the investment.

The following equation is used to find the annual carrying charges in year k, CC_k:

$$CC_k = d_{B_k} + [(1 - \lambda)e_a + \lambda i_b] \cdot UI_k + T_k, \qquad (7\text{-}5)$$

where d_{B_k} = book depreciation taken in year k, $1 \leq k \leq N$,

$\qquad \lambda$ = fraction of borrowed money in a utility's total capitalization,

$\qquad e_a$ = return to equity capital (as a decimal),

$\qquad i_b$ = cost of borrowed capital (as a decimal),

$\qquad UI_k$ = unrecovered investment at the beginning of year k,

$$UI_k = \begin{cases} I \text{ (initial investment)}, k = 1 \\ UI_{k-1} - d_{B_{k-1}}, 2 \leq k \leq N, \end{cases}$$

$\qquad T_k$ = income taxes paid in year k.

Since depreciation claimed for income tax purposes and interest paid on debt are tax deductible, the income tax in any given year is determined by the following equation:

$$T_k = t(CC_k - \lambda \cdot i_b \cdot UI_k - d_{Tk}), \qquad (7\text{-}6)$$

where d_{T_k} is the tax depreciation for income tax purposes in year k and t is the effective income tax rate.

Note that carrying charges (CC_k) are a function of income taxes (T_k) in Eq. 7-5 and that income taxes (T_k) are a function of carrry charges (CC_k) in Eq. 7-6. This can be seen clearly in Fig. 7-1. The revenue requirement can be determined if the income taxes are known and, similarly, the income taxes can be computed if the revenue requirement is known. There are two equations and two unknowns (i.e., CC_k and T_k). Solving for T_k, we find

$$T_k = [t/(1 - t)]\left[(1 - \lambda)e_a \cdot UI_k + d_{B_k} - d_{T_k}\right]. \tag{7-7}$$

The revenue requirement in year k, (RR_k), is

$$RR_k = CC_k + C_k, \tag{7-8}$$

where C_k represents all recurring annual expenses in year k.

The following assumptions are common when using the revenue requirement method:

1. The total investment in an asset during any year is equal to its beginning-of-year book value.

2. The amount of debt capital invested in an asset during any year is a constant fraction of its book value during that year, and this fraction remains constant throughout the asset's life.

3. Equity and debt capital involve constant rates of return throughout the life of the project.

4. Book depreciation charges are used to retire capital stock and bonds each year in proportion to the debt–equity mix of financing employed.

5. The effective income tax rate is constant over the life of the project.

7.4.1 Utility Rate Regulation[2]

Utility rates are established during a regulatory rate proceeding. When changes in a utility's cost or income occur because of a change in the company's physical plant, a regulatory rate proceeding takes place to consider whether a new rate is warranted. First, an acceptable return on investors' equity is determined based on factors such as what is required to instill financial confidence in the utility, what other utilities are allowed when operating in the same business risk environment, and what is fair and reasonable. Revenues are then calculated to yield the required return on equity.

H. G. Stoll distinguishes these two criteria for electric rate regulation: the return on equity and the return on the rate base[3]. The return on common equity criterion calculates the ratio of net income available for common stock (from the utility's income statement) to the average or end-of-year common equity (from the utility's balance sheet). The revenue requirement is then increased or decreased so that the return-on-

---

[2] Inflation-adjusted interest rates are required in this section. A fairly thorough discussion of inflation and its effect on capital investment analysis is provided in Chapter 18.

[3] H. G. Stoll, *Least-Cost Electric Utility Planning* (New York: Wiley, 1987).

equity target is achieved. Alternatively, return on the rate base is a more traditional rate regulation criterion. The rate base is defined as follows:

Rate base = total plant in service
 − accumulated depreciation reserve
 + materials and supplies (optional)
 + fossil fuel inventory (optional)
 + working capital allowance (optional)
 − deferred income taxes (optional)
 − deferred investment tax credit (optional)
 + construction work in progress (optional).

Because of the role and importance of a utility's cost of capital and capitalization structure to the minimum revenue requirement, we discuss selected aspects of an investor-owned utility's financing operation at this point. First, we observe that the interest paid on borrowed capital (debt) is tax deductible. Therefore, the after-tax cost of debt, i'_a, is

$$i'_a = i'_b - t \cdot i'_b$$
$$= (1-t)[(1+i_b)(1+\bar{f}) - 1],$$

(7-9)

where i'_b = inflation-adjusted cost of borrowed capital
$$= [(1+i_b)(1+\bar{f}) - 1],$$
t = effective income tax rate,
$\bar{f}$ = average annual inflation rate.

Second, the firm's cost of capital depends on the proportion and cost of both debt and equity capital. The *after-tax cost of capital,* K'_a including, an adjustment for inflation, is

$$K'_a = \lambda i'_a + (1-\lambda)e'_a$$
$$= \lambda(1-t)i'_b + (1-\lambda)e'_a,$$

(7-10)

where λ = fraction of borrowed money in the utility's total capitalization,
$(1-\lambda)$ = fraction of equity capital in total capitalization,
e'_a = inflation-adjusted equity rate = $[(1+e_a)(1+\bar{f}) - 1]$.

The real (inflation-free) after-tax cost of capital, K_a, is

$$K_a = \frac{1+K'_a}{1+\bar{f}} - 1$$
$$= \frac{\lambda(1-t)i_b + (1-\lambda)e_a - \lambda t\bar{f}}{1+\bar{f}},$$

(7-11)

where e_a is the real equity rate.

7.4.2 Illustrations of the Revenue Requirement Method

Using a tabular form to calculate the annual revenue requirements for a utility project offers an easy-to-manipulate and understandable computational format. The analyst can use tabular columns, as required by a given problem, to account for the various components of revenue requirement shown in Fig. 7-1.

Example 7-5

This example, evaluating a single investment project, utilizes the following project data and the column-by-column operations shown in Table 7-1:

> Project life, N = book life = 4 years
> Initial investment, I = \$7,500
> Salvage value, S = \$1,500
> Annual O&M expenses, C = \$500
> Real (inflation-free) cost of borrowed money, i_b = 5%
> Real (inflation-free) return on equity, e_a = 16.07%
> Debt ratio, λ = 0.3
> Effective income tax rate, t = 50%
> Book depreciation method = straight line
> Tax depreciation method = straight line
> Average annual inflation rate, $\bar{f}$ = 0%

For each operating year, k, $1 \leq k \leq 4$, the revenue requirement in year k, RR_k, is calculated by using Eq. 7-8. One column is reserved for each term of the carrying charges (see Eq. 7-5), and an additional column is used for the recurring annual expenses associated with the project.

For instance, RR_2 is calculated as follows:

$$= \$7,500 - \$1,500 = \$6,000$$

Columns 2 $d = (I - S)/N$
and 3: $= (\$7,500 - \$1,500)/4 = \$1,500$

Column 4: $\lambda i_b \cdot UI_2 = 0.3(0.05)(\$6,000) = \$90$

Column 5: $(1 - \lambda)e_a \cdot UI_2 = 0.7(0.1607)(\$6,000) = \$674.94$

Column 6: $T_2 = [t/(1 - t)][(1 - \lambda)e_a \cdot UI_2 + d_{B_2} - d_{T_2}]$
             $= [0.5/(1 - 0.5)] \cdot [0.7 \cdot 0.1607 \cdot \$6,000 + \$1,500 - \$1$
             $= \$674.94$

Column 7: $C_2 = \$500$

Column 8: $RR_2 = \$1,500 + \$500 + \$90 + \$674.94 + \$674.94$
             $= \$3,439.88$

TABLE 7-1 Annual Revenue Requirements for Example 7-5

	(1) Unre-covered invest., UI_k	(2) Book depr., d_{B_k}	(3) Tax depr., d_{T_k}	(4) Debt return $\lambda i_b UI_k$	(5) Equity return, $(1-\lambda)e_a UI_k$	(6) Income tax, T_k	(7) Annual expenses, C_k	(8) $RR_k =$ (2) + (4) + (5) + (6) + (7)
Year, k								
1	\$7,500	\$1,500	\$1,500	\$113	\$844	\$844	\$500	\$3,780
2	6,000	1,500	1,500	90	675	675	500	3,440
3	4,500	1,500	1,500	68	506	506	500	3,080
4	3,000	1,500	1,500	45	337	337	500	2,720

Calculations for the remaining years are performed similarly. Table 7-1 provides a summary of the results in Example 7-5. Note that no unrecovered investment is left after the end of year 4.   ■

> It is customary to express the yearly revenue requirements (column 8) as a single measure of worth for the project under study.

Cumulative present worth, equivalent annual worth (also called *levelized revenue requirement, $\overline{RR}$*), and capitalized worth are the three measures most often used by utilities to report the economic merit of a given project. To calculate these quantities, a discounting factor is needed to account for the time value of money. The utility's real after-tax cost of capital, K_a, is usually the interest rate employed for such calculations.

In the preceding example, K_a is determined using Eq. 7-11 with an inflation rate of $\bar{f} = 0$:

$$K'_a = 0.3 \cdot (1 - 0.5) \cdot 0.05 + (1 - 0.3) \cdot 0.1607 - 0.3 \cdot 0.5 \cdot 0/(1 + 0)$$
$$= 0.12.$$

Hence, the present worth of RR (PWRR) as a function of K_a is

$$PWRR(K_a) = \sum_{k=1}^{N} RR_k \cdot (P/F, K_a\%, k)$$
$$= [\$3,799.86(P/F, 12\%, 1) + \$3,439.88(P/F, 12\%, 2)$$
$$+ 3,079.92(P/F, 12\%, 3) + \$2,719.94(P/F, 12\%, 4)]$$
$$= \$10,055.59.$$

The levelized revenue requirement $(\overline{RR})$ is then

$$\overline{RR}(K_a) = PWRR(K_a) \cdot (A/P, K_a\%, N)$$
$$= \$10,055.59 \cdot (A/P, 12\%, 4)$$
$$= \$3,310.70.$$

Finally, the capitalized revenue requirement (CRR) is

$$CRR(K_a) = \overline{RR}(K_a) \div K_a$$
$$= \$3,310.70 \div 0.12$$
$$= \$27,589.17.$$

In selecting among alternative investment projects, the three quantitative measures above are equivalent. The alternative that minimizes the selected revenue requirement measure represents the most economical choice. Because the utility is obligated to provide services to the public, a request for a rate increase can be presented to the regulatory commission if the investors feel that revenues generated from a project are unsatisfactory.

Example 7-6

A public utility must extend electric power service to a new shopping center. A decision must be made as to whether a pole-line or an underground system should be used. The pole-line system would cost only $158,000 to install, but because of numerous changes that are anticipated in the development and use of the shopping center, it is estimated that annual maintenance expenses would be $29,000. An underground system would cost $315,000 to install, but the annual maintenance expenses would not exceed $5,500. Annual property taxes are 1.5% of the first cost. The company operates with 33% borrowed capital, on which it pays an interest rate of 8%. Capital should earn about 11% after taxes. For this problem, the after-tax return of 11% is interpreted as the value of K_a. A 20-year study period is to be used, and the effects of inflation on cash flows are to be ignored. The straight-line method is to be used for both book and tax depreciation purposes. Finally, the effective income tax rate is 39.94%.

Before column 5 (equity return) in the tabular procedure can be computed, the value of the return on equity, e_a, has to be determined. By using Eq. 7-11, one can write

$$
\begin{aligned}
e_a &= \{K_a - \lambda \cdot [(1 - t) \cdot i_b - t \cdot \bar{f} \cdot (1 - \bar{f})]\} / (1 - \lambda) \\
&= \{0.11 - 0.33 \cdot [(1 - 0.3994) \cdot 0.08 - 0.3994 \cdot 0 \cdot (1 - 0)]\} / (1 - 0.33) \\
&= 0.1405.
\end{aligned}
$$

Tables 7-2 and 7-3 display the yearly RR results for the pole-line and underground systems, respectively. The values of $\overline{RR}$ are given for each alternative at the bottom of the corresponding RR table. Accordingly, the pole-line system shows a lower $\overline{RR}$ and is, therefore, the system to be chosen based on financial considerations only.   ■

7.5 Immediate Versus Deferred Investment

Because utilities must always be prepared to meet the demands for service placed on them, many engineering economy studies in utility companies involve immediate versus deferred investment to meet future demands. The following is an example.

Example 7-7

A water company must decide whether to install a new pumping plant now and abandon a gravity-feed system, which has been fully depreciated, or wait 5 years to install the new plant because of deteriorated piping in the gravity-feed system. Annual O&M expenses and taxes for the gravity system are $45,000. The pumping plant will cost $375,000 to install, and it is estimated that it would have a salvage value of 5% of its initial cost at the time of removal from service 20 years hence, when a new and larger system will be installed. Annual O&M expenses and property taxes for the proposed plant would be $30,000. The gravity-feed system has no salvage value now or later.

If the pumping plant is installed now, it would have a useful life of 20 years. If installed 5 years hence, its useful life would be only 15 years, but its salvage value would still be 5% of its initial cost. Using the revenue requirement method, determine which alternative is better. Straight-line depreciation is assumed for both book and tax purposes. The company operates with 50% borrowed capital, on which it pays an interest rate of 7%. The equity rate is expected to be about 14%, and the company pays an effective income tax rate of 50%.

Solution   First, we determine K_a from Eq. 7-11 to be $0.5[(1 - 0.5)(0.07)] + 0.5(0.14) = 0.0875$. Next, from Table 7-4, the levelized revenue requirement of the new pumping plant using the tax-adjusted cost of capital is found to be

$$\overline{RR}(8.75\%) = \$92,135.$$

TABLE 7-2 Pole-Line System Calculations for Example 7-6

Year, k	(1) Unrecovered investment	(2) Book depreciation	(3) Tax depreciation	(4) Debt return	(5) Equity return	(6) Income taxes	(7) Annual expenses	(8) $RR_k =$ (2) + (4) + (5) + (6) + (7)
1	$158,000	$7,900	$7,900	$4,171	$14,875	$9,892	$31,370	$68,208
2	150,100	7,900	7,900	3,963	14,131	9,397	31,370	66,761
3	142,200	7,900	7,900	3,754	13,387	8,902	31,370	65,313
4	134,300	7,900	7,900	3,546	12,644	8,408	31,370	63,867
5	126,400	7,900	7,900	3,337	11,900	7,914	31,370	62,420
6	118,500	7,900	7,900	3,128	11,156	7,419	31,370	60,974
7	100,600	7,900	7,900	2,920	10,412	6,924	31,370	59,526
8	102,700	7,900	7,900	2,711	9,669	6,430	31,370	58,080
9	94,800	7,900	7,900	2,503	8,925	5,935	31,370	56,632
10	86,900	7,900	7,900	2,294	8,181	5,440	31,370	55,185
11	79,000	7,900	7,900	2,086	7,437	4,946	31,370	53,739
12	71,000	7,900	7,900	1,877	6,694	4,452	31,370	52,292
13	63,200	7,900	7,900	1,668	5,950	3,957	31,370	50,845
14	55,300	7,900	7,900	1,460	5,206	3,462	31,370	49,399
15	47,400	7,900	7,900	1,251	4,462	2,967	31,370	47,951
16	39,500	7,900	7,900	1,043	3,719	2,473	31,370	46,504
17	31,600	7,900	7,900	834	2,975	1,978	31,370	45,057
18	23,700	7,900	7,900	626	2,231	1,484	31,370	43,611
19	15,800	7,900	7,900	417	1,487	989	31,370	42,163
20	7,900	7,900	7,900	209	744	495	31,370	40,717

$$\overline{RR}(K_a) = \$59,496.77$$

TABLE 7-3 Underground System Calculations for Example 7-6

Year, k	(1) Unrecovered investment	(2) Book depreciation	(3) Tax depreciation	(4) Debt return	(5) Equity return	(6) Income taxes	(7) Annual expenses	(8) $RR_k =$ (2) + (4) + (5) + (6) + (7)
1	$315,000	$15,750	$15,750	$8,316	$29,655	$19,721	$10,225	$83,667
2	299,250	15,750	15,750	7,900	28,173	18,735	10,225	80,783
3	283,500	15,750	15,750	7,484	26,690	17,749	10,225	77,899
4	267,750	15,750	15,750	7,069	25,207	16,763	10,225	75,014
5	252,000	15,750	15,750	6,653	23,724	15,777	10,225	72,129
6	236,250	15,750	15,750	6,237	22,242	14,791	10,225	69,245
7	220,500	15,750	15,750	5,821	20,759	13,805	10,225	66,360
8	204,750	15,750	15,750	5,405	19,276	12,819	10,225	63,476
9	189,000	15,750	15,750	4,990	17,793	11,832	10,225	60,590
10	173,250	15,750	15,750	4,574	16,310	10,846	10,225	57,705
11	157,500	15,750	15,750	4,158	14,828	9,861	10,225	54,822
12	141,750	15,750	15,750	3,742	13,345	8,874	10,225	51,936
13	126,000	15,750	15,750	3,326	11,862	7,888	10,225	49,052
14	110,250	15,750	15,750	2,911	10,379	6,902	10,225	46,167
15	94,500	15,750	15,750	2,495	8,897	5,917	10,225	43,283
16	78,750	15,750	15,750	2,079	7,414	4,930	10,225	40,398
17	63,000	15,750	15,750	1,663	5,931	3,944	10,225	37,513
18	47,250	15,750	15,750	1,247	4,448	2,958	10,225	34,629
19	31,500	15,750	15,750	832	2,966	1,972	10,225	31,744
20	15,750	15,750	15,750	416	1,483	986	10,225	28,859

$$\overline{RR}(K_a) = \$66,304.74$$

TABLE 7-4 Install New Pumping Plant Now for Example 7-7

Year, k	(1) Unrecovered investment	(2) Book depreciation	(3) Tax depreciation	(4) Debt return	(5) Equity return	(6) Income taxes	(7) Annual expenses	(8) $RR_k =$ (2) + (4) + (5) + (6) + (7)
1	$375,000.00	$17,812.50	$17,812.50	$13,125.00	$26,250.00	$26,250	$30,000	$113,438
2	357,187.50	17,812.50	17,812.50	12,501.56	25,003.13	25,003	30,000	110,320
3	339,375.00	17,812.50	17,812.50	11,878.13	23,756.25	23,756	30,000	107,203
4	321,562.50	17,812.50	17,812.50	11,254.69	22,509.38	22,509	30,000	104,086
5	303,750.00	17,812.50	17,812.50	10,631.25	21,262.50	21,263	30,000	100,970
6	285,937.50	17,812.50	17,812.50	10,007.81	20,015.62	20,016	30,000	97,852
7	268,125.00	17,812.50	17,812.50	9,384.38	18,768.76	18,769	30,000	94,735
8	250,312.50	17,812.50	17,812.50	8,760.94	17,521.88	17,522	30,000	91,618
9	232,500.00	17,812.50	17,812.50	8,137.50	16,275.00	16,275	30,000	88,501
10	214,687.50	17,812.50	17,812.50	7,514.06	15,028.12	15,028	30,000	85,383
11	196,875.00	17,812.50	17,812.50	6,890.63	13,781.26	13,781	30,000	82,266
12	179,062.50	17,812.50	17,812.50	6,267.19	12,534.38	12,534	30,000	79,148
13	161,250.00	17,812.50	17,812.50	5,643.75	11,287.50	11,288	30,000	76,033
14	143,437.50	17,812.50	17,812.50	5,020.31	10,040.62	10,041	30,000	72,915
15	125,625.00	17,812.50	17,812.50	4,396.88	8,793.76	8,794	30,000	69,798
16	107,812.50	17,812.50	17,812.50	3,773.44	7,546.88	7,547	30,000	66,680
17	90,000.00	17,812.50	17,812.50	3,150.00	6,300.00	6,300	30,000	63,563
18	72,187.50	17,812.50	17,812.50	2,526.56	5,053.12	5,053	30,000	60,446
19	54,375.00	17,812.50	17,812.50	1,903.31	3,806.26	3,806	30,000	57,328
20	36,562.50	17,812.50	17,812.50	1,279.69	2,559.38	2,559	30,000	54,210

$$\overline{RR}(K_a) = \$92,135$$

TABLE 7-5 Defer Installation of New Pump for 5 Years for Example 7-7

Year, k	(1) Unrecovered investment	(2) Book depreciation	(3) Tax depreciation	(4) Debt return	(5) Equity return	(6) Income taxes	(7) Annual expenses	(8) $RR_k =$ (2) + (4) + (5) + (6) + (7)
1	$ 0	$ 0	$ 0	$ 0	$ 0	$ 0	$45,000	$45,000
2	0	0	0	0	0	0	45,000	45,000
3	0	0	0	0	0	0	45,000	45,000
4	0	0	0	0	0	0	45,000	45,000
5	0	0	0	0	0	0	45,000	45,000
6	375,000	23,750	23,750	13,125.00	26,250	26,250	30,000	119,375
7	351,250	23,750	23,750	12,293.75	24,588	24,588	30,000	115,220
8	327,500	23,750	23,750	11,462.50	22,925	22,925	30,000	111,063
9	303,750	23,750	23,750	10,631.25	21,263	21,263	30,000	106,907
10	280,000	23,750	23,750	9,800.00	19,600	19,600	30,000	102,750
11	256,250	23,750	23,750	8,968.75	17,938	17,938	30,000	98,595
12	232,500	23,750	23,750	8,137.50	16,275	16,275	30,000	94,438
13	208,750	23,750	23,750	7,306.25	14,673	14,673	30,000	90,402
14	185,000	23,750	23,750	6,475.00	12,950	12,950	30,000	86,125
15	161,050	23,750	23,750	5,643.75	11,288	11,288	30,000	81,970
16	137,500	23,750	23,750	4,812.50	9,625	9,625	30,000	77,813
17	113,750	23,750	23,750	3,981.25	7,963	7,963	30,000	73,657
18	90,000	23,750	23,750	3,150.00	6,300	6,300	30,000	69,500
19	66,250	23,750	23,750	2,318.75	4,638	4,638	30,000	65,345
20	42,500	23,750	23,750	1,487.50	2,975	2,975	30,000	61,188

$$\overline{RR}(K_a) = \$74,876$$

From Table 7-5, the levelized revenue requirement of the deferred installation is

$$\overline{RR}(8.75\%) \ = \ \$74,876.$$

Finally, a comparison of the levelized revenue requirements for both alternatives shows that it is more economical to defer the new pumping plant for 5 years.   ∎

7.6 Summary

This chapter has addressed many unique topics associated with capital investment analyses in government agencies and public (investor-owned) utilities. The benefit–cost ratio method was shown to provide investment recommendations identical to other theoretically correct methods (e.g., PW).

The revenue requirement method was introduced as an appropriate economic evaluation technique for public utility projects. A fundamental principle underlying utility rate regulation is the fact that revenues should be generated so as to just cover the utility service expenses and provide a fair return on equity to investors. The revenue requirement method is equivalent to an after-tax PW or AW analysis of competing alternatives. Only the perspective is different. That is, PW and AW methods evaluate the project from the shareholder's viewpoint, while the revenue requirement method uses the utility customer's viewpoint because rates are regulated by representatives of the public.

REFERENCES

COMMONWEALTH EDISON COMPANY, *Engineering Economics.* Chicago: Commonwealth Edison Company, 1975.

Jeynes, P. H., *Profitability and Economic Choice.* Ames, IA: Iowa State University Press, 1968.

MAYER, R. R., "Finding Your Minimum Revenue Requirements," *Industrial Engineering* 9, no. 4 (April 1977): 16–22.

STEVENS, G. T., *Economic and Financial Analysis of Capital Investments.* New York: Wiley, 1979.

STOLL, H. G., *Least-Cost Electric Utility Planning.* New York: Wiley, 1987.

Ward, T. L., and W. G. Sullivan, "Equivalence of the Present Worth and Revenue Requirements Method of Capital Investment Analysis," *AIIE Transactions* 13, no. 1 (March 1981), pp. 29–40.

PROBLEMS

7-1. Consider the two independent alternatives given below. Using the modified benefit–cost ratio, which (if either) should be selected?

	Alternatives	
	I	II
First cost:	$20,000	$20,500
Life:	10 yr	10 yr
Salvage value:	$1,000	$1,500
Annual receipts:	$4,740	$4,800
MARR = 10%		
Study period = 10 yr		

7-2. A construction firm is considering leasing a crane for 4 years for $200,000 payable now. As an alternative, the crane can be purchased for $250,000 and sold for $100,000 at the end of 4 years. Annual maintenance costs for ownership are expected to be $10,000 per year for the first 2 years and $15,000 per year for the last 2 years. Using a study period of 4 years and a MARR of 10%, perform a B/C analysis to determine the preferred alternative.

7-3. Perform a conventional B/C analysis for the following six mutually exclusive alternatives. Also, compute the conventional B/C values for each individual alternative; compare the values obtained with the modified B/C values. MARR =10%.

	Alternative project					
	A	B	C	D	E	F
Investment:	$1,000	$1,500	$2,500	$4,000	$5,000	$7,000
Annual savings in cash disbursements:	150	375	500	925	1,125	1,425
Salvage value:	1,000	1,500	2,500	4,000	5,000	7,000

7-4. A city is considering the elimination of a railroad grade crossing by building an overpass. The overpass would cost $1,000,000 and is estimated to have a useful life of 40 years and a $100,000 salvage value. Approximately 2,000 vehicles per day are delayed an average of 2 minutes each due to trains at the grade crossing. Trucks comprise 40% of the vehicles, and the opportunity cost of their delay is assumed to average $20 per truck-hour. The other vehicles are cars having an assumed average opportunity cost of $4.00 per car-hour. The installation will save the railroad an annual expense of $30,000 for lawsuits and maintenance of crossing guards, but it will not save the railroads anything for time to pass through the intersection. It is estimated that the new overpass will save the city approximately $4,000 per year in expenses directly due to accidents. Should the overpass be built by the city if it is to be the owner and the opportunity cost of its capital is 8%? How much should the railroad reasonably be asked to contribute toward construction of the bridge if its opportunity cost of capital is assumed to be 15%?

7-5. A steel bridge cost $400,000 when built by the state 15 years ago. Average annual maintenance costs are expected to be $10,000 per year; therefore, it has been proposed to replace the steel bridge with a concrete bridge of comparable capacity. A new concrete bridge would cost $700,000, but it would require only $2,000 per year maintenance cost and would last 60 years and cost $40,000 to remove then. The present realizable salvage value for the steel bridge is $50,000 now, which means that the new capital required for the new bridge would be only $650,000. If kept, the old steel bridge would be expected to last 25 more years and have a $15,000 salvage value at that time.

   The state has sufficient tax funds on hand to finance the new bridge without borrowing, but some contend that money invested in highway improvements has an opportunity cost to taxpayers of at least 8%. The following is an annual cost economic comparison provided by an engineer.

Keep Present Bridge

Depreciation (based on 15-year life) = $400,000/15 =		$26,667
Maintenance		$10,000
	Total annual cost	$36,667

Replace with Proposed Bridge

Depreciation (based on 60-year life) = $650,000/60 =		$10,833
Maintenance		$2,000
	Total annual cost	$12,833

List any errors you perceive in the engineer's analysis. Make an annual cost comparison that is more valid. Use any assumptions you think are reasonable.

7-6. Five mutually exclusive alternatives are being considered for providing a sewage treatment facility. The costs and estimated benefits of the alternatives are as follows:

	Annual equivalents (in thousands)	
Alternative	Cost	Benefits
A	$1,050	$1,110
B	900	810
C	1,230	1,390
D	1,350	1,500
E	990	1,140

Which plan should be adopted, if any, if the Sewage Authority wishes to invest if, and only if, the B/C ratio on any cost is at least 1.0?

7-7. A government highway department is considering the economics of installing a four-lane highway versus a six-lane highway in a certain location now. A four-lane highway would cost $3,000,000 now and would be sufficient to meet requirements for 10 years, at which time it is projected that the highway would have to be expanded by two lanes at a cost of $1,000,000. Alternatively, the full six-lane highway could be built now for $3,400,000, and this would be sufficient for needs into the indefinite future. Average annual maintenance costs would be $100,000 for the four-lane highway and $120,000 for the six-lane highway. If money used by the highway department is assumed to cost 8%, which is the better choice based on an annual cost comparison?

7-8. How much could a city afford to pay now for a new sewage treatment facility that would eliminate contractor charges of $100,000 each year starting this year and would also result in pollution reduction benefits judged to be worth approximately $200,000 beginning at the end of 5 years from now and increasing by $10,000 at the end of each year thereafter until the end of 25 years from now? Assume capital used by the city has an opportunity cost of 10% per year.

7-9. Alternative methods I and II are proposed for a municipal plant operation. The following is comparative information:

	Method I	Method II
First cost:	$10,000	$40,000
Life:	5 yr	10 yr
Salvage value:	$1,000	$5,000
Annual disbursements:	$14,150	$7,000
MARR = 20%		
Study period = 10 yr		

Using a modified B/C analysis to determine the preferred alternative, compare these mutually exclusive alternatives.

7-10. Two alternative machines will produce the same product, but one will produce higher-quality items that can be expected to return greater revenue. Given the following data, determine which machine is better.

	Machine A	Machine B
First cost:	$ 20,000	$ 30,000
Salvage value:	2,000	0
Annual receipts:	150,000	180,000
Annual disbursements:	138,000	170,000
MARR = 15%		
Study period = 10 yr		

Determine the B/C values for each machine by using both the conventional and the modified formulations.

7-11. Five mutually exclusive machines are being considered for a particular job. Each is expected to have a salvage value of 100% of the investment amount at the end of the 4-year study period. Using a B/C analysis, which machine should be selected on the basis of the following data?

			Alternatives		
	A	B	C	D	E
Investment:	$2,100	$3,400	$1,000	$2,700	$1,400
Net cash flow per yr:	280	445	110	340	180
MARR = 12%					

7-12. Given the following mutually exclusive alternatives, show which is best considering a savings in operating disbursements as ΔB.

		Incinerator		
	A	B	C	D
First cost:	$3,000	$3,800	$4,500	$5,000
Life:	10 yr	10 yr	10 yr	10 yr
Salvage value:	$0	$0	$0	$0
Annual operating disbursements:	$1,800	$1,770	$1,470	$1,380
MARR = 10%				

7-13. Briefly summarize the basic characteristics that distinguish investor-owned utilities from nonregulated industries such as steel, automobile, and chemical manufacturing.

7-14. Why are most utilities heavily financed with borrowed capital? What characteristics of this industry make it possible to attract large amounts of borrowed capital, and what advantages (disadvantages) are associated with the use of borrowed money?

7-15. Explain why it may be in the best interest of the consuming public for a regulatory agency to permit a utility to charge sufficiently high rates to allow it to earn an adequate return on its capital.

7-16. a. In a certain state a member of the Public Utilities Commission said, "I will oppose all rate increases. I am interested only in the rates the customers have to pay today." Comment on the results that could follow if all members of the commission rigidly followed this concept.

    b. Comment on this statement "No company that provides an exclusive and required service, such as electric power, should be permitted to make a profit."

7-17. Is there justification for a privately owned, regulated utility being permitted to include in its rates the cost of advertising that encourages the public to increase utilization of its service?

Note: Solve the remaining problems by using the after-tax cost of capital, K_a (or K_a').

7-18. A telephone company can provide certain facilities having a ten-year life and zero salvage value by either of two alternatives. Alternative A requires a first cost of $70,000 and $3,000 per year for maintenance. Alternative B will have a first cost of $48,000 and will require $6,000 annually for maintenance. Property taxes and insurance would be 4% of the first cost per year for either alternative. The after-tax cost of capital is 10%, with 30% being borrowed at a 6% interest rate. The effective income tax rate is 50%. Which alternative will provide the lower annual equivalent revenue requirement? Depreciation for income tax purposes is MACRS (recovery period = 5 years), and book depreciation is computed with the straight-line method over 10 years.

7-19. A gas company must decide to build a new meter-repair and testing facility now or wait three years before doing so. It estimates that until the new facility is built, its annual costs for these functions will be $90,000 greater than when the new facility is completed. The new facility would cost $900,000 and would not be needed after 20 years. The ultimate salvage value would be $200,000 at that time. The company uses 40% borrowed capital, paying 8% interest (before taxes) for it, and the regulatory body permits it to earn 13.8% on its equity capital. Assuming that the company has a 40% effective income tax rate, determine the equivalent annual revenue requirement for both options and recommend which is better. Assume that book (and tax) depreciation over 20 years is computed with the straight-line method.

7-20. A utility company can construct a modern power plant that can generate power at 24 mils ($0.024) per kWh at a 70% load factor. The 24 mils covers all costs, including profit on capital and also income taxes. A large, industrially owned power plant will soon make wholesale power available. To take advantage of the wholesale power, it will cost $180 per kilowatt of capacity to build the necessary transmission line, which will experience a 70% load factor. Annual maintenance costs for this line will be $0.90 per kilowatt of capacity, and it is to be fully depreciated for book purposes over a 30-year period. A 15-year MACRS recovery period will be utilized for calculating depreciation for income tax purposes. The cost of money for the company is 12%, with 40% borrowed capital at an interest rate of 7%. The income tax rate is 50%. At what price must the company be able to purchase the power in order to be as economical as generating with the new modern plant?

7-21. Determine the annual revenue requirements for the following proposed 280-KVA transformer bank.

Installed cost	= $240,000
Property taxes and insurance/yr	= 2% of installed cost
Salvage value	= 0
Tax life = book life	= 4 years
Depreciation method (for book purposes)	= Straight line
Depreciation method (for tax purposes)	= MACRS (3-year recovery period)
Effective income tax rate	= 0.40 (40%)
Cost of equity capital	= 20%, $(1 - \lambda) = 0.60$
Cost of borrowed funds	= 12%, $\lambda = 0.40$

Use the format in Tables 7-2 and 7-3 for completing this problem.

7-22. A telephone company must provide a direct current battery unit to a new service region in 1997. The expected useful life of the equipment is seven years. Alternative A requires a first cost of $75,000 and has O&M costs of $8,000 per year. The salvage value for tax purposes is zero, and this is also the expected realizable market value. A tax life of 5 years will be claimed, and MACRS depreciation will be utilized for tax purposes. However, straight-line depreciation over seven years will be taken for rate-setting purposes (i.e., book depreciation).

The after-tax cost of capital (K_a') is 12%, with 40% being borrowed at 8% per year. The effective income tax rate is 40%, and the general inflation rate is 6% per year. Only O&M costs are affected by inflation, and the costs of capital given previously include an allowance for anticipated inflationary pressures in the economy.

For alternative A, answer the following questions. Be sure to state any assumptions you feel are appropriate and necessary.

a. What is the actual dollar ATCF in year 5 of this alternative's useful life?

b. What is the *income tax* entry in an RR table for year 5?

7-23. In 1998 the installed cost of a new transformer at the CEPO Utility Company is $50,000. Annual maintenance costs, which are expected to escalate by 5% each year, are $1,500 in today's dollars. A five-year MACRS recovery period is to be used for tax depreciation purposes, and the expected life of the transformer is eight years. The terminal MV is negligible. Straight-line depreciation is used for determining BV for the rate-setting purposes. Borrowed capital represents 40% of the company's capitalization, and it costs 10% per year before taxes. The return to equity is approximately 15% per year.

a. If the firm's effective income tax rate is 40%, calculate the RR in year 3.

b. If the firm's effective income tax rate is 50%, by how much does the RR in year 3 increase?

7-24. An electric utility company has an opportunity to build a small hydroelectric generating plant of 20,000-kW capacity, on a mountain stream where the flow is seasonal. As a consequence, the annual output of energy would be only 40,000,000 kWh. The initial cost would be $2,000,000, and it is estimated that the annual operation and maintenance costs would be $32,000 during its estimated 30-year economic life. It is believed that the property would have a salvage value of $200,000 at the end of the 30-year period. An alternative is to build a geothermal generating plant, which would have the same annual capacity, at a cost of $1,600,000. Because the company would have to pay the owners of the property for the geothermal steam, the estimated annual cost for the steam and operation and maintenance is $120,000. A 30-year contract can be obtained on the steam supply, and it is believed that this period is realistic for the useful life of the plant but that the salvage value at that time would be little more than zero. Property taxes and insurance on either plant would be 2% of the first cost per year. The company employs 40% borrowed capital, for which it pays 8.5% interest. It earns 13% after taxes on total capital, and it has a 50% effective income tax rate. Which development should be undertaken? Use the RR method.

7-25. In making its forecast of requirements in a certain area for the next 30 years, a telephone company has determined that a 600-pair cable is required immediately and a total of 1,000 pairs will be required by the end of 15 years. An underground conduit of sufficient size to handle the cable needs is being installed now at a cost of $10,000. If a 1,000 pair-cable is installed now, it will cost $30,000. As an alternative, the company can install a 600-pair cable immediately at a cost of $20,000 and install an additional 400-pair cable at the end of 15 years at an estimated cost of $16,000. Because of technical obsolescence, it is company policy to consider the useful life of either installation to be 30 years from the present time. Annual property taxes on either alternative would be 2% of the installed cost, and the salvage value of all cable and conduit at the end of the 30-year period is estimated to be 10% of the first cost. The company uses 40% borrowed capital, for which it pays 8%. It earns 12% after taxes on total capital and has a 50% effective income tax rate. Which alternative would you recommend? Assume that depreciation for tax and book purposes is computed with the straight-line method over 15 years for both alternatives. Use the RR method.

7-26. Use the RR method to compare alternatives A and B in Problem 7-18 when the average inflation rate on maintenance is 6% per year. Assume that property taxes do not respond to inflation, and adjust the cost of capital to account for inflation.

8
Replacement Analyses

8.1 Introduction

Replacement studies are of two general types. The first type involves studies on whether to keep an old asset (called the *defender*) or to replace the old with a new asset (called the *challenger*) at a given point in time. The second general type involves determining, in advance, the economic service life of an asset. The latter type of replacement study is treated briefly in this chapter.

The economics of replacement can generally be studied by any of the methods used for economic analyses of alternatives; e.g., rate of return, annual worth, present worth, future worth, or benefit–cost ratio.

8.2 Importance of Replacement Studies

The formulation of a replacement policy plays a major part in the determination of the basic technological and economic progress of a firm. Undue or hasty replacement can leave a firm pressed for capital that may be needed for other beneficial uses. Furthermore, if replacement is postponed beyond a reasonable time, the firm may find that its production costs are rising while the costs of its competitors who are using more modern equipment are declining. This can result in the firm's loss of ability to meet price competition and in a technological and economic trap of major proportions.

8.3 Causes of Retirement

Property retirement for economic study purposes is said to occur whenever the asset is physically removed, abandoned, or reassigned to a secondary service function. The following are common causes of retirement:

1. unsatisfactory functional characteristics, such as deterioration or inadequacy to meet requirements for safety, capacity, style, or quality;
2. end of need for output capability of asset; and
3. the old asset becomes uneconomical due to the appearance of improved assets with lower operating costs.

8.4 Replacement Considerations and Assumptions

Several important classes of considerations and assumptions inherent in replacement analyses (as well as in most analyses of nonreplacement alternatives) are discussed below.

1. The *planning horizon* is the farthest time in the future considered in the analysis. Often, an infinite planning horizon is used if it is difficult or impossible to predict when the activity under consideration will be terminated. Whenever it is clear that the project will have a definite and predictable duration, it is more realistic to base the study on a finite planning horizon.
2. The *technology* is important with respect to the characteristics of machines that are candidates to replace those under analysis. If it is assumed that all future assets will be the same as those presently in service, this implies that there will be no technological progress for that type of asset. It is probably more realistic to expect some obsolescence of old assets with respect to available new assets.
3. *Cost and return patterns over asset life* can take an infinite variety of forms. It is fairly common to assume they are uniform (constant) for lack of ability or willingness to try to estimate more closely. Common alternatives are to assume an increasing or decreasing pattern according to some function of time.
4. The *availability of capital* can be quite important to any replacement analysis, for the alternatives usually involve keeping the old asset so that little or no additional capital is needed as opposed to investing in some replacement asset with a marked capital outlay. Many studies are made assuming infinite capital available at some specified minimum rate of return. On the other hand, it may be desirable to consider some limitation on capital available, at least in the choice between alternative projects in a given time period.

PW (New-Old)   - Neg - New cost is worth more - stay w/ old
                  Pos - Old cost is more - replace w/ new

8.5 Market Value of Old Asset

When equipment is replaced, the old equipment is often accepted as partial payment for its replacement. If the vendor offers an allowance of, say, $15,000 on an older truck when traded for a newer model listed at, say, $25,000, the exchange price ($10,000) is then known, but the true market value of the older truck is not necessarily known. It is only by knowing the cash price of the newer model, say $21,000, that we can have a true estimate of the actual market value of the old truck ($21,000 – $10,000 = $11,000). Thus, the difference between the "apparent" market value based on trade-in quotation and the true market value is $15,000 – $11,000 = $4,000. The erroneous overstatement of challenger and defender costs by this $4,000 does not result in compensating errors whenever the expected life of the challenger differs from that for the defender and the conventional approach (to be explained later) is used. The added stated investment for the challenger then is spread over a different number of years than for the defender.

8.6 Specifying the Planning Horizon for a Replacement Study

As pointed out in Chapter 3, the determination of the study period or planning horizon is an important step in comparing investment alternatives. In this section, aspects of that determination peculiar to replacement studies are examined.

In a replacement study it is very common for the anticipated remaining life of the defender to be different from the anticipated lives of the challengers. In fact, it is usually the case that the remaining life for the defender is less than the useful life of the challenger. For this reason, it is very common to specify a planning horizon equal ✓to the anticipated remaining life of the defender.

If a particular situation dictates that the planning horizon should be greater than the anticipated remaining life of the defender, then forecasts are required for potential future replacements of the defender. Although this is normally the case in comparing any investment alternative, it is commonly assumed that replacements will have identical cash flow profiles. (Recall that the *repeatability* assumptions were used in several of the examples in previous chapters.) Such assumptions are questionable because the defender normally would not last more than one life cycle. For this reason, it is often assumed that the defender will be replaced at the end of its useful life with an asset having a cash flow profile equal to or better than the best challenger available currently.

Example 8-1

A machine has been in use for some time and can be continued in service for at most four more years. A new machine can be purchased for $18,000, and at the end of the 10-year planning horizon it will have a market (and book) value of $1,500. The defender has a market value of $1,800 currently; at the end of 4 years it will have a negligible market value. The annual expenses for the defender equal $3,500 and for the challenger equal $2,100.

If the defender remains in service for 4 years and is replaced at that time, the replacement is anticipated to cost $18,000 initially, have annual expenses of $2,100, and have a market value of $X after 6 years of service. A before-tax study and a MARR of 15% are deemed appropriate. Further suppose that the market value of the defender is used to offset the investment cost of the challenger.

   Solution   Two approaches come to mind in estimating the value of X. One approach is to specify a value of X such that the capital recovery cost for the replacement is equal to the capital recovery cost of the challenger. While such an approach may, on the surface, appear reasonable, it assumes implicitly that X equals the book value after 6 years using the sinking fund depreciation method, which may not adequately describe the way in which most equipment truly depreciates in value. Hence, it does not seem proper to select X in this manner. Instead, the proper approach is to forecast the value of X directly. For the purpose of this example, the straight-line depreciation book value based on a 10-year life will be used; hence, $X = \$8,100$. In this case, the capital recovery cost of the replacement for the defender will be larger than that of the challenger.

Year	Defender	Challenger
0	$0	$-\$18,000 + \$1,800$
1–4	$- 3,500$	$-2,100$
4	$-18,000$	0
5–10	$- 2,100$	$-2,100$
10	$8,100$	$1,500$

$$\text{PW (Defender)} = -\$3,500(P/A, 15\%, 4) - \$18,000(P/F, 15\%, 4)$$
$$- \$2,100(P/A, 15\%, 6)(P/F, 15\%, 4)$$
$$+ \$8,100(P/F, 15\%, 10)$$
$$= -\$22,826$$
$$\text{PW (Challenger)} = -\$16,200 - \$2,100(P/A, 15\%, 10) + \$1,500(P/F, 15\%, 10)$$
$$= -\$26,369$$

   The defender has the lower discounted cost, even though a somewhat conservative approach was used to account for the replacement of the defender at the end of its useful life. Technological improvements may well produce replacement candidates in the future that are superior to those available currently.   ■

8.7 Cash Flow Approach and Conventional Approach

In performing replacement studies a cash flow approach can usually be employed directly. If such an approach is used, the trade-in or current market value of the defender would be shown as a positive cash flow for the challenger, as in the previous example, because the positive cash flow will be realized if the defender is replaced. The cash flow approach allows a direct treatment of a situation in which the current market value for the defender can be different for different challengers. Finally, tax aspects involving capital gains and losses are easily handled when a cash flow approach is used.

Example 8-2

An industrial lift truck has been in service for several years and management is contemplating replacing it. Two replacement candidates (A and B) are under consideration. If new truck A is purchased, the old lift truck will have an actual market value of $2,000; if new truck B is purchased, the defender will have an actual market value of $1,500.

A planning horizon of 5 years is to be used. If the defender is retained, it is anticipated to have annual operating and maintenance costs of $7,300; it will have a zero market value at the end of 5 additional years of service.

Challenger A will cost $12,000 and will have operating and maintenance costs of $4,400; at the end of the planning horizon, it will have a market value of $4,000. Challenger B will cost $10,000 and will have operating and maintenance costs of $5,100; at the end of the planning horizon it will have a market value of $2,500.

A before-tax analysis is to be used to determine the preferred alternative with the cash flow approach. A present worth comparison and a minimum attractive rate of return of 20% are to be used.

Solution   The following cash flow profiles exist for the three alternatives:

| End of | | Challenger | |
year	Defender	A	B
0	$0	−$12,000 + $2,000	−$10,000 + $1,500
1–5	−7,300	−4,400	−5,100
5	0	4,000	2,500

$$PW \text{ (Defender)} = -\$7,300(P/A, 20\%, 5) = -\$21,831$$
$$PW \text{ (A)} = -\$10,000 - \$4,400(P/A, 20\%, 5) + 4,000(P/F, 20\%, 5)$$
$$= -\$21,553$$
$$PW \text{ (B)} = -\$8,500 - \$5,100(P/A, 20\%, 5) + 2,500(P/F, 20\%, 5)$$
$$= -\$22,749$$

Because challenger A has the lower equivalent present cost ($21,553), it would be recommended that challenger A be purchased.   ■

Example 8-3

Suppose that in the previous example an after-tax analysis is to be performed. Assume the defender has a current book value of $750 and is being depreciated using MACRS depreciation with depreciation charges of $500 and $250 remaining. Capital gains are taxed at 40% and ordinary income is taxed at 40%.

The challenger lift trucks will be depreciated using MACRS depreciation over a 5-year period. An after-tax minimum attractive rate of return of 12% is to be used in the analysis. Again, we employ the cash flow approach.

Solution

Defender: Cash Flow Approach

Year	BTCF	Depreciation deduction	Taxable income	Cash flow for income taxes ($t = 0.40$)	ATCF
0	$0	$0	$0	$0	$0
1	−7,300	500	−7,800	3,120	−4,180
2	−7,300	250	−7,550	3,020	−4,280
3	−7,300	—	−7,300	2,920	−4,380
4	−7,300	—	−7,300	2,920	−4,380
5	−7,300	—	−7,300	2,920	−4,380
				PW (12%) =	−$15,531

Challenger A: Cash Flow Approach

Year	BTCF	MACRS depreciation	Book value	Taxable income	Income tax ($t = 0.40$)	ATCF
0	−$10,000	$0	$12,000	$1,250*	−$500	−$10,500
1	−4,400	2,400	9,600	−6,800	2,720	−1,680
2	−4,400	3,840	5,760	−8,240	3,296	−1,104
3	−4,400	2,304	3,456	−6,704	2,682	−1,718
4	−4,400	1,382	2,074	−5,782	2,313	−2,087
5	−4,400	691	1,382	−5,091	2,037	−2,363
5	−4,000	—	—	2,618†	−1,047	2,953
					PW (12%) =	−$15,095

*$1,250 = $2,000 MV − $750 BV

†MV − BV = $4,000 − $1,382 = $2,618 (assuming half-year convention at time of disposal)

Challenger B: Cash Flow Approach

Year	BTCF	MACRS depreciation	Book value	Taxable income	Income tax ($t = 0.40$)	ATCF
0	−$8,500	$0	$10,000	$750*	−$300	−$8,800
1	−5,100	2,000	8,000	−7,100	2,840	−2,260
2	−5,100	3,200	4,800	−8,300	3,320	−1,780
3	−5,100	1,920	2,880	−7,020	2,808	−2,292
4	−5,100	1,152	1,728	−6,252	2,501	−2,599
5	−5,100	576	1,152	−5,676	2,270	−2,830
5	2,500	—	—	1,348†	−539	1,961
					PW (12%) =	−$16,013

*$750 = $1,500 MV − $750 BV

†MV − BV = $2,500 − $1,152 = $1,348 (assuming half-year convention at time of disposal)

On the basis of an after-tax analysis, the recommendation is made to replace the defender with challenger A. Notice that the cash flow approach accounts for the market value of the defender over the study period (often useful life) of the *challenger*. The conventional approach illustrated in Example 8-4 incorporates the market value of the defender as an opportunity cost of keeping the defender rather than selling it. Thus, the before-tax investment cost of the defender is its market value. The cash flow approach and the conventional approach provide identical recommendations when the study periods are the same. Under the conventional approach when the life of the defender is less than that of the challenger, the market value of the defender is recovered over a shorter life—which may bias against keeping the defender. This conservative method of making comparisons involving different lives of the defender and challenger is favored by most analysts, leading to the heavy use of the conventional approach in replacement studies.   ∎

Example 8-4

In the previous example what would be the PWs for the alternatives using the conventional approach of treating the current market value of the defender as a cost of continuing it in service?

   Solution   Since the old lift truck will have a market value of $2,000 if traded in for new truck A and a market value of $1,500 if traded in for new truck B, a current market value of $2,000 will be used as the opportunity cost of continuing the old truck in service. Consequently, the before-tax cash flow for the defender will be −$2,000 in year 0; for challenger A a value of −$12,000 will apply in year 0; and for challenger B values of −$10,000 and −$500 are entered for year 0. The latter entry would represent the difference in market values (i.e., opportunity foregone for not having taken the higher trade-in value).

   The treatment of recaptured depreciation will be handled differently. Namely, a tax savings can be applied to the defender; in such a case, for each alternative one would add the amount 0.40($2,000 − $750) = $500 in year 0 to the entry denoting the cash flow for income taxes.

   Regardless of the market value used to assess the opportunity cost of retaining the defender and the method of assessing the tax on depreciation recapture, the differences in the present worths for the alternatives remain unchanged. Using a $2,000 market value and assessing the tax savings as indicated above, the reader may wish to confirm that the following after-tax present worths result:

Defender: Conventional Approach

EOY	BTCF	Depreciation	Taxable income	Income tax	ATCF
0	−$2,000	—	−$1,250	$500	−$1,500
1	−7,300	$500	−7,800	3,120	−4,180
2	−7,300	250	−7,550	3,020	−4,280
3	−7,300	—	−7,300	2,920	−4,380
4	−7,300	—	−7,300	2,920	−4,380
5	−7,300	—	−7,300	2,920	−4,380
				PW (12%) = −$17,031	

Challenger A: Conventional Approach

EOY	BTCF	MACRS depreciation	Book value	Taxable income	Income tax	ATCF
0	-$12,000	$0	$12,000	$0	$0	-$12,000
1	-4,400	$2,400	9,600	-6,800	2,720	-1,680
2	-4,400	3,840	5,760	-8,240	3,296	-1,104
3	-4,400	2,304	3,456	-6,704	2,682	-1,718
4	-4,400	1,382	2,074	-5,782	2,313	-2,087
5	-4,400	691	1,382	-5,091	2,037	-2,363
5	4,000	—	—	2,618	1,047	-2,953
						PW (12%) = -$16,595

Challenger B: Conventional Approach

EOY	BTCF	MACRS depreciation	Book value	Taxable income	Income tax	ATCF
0	-$10,500	$0	$10,000	-$500	$200	-$10,300
1	-5,100	$2,000	8,000	-7,100	2,840	-2,260
2	-5,100	3,200	4,800	-8,300	3,320	-1,780
3	-5,100	1,920	2,880	-7,020	2,808	-2,292
4	-5,100	1,152	1,728	-6,252	2,501	-2,599
5	-5,100	576	1,152	-5,676	2,270	-2,830
5	2,500	—	—	1,348	-539	1,961
						PW (12%) = -$17,513

Notes:

a. Half-year convention applied in year 5; i.e., half of usual depreciation is allowed.

b. Assume that tax credits are offset against tax liabilities of the company.

c. PW values are calculated as present worth of cash flow stream from year 0 to year 5.

d. Challenger B has a cost penalty of $500 (added to the cost of challenger B).

e. Assumed that BV of challenger B = $10,000, and $500 ($\Delta$MV) is treated as an expense.

f. The effective income tax rate (*t*) is 40%.

The decision is again to choose challenger A to minimize the present worth of costs. Notice that the *differences* in present worths are constant between alternatives for both the cash flow approach (Example 8-3) and the conventional approach (Example 8-4). This will always be true when identical study periods are used in the comparisons:

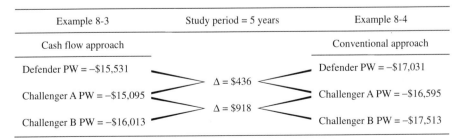

Example 8-3	Study period = 5 years	Example 8-4
Cash flow approach		Conventional approach
Defender PW = -$15,531	Δ = $436	Defender PW = -$17,031
Challenger A PW = -$15,095		Challenger A PW = -$16,595
	Δ = $918	
Challenger B PW = -$16,013		Challenger B PW = -$17,513

8.8 Asset Life Types

Economic studies, because they deal with the future, involve estimated rather than observed lives of properties. The term *life* can have a number of meanings; hence, the following defining distinctions are given:

1. *Economic life (service life)* is the period of time from the date of installation to date of retirement from the primary intended service of the asset. An estimate of the economic life of an asset may be based on the period that maximizes the annual worth of the proposed asset, assuming that the asset's service will be needed for that long and that no superior alternatives become available during that period. Retirement, however, is signaled by a future economic study when the annual worth of a prospective new asset becomes greater than the annual worth of retaining the present asset for one or more years. Retirement may constitute either disposal or demotion of the asset to a lower, less useful grade of service than was originally intended (such as for standby service).

2. *Ownership life* is the period of time from the date of installation to the date of actual disposal by a specific owner. The ownership life of an asset may well consist of one or more periods of secondary (downgraded) service in addition to the period of primary service, which constitutes its economic life.

3. *Physical life* is the period of time from the date of installation by the original owner until the asset is ultimately disposed of by its final owner.

There are several reasons underlying the difference between physical life and economic life. The first reason is the technological improvements in reliability or uniformity, the increased output capacities, and the reduced operating costs of successive new models. Second, the old equipment may show a pattern of increasing maintenance and operating costs as well as a deterioration in the quality of the goods or services produced. While the above reasons underlie the need for replacement in general, they also should dictate the pattern, if any, of physical life in addition to economic life. It should be kept in mind that we are generally concerned with economic lives in replacement studies.

8.9 Calculation of Economic Life

The second general type of replacement study mentioned at the beginning of this chapter is the determination, in advance, of how many years an asset should be kept for most economical service. In determining the economic life of a prospective new asset, there are three main types of costs to be taken into account: (1) costs of capital recovery, (2) costs inherent to the asset, and (3) costs relative to available improved models. Generally, the longer an asset is kept, the lower will be its annual cost of capital recovery and the higher will be its inherent and relative operating costs. An explanation of the meaning and calculation of each of these three types of costs is given in the following sections.

8.9.1 Costs of Capital Recovery

A realistic year-by-year cost of capital recovery can be determined only by estimating the net market (resale) value at the end of each year. The capital recovery cost for any year can be calculated as the decline in market value plus the interest on the asset value for that year. For example, the expected year-by-year market values of an asset X costing $23,000 and the calculation of yearly capital recovery costs, assuming interest at 10%, are shown in Table 8-1.

TABLE 8-1 Yearly Capital Recovery Cost Data for Asset X

Year	Market value at end of year	Decrease in market value during year	Interest on investment at beginning of year (@10%)	Capital recovery cost for year
1	$15,000	$8,000	$2,300	$10,300
2	8,500	6,500	1,500	8,000
3	3,000	5,500	850	6,350
4	1,000	2,000	300	2,300
5	1,000	0	100	100

8.9.2 Costs Inherent to the Asset

Asset inherent costs include the cost of operation and maintenance and the costs of capacity decreasing and quality declining compared to when the asset was new. Decreasing capacity cost is the loss in output due to downtime and reduced operating rate. Declining quality cost is the cost of lost sales, and reduced sales price or scrap rework due to the reduced quality capability of the asset. It should be kept in mind that the costs of decreasing capacity and declining quality are compared to the asset when it was new and not compared to possible improved models of the asset. Consideration of the effect of improved models is discussed in Section 8.9.3.

For example, the inherent costs of asset X for which capital recovery data were given in Table 8-1 is shown in Table 8-2.

TABLE 8-2 Yearly Inherent Cost Data for Asset X

Year	Cost of operation and maintenance	Cost of decreasing capacity	Cost of declining quality	Total inherent costs for year
1	$ 6,000	$ 0	$ 0	$ 6,000
2	6,000	0	0	6,000
3	6,000	500	500	7,000
4	8,000	1,000	4,000	13,000
5	10,000	2,000	5,000	17,000

8.9.3 Costs Relative to Improved Models

It is common that improved models of an asset will produce at lower operating costs and/or will produce a higher-quality product than the original models of that asset. There is thus an increase in operating costs and an increase in the cost of declining quality in the old asset relative to the new improved model of the asset. These increases in costs are alternative or opportunity costs. Another relative cost is the decreasing net income caused by obsolescence affecting the demand for the asset's products or services.

For example, the relative costs for the same asset X as in the preceding two sections are shown in Table 8-3.

Table 8-4 summarizes the data of Tables 8-1, 8-2, and 8-3 to provide for the calculation of total costs for each year and total equivalent annual costs for purposes of determining the most economical life for asset X.

Thus, from Table 8-4 it can be seen that the equivalent annual cost is at a minimum if asset X is retired at the end of three years. It can also be observed that the total marginal cost (shown in the fifth column as "Total cost for year") exceeds the equivalent annual cost after the third year. This illustrates the general principle for the determination of economic life, which may be stated as follows for the general case in which year-by-year costs are increasing:

Replace at the end of any period for which the total cost in the next period exceeds the average cost up to that period. Do not replace as long as the total cost in a period does not exceed the average cost to the end of that period.

TABLE 8-3   Yearly Relative Cost Data for Asset X

Year	Operating cost inferiority	Quality inferiority	Obsolescence	Total relative costs for year
1	$ 0	$ 0	$ 0	$ 0
2	500	0	0	500
3	800	850	0	1,650
4	2,000	3,000	0	5,000
5	2,500	3,500	1,000	7,000

TABLE 8-4 Economic Life Calculation for Asset X

Year	Capital recovery cost for year	Total inherent cost for year	Total relative cost for year	Total (marginal) cost for year	$(P/F,10\%,N)$	PW of cost for year [Total cost $\times$ $(P/F,10\%,N)$]	Cumulative PW of cost since placed in service $(\Sigma\,PW)$	$(A/P,10\%,N)$	Equivalent annual cost if retired at end of year [$\Sigma\,PW \times (A/P,10\%,N)$]
1	$10,300	$ 6,000	$ 0	$16,300	0.909	$14,790	$14,790	1.100	$16,300
2	8,000	6,000	500	14,500	0.826	12,700	27,400	0.576	15,800
3	6,350	7,000	1,650	15,000	0.751	11,300	38,700	0.402	15,570 (minimum)
4	2,300	13,000	5,000	20,300	0.683	13,900	52,600	0.315	17,800
5	100	17,000	7,000	24,100	0.621	15,000	67,600	0.264	17,850

Example 8-5

As an additional example of calculation of economic life without consideration of income taxes, suppose a certain machine has an installed price of $10,000 and the projected year-by-year operating costs and market values shown in the table below.

Year	Total annual inherent and relative operating cost	Market value
1	$3,000	$6,000
2	3,500	3,000
3	8,000	1,000

Neglecting income taxes and assuming an interest rate of 0%, calculations for determining the most economic replacement interval are as follows.

AC (for 1-year interval):

$$(\$10,000 - \$6,000)(A/P, 0\%, 1) + \$6,000(0\%) + \$3,000 = \$7,000$$

AC (for 2-year interval):

$$(\$10,000 - \$3,000)(A/P, 0\%, 2) + \$3,000(0\%)$$
$$+ [\$3,000 + 3,500(P/F, 0\%, 2)](A/P, 0\%, 2) = \$6,750$$

AC (for 3-year interval):

$$(\$10,000 - \$1,000)(A/P, 0\%, 3) + (\$1,000)(0\%) + [\$3,000(P/F, 0\%, 1)$$
$$+ \$3,500(P/F, 0\%, 2) + \$8,000(P/F, 0\%, 3)](A/P, 0\%, 3) = \$7,833$$

Thus, replacement at two-year intervals is apparently slightly more economical than one-year or three-year intervals. ∎

8.10 Calculation of Remaining Economic Life—Existing Asset

In this section we turn our attention to determining the economic life of a defender asset.

Example 8-6

As an example of replacement analysis involving determination of the optimal remaining life of an old asset, suppose the replacement of a spray system is being considered by the Hokie Metal Stamping Company. The new improved system will cost $60,000 installed and will have an estimated economic life of 12 years and a $6,000 market value. Further, it is estimated that annual operating and maintenance costs will average $32,000 per year for the new system and that straight-line depreciation will be used. The present system has a book value of $12,000 and a present realizable market value of $8,000. Its estimated costs and market and book values for the next three years are as follows:

Year	Market value at end of year	Book value at end of year	Inherent and relative operating costs during year
1	$6,000	$9,000	$40,000
2	5,000	6,000	50,000
3	4,000	3,000	60,000

Table 8-5 shows calculations to determine relevant after-tax cash flows for this problem. It is assumed that the ordinary income tax rate is 50% and that any gain or loss on disposal of the old asset affects income taxes at the full ordinary rate.

After-tax annual cost calculations for each alternative utilizing the results in column (6) of Table 8-5 (with the signs reversed) and a 15% minimum after-tax rate of return are shown below.

New system:

$(\$60,000 - \$6,000)(A/P, 15\%, 12) + \$6,000(15\%)$

$$+ \$13,750 = \$24,613$$

Old system, keep one year:

$(\$10,000 - \$7,500)(A/P, 15\%, 1) + \$7,500(15\%)$

$$+ \$18,500 = \$22,500$$

Old system, keep two years:

$(\$10,000 - \$5,500)(A/P, 15\%, 2) + \$5,500(15\%)$

$$+ [\$18,500(P/F, 15\%, 1)$$

$$+ \$23,500(P/F, 15\%, 2)](A/P, 15\%, 2) = \$24,418$$

Old system, keep three years:

$(\$10,000 - \$3,500)(A/P, 15\%, 3) + \$3,500(15\%)$

$$+ [\$18,500(P/F, 15\%, 1) + \$23,500(P/F, 15\%, 2)$$

$$+ \$28,500(P/F, 15\%, 3)](A/P, 15\%, 3) = \$26,408$$

On the basis of the above analysis, one would tend to say that the old system should be kept at least one more year. However, in this situation one should examine marginal costs. The valid economic criterion when operating costs are increasing over time is to keep the old system as long as the marginal cost of an additional year of service is less than the equivalent annual cost of the new system. The marginal cost of keeping the old system for the first year is the $22,500 previously computed. This $22,500 is less than the $24,613 average annual cost of the new system, thus justifying keeping the old system for the first year.

The marginal cost of keeping the old system for the second year is ($7,500 – $5,500) + 7,500(15%) + $23,500 = $26,625. (*Note:* These numbers are from Table 8-5, column 6, for the difference between "keep 2 years" and "keep 1 year" and ordered to reflect depreciation, interest, and operating costs, respectively.) The $26,625 is slightly greater than the $24,613 average

TABLE 8-5 Calculation of After-Tax Cash Flows for Example Replacement Analysis

(1) Year	(2) BTCF	(3) Depreciation deduction	(4) = (2) + (3) Taxable income	(5) = −(4) × 50% Cash flow for income taxes	(6) = (2) + (5) ATCF
New system					
0	− $60,000				− $60,000
1–12	− 32,000	−$4,500	− $36,500	+ 18,250	− 13,750
12	+ 6,000				+ 6,000
Old system, keep 1 yr					
0	(−)$ 8,000		(−)($ 8,000 − $12,000)	(−)$ 2,000	(−)$10,000
1	− 40,000	−$3,000	− 43,000	+ 21,500	− 18,500
1	+ 6,000		+ 6,000 − 9,000	+ 1,500	+ 7,500
Old system, keep 2 yr					
0	(−)$ 8,000		(−)($ 8,000 − $12,000)	(−)$ 2,000	(−)$10,000
1	− 40,000	−$3,000	− 43,000	+ 21,500	− 18,500
2	− 50,000	− 3,000	− 53,000	+ 26,500	− 23,500
2	+ 5,000		+ 5,000 − 6,000	+ 500	+ 5,500
Old system, keep 3 yr					
0	(−)$ 8,000		(−)($ 8,000 − $12,000)	(−)$ 2,000	(−)$10,000
1	− 40,000	−$3,000	− 43,000	+ 21,500	− 18,500
2	− 50,000	− 3,000	− 53,000	+ 26,500	− 23,500
3	− 60,000	− 3,000	− 63,000	+ 31,500	− 28,500
3	+ 4,000		+ 4,000 − 3,000	− 500	+ 3,500

Note: Negative signs in parentheses represent the result of "opportunity foregone"—i.e., if the old system were sold, a certain cash flow would result, but by keeping it the opportunity for the cash flow is foregone; hence, the reversal of cash flow sign.

annual cost of the new system, thus indicating that the old system should not be kept the second year, but rather that it be replaced at the end of the first year. (Incidentally, the reader may want to see that the marginal cost of keeping the old system for the third year is ($5,500 – $3,500) + $5,500(15%) + $28,500 = $31,525.)   ∎

The preceding example assumes that there is only one new asset (challenger) alternative available and that the net annual cost of the new asset will be the same regardless of when it replaces the defender. It shows the general relationship that if the old asset (defender) is retained beyond the break-even point, its costs continue to grow and replacement becomes more urgent, as illustrated in Fig. 8-1.

Figure 8-2 illustrates the effect of improved new challengers in the future. If an improved challenger X becomes available before replacement with the new asset of Fig. 8-1, then a new replacement study probably should take place to consider that improved challenger. If there is a possibility of a further improved challenger Y as of, say, four years later, it may be still better to postpone replacement until that challenger becomes available. Thus, retention of the old asset beyond the break-even point has a cost that may well grow with time, but this cost of waiting can, in some instances, be worthwhile if it permits purchase of an improved asset having economies that offset the cost of waiting. Of course, a decision to postpone a replacement may also "buy time and information." Because technological change tends to be sudden and dramatic rather than uniform and gradual, new challengers with significantly improved features can arise sporadically and can change replacement plans substantially.

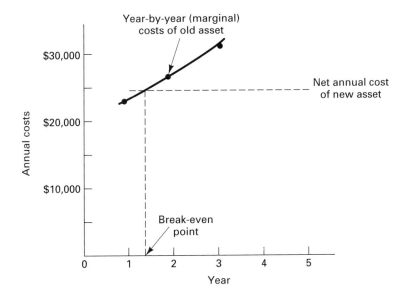

Figure 8-1   Old versus new asset costs.

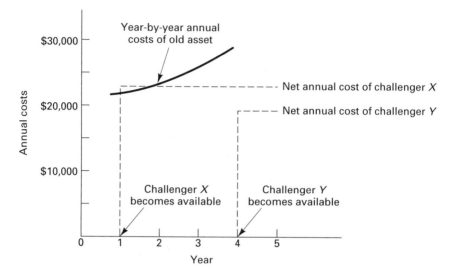

Figure 8-2 Old versus new asset costs with improved challengers becoming available in future.

8.11 Intangibles in Replacement Problems

As in other management decisions, replacement problems have intangible aspects that may have sufficient weight to control the decision. Intangibles in replacement analyses may include competing needs for capital, future uncertainties in market and costs, possibilities of product change, financial condition of the business, attitudes and limitations of personnel, labor shortage or surplus, ethical and social problems, and so forth. Intangibles are notably reflected in two of the quantitative factors or elements used in the previous illustrations, namely the study period and the minimum required rate of return on investment.

8.12 Replacement of Assets Subject to Sudden Failure

Up to this point, we have considered the replacement of assets that deteriorate gradually over time. In this section assets that are subject to sudden failure are modeled.

Some assets, such as light bulbs, electronic parts, and missiles, do not exhibit significant deterioration in capabilities over time, but they are usually subject to increasing opportunities for failure as usage or age increases. The usual objective in this type of problem is to determine the amount and timing of replacement (or maintenance) of such assets. Three types of problems are as follows:

1. Should a group of such assets be replaced in entirety, or should they be individually replaced upon failure?

   2. If a group replacement is the best policy, what is the optimum group-replacement interval?

   3. How much and how often should preventive maintenance be performed?

   To solve the first type of problem, the cost of group replacement can be compared with that of individual replacement upon failure. Whenever group replacement can lead to reduced costs (through labor savings, materials discounts, etc.), the optimum group-replacement interval can be computed using calculus or arithmetic approximation methods. The same methods may be used to solve preventive-maintenance problems. Group-replacement and preventive-maintenance interval problems can also be solved by dynamic programming.

8.12.1 An Illustrative Model

   Let us assume that failures occur only at the end of a period. Thus, replacements of failures that occur at the end of, say, the third period will be age zero at the beginning of the fourth period. During the first $(N - 1)$ time intervals, all failures are replaced as indicated in the foregoing. At the end of the Nth time interval, all units are replaced regardless of their ages. The problem is to find the value of N that minimizes total cost per period. If it is assumed that the entire replacement interval in question is of short duration such that the time value of money can be neglected, then the total cost from time of group installation until the end of N periods is given by

$$K(N) = QC_1 + C_2 \sum_{x=1}^{N-1} f(x), \qquad (8\text{-}1)$$

where

$$\begin{aligned}
K(N) &= \text{expected total cost for } N \text{ periods,}\\
C_1 &= \text{unit cost of replacement in a group,}\\
C_2 &= \text{unit cost of individual replacement after failure,}\\
f(x) &= \text{expected number of failures in the } x\text{th period,}\\
Q &= \text{number of units in the group.}
\end{aligned}$$

The objective is to find the value of N that minimizes $K(N)/N$. If an individual unit fails during the jth period with probability $P(j)$, then $f(x)$ is obtained as follows:

$$f(1) = QP(1)$$
$$f(2) = QP(2) + f(1)P(1)$$
$$f(3) = QP(3) + f(1)P(2) + f(2)P(1)$$
$$\vdots$$
$$f(x) = QP(x) + f(1)P(x - 1) + f(2)P(x - 2) + \cdots + f(x - 1)P(1)$$

or

$$f(x) = QP(x) + \sum_{k=1}^{x-1} f(k)P(x - k). \qquad (8\text{-}2)$$

Equation 8-2 is based on the assumption that failures occur independently. If the failure of a unit affects the probability of another unit's failing, then $f(x)$ cannot be computed using Eq. 8-2.

Costs are minimized for a policy of group replacing after $\hat{N}$ periods if

$$\frac{K(\hat{N})}{\hat{N}} \leq \frac{K(\hat{N}+1)}{\hat{N}+1} \quad \text{and if} \quad \frac{K(\hat{N})}{\hat{N}} \leq \frac{K(\hat{N}-1)}{\hat{N}-1}. \tag{8-3}$$

Considering the first condition, from Eq. 8-1 it is seen that

$$\frac{QC_1}{\hat{N}} + \frac{C_2 \sum\limits_{x=1}^{\hat{N}-1} f(x)}{\hat{N}} \leq \frac{QC_1}{\hat{N}+1} + \frac{C_2 \sum\limits_{x=1}^{\hat{N}} f(x)}{\hat{N}+1}$$

reduces to

$$QC_1(\hat{N}+1) + C_2(\hat{N}+1)\sum_{x=1}^{\hat{N}-1} f(x) \leq QC_1\hat{N} + C_2\hat{N}\sum_{x=1}^{\hat{N}-1} f(x) + C_2\hat{N}f(\hat{N}).$$

Cancelling like terms on both sides of the inequality gives

$$QC_1 + C_2 \sum_{x=1}^{\hat{N}-1} f(x) \leq C_2\hat{N}f(\hat{N}),$$

or

$$\frac{K(\hat{N})}{\hat{N}} \leq C_2 f(\hat{N}). \tag{8-4}$$

Similarly, the second condition gives

$$\frac{K(\hat{N}-1)}{(\hat{N}-1)} \geq C_2 f(\hat{N}-1). \tag{8-5}$$

Hence, the following interpretation is obtained: *Group replacement should be performed at the end of period* $\hat{N}$ *if the average cost per period to date is less than the cost of individual replacements during the period. So long as the average cost per period is greater than the cost of replacing individual units during the period, do not group replace.*

If $\hat{N}$ is the optimum group-replacement interval, then $K(\hat{N})/\hat{N}$ should be compared with the expected cost of the policy of replacing only individual units. Namely, group replacement should occur if and only if

$$\frac{K(\hat{N})}{\hat{N}} < \frac{C_2 Q}{E[N]},$$

where $E[N]$ is the expected service life of an individual unit, i.e.,

$$E[N] = \sum_{N=1}^{\infty} NP(N).$$

Example 8-7

As an illustration of the calculations of the optimal group-replacement interval for a situation that fits the model represented by Eq. 8-1, suppose that for a group of 10,000 electronic parts subject to sudden failure the unit cost of group replacement is \$0.50 and the unit cost of individual replacement is \$2.00. Further, the failure probabilities are as shown in Table 8-6. Table 8-7 shows the calculations for $f(x)$. As shown in Table 8-8, the lowest cost is obtained if group replacement is performed every three periods. The expected cost per period with individual replacements alone is found to be

$$\frac{C_2 Q}{E[N]} = \frac{\$2(10,000)}{4} = \$5,000 > \$2,683.33,$$

where, from Table 8-6, $E[N] = 1(0.05) + \cdots + 7(0.05) = 4$. Therefore, the group-replacement policy is an optimum policy.

Table 8-6 Failure
Probabilities

j	$P(j)$
1	0.05
2	0.10
3	0.20
4	0.30
5	0.20
6	0.10
7	0.05

TABLE 8-7 Computation of $f(x)$ Values

x	$f(x)$	
1	500	= 500.00
2	1,000 + 25	= 1,025.00
3	2,000 + 50 + 51.25	= 2,101.25
4	3,000 + 100 + 102.50 + 105.06	= 3,307.56
5	2,000 + 150 + 205.00 + 210.12 + 165.38	= 2,730.50
6	1,000 + 100 + 307.50 + 420.25 + 330.76 + 136.52	= 2,295.03

TABLE 8-8 Determination of the Optimum
Group-Replacement Interval

N	$\sum_{x=1}^{N-1} f(x)$	$K(N)$	$K(N)/N$
1	\$0	\$5,000.00	\$5,000.00
2	500.00	6,000.00	3,000.00
3	1,525.00	8,050.00	2,683.33
4	3,626.25	12,252.50	3,063.13
5	6,933.81	18,867.62	3,773.52
6	9,664.31	24,328.62	4,054.77

In a number of cases the values of $P(j)$ are not available. Rather, failure data are collected such that $f(x)$ values are obtained directly. This situation is commonly encountered when a group-replacement policy is already in use.

Example 8-8

As an illustration of the determination of the optimum group-replacement interval when $f(x)$ values are given directly, consider a situation involving 50 cutting tools on a number of numerically controlled milling machines. Data collected on the number of replacements of cutting tools per day are given in Table 8-9. A unit cost of $10.00 occurs when group replacement is performed; a unit cost of $50.00 applies to individual replacement. The calculations shown in Table 8-10 indicate the optimum replacement interval is 3 days. (Notice that there are two local minima at $N = 3$ and $N = 8$.)

TABLE 8-9 Failure Data
for the Example

x	$f(x)$	x	$f(x)$
1	2	6	7
2	4	7	6
3	7	8	9
4	7	9	7
5	8	10	8

TABLE 8-10 Computation of the Optimum
Replacement Interval

N	$\sum\limits_{x=1}^{N-1} f(x)$	$K(N)$	$K(N)/N$
1	$0	$500	$500
2	2	600	300
3	6	800	267
4	13	1,150	288
5	20	1,500	300
6	28	1,900	317
7	35	2,250	321
8	41	2,550	319
9	50	3,000	333
10	57	3,350	335

■

8.13 Summary

The early part of this chapter illustrated replacement analyses concerned with whether and when to replace existing equipment for commonly assumed constant conditions. The latter part of the chapter provides only a brief sampling of types of models for determination, in advance, of economical replacement intervals under various circumstances. Models such as these should be used only when the inherent assumptions mesh sufficiently closely with the estimated situation under study. In general, the closer such models approach reality, the more complex the construction and use of those models. The greatest strength in using such models is that they tend to force the analyst to examine objectively the nature of the estimates and assumptions that must be made in the process of analysis for decision making.

REFERENCES

BERNHARD, R. H., "Improving the Economic Logic Underlying Replacement Age Decisions for Municipal Garbage Trucks: Case Study," *The Engineering Economist* 35, no. 2 (Winter 1990): 129–147.

LAKE, D. H., AND A. P. MUHLEMANN, "An Equipment Replacement Problem," *Journal of the Operational Research Society* 30, no. 5 (1979): 405–411.

LEUNG, L. C., AND J. M. A. TANCHOCO, "Multiple Machine Replacement within an Integrated Systems Framework," *The Engineering Economist* 32, no. 2 (1987): 89–114.

MATSUO, H., "A Modified Approach to the Replacement of an Existing Asset," *The Engineering Economist* 33, no. 2 (Winter 1988): 109–120.

OAKFORD, R. V., J. R. LOHMANN, AND A. SALAZAR, "A Dynamic Replacement Economy Decision Model," *IIE Transactions* 16, no. 1 (1984): 65–72.

PROBLEMS

8-1. A firm is contemplating replacing a computer it purchased 3 years ago for $450,000. Operating and maintenance costs have been $85,000 per year. Currently, the computer has a trade-in value of $300,000 toward a new computer that costs $650,000 and has a life of 5 years, with a value of $200,000 at that time. The new computer will have annual operating and maintenance costs of $80,000.

If the current computer is retained, another small computer will have to be purchased in order to provide the required computing capacity. The smaller computer will cost $300,000, have a value of $50,000 in 5 years, and have annual operating and maintenance costs of $55,000.

Using a before-tax analysis, with a minimum attractive rate of return of 30%, determine the preferred course of action.

8-2. A firm owns a pressure vessel that it is contemplating replacing. The old pressure vessel has annual operating and maintenance costs of $60,000 per year, and it can be kept for 5 more years, at which time it will have zero salvage value.

The old pressure vessel can be traded in on a new one; the trade-in value is $30,000; the purchase price for the new pressure vessel is $120,000. The new pressure vessel will have a market value of $50,000 in 5 years and will have annual operating and maintenance costs of $30,000 per year. Using a minimum attractive rate of return of 20% and a before-tax analysis, determine whether or not the pressure vessel should be replaced.

8-3. A building supplies distributor purchased a gasoline-powered forklift truck 5 years ago for $10,000. At that time, the estimated useful life was 10 years with a market value of $1,000. The truck can now be sold for $2,500. For the old truck, annual average operating expenses for year j have been in accordance with

$$C_j = \$2,000(1.10)^{j-1}.$$

The distributor if considering replacing the lift truck with a smaller, battery-powered truck costing $8,000. The estimated life is 10 years with the market value decreasing by $600 each year. Annual average operating expenses are expected to be $2,000. If a MARR of 10% is assumed and a 5-year planning horizon is adopted, should the replacement be made?

8-4. The Ajax Corp. has an overhead crane that has an estimated remaining life of ten years. The crane can be sold for $8,000. If the crane is kept in service, it must be overhauled immediately at a cost of $4,000. Operating and maintenance costs will be $3,000 per year after the crane is overhauled. After the crane is overhauled, it will have zero market value at the end of the ten-year period. A new crane will cost $18,000, will last for 10 years, and will have a $4,000 market value at that time. Operating and maintenance costs are $1,000 for the new crane. The company uses an interest rate of 10% in evaluating investment alternatives. Should the company replace the old crane?

8-5. A firm is considering replacing a compressor that was purchased 4 years ago for $50,000. Currently, the compressor has a book value of $30,000 based on straight-line depreciation. If the compressor is retained, it will probably be used for 4 more years, at which time it is estimated to have a market value of $10,000. If the old compressor is retained for 8 more years, it is estimated that its market value will be negligible at that time. Operating and maintenance costs for the compressor have been increasing at a rate of $1,000 per year, with the cost during the past year being $8,000.

A new compressor can be purchased for $60,000. It is estimated to have uniform annual operating and maintenance costs of $8,000 per year. The market value for the compressor is estimated to be $30,000 after 4 years and $15,000 after 8 years. If a new compressor is purchased, the old compressor will be traded in for $20,000.

Using an after-tax analysis with straight-line depreciation and a MARR of 10%, determine the preferred alternative using a planning horizon of (a) 4 years and (b) 8 years. Assume a 40% ordinary income tax rate and a 40% capital gain or loss tax rate.

8-6. A new numerically controlled drill press can be used to replace three manual drill presses in use. The N/C drill press will cost $55,000, will last 10 years and have a negligible market value, and will have annual operating and maintenance costs of $10,000 per year.

The manual drill presses were purchased 5 years ago for $11,000 each; they were estimated to have lives of 10 years, with negligible market values. Currently, the manual drill presses are worth $1,500 each. Annual operating and maintenance costs are $6,000 per drill press.

A study period of 5 years is to be used in the replacement study. A market value of $15,000 is estimated for the N/C drill press at the end of the study period. An after-tax MARR of 15% and a tax rate of 40% are to be used. Any capital losses will be assumed to be deducted from taxable income in the year they occur. MACRS depreciation with a 5-year recovery period is to be used for both alternatives. What action do you recommend?

8-7. Machine X has been used for 10 years and currently has a book value of $20,000. A decision must be made concerning the most economic action to take: keep X; replace X with Y; or replace X with Z.

If machine X is continued in service, it can be used for 6 years and scrapped at zero value. Annual operating and maintenance costs will equal $95,000.

If machine X is replaced with machine Y, a trade-in allowance of $25,000 will be provided for X. The original purchase price for Y, excluding the trade-in allowance, is $120,000. At the end of the 6-year planning horizon, Y will have a market value of $30,000. Annual operating and maintenance costs will total $80,000.

If machine X is replaced with machine Z, no trade-in allowance will be provided for X. The purchase price for Z is $150,000. At the end of the 6-year horizon, Z will have a market value of $50,000. Annual operating and maintenance costs will total $60,000.

Using a MARR of 15% and a before-tax analysis, determine the preferred course of action.

8-8. An economic life analysis has been performed on a challenger and it was found to have the total equivalent uniform annual cost (EUAC) values shown below. A defender can be sold for $15,000 today, $3,000 in one year, and $1,500 in two years. If kept, the defender will have operating and maintenance costs of $7,500 next year, and $9,000 the following year. Should the defender be replaced based on a before-tax study? The MARR is 20%.

N	EUAC
1	$21,760
2	21,153
3	20,833
4	19,185
5	19,684

8-9. A man owns a side business he purchased 10 years ago for $46,000. Straight-line depreciation has been charged assuming a 25-year life and $6,000 estimated salvage value. He now has an offer to sell the business for $50,000. He estimates that if he does not sell now, he will hold the property for another 7 years and sell it at that time for $45,000. If he keeps the business, he estimates he will pay taxes at the rate of 40%. (Any capital gain or capital loss will affect taxes at the rate of 40%. The business will have an annual net cash flow of $6,600, from which $1,600 depreciation is deducted.)

a. Compute the before-tax rate of return from continued ownership of the side business.

b. If he can get 8% after taxes from the use of his capital in some other venture of comparable risk, should he sell now, or wait 7 years? Compute the after-tax internal rate of return on continued ownership.

8-10. You have a machine that cost $30,000 two years ago; it has a present MV of $5,000. Operating costs total $2,000 per year as long as the machine is in use. In two more years it will no longer be useful, and you can sell it for $500 scrap value. You are considering replacing this machine with a new model incorporating the latest technology; this new model will cost you $20,000 now. It has an annual operating cost of $1,000 with a useful life of 8 years and negligible salvage value. If you delay the purchase of the new model, the cost will be $24,000 because of the installation difficulties that do not exist now. When your present machine is retired, you have no hopes of getting another one like it, since it was a very limited model. If money is worth 10% before taxes, what should you do?

8-11. A construction firm currently owns a heavy-duty tractor that has a present MV of $80,000. Estimates of the tractor's O&M expenses and MV at the end of each of the remaining six years of life are as follows:

	\multicolumn{6}{c}{End of year k}
	1	2	3	4	5	6
O&M expenses:	$20,000	$25,000	$38,000	$45,000	$47,000	$50,000
MV:	70,000	60,000	50,000	40,000	30,000	20,000

The firm is considering a new heavy-duty tractor to replace the one presently owned. The new tractor's purchase price is $220,000, and its estimated O&M and MV for each of the next 6 years of the study period are these:

	\multicolumn{6}{c}{End of year k}
	1	2	3	4	5	6
O&M expenses:	$10,000	$12,000	$16,000	$17,000	$20,000	$22,000
MV:	180,000	150,000	120,000	100,000	90,000	85,000

If the MARR = 0%, should the new tractor be purchased? If so, when?

8-12. A 3-year-old asset that was originally purchased for $4,500 is being considered for replacement. The new asset under consideration would cost $6,000. The engineering department has made the following estimates of the O&M costs of the two alternatives.

Year	Old asset	New asset
1	$2,000	$ 500
2	2,400	1,500
3	—	2,500
4	—	3,500
5	—	4,500

The dealer has agreed to place a $2,000 market value on the old asset if the new one is purchased now. It is estimated that the residual value for either of the assets will be zero at any time in the future. If the before-tax MARR is 12%, make an equivalent AC analysis of this situation and recommend which course of action should be taken now. Use the repeatability assumption and give a short written explanation of your answer.

8-13. Use the PW method to select the better alternative shown below. Assume that the defender was installed in 1989 and that its MACRS recovery period is 7 years. The time of this comparison is 1994. Let the MARR equal 10% after taxes, and let the effective income tax rate be 40%. Capital gains (losses) are taxed at 40%. The firm is known to be profitable in its overall operation.

Data	Alternative A	Alternative B
Labor per year	$300,000	$250,000
Material cost per year	$250,000	$100,000
Insurance and property taxes per year	4% of first cost	None
Maintenance per year	$8,000	None
Rental cost per year	None	$100,000

Alternative A: Retain an already owned machine in service for 8 more years.
Alternative B: Sell the old machine and rent a new one for 8 years.

Alternative A:

Cost of old machine 5 years ago = $500,000.

BV now = $111,500.

Depreciation with MACRS.

Estimated market value at end of useful life = $50,000.

Present MV = $150,000.

8-14. a. Find the economic life of an asset having the following projected cash flows. The MARR is 0% per year.

Initial investment = $5,000.

MV = 0 (at all times).

Operating costs = $3,000 (EOY 1),
                    $4,000 (EOY 2),
                    $5,000 (EOY 3),
                    and $6,000 (EOY 4).

[Answer: $N^* = 3$ years]

b. Find the economic life of another asset having these cash flow estimates. The MARR is 12% per year.

Initial investment = $10,000.

MV = $10,000 (at all times).

Operating costs = $3,000 (EOY 1),
                    $4,000 (EOY 2),
                    $5,000 (EOY 3),
                    and $6,000 (EOY 4).

[Answer: $N^* = 1$ year]

c. Repeat (b), except that MV = $0 at all times.

[Answer: $N^* = 4$ years]

8-15. Consider a piece of equipment that initially costs $8,000 and has these estimated annual operating expenses and market values:

End of year	Operating expenses for the year	MV at end of year
1	$3,000	$4,700
2	3,000	3,200
3	3,500	2,200
4	4,000	1,450
5	4,500	950
6	5,520	600
7	6,250	300
8	7,750	0

If the interest rate is 8% per year, before taxes, determine the most economical time to replace this equipment.

8-16. A manufacturer moves pallets of materials with a forklift truck. He has consistently used the same make and model of forklift over the past several years. Their purchase price has been relatively constant at $8,000 each. Records of O&M indicate average expenses per year as a function of age of the vehicle as shown in the following table. The market values (MVs) of the forklift are reasonably well known as a function of age. Find the best time to replace a forklift truck if MARR = 15% per year.

Year	1	2	3	4	5
Operating expense (O&M):	$3,000	$3,000	$3,500	$4,000	$4,500
MV:	6,000	5,000	4,000	2,500	1,250

8-17. An existing machine (defender) has a current market value of $4,000, and its future market values are expected to be negligible. The projected operating and maintenance costs are given by the following cash flow diagram.

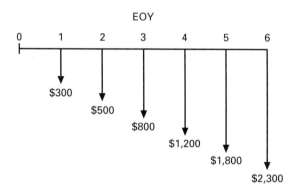

a. If the MARR is 0%, what is the before-tax economic life of this machine?

b. If the MARR is 15%, what is the before-tax economic life of this machine?

8-18. Suppose there are 1,000 identical independent electronic components, the failure rate of which as a function of days of use, t, is $f(t) = 50 + 60t$. The unit cost of replacement of the entire group is $0.30, while the unit cost of individual replacement after failure is $2.00. Find the most economical replacement interval.

8-19. The probability that a certain critical machine component will last t hours is given by

		t		
	20	40	60	80
$P(t)$	0.90	0.70	0.45	0.15

If the component fails during production, the machine is stopped and a cost of $400 is incurred for downtime and repairs. Preventive replacement of the component costs approximately $50. Find the replacement policy that will minimize the total expected costs.

8-20. The manufacturer of "Surefire" solenoids has life-tested 1,000 of these devices and has released the information:

Life (in days)	0–20	21–40	41–60
Cumulative no. of failures	500	800	1,000
Probability of failure	0.5	0.3	0.2

Suppose your company has 100 of these solenoids in service at the present time. Management has specified that all failures must be replaced as soon as they are detected. If the solenoid fails in service the cost of individual replacement is $100, but if the device is replaced before it fails (e.g., group replacement) the unit cost is only $60.

  a. Determine the cost of a "fix as they go bad" replacement policy.

  b. If a group-replacement strategy is justified, what time interval between group replacements should be recommended to management?

  c. What assumptions did you make in (a) and (b)?

8-21. The probability that a certain personal computer component will last x hours is given by

x	100	200	300	400	500
Survival rate	0.90	0.80	0.60	0.45	0.15

If this component fails when the computer is in use, the contents of the hard drive will be erased (lost). In this case, the cost of restoring the hard drive to a functional status is $1,000. However, preventive maintenance of the component costs only $100. What preventive-maintenance replacement policy should the computer manufacturer recommend to minimize total expected costs? Assume there are thousands of components in service.

8-22. Refer to Section 8.12.1 and find N under these conditions:

$$Q = 1,000,$$
$$f(x) = 10 + 15x,$$
$$C_1 = \$20.00,$$
$$C_2 = \$100.00.$$

9
Capital Planning and Budgeting

9.1 Introduction

Proper capital planning, budgeting, and management represent the basic top management function of an enterprise and are crucial to its welfare. *Capital budgeting* is commonly understood to be a function that takes place at the highest levels of management, such as office of the controller or corporate executive committee, but it should be recognized that decisions made at the lower levels in the management hierarchy directly affect those proposals that are ultimately considered as contenders in the overall capital budget. For example, before a major project is considered in top management's capital budget, usually many subalternatives of design and specification will have been considered and the related decisions virtually made as part of the recommended project package. Appropriate procedures for evaluation of these subalternatives should be available and uniformly applied to ensure that economic consequences are considered at all levels.[1]

Capital budgeting may be defined as the series of decisions by individual economic units as to how much and where resources will be obtained and expended for future use, particularly in the production of future goods and services. The scope of capital budgeting encompasses

1. how the money is acquired and from what sources,
2. how individual capital project opportunities (and combination of opportunities) are identified and evaluated,
3. how minimum requirements of acceptability are set,

[1] Mathematical programming formulations of capital budgeting problems are dealt with in Chapter 16.

4. how final project selections are made, and

5. how postmortem reviews are conducted.

These facets of capital budgeting are highly interrelated and will be discussed in turn in the next several sections.

9.2 Sources of Funds

The determination of how much capital (and from what sources) should be applied toward project investments is a function of the individual project investment require-ments; prospective profitability, type, and risk; the amount, conditions, and prices of funds to be obtained from internal or external sources; and the firm's financial policies and condition.

Most investment funds are obtained from internal sources; retained earnings and reinvested depreciation reserves. When outside sources are used, it can be to the disadvantage of the present owners to the extent that present ownership control is diluted, as in the case of new equity stock issuance; or to the extent that the company is burdened with new fixed monetary obligations and operating restrictions, as in the case of long-term borrowing. However, outside sources are often used when it is judged to be in the best interests of the existing stockholders.

In general, the more attractive the investment proposal available, the more the company will be willing to go to outside sources to obtain capital in order to take advantage of more investments. However, this has to be balanced against the cost of obtaining outside capital. The more outside capital the company obtains by borrow-ing, the higher the cost—in terms of both interest and risk—is likely to be.

9.3 Identification and Evaluation of Opportunities ✓

All levels of the organization—operating staff, supervisory, and engineering, as well as top management—should be encouraged to develop proposals for capital invest-ment projects. For example, the research section may discover new products and pro-cesses. The engineering section may create improved designs in product, packaging, or methods. The manufacturing section may propose the installation of more efficient facilities. The marketing section may propose programs of advertising, sales, or inventory expansion for the development of new markets or expansion of existing ones. Finally, top management may, for example, develop plans for major acquisi-tions leading to integration or diversification.

The importance of identifying all opportunities reasonably worth consideration can hardly be overemphasized. It doesn't matter how thorough and accurate are the evaluations and final selection procedures of a firm; if a project(s) that would have been superior to others accepted is (are) never even considered, then the firm will have suboptimized. That is, the firm will have invested in the project(s) considered best, but it will have given up the opportunity to have generated even greater benefits that could have been obtained had the limited capital been invested in the superior opportunity that ". . . might have been."

A dearth of good investment proposals within a firm indicates that the firm lacks a healthy climate for encouraging the search for investment opportunities. This climate should exist in order for the firm to create the best economic opportunities for itself. Indeed, the development of good investment proposals can even become a question of the firm's survival.

Evaluation of opportunities is undertaken to determine which of the opportunities is (are) best and, sometimes, also to determine whether the "best" is (are) good enough. This book has been concerned primarily with methods of evaluation on the basis of monetary criteria, supposedly leading to profit maximization or cost minimization. But it must be recognized again that the objectives of a firm are not necessarily solely, or even dominantly, based on monetary criteria.

9.4 Minimum Requirements of Acceptability

The determination of the *minimum acceptable rate of return* (MARR), sometimes also called *cost of capital,* for the project proposals of a firm is generally controversial and difficult. From a purely monetary viewpoint this minimum rate of return should be selected to maximize the economic well-being of the present owners. The outward manifestation of this viewpoint is that an investment should be undertaken as long as the present value of the existing owner's equity in the firm is enhanced. Even with agreement on this, there are many viewpoints on just how the minimum rate of return should be determined. Several of these viewpoints will be discussed ahead.

An easy-to-compute method for determining what is alleged to be a "minimum rate of return" is to determine the rate of cost of each source of funds and to weight these by the proportion that each source constitutes of the total. For example, if one-third of a firm's capital is borrowed at 6% and the remainder of its capital is equity earning 12%, then the alleged minimum rate of return is $\frac{1}{3} \times 6\% + \frac{2}{3} \times 12\% = 10\%$.

Another school of thought maintains that if particular projects are to be undertaken using borrowed funds, then the minimum rate of return should be based on the rate of cost of those borrowed funds alone. Yet another school of thought, as exemplified by Solomon,[2] maintains that the minimum rate of return should be based on the cost of equity funds alone, on the grounds that firms tend to adjust their capitalization structure to the point at which the real costs of new debt and new equity capital are equal.

Great stimulation to thinking on the determination of the minimum rate of return (i.e. cost of capital) was caused by Modigliani and Miller.[3] They developed a theory that essentially asserts that the average cost of capital to any firm is completely independent of its capital structure and is strictly the capitalization rate of future equity earnings. Since Modigliani and Miller's article, there have been many articles criticizing their contention on the grounds of oversimplification and unfounded postulation.[4]

---

[2] Ezra Solomon, *The Management of Corporate Capital* (New York: The Free Press, 1959), p. 136.

[3] Franco Modigliani and Menton H. Miller, "The Cost of Capital, Corporation Finance, and the Theory of Investment," *American Economic Review* (June 1958):261–297.

[4] See principally D. Durand, "The Cost of Capital in an Imperfect Market: A Reply to Modigliani and Miller," *American Economic Review* (Sept. 1959):639–655.

There has been no clearcut settlement of this issue, and the problem of determining the cost of capital is still one of open controversy in both theory and practice.

Another viewpoint on the determination of the minimum rate of return commonly overlooked (and which we feel is most sound) is the *opportunity cost* viewpoint; it comes as a direct result of the phenomenon of "capital rationing." *Capital rationing* describes what is necessary when there is a limitation of funds relative to prospective proposals to use the funds. This limitation may be either internally or externally imposed. Its parameter is often expressed as a fixed sum of capital; but when the prospective returns from investment proposals together with the fixed sum of capital available to invest are known, then the parameter can be expressed as a minimum acceptable rate of return, or cut-off rate.

Ideally, the cost of capital by the opportunity cost principle can be determined by ranking prospective projects according to a ladder of profitability and then establishing a cut-off point where the capital is used on the better projects. The rate of return earned by the last project before the cut-off point is the cost of capital or minimum rate of return by the opportunity cost principle.

To illustrate the above, Fig. 9-1 ranks projects according to prospective rate of return and the cumulative investment required. For purposes of illustration, the amount of capital shown available is $4 million. By connecting up (to the next whole project within the $4 million) and across, one can read the minimum rate of return under the conditions, which turns out to be 25%.

It is not uncommon for firms to set two or more MARR levels according to risk categories. For example, one major industrial firm defines risk categories for income-producing projects and "normal" MARR standards for each as follows:

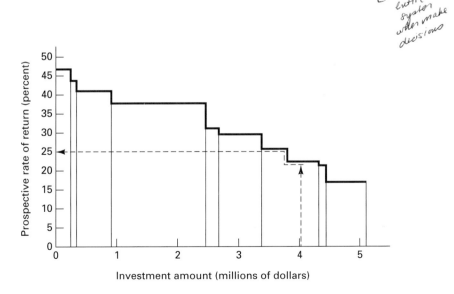

Figure 9-1   Schedule of prospective returns and investment amounts.

1. *High Risk* (MARR = 40%)
   New products
   New business
   Acquisitions
   Joint ventures
2. *Moderate Risk* (MARR = 25%)
   Capacity increase to meet forecasted sales
3. *Low Risk* (MARR = 15%)
   Cost improvements
   Make versus buy
   Capacity increase to meet existing orders

To illustrate how the above set of MARR standards could be determined, the firm could rank prospective projects in each risk category according to prospective rates of return and investment amounts. After tentatively deciding how much investment capital should be allocated to each risk category, the firm could then determine the MARR for each risk category as illustrated in Fig. 9-1 for a single category. Of course, the firm might reasonably shift its initial allocation of funds according to the opportunities available in each risk category, thereby affecting the MARR for each category allocated more or fewer investment funds.

✓In principle, it would be desirable for a firm to invest additional capital as long as the return from that capital were greater than the cost of obtaining that capital. In such a case, the opportunity cost would equal the marginal cost (in interest and/or stockholder returns) of the last capital used. In practice, however, the amount of capital actually invested is more limited due to risk and conservative money policies; thus, the opportunity cost is higher than the marginal cost of the capital.

It is often reasonable to argue that since one cannot truly know the opportunity cost for capital in any given period (such as a budget year), it is useful to proceed as if the MARR for the upcoming period were the same as in the previous period. In addition to the normal difficulty of projecting the profitability of future projects and the availability of capital, there also may be pressures to manipulate the standards of acceptability to permit the approval of favored, even if economically undesirable, projects or classes of projects.

9.5 Project Selection

To the extent that project proposals can be justified through profitability measures, the most common basis of selection is to choose those proposals that offer the highest prospective profitability subject to allowances for intangibles or nonmonetary considerations, risk considerations, and limitations on the availability of capital. If the minimum acceptable rate of return has been determined correctly, one can choose proposals according to the rate of return method, annual worth method, or present worth method.

For certain types of project proposals, monetary justification is not feasible—or at least any monetary return is of minor importance compared to intangible or nonmonetary considerations. These types of projects should require careful judgment and analysis, including how they fit in with long-range strategy and plans. Factor-weighting methods such as those in Chapter 19 are particularly suited to projects for which monetary justification is not feasible.

The capital budgeting concepts discussed in this chapter are based on the presumption that the projects under consideration are *not* mutually exclusive (i.e., the adoption of one does not preclude the adoption of others, except with regard to the availability of funds). Whenever projects are mutually exclusive, the alternative chosen should be based on justification through the incremental return on any incremental investment(s) as well as proper consideration of nonmonetary factors.

9.5.1 Classifying Investment Proposals

For purposes of study of investment proposals, there should be some system or systems of classification into logical, meaningful categories. Investment proposals have so many facets of objective, form, and competitive design that no one classification plan is adequate for all purposes. Several possible classification plans are:

1. according to the kinds and amounts of scarce resources used, such as equity capital, borrowed capital, available plant space, the time required of key personnel, etc.;

2. according to whether the investment is tactical or strategic: a *tactical investment* does not constitute a major departure from what the firm has been doing in the past and generally involves a relatively small amount of funds; _strategic investment_ decisions, on the other hand, may result in a major departure from what a firm has done in the past and may involve large sums of money;

3. according to the business activity involved, such as marketing, production, product line, warehousing, etc.;

4. according to priority, such as absolutely essential, necessary, economically desirable, or general improvement;

5. according to type of benefits expected to be received, such as increased profitability, reduced risk, community relations, employee benefits, etc.;

6. according to whether the investment involves facility replacement, facility expansion, or product improvement;

7. according to the way benefits from the proposed project are affected by other proposed projects; this is generally a most important classification consideration, for there quite often exist interrelationships or dependencies between pairs or groups of investment projects.

Of course, all of the above classification systems probably are not needed or desirable. As an example, one major corporation uses the following four major categories for higher management screening:

1. expanded facilities;

2. research and development;

3. improved facilities—for process improvement, cost savings, or quality improvement; *new business*

4. necessity—for service facilities, emergency replacements, or for the removal or avoidance of a hazard or nuisance. *Maintenance*

9.5.2 Degrees of Dependency Between Projects

Several main categories of dependency between projects are briefly defined in Table 9-1. Actually, the possible degrees of dependency between projects can be expressed as a continuum from "prerequisite" to "mutually exclusive," with the degrees "complement," "independent," and "substitute" between these extremes, as shown in Fig. 9-2.

In developing a project proposal to be submitted for review and approval, the sponsor should include whatever complementary projects seem desirable as part of a single package. Also, if a proposed project will be a partial substitute for any projects to which the firm is already committed or that are under consideration, this fact should be noted in the proposal.

In cases where choices involved in planning a proposed project are considered sufficiently important so that the final decision should be made by higher levels of management, the project proposal should be submitted in the form of a set of mutually exclusive alternatives. For example, if it is to be decided whether to move a plant to a new location and several alternative sites are possible, then separate proposals should be made for each site so as to facilitate the choice of which site, if any, should be chosen.

Whenever capital budgeting decisions involve several groups of mutually exclusive projects and independent projects to be considered within capital availability constraints, the mathematical programming models of Chapter 16 can be useful for selecting the optimal combination of projects.

TABLE 9-1 Degrees of Dependence Between Pairs of Projects

"If the results of the first project would _____ by acceptance of second project . . .	. . . then the second project is said to be _____ _____ the first project."	Example
be technically possible or would result in benefits only	a prerequisite of	Car stereo purchase feasible only with purchase of car
have increased benefits	a complement of	Additional hauling trucks more beneficial if automatic loader purchased
not be affected	independent of	A new engine lathe and a fence around the warehouse
have decreased benefits	a substitute for	A screw machine that would do part of the work of a new lathe
be impossible or would result in no benefits	mutually exclusive with	A brick building or a wooden building for a given need

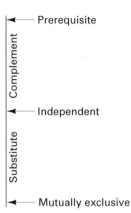

Figure 9-2 Continuum of degrees of dependence between pairs of projects.

9.5.3 Organization for Capital Planning and Budgeting

In most large organizations project selections are accomplished by sequential review through various levels of the organization. The levels required for approval should depend on the nature and importance of the individual project, as well as the particular organizational makeup of the firm. In general, a mix of central control and coordination together with authority to make project commitments delegated to operating divisions is considered desirable. Three typical basic plans for delegating investment decisions are listed below.

1. Whenever proposals are clearly "good" in terms of economic desirability according to operating division analysis, the division is given the power to commit as long as appropriate controls can be maintained over the total amount invested by each division and as long as the division analyses are considered reliable.

2. Whenever projects represent the execution of policies already established by headquarters, such as routine replacements, the division is given the power to commit within the limits of appropriate controls.

3. Whenever a project requires a total commitment of more than a certain amount, this request is sent to higher levels within the organization. This is often coupled with a budget limitation regarding the maximum total investment that a division may undertake in a budget period.

To illustrate the concept of larger investments requiring higher administrative approval, the limitations for a medium-sized firm might be as follows:

| If the total investment is . . . | | then approval is required through |
more than	but less than	
$ 50	$ 1,000	Plant manager
1,000	10,000	Division vice-president
10,000	25,000	President
25,000	—	Board of directors

9.5.4 Communication

The importance of effective communication of capital investment proposals is often overlooked. No matter how great are the merits of a proposed project, if those merits are not communicated to the decision maker(s) in understandable terms with emphasis on the proper matters, that proposal may well be rejected in favor of less desirable, though better communicated, proposals. Klausner[5] provides good insight into this problem with emphasis on the differing perspectives of engineers responsible for technical design and proposal preparation, and management decision makers responsible for monitoring the firm's capital resources. The proposal preparer should be as aware as possible of the decision maker's perspective and related informational needs. For example, in addition to basic information such as investment requirements and measures of merit and other expected benefits, the decision maker may well want clearly presented answers to such questions as:

1. What bases and assumptions were used for estimates?
2. What level of confidence does the proposer have regarding these estimates?
3. How would the investment outcome be affected by variations in these estimated values?

If project proposals are to be transmitted from one organizational unit to another for review and approval, there must be effective means for communication. The means can vary from standard forms to personal appearances. In communicating proposals to higher levels, it is desirable to use a format that is as standardized as possible to help assure uniformity and completeness of evaluation efforts. In general, the technical and marketing aspects of each proposal should be completely described in a format that is most appropriate to each individual case. However, the financial implications of all proposals should be summarized in a standardized manner so that they may be uniformly evaluated.

9.6 Postmortem Review

The provision of a system for periodic postmortem reviews (post-audits) of the performance of consequential projects previously authorized is an important aspect of a capital budgeting system. That is, the earnings or costs actually realized on each such project undertaken should be compared with the corresponding quantities estimated at the time the project investment was committed.

This kind of feedback review serves at least three main purposes, as follows:

1. It determines if planned objectives have been obtained.
2. It determines if corrective action is required.
3. It improves estimating and future planning.

---

[5] R. F. Klausner, "Communicating Investment Proposals to Corporate Decision Makers," *The Engineering Economist* (Fall 1971):45.

Postmortem reviews should tend to reduce biases in favor of what individual divisions or units preparing project proposals see as their own interests. When divisions of a firm have to compete with each other for available capital funds, there is a tendency for them to evaluate their proposals optimistically. Estimating responsibilities can be expected to be taken more seriously when the estimators know that the results of their estimates will be checked. However, this checking function should not be overexercised, for there is a human tendency to become overly conservative in estimating when one fears severe accountability for unfavorable results.

It should be noted that a postmortem audit is inherently incomplete. That is, if only one of several alternative projects is selected, it can never be known exactly what would have happened if one of the other alternatives had been chosen. "What might have been if . . ." is at best conjecture, and all postmortem audits should be made with this reservation in proper perspective.

9.7 Budget Periods

The approved capital budget is limited typically to a one- or two-year period or less, but this should be supplemented by a long-range capital plan with provision for continual review and change as new developments occur. The long-range plan (or plans) can be for a duration of from two years to twenty years, depending on the nature of the business and the desire of management to force preplanning.

Even when the technological and market factors in the business are so changeable that plans are no more than guesses to be continually revised, it is valuable to plan and budget as far ahead as possible. Planning should encourage the search for investment opportunities, provide a basis for adjusting other aspects of management of the firm as needed, and sharpen management's forecasting abilities. Long-range budget plans also provide a better basis for establishing minimum rate of return figures that properly take into account future investment opportunities.

9.8 Timing of Capital Investments and Management Perspective

An aspect of capital budgeting that is difficult and often important is deciding how much to invest now as opposed to later. If returns are expected to increase for future projects, it may be profitable to withhold funds from investment for some time. The loss of immediate return, of course, must be balanced against the anticipation of higher future returns.

In a similar vein, it may be advantageous to supplement funds available for present projects whenever returns for future projects are expected to become less than those for present projects. Funds for present investment can be supplemented by the reduction of liquid assets, the sale of other assets, and the use of borrowed funds.

The procedures and practices discussed in this book are intended to aid management in making sound investment decisions. Management's ability to sense the opportunities for growth and development and to time their investments to achieve optimum advantage is a primary ingredient of success for an organization.

9.9 Leasing Decisions

By the term *lease,* we normally are referring to the financial type of lease; that is, a lease in which the firm has a legal obligation to continue making payments for a well-defined period of time for which it has use, but not ownership, of the asset(s) leased. Financial leases usually have fixed durations of at least one year. Many financial leases are very similar to debt and should be treated in essentially the same manner as debt. A significant proportion of assets in some firms is acquired by leasing, thus making the firm's capital available for other uses. Buildings, railroad cars, airplanes, production equipment, and business machines are examples of the wide array of facilities that may be leased.[6]

Lease specifications are generally detailed in a formal written contract. The contract may contain such specifications as the amount and timing of rental payments; cancellability and sublease provisions, if any; subsequent purchase provisions; and lease renewability provisions.

Although the financial lease provides an important alternative source of capital, it carries with it certain subtle, but important, disadvantages. Its impact is similar to that of added debt capital. The acquisition of financial leases or debt capital will reduce the firm's ability to attract further debt capital and will increase the variability (leverage) in prospective earnings on equity (owner) capital. Higher leverage results in more fixed charges for debt interest and repayment, and thus will make good conditions even better and poor conditions even worse for the equity owners.

Confused reasoning may be introduced into economic studies in which calculations appropriate for justifying long-run economy of proposed investments in physical assets are combined with calculations related to the financing of those assets. It is recommended that analyses (and decisions) such as these be made *separately,* not mixed, whenever possible.

Figure 9-3 depicts the types of analyses that should be made for lease-related decisions and also shows what conclusion (final choice) should be made for various combinations of conditions (analysis outcomes). The "buy" versus "status quo" (do nothing) decision is to determine long-run economy or feasibility and is sometimes referred to as an *equipment* decision. In general practice, only if this is investigated and "buy" is found to be preferable should one be concerned with the "buy" versus "lease" question, which is considered a financing decision. However, we cannot always separate the *equipment* and *financing* decisions. For example, if the *equipment* decision results in "status quo" being preferable and yet it is possible that the equipment might be favorable if leased, then we should compare "status quo" with "lease." This, by definition, is a *mixed* decision involving both *equipment* and *financing* considerations.

The main point one should retain from the above is that one should not merely compare "buy" versus "lease" alternatives; one should also compare, if possible, against the "status quo" alternative to determine if the asset is justified under any financing plan.

A major factor in evaluating the economics of leasing versus buying is the tax deductions (reductions in taxable income) allowable. In the case of leasing, one is allowed to deduct the full cost of normal financial lease payments. In the case of buying

---

[6] So-called leases with "an option to buy" are treated as conditional sales contracts and are not considered to be true financial leases.

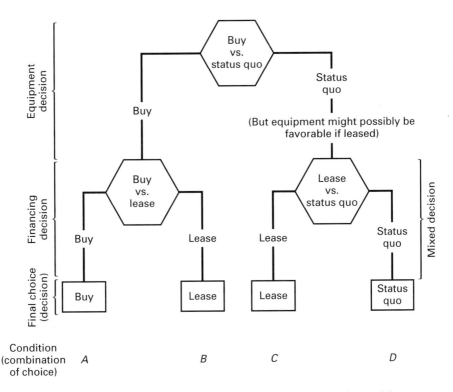

Figure 9-3 Separability of "buy" versus "lease" versus "status quo" decisions, and the choices that should result.

(ownership), only depreciation charges and interest payments, if any, are deductible. Of course, other disbursements for operating the property are tax deductible under either financing plan.

Example 9-1

Suppose a firm is considering purchasing equipment for $200,000 that would last for 5 years and have zero market value. Alternatively, the same equipment could be leased for $52,000 at the *beginning* of each of those 5 years. The composite income tax rate for the firm is 55% and straight-line depreciation is used. The net before-tax cash benefits from the equipment are $56,000 at the end of each year for 5 years. If the after-tax MARR for the firm is 10%, use the present worth method to show whether the firm should buy, lease, or maintain the status quo.

Solution

Alternative	Yr	Before-tax cash flow	Depreciation	Taxable income	Income taxes @ 55%	After-tax cash flow
Buy	0	−$200,000				−$200,000
	1–5		−$40,000	−$40,000	+$22,000	+ 22,000
Lease	0	− 52,000		− 52,000	+ 28,600	− 23,400
	1–4	− 52,000		− 52,000	+ 28,600	− 23,400
Status quo	{ 1–5	− 56,000		− 56,000	+ 30,800	− 25,200

Thus, the present worths (after taxes) for the three alternatives are:

$$\text{Buy:} \quad -\$200,000 + \$22,000(P/A, 10\%, 5) = -\$116,600,$$
$$\text{Lease:} \quad -\$23,400 + \$23,400(P/A, 10\%, 4) \ \ = -\$97,600,$$
$$\text{Status quo:} \ -\$25,200(P/A, 10\%, 5) \qquad\qquad = -\$95,500.$$

If the sequence of analysis follows that in Fig. 9-3, we would first compare "buy" (PW = −\$116,600) against "status quo" (PW = −\$95,500) and find the status quo to be better. Then we would compare "lease" (PW = −\$97,600) against "status quo" (PW = −\$95,500) and find the "status quo" to be better. Thus, "status quo" is the better final choice. It should be noted that had we merely compared "buy" versus "lease," "lease" would have been the choice. But the final decision should not have been made without comparison against "status quo."   · ■

Another way the problem could have been solved would have been to include the +\$56,000 before-tax cash benefits with both the "buy" and the "lease" alternatives. Then the two alternatives being considered would have been "buy rather than status quo" (PW = −\$116,600 + \$95,500 = −\$21,100) and "lease rather than status quo" (PW = −\$97,600 + \$95,500 = −\$2,100). Of these alternatives, "lease rather than status quo" is the better, but "lease" is still not justified because of the negative PW, indicating that the costs of "lease" are greater than the benefits.

9.10 Capital Expenditure Practices in the United States

There have been a number of extensive surveys of the capital expenditure practices used by mostly large industrial firms. The National Industrial Conference Board published a definitive work[7] that included good examples of procedures and forms used by a variety of firms. This was updated somewhat in 1974,[8] with emphasis on the methods used for evaluating prospective projects as well as the policies and procedures for screening and approval.

Abdelsamad[9] summarized the results for an early 1970s survey of capital expenditure practices of some 200 of the largest industrial corporations in the United States. The results confirmed other surveys by showing that there is considerable growing sophistication in the use of valid proposal evaluation techniques among the larger industrial firms. Of particular interest is the following list of six problem areas that were considered most important in the evaluation of capital expenditure proposals by the respondent firms:

1. forecasting in general;
2. disclosure of alternatives;
3. inability of the accounting department to later confirm or disprove accuracy of original cash flow estimates;

[7] Norman E. Pflomm, *Managing Capital Expenditures,* National Industrial Conference Board Studies in Business Policy, no. 107, 1964.

[8] Patrick J. Davey, *Capital Investments: Appraisals and Limits,* The Conference Board, Report no. 641, 1974.

[9] M. Abdelsamad, *A Guide to Capital Expenditure Analysis,* AMAACOM, Division of American Management Association, 1973.

4. qualitative information not subject to quantitative analysis;

5. overestimation of benefits;

6. lack of standardization in methods and assumptions used by analysts and evaluators.

Petty, Scott, and Bird[10] reported on a thorough survey that obtained responses from over 100 of the 500 largest industrial firms in the United States. The following are some highlights:

(a) The most important financial objectives were ranked (in decreasing order of importance) as follows:

    1. target earnings per share growth rate;

    2. maximize percent return on asset investment;

    3. maximize aggregate dollar earnings;

    4. maximize common stock price.

(b) The most-used evaluation techniques were ranked as follows:

    1. internal rate of return;

    2. accounting return on investment $\left(= \dfrac{\text{accounting profit/yr}}{\text{book value of investment}} \right)$;

    3. payback period;

    4. net present value.

(c) The most important qualitative (nonmonetary) factors affecting investment decisions were ranked as follows:

    1. legal;

    2. image with industry, investor, and customer;

    3. environmental responsibility;

    4. employee safety.

(d) The most-used definitions of risk were ranked as follows:

    1. probability of not achieving a target return (in percent or dollars);

    2. variation in returns;

    3. payback period uncertain.

(e) The most-used methodologies for analyzing the riskiness of a capital investment were ranked as follows:

    1. payback period;

    2. risk-adjusted discount rate;

    3. measuring expected variation in returns;

    4. simulation.

(f) The minimum return standard (method of determining MARR) was ranked as follows:

    1. management-determined target RR;

    2. weighted cost (average) of sources of funds;

    3. cost of a specific source of funds;

    4. firm's historical RR.

[10] Petty, J. W. Scott, and M. M. Bird, "The Capital Expenditure Decision-Making Process of Large Corporations," *The Engineering Economist* (Spring 1975):159.

Another excellent survey of U. S. capital budgeting practices was reported by Gurnani in 1984.[11] In addition to referencing most previous studies of capital budgeting theory versus actual practice, the 1984 article confirms a continuing and growing trend for firms to utilize quantitative techniques for considering risk and uncertainty in the strategic evaluation of large capital investments.

9.11 Capital Budgeting Forms Used by a Large Corporation[12]

The corporate guidelines and forms illustrated in this section are well-designed analysis and control procedures. (However, they should not be accepted as the best available or as necessarily appropriate for use in companies searching for "cookbook" procedures.) Clearly, an organization's capital allocation guidelines should be designed and written around its own set of business goals and objectives and long-term strategy. The materials for this large company typify fundamental calculations required to develop and analyze cash flows associated with capital investment proposals. It is our intent to demonstrate that previous topics covered in this book are integral components of the process used by top management to execute their responsibility for judiciously allocating capital in building a company's future.

"Corporation A" is a Fortune 100 company known for its thorough and careful evaluation of proposed capital investment opportunities. The capital allocation procedure summarized ahead requires information concerning estimated project cash flow profiles and nonmonetary considerations, sensitivity analyses of key parameters, inflation adjustments, business history of the project involved, and so forth. Be sure to observe the sequence of steps comprising this company's procedure and the timing of various approvals required to move a capital investment proposal forward to final acceptance and funding. This company requires that a plant appropriation request (PAR) be prepared for capital expenditures that exceed $50,000. Who makes the final approvals for PARs then depends on the amount of capital requested. For example, when $20 million or more is being sought, the Board of Directors must grant final approval.

I. Application of the Procedure.   Responsibilities for initiation, preparation, review, and approval, as well as the routine to be followed in processing a plant appropriation request (PAR) requiring Corporate Executive Office (CEO) or Board of Directors (BOD) approval are specified in this Company Procedure. Principal topics covered in this Procedure include

- approvals in conjunction with acquisition/disposition proposals,
- notification of initiation (NOI),
- amount of PAR and scope of project,

---

[11] C. Gurnani, "Capital Budgeting: Theory and Practice," *The Engineering Economist* (Fall 1984): 19–46.

[12] Reprinted with permission of the General Electric Company. A review of capital budgeting methods used by other manufacturing firms is provided in M. D. Proctor and J. R. Canada, "Past and Present Methods of Manufacturing Investment Evaluation: A Review of the Empirical and Theoretical Literature," *The Engineering Economist* 38, no. 1 (Fall 1992):45–58.

- PAR reviews,
- approval, implementation, and follow-up.

II. Transactions Requiring a PAR.   Any (1) facility investment, (2) purchase or sale of land, (3) sale or lease of company-owned facilities, (4) plant relocation, (5) sublease of facilities to external parties (unless such transactions are normally part of the component's business), and/or (6) lease commitment requiring CEO or BOD approval must be supported by an approved PAR before any commitment is made. The calculations to determine approval amounts are discussed in the PAR instructions provided in Section VI.

III. Purpose of the Notification of Initiation.   The use of the notification of initiation (NOI) is mandatory for PARs requiring CEO or BOD approval. The principal purpose of the NOI is for the use of the initiating manager in securing the concurrence of the approving authority prior to proceeding with the preparation and review of a PAR. The NOI establishes the time schedule for review and specifies the Functional reviews required by the approving authority.

The initiating manager provides basic information pertaining to a project as follows:

- a summary description of the proposed project including a preliminary evaluation of the compatibility with strategic plan;
- *estimated* expenditures (investment and expense);
- proposed target dates for endorsement or approval by the approving authorities;
- recommendation of functional reviews and critical issues to be considered.

The initiating manager will then forward the NOI through channels to the final approving authority for completion of

- approval or modification of the initiating manager's proposed schedule and functional reviews;
- specifications for additional functional reviews and/or additional instructions.

The primary purpose of the preliminary evaluation of compatibility with strategic plan is to confirm that the strategic plan, of which the proposed investment is a part, remains current and viable. It is not intended as a detailed strategic review of the PAR or as a detailed evaluation of the merits of the project. In the case of projects not included in an approved strategic plan, the scope of the preliminary evaluation of compatibility will usually be broadened to a more detailed critique of the supporting assumptions and documentation of the consistency of the proposed investment with the approved strategic plan.

IV. Timing of NOI.   An NOI should be prepared after the preliminary analyses have been completed and immediately upon the decision to seek funds, but *prior* to the preparation of the PAR.

Return of the signed NOI by the responsible executive officer, through channels, constitutes approval for the initiating manager to proceed. The initiating manager will then provide copies of the approved NOI to the designated reviewers.

V. Amount of PAR and Scope of Project.   In defining the total amount to be requested and the scope of a project, all purchases, costs, activities, and personnel essential to plan and complete the undertaking must be documented, identifying all items of expense. The project scope should be such that the proposal can be considered independently without regard to other plant investment needs. Also, the evaluation of the financial benefits must be based solely on the benefits achieved by the project itself.

VI. What Is Required in a Plant Appropriation Request.   All PARs should include the following:

1. *Executive Summary*—should be limited to two pages, and in most instances should cover the items outlined below in roughly the order suggested. This is not an all-inclusive list, however, and the approach should be modified as the facts warrant.
   (a) *Narrative Description of Project*—including the following where applicable:
      • organizational unit requesting, project line involved;
      • category—for example: cost reduction, capacity, pollution control, etc.;
      • project description and location;
      • amount, amounts previously requested, timing;
      • how project will be implemented (for example: purchase equipment, lease facility, rearrange, etc.);
      • why project is required;
      • expected results;
      • project phases;
      • consistency with strategic plan and budget;
      • degree of flexibility, for example:
         — if lease, minimum payment commitment;
         — if purchase, possible alternate uses of proposed equipment.
   (b) *Financial Benefits*—brief description of the basis for determining project financial benefits (for example: incremental sales, cost improvements, etc.). Identify key assumptions on which benefits are based and sensitivity of benefits to changes in these assumptions. State the discounted funds flow rate of return (DCRR) and the payback period. Comment on upside and downside potential. If financial benefits cannot be quantified, state other benefits on which approval should be based.
   (c) *Alternatives*—other alternatives considered and reason for selection of recommended approach, including relative financial benefits.
   (d) *Risks*—summary and assessment of key risks to project success. Quantify impact on DCRR.
   (e) *Conclusion*—brief summary of financial and nonfinancial reasons for approval.

2. Exhibits A through C

    (a) Exhibit A—contains the basic financial data required for PARs.

    (b) Exhibit B, Incremental Funds Flow Worksheet—facilitates calculation of the funds flow on reported and inflation adjusted bases.

    (c) Exhibit C, Key Assumptions, Evaluation of Alternatives, and Sensitivity Analysis—the main purpose of the exhibit is to facilitate the communication, understanding, and appraisal of information for a proposed project with regard to

- alternative courses of action considered;
- principal advantages and disadvantages of this request and the alternatives considered, as well as reasons for rejecting the alternatives;
- key assumptions basic to achieving project objectives;
- key assumptions upon which individual alternatives are based;
- evaluation of risk associated with this request.

VII. PAR Reviews.   PAR reviews are classified into three main categories: financial, strategic, and functional.

A. Financial Review

A financial review is mandatory in addition to any other accounting or treasury participation in the PAR review. The purpose of the financial review is to provide a financial evaluation of the attractiveness of the proposal and to verify that the financial information is completed and properly presented and meets the requirement for a detailed financial expression of the proposed investment.

B. Strategic Review

A strategic review is mandatory. The purpose is to determine that the proposal is compatible with the most recently approved strategic plan. The strategic review will be performed by Corporate Strategy Review. After receipt of the financial and any functional review letters, the strategic reviewer will prepare a strategic review report. The strategic review report will include an assessment of the purpose/need for the proposed project and its compatibility with the approved strategic plan, an evaluation of the principal benefits, issues and risks associated with the project, and conclusions and recommendations.

C. Functional Reviews

The purpose of any functional review approved by an executive officer is to ensure the effectiveness and practicability of the proposed investment (as described in the PAR) to produce intended results within the estimated project cost. Functional reviews may be designated to cover such areas/considerations as

- employee relations,
- engineering,
- legal,
- manufacturing,
- marketing,
- research and development,

Plant appropriation request (PAR)   No. _____

1. . . . . .Sector
   . . . . .Group
   . . . . .Division
   . . . . .Department
   . . . . .Product line
   . . . . .Project location

   Approval required

   . . . .
   (Dollar amounts in thousands)

2. Summary Description of proposed project

3. Project expenditures

	This request	Previously approved	Future requests	Total project
Basis for approval.............................				
Related expense.................................	_____	_____	_____	_____
Total..	══════	══════	══════	══════

4. Key financial measurements

	Reported	Inflation adjusted
DCRR*	_____%	_____%
Payback period (years)	_____	_____

Chart of cumulative funds flow

——Reported
----Inflation adjusted

Amount

0

19   19   19   19   19   19   19   19   19   19   19

*Discounted funds flow rate of return (DCRR) is synonymous with IRR.

Exhibit A

Incremental funds flow worksheet

(in thousands)

19_ 19_ 19_ 19_ 19_ 19_ 19_ 19_ 19_ 19_

REPORTED FUNDS FLOW
Incremental increase/decrease in net income
  resulting from the project based on:

    Cost reduction......................
    Incremental volume..................
    Changes in average selling price.........
    Investment credit...................
       Total–a).........................

Provision for incremental depreciation expense.

Increased/decreased incremental investment in:

    Plant and equipment–b)...............
    Inventories........................
    Receivables.......................
    All other..........................

Reported annual funds flow.............

Reported cumulative funds flow.........

INFLATION ADJUSTED FUNDS FLOW
    Reported annual funds flow.............
    CPI (1st year of investment = 100.0)......

Inflation adjusted annual funds flow......

Inflation adjusted cumulative funds flow...

Residual investment value at the end of funds
  flow period: Reported _____ Inflation adjusted _____

(a–Must agree with incremental net income as shown in Section 6 (c) of Exhibit A.
(b–Gross expenditures only, exclusive of provision for depreciation expense.

Exhibit B

<u>KEY ASSUMPTIONS</u>
<u>EVALUATION OF ALTERNATIVES</u>
<u>AND</u>
<u>SENSITIVITY ANALYSIS</u>

1. <u>Brief description of major alternatives</u>

   This request –

   Alternative I –

   Alternative II –

   Alternative III –

2. <u>Principal advantages/disadvantages and reasons for rejecting alternatives</u>

   This request –

   Alternative I –

   Alternative II –

   Alternative III –

3. <u>Key assumptions basic to achieving project objectives upon which this request is based:</u>

   a.

   b.

   c.

   d.

   e.

Exhibit C

<u>KEY ASSUMPTIONS</u>
<u>EVALUATION OF ALTERNATIVES</u>
<u>AND</u>
<u>SENSITIVITY ANALYSIS</u>

4. Key assumptions upon which individual alternatives are based

Alternative I   a.

              b.

              c.

Alternative II   a.

              b.

              c.

Alternative III   a.

              b.

              c.

5. <u>Sensitivity Analysis</u> – Evaluate the project <u>downside risk</u> by identifying "worst case" changes in key variables. The project's upside potential, if relevant, may also be identified here, or in the text of the appropriation.

Variable (selling prices, market share, material costs, etc.)	Assumption reflected in project funds flow	Management assessment of "worst case" result	Downside DCRR	
			Reported	Inflation adjusted

Exhibit C continued

- international considerations,
- real estate and construction operation,
- treasury.

D. Endorsement Prior to Final Approval

When this work is completed, the strategic reviewer will assemble and forward to the cognizant executive officer the PAR review package (strategic review report and review letters) to obtain the endorsement for transmittal of the PAR package to those senior vice-presidents, Corporate Components, designated by the CEO for review and comment. The position of the designated senior vice-presidents will be documented in a brief letter addressed to the CEO with copies to the manager of Corporate Strategy Review.

VIII. Notification of Final Approval. For PARs that require CEO approval, the documents provided by Corporate Strategy Review will enable the CEO to return a signed copy that authorizes the project. Corporate Strategy Review will notify the initiating manager and others who need to know upon receipt of the signed approval document.

For PARs that require BOD approval, the secretary of the BOD will notify Corporate Finance staff of the action taken by the BOD. Corporate finance staff, in turn, will notify the initiating manager and others who need to know.

IX. Implementation and Follow-up.

A. Change in Scope or Results, Requirements for Additional Funds

During the implementation of an approved PAR, it is the responsibility of the Component to inform the approving authority in writing of any major change in the scope of the project, such as site location, reduced profit forecast, or cost overrun. In these instances, further commitment of funds must receive the same careful planning and correlation of investment opportunities with the Strategic Business Unit's strategic plans as the original proposal.

B. Follow-Up Reporting

The following two reports must be prepared periodically and submitted to Corporate Finance staff.

1. Status reports on projects covered by open appropriations approved by the CEO or BOD. These reports must reflect and explain deviations from the approved PARs with respect to estimates of timing, amount of expenditure, and benefits realized.

2. Closed appropriation reports on projects approved by the CEO or BOD after the appropriations are closed and results can be appraised. These reports must summarize and explain variances between estimated and actual expenditures and benefits.

PROBLEMS

9-1. Discuss the scope of capital budgeting with particular emphasis on how the various main considerations in capital budgeting are interrelated.

9-2. In your opinion, what is the most valid philosophy on how the minimum acceptable rate of return should be determined?

9-3. Under what circumstances is it reasonable to have more than one minimum acceptable rate of return for a given firm?

9-4. Select the classification systems for investment proposals you think would be most useful for the typical small (say, fewer than 200 employees) enterprise. Do the same for the typical large (say, more than 2,000 employees) enterprise. Explain the reasonableness of any differences in your selections for the two size groups.

9-5. What is the purpose of a postmortem review? Can it be a means of correcting unwise commitments made as a result of past project analyses?

9-6. Explain the various degrees of dependency between two or more projects. If one or more projects seem desirable and are complementary to a given project, should those complementary projects be included in a proposal package or kept separate for review by management?

9-7. A 4-year-old truck has a present net realizable value of $6,000 and is now expected to have a market value of $1,800 after its remaining 3-year life. Its operating disbursements are expected to be $720 per year.

An equivalent truck can be leased for $0.40 per mile plus $30 a day for each day the truck is kept. The expected annual utilization is 3,000 miles and 30 days. If the before-tax MARR is 15%, find which alternative is better by comparing before-tax equivalent annual costs

   a. using only the above information;

   b. using further information that the annual cost of having to operate without a truck is $2,000.

9-8. Work Problem 9-7 by comparing after-tax equivalent present worths if the effective income tax rate is 40%, the present book value is $5,000, and the depreciation charge is $1,000 per year if the firm continues to own the truck. Any gains or losses on disposal of the old truck affect taxes at the full 40% rate, and the after-tax MARR is 5%.

9-9. A lathe costs $56,000 and is expected to result in net cash inflows of $20,000 at the beginning of each year for 3 years and then have a market value of $10,000 at the end of the third year. The equipment could be leased for $22,000 a year, with the first payment due immediately.

   a. If the organization does not pay income taxes and its MARR is 10%, show whether the organization should lease or purchase the equipment.

   b. If the lathe is thought to be worth only, say, $18,000 per year to the organization, what is the better economic decision?

9-10. The Shakey Company can finance the purchase of a new building costing $2 million with a bond issue, for which it would pay $100,000 interest per year, and then repay the $2 million at the end of the life of the building. Instead of buying in this manner, the company can lease the building by paying $125,000 per year, the first payment being due one year from now. The building would be fully depreciated for tax purposes over an expected life of twenty years. The income tax rate is 40% for all expenses and capital gains or losses, and the firm's after-tax MARR is 5%. Use annual worth analysis based on equity (nonborrowed) capital to determine whether the firm should borrow and buy or lease if at the end of 20 years the building has the following market values for the owner: (a) nothing, (b) $500,000. Straight-line depreciation will be used, but is allowable only if the company purchases the building.

9-11. The Capitalpoor Company is considering purchasing a business machine for $100,000. An alternative is to rent it for $35,000 at the beginning of each year. The rental would include all repairs and service. If the machine is purchased, a comparable repair and service contract can be obtained for $1,000 per year.

       The salesperson of the business machine firm has indicated that the expected useful service life of this machine is five years, with zero market value, but the company is not sure how long the machine will actually be needed. If the machine is rented, the company can cancel the lease at the end of any year. Assuming an income tax rate of 25%, a straight-line depreciation charge of $20,000 for each year the machine is kept, and an after-tax MARR of 10%, prepare an appropriate analysis to help the firm decide whether it is more desirable to purchase or rent.

PART TWO

Capital Investment Evaluation Under Risk and Uncertainty

10

Introduction to Risk and Uncertainty

→ Assigns ownership
0 to risk

All of the economic study methods and illustrations in Part I were for conditions of "assumed certainty"; i.e., all elements (parameters) considered were estimated or specified by a single figure. Generally, such elements as life, salvage value, and periodic incomes and costs are random variables rather than known constants. Hence, in many economic studies it is necessary or desirable to extend the results of assumed certainty analyses by directly considering the risk and uncertainty involved due to variability in the outcome of elements. Part II of this book treats the subject of capital investment evaluation under risk and uncertainty conditions.

10.1 The Difference Between Risk and Uncertainty

The classical distinction between risk and uncertainty is that an element or analysis ✓ involves *risk* if the probabilities of the alternative, possible outcomes are known, while it is characterized by *uncertainty* if the frequency distribution of the possible outcomes is not known. The distinction between conditions of assumed certainty, risk, and uncertainty for a given element such as project life is portrayed graphically in Fig. 10-1.

Another less restrictive distinction between risk and uncertainty is that risk is ✓ the dispersion of the probability distribution of the element being estimated or calculated outcome(s) being considered, while uncertainty is the degree of lack of confidence that the estimated probability distribution is correct. The word *risk* can be used to apply to the outcome of any element or measure of merit. Colloquially, "risk" is often used merely to denote variability of outcome, and often the only variability of concern is variability in an unfavorable direction.

goal is to move away from risk to uncertainty

Risk Event Status = probability × amt at stake

269

Figure 10-1   Illustrations of assumed certainty, risk, and uncertainty as applied to life of a project.

There are several combinations of risk and assumed certainty that can specify a given element estimate over time. For example, Fig. 10-2 represents an assumed-certain outcome amount (e.g., cash flow) at an assumed-certain point in time, while Fig. 10-3 shows a risk amount at an assumed-certain point in time. Figure 10-4 represents an assumed-certain amount at risk (discrete) points in time, while Fig. 10-5 represents random amounts at each risk (discrete) point in time.

10.2 Causes of Risk and Uncertainty

Risk and uncertainty in project investment decisions are attributable to many possible sources. A brief description of some main causes follows.

> **1.** *Insufficient number of similar investments.* In general, a firm will have only a few investments of a particular type. This means that there will be insufficient opportunity for the results of a particular investment type to "average out," i.e., for the effect of unfavorable outcomes to be virtually cancelled by favorable outcomes. This type of risk is dominant when the magnitude of

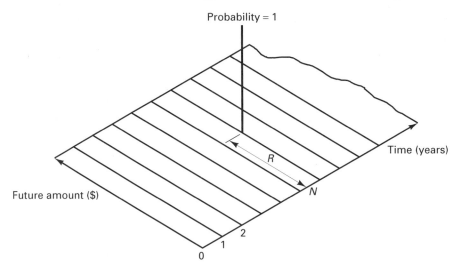

Figure 10-2   Assumed-certain outcome at assumed-certain point in time.

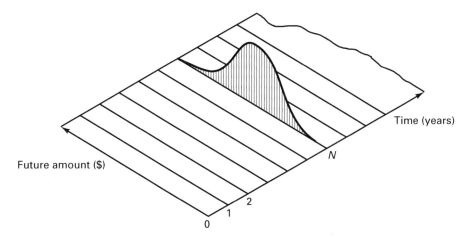

Figure 10-3   Risk amount at assumed-certain point in time.

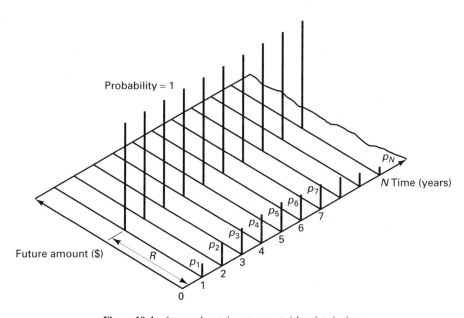

Figure 10-4   Assumed-certain outcome at risk points in time.

the individual investment commitment is large compared to the financial resources of the firm.

2. *Bias in the data and its assessment.* It is common that individuals making or reviewing economic analyses have biases of optimism or pessimism or are unconsciously influenced by factors that should not be a part of an objective study. A pattern of consistent undue optimism or pessimism on the part of an analyst should be recognized through analysis review procedures.

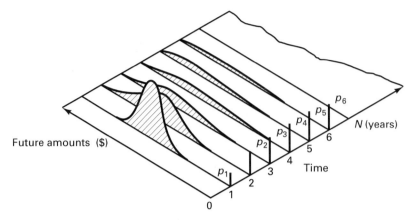

Figure 10-5   Risk amounts at risk points in time.

3. *Changing external economic environment, invalidating past experience.* Whenever estimates are made of future conditions, the usual bases are past results for similar quantities, whenever available. While the past information is often valuable, there is risk in using it directly without adjustment for expected future conditions.

4. *Misinterpretation of data.* Misinterpretation may occur if the underlying factors behind elements to be estimated are so complex that the relationship of one or more factors to the desired elements is misunderstood.

5. *Errors of analysis.* Errors can occur either in analysis of the technical operating characteristics of a project or in the analysis of the financial implications of a project.

6. *Managerial talent availability and emphasis.* The performance of an industrial investment project or set of projects usually depends in substantial part on the availability and application of managerial talent once the project has been undertaken. In general, management talent is a very limited resource within a firm; hence, it follows that the results of some projects are going to suffer compared to the results of other projects. Thus, there is risk due to lack of availability or neglect in needed managerial talent applied to investment projects.

7. *Salvageability of investment.* Of prime consideration in judging risk is the relative recoverability of investment commitments if a project, for performance considerations or otherwise, is to be liquidated. For example, an investment in special-purpose equipment that has no value to other firms entails more risk than an investment in general-purpose equipment that would have a high percentage salvage value if sold because of poor operating results. A descriptive synonym is "bailoutability."

8. *Obsolescence.* Rapid technological change and progress are characteristic of our economy. Not only do products become superseded, thus rendering those products' productive facilities less needed or useless, but also changes in process technology can render existing facilities obsolete.

10.3 Weakness in Probabilistic Treatment of Project Analyses Involving Risk

While the use of probabilities is freely made in analyses of projects involving risk, it should be pointed out that these probabilities are not generally objectively verifiable, and hence are generally *subjective* (sometimes called *personal*) probabilities. A further weakening fact is that the evidence supporting any given probability in an analysis may differ markedly in both quality and quantity from that for any other probability.

When probabilities are used, the risk and uncertainty concerning outcomes in question are not eliminated, but rather the uncertainty then becomes uncertainty connected with the probabilities on which the analysis is based. Nevertheless, it is often worthwhile to express degree of confidence in estimates through the use of probability distributions rather than through subjective verbal expressions.

10.4 Ways to Change or Influence Degree of Uncertainty

It is usually possible for the firm to take actions that will decrease the degree of uncertainty to which it is subject as a result of investment project selection. Several notable ways are

1. by increasing information obtained before decision, such as through additional market research or investigation of technical performance characteristics,

2. by increasing size of operations so as to have enough different investment projects to increase expectation that results will "average out,"

3. by diversifying products, particularly by choosing product lines for which sales are affected differently by changes in business activity (i.e., when sales of some products decrease, then sales of other products can be expected to increase).

10.5 Return, Risk, and Choice

It is generally accepted that the riskier a project, the higher the apparent return it must promise to warrant its acceptance. It would be desirable to determine differential risk allowances that would reduce all projects to a common basis. This cannot be done precisely, however, for the statement of differential risk allowances is very much a matter of subjective judgment.

Before a firm can make investment decisions to include allowances for risk, the firm's policy toward risk should be determined. The amount of risk a firm is prepared to undertake to secure a given actual or apparent monetary return is a general question of values. There is no rational or logical criterion by which the choice can be made. Rather, this is largely a function of the preferences of the decision makers of the firm and the amount of risk to which the firm is already exposed.

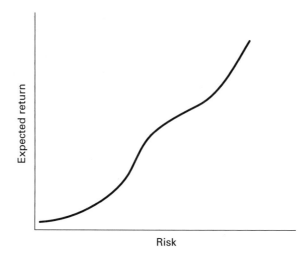

Figure 10-6   Relationship between return and risk.

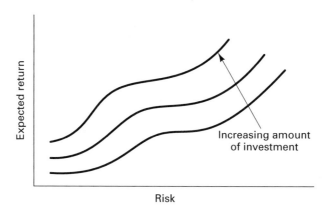

Figure 10-7   Relationship between return and risk, considering amount of investment.

In general, the relationship between expected return and risk (degree of variability of the return) can be represented as in Fig. 10-6. A refinement of the concept of Fig. 10-6 can be made if risk is subcategorized into the quantitative components of variability of returns and amount of investment. This is shown in Fig. 10-7.

10.6 Decision Guides on When and How Much to Consider Risk and Uncertainty

The question of the extent to which risk and uncertainty should be directly considered in economic analyses is of great concern and cannot be answered categorically. The concern stems from the fact that the risk and uncertainty of the future pervade most capital projects for which analyses are made. The impossibility of categorical answers stems from the fact that there is an infinite variety of sources of risk and relative degrees of risk for various projects and firms.

It is sometimes felt that the risk and uncertainty inherent in most investment decisions make it not worthwhile to engage in any complex or "sophisticated" methods of analysis. While this may be true in particular situations, the position is strongly suggested that it is, in general, very worthwhile to supplement judgment quantitatively and explicitly in the analysis of risk and uncertainty for investment projects. Analysis is needed to ensure that the implementation of judgment is not accompanied by errors in quantification and omissions in factors considered. Indeed, with the increasing complexity and economic size of individual projects compared to the worth of the firm as a whole, quantitative consideration of risk and uncertainty in capital project analyses becomes increasingly important. This section provides some qualitative decision guides regarding this question.

A conceptual answer to the question of when and how much risk and uncertainty should be considered can be reduced to simple economics—that is, put more study effort into the analysis as long as the savings from further study is greater than the cost of further study (i.e., as long as marginal savings is greater than marginal cost). Since the marginal savings (and possibly marginal cost also) for a given amount of added study is a variable, it is necessary to modify the rationale. A reasonable modification is suggested then to be the consideration of expected values. Thus the rationale can be restated as: put more study effort into the analysis as long as the expected savings from further study is greater than the expected cost of that further study.

The great problem in applying this rationale in practice is that it is quite difficult to estimate the expected savings from further study. In economic analyses of mutually exclusive projects (i.e., when at most one project can be chosen), savings from further study occurs if the further study correctly causes a reversal or change in decision as to the project accepted. In economic analyses of nonmutually exclusive projects (i.e., when the choice of one project does not affect the desirability of choice of any other project), savings from further study occurs if the further study correctly causes the decision maker to drop one or more projects previously accepted and/or correctly to add one or more projects not previously accepted. By "correctly" is meant "with favorable consequences." Other savings can be created by the added study. For example, the added study may provide information that will prove useful in future operating decisions and/or investment analyses.

The savings from further study can be conceptually determined as the discounted present worth of the new project(s) accepted after the further study minus the present worth of the project(s) accepted before the further study. However, the practical problem of determination of the expected savings from further study, as based on the amount of savings and the likelihood or probability of those savings, is generally quite difficult. It should be noted that the expected savings from added study may well not be a continuous function of the amount of the added study, but rather it is likely to change in discrete steps.

The expected cost of added study is more readily determinable than the expected savings from that study; nevertheless, it is not always apparent. Two common viewpoints on this cost are that it is equal to the direct cost of the resources devoted to the added study or that it is essentially zero on the grounds that the resources are available and paid for regardless of whether or not they are used on that added study. The most defensible cost of added study is based on the opportunity cost principle; that is, the cost of the added study should be determined by the value to the company of those study resources if put to best productive use on work other than that added study. While this opportunity cost is often hard to evaluate, it seems reasonable that in a well-managed company the cost will be at least as great as the direct cost of those resources.

Figure 10-8 show a flow diagram that depicts a general recommended sequence of steps in making economic analyses and shows qualitative test points regarding the extent of the analysis. This sequence would be applicable to analyses of groups of either mutually exclusive or nonmutually exclusive projects. Note that the recommended sequence shown in the figure shows four different points at which the decision could be made concerning which project(s) to accept. Also, there are four stages at which provision is made for dropping from further consideration projects that analysis indicates are clearly not contenders worthy of further study.

The meaning of the test points included is worthy of discussion. The test points are depicted as diamond shapes and are numbered in parentheses. Test point 1 considers the magnitude of the fixed monetary commitments involved in the decision for purposes of deciding whether further study is justified. The relevant amount of money to consider is the total present worth of the nonsalvageable investment costs as well as other fixed costs the company would incur if it should accept that project. If the magnitude of the fixed commitments for each of the projects being considered is low compared to the cost of further study, then it may be decided that further study is not justified and that the choice(s) should be made. The break-even point concerning the size of fixed commitments to use as this criterion is rather subjective. Determined intuitively, it appears that this point would be related to the company's financial health, the size of the projects usually considered, and the availability of resources for further analysis.

Test point 2 in Fig. 10-8 considers how close is the choice between projects. In this case, "close" can be defined as the nearness of the measure of merit for the most preferred alternative to the next most preferred alternative. If the assumed certainty analysis results up to that point show that the decision is not at all close [i.e., the choice(s) is (are) apparent], then further study is hardly justified and the choice(s) should be made.

Test point 3 is concerned with the decision of whether the results of an initial analysis considering variation of elements (which would be essentially a risk analysis such as described in the following chapters) is based on sufficient study considering the economic importance of the decision and the closeness of the analysis results for the projects considered. If the decision is important enough in terms of worth of the fixed commitments for the projects considered and the analysis results are somewhat close, then further study should be performed before the choice(s) is (are) made. The further study would take the form of closer estimations of elements and sensitivity analyses.

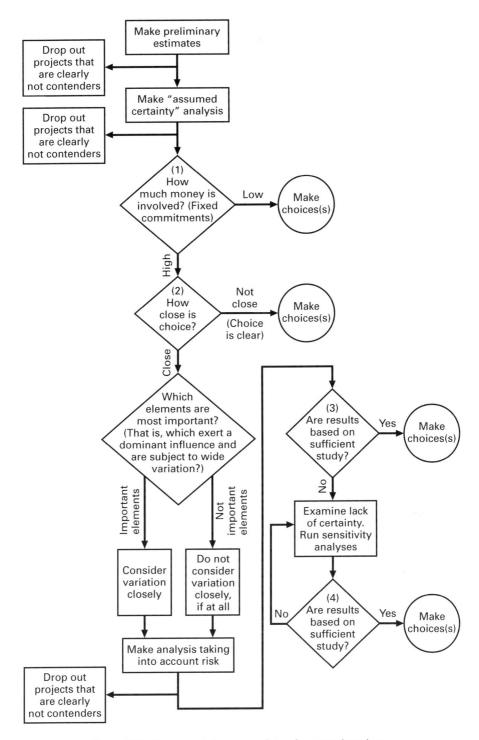

Figure 10-8   Recommended sequence of steps for economic analyses.

Test point 4 in Fig. 10-8 is repetitious of test point 3. It shows that closer estimations and sensitivity analyses would be continued until it is decided that the results of the analyses are based on sufficient study and the choice(s) can be made.

The sequence suggested in Fig. 10-8 is subject to shortcuts in cases where warranted. For example, if a given analysis involves projects that are of extreme importance to the future of the company, it may be decided to perform directly a risk analysis that considers variation of multiple elements without bothering to perform an initial "assumed certainty" study.

The sequence of steps shown in Fig. 10-8 provides a conceptual basis for determining the extent to which economic analyses should be performed. The decisions in the sequence are rather intangible; nevertheless, the sequence of steps represents a formalized structure for thinking which the analyst or decision makers can follow in determining the extent to which analyses should be carried out in particular situations. It should be recognized that all of these steps of analyses involving a particular set of alternatives are not necessarily performed at a single level in the organization. In fact, the more money involved, the higher the level in the organization to which the analysis is referred, in general, before a decision is made.

In summary, the justifiable extent of an economic analysis depends on the economic importance of the study. When the risks or potentials are substantial, systematic procedures are needed to reduce the uncertainties inherent and to give them appropriate weight in arriving at decisions.

10.7 General Model for Risk and Uncertainty Problems

In order to provide a framework for the subsequent discussion, it is useful to employ a general model of decision problems in which there are various possible outcomes (called *states of nature*) in combination with several alternative actions. As depicted in Table 10-1, m mutually exclusive investment alternatives, $\{A_i\}$, and k mutually exclusive and collectively exhaustive states of nature, $\{S_j\}$, have been identified. The combination of action A_i and state of nature S_j yields a net result R_{ij}. The outcome, R_{ij}, is normally expressed in equivalent returns or costs, but it can be in any measure. (In fact, it may well be that the outcome is multidimensional; however, it is generally assumed that, in such cases, a one-dimensional utility measure can be obtained.) This tabular model is often descriptively called a *payoff table.* *(matrix)*

When the decision problem is considered a decision under risk, then for $j = 1$, ..., k, estimates are provided of $P(S_j)$, which is the probability of state of nature S_j occurring. In the absence of such probabilities, the decision is considered to be a decision under uncertainty.

The general model is very useful in the case of a single decision. However, when sequential decisions are made, with each subsequent decision being influenced by a partial realization of the future state, then an alternative model is generally used. In Chapter 14 a decision tree model is used to represent decision problems under risk that involve sequential decisions.

TABLE 10-1   General Model for Risk and Uncertainty Problems

Alternative	State of nature (Probability of state)			
	S_1 $P(S_1)$	$S_2 \ldots$ $P(S_2) \ldots$	$S_j \ldots$ $P(S_j) \ldots$	S_k $P(S_k)$
A_1	R_{11}	$R_{12} \ldots$	$R_{1j} \ldots$	R_{1k}
A_2	R_{21}	$R_{22} \ldots$	$R_{2j} \ldots$	R_{2k}
.	.	.	.	.
.	.	.	.	.
A_i	R_{i1}	$R_{i2} \ldots$	$R_{ij} \ldots$	R_{ik}
.	.	.	.	.
.	.	.	.	.
A_m	R_{m1}	$R_{m2} \ldots$	$R_{mj} \ldots$	R_{mk}

10.8 Estimating in Terms of Probability Distributions

It is reasonable to estimate many element outcomes in terms of subjective, usually continuous, probability distributions. This can be most useful, either for purposes of calculating measures of merit that directly take these distributions into account or for purposes of merely judging the degree and effect of probable outcome variation. When it is desired to estimate the subjective probability distribution of an element and that element is not thought to fit one of the computationally convenient distributions, one good way is to estimate in terms of a cumulative probability distribution function and then convert the results to other forms if needed. For example, suppose you desire to estimate the life of a project (such as the length of time before your car will have a major breakdown). After considering the experience records for similar cars and making your subjective adjustments for future conditions, you may decide that, say, there is practically nil probability that the life will be equal to or less than 2 years; 0.10 probability that the life will be equal to or less than 3 years; and you keep making similar estimates until the life is reached at which you feel there is 100% chance that the life will not be exceeded. A complete set of estimates for this example is given in Table 10-2 and then graphed, assuming a continuous distribution, in Fig. 10-9. Figure 10-10 shows the same estimates converted into the more commonly portrayed probability density form. The reader may recall that the probability density function (height of the curve) for a continuous distribution equals the slope of the cumulative probability distribution function over the entire range of the element estimated.

Another way to estimate in terms of a cumulative probability distribution, and which is probably even easier to do than that just depicted, is to estimate median, quartile, and extreme values. Below are typical suggested questions and example answers for estimating, say, the labor cost for a new product.

TABLE 10-2   Example Estimates Expressed
in Cumulative Probability Form

Life (yr)	Probability that life will be equal to or less than life given
2	0.0
3	0.1
4	0.3
5	0.7
6	0.9
7	1.0

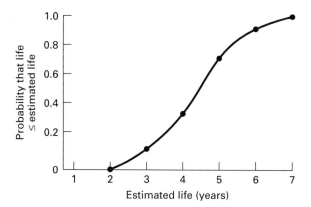

Figure 10-9   Cumulative probability distribution representation of data in Table 10-2.

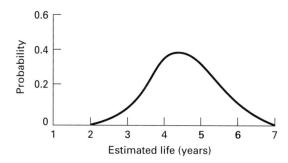

Figure 10-10   Alternative form of probability distribution (probability density) for data given in Table 10-2.

For Median (50% Cumulative Probability):
At what value is the labor cost as likely to be above as below that value? (*Ex. Ans.:* $18)

For Upper Quartile (75% Cumulative Probability):
Given that the labor cost is above the median, $18, at what value is the labor cost as likely to be above as below that value? (*Ex. Ans.:* $28)

For Lower Quartile (25% Cumulative Probability):
Given that the labor cost is below the median, $18, at what value is the labor cost as likely to be above as below that value? (*Ex. Ans.:* $12)

For Upper Extreme (99% Cumulative Probability):
What value would the labor cost exceed only 1% of the time? (*Ex. Ans.:* $45)

For Lower Extreme (1% Cumulative Probability):
What value would the labor cost be lower than only 1% of the time? (*Ex. Ans.:* $8)

After having obtained the above answers, one can either plot directly on a cumulative probability distribution or first check the answers for consistency and make adjustments by asking questions such as:

1. Is the labor cost more likely to be within the two quartiles (i.e., $12 to $28) or outside the two quartiles (i.e., < $12 or > $28)?

2. Is the labor cost more likely to be less than the lower quartile estimate, $12, or greater than the upper quartile estimate, $28?

If the answers to the above questions are not ". . . equally likely . . . ," then adjustments in one or more of the original estimates should be made until one is satisfied that the estimates represent the best judgments that can be made within the time and talent resources available. Let us suppose that after such adjustments the final estimates are as follows:

Median (M)	$19
Upper quartile (UQ)	30
Lower quartile (LQ)	12
Upper extreme (UE)	48
Lower extreme (LE)	8

Figure 10-11 shows these results on a cumulative probability graph.

The following two sections give simplified approximation procedures for estimating parameters of elements thought to be distributed according to the Beta distribution and to the normal distribution.

10.8.1 Beta II Distribution

The Beta II distribution is of interest because it can describe a wide range of left-skew and right-skew conditions of differing variances.[1]

The following Beta estimation procedure is based on a system developed for the PERT network planning and scheduling technique. It involves first making an "optimistic" estimate, a "pessimistic" estimate, and a "most likely" estimate for the element. These estimates are to correspond to the lower (or upper) bound, upper (or lower) bound, and mode, respectively, of the assumed Beta II distribution describing the element. Figure 10-12 shows an assumed Beta distribution for a typical element

---

[1] Whereas the Beta I (normally called just Beta) distribution applies to variables ranging between 0 and 1, the Beta II distribution applies to variables ranging over any set of outcomes.

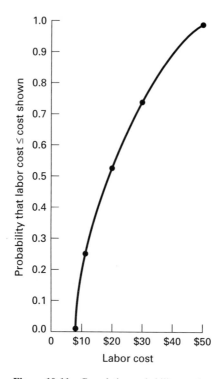

Figure 10-11   Cumulative probability graph.

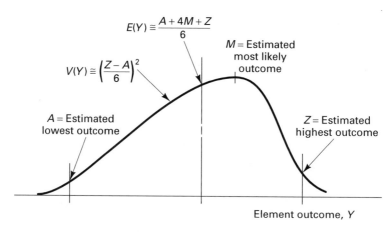

Figure 10-12   Demonstration of estimates with Beta distribution.

together with the meaning of the above types of estimates. In this case, the distribution happens to be left-skewed.

Once the three estimates of element outcome have been made, the approximate mean and variance of the Beta distribution for the element may be calculated as

$$E[Y] \approx \frac{A + 4M + Z}{6} \Bigg] \qquad (10\text{-}1)$$

and

$$V[Y] \approx \left(\frac{Z - A}{6}\right)^2, \qquad (10\text{-}2)$$

where

$E[Y]$ = estimated expected outcome,
$V[Y]$ = estimated variance of outcome,
A = estimated lowest outcome,
M = estimated most likely outcome (mode),
Z = estimated highest outcome.

It is worthy of note that the difference between the approximate expected values as calculated by Eq. 10-1 and the exact formula is relatively small for a wide range of Beta II distribution conditions. On the other hand, the difference between the approximate variance as calculated by Eq. 10-2 and the exact formula can be quite high, and the difference usually is the direction of underestimation of the exact value.

If several elements, as estimated by the above procedure, are assumed to be independent and are added together, the distribution of the total outcome so obtained is approximately normal according to the central limit theorem. The mean of this total outcome distribution can be calculated by adding the means of the individual elements. Further, the variance of this total outcome distribution can be calculated by adding the variances of the distributions of the individual elements.

10.8.2 Normal Distribution

Quite often the best subjective estimate of the shape of the distribution of an element that can be made in practice is that the distribution is normal. It can be observed from tables of area under the normal distribution that the middle 50% of a normal distribution (i.e., between the upper quartile and the lower quartile) is within ± 0.675 standard deviations of the mean of that distribution, as shown in Fig. 10-13. Thus, for a normally distributed element, if one is willing to estimate the smallest range $r\ (= \mathrm{UQ} - \mathrm{LQ})$ within which that variable is expected to occur with 50% probability, then the standard deviation σ for that variable can be calculated by the relation $0.675\sigma = r/2$. In practice, it is generally sufficiently close to approximate the 0.675 with the fraction $\frac{2}{3}$.

This same idea for estimating the variance for normally distributed variables could be applied using any other number of standard deviations and the associated probabilities. The values suggested above, however, are probably most useful because of the relative ease of visualizing the minimum range that would include 50% probability of occurrence.

As an example, suppose that the investment for a project is estimated to be normally distributed, and it is thought that there is a 50% chance it will be between $9,000 and $12,000. The standard deviation for this distribution is calculated as $\frac{2}{3}\sigma =$ ($12,000 − $9,000)/2, or $\sigma = \$2,250$. The mean for the distribution is, of course, ($12,000 + $9,000)/2 = $10,500.

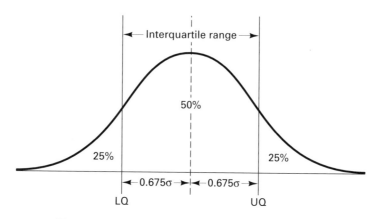

Figure 10-13 Estimating relationships for normal distribution.

10.9 Human Problems in Reflecting Degree of Uncertainty in Estimates

An interesting, but troublesome, phenomenon that should be taken into account when estimating variability is a human tendency to reflect greater certainty (i.e., less variability or spread) in estimates than is justified. This has been shown in numerous empirical studies, perhaps the most notable of which used thousands of Harvard MBA students as subjects over many years. Each student was asked to estimate the values of numerous physical phenomena, which they presumably did not know for certain, in terms of upper quartile and lower quartile values. When the exact values (outcomes) of the phenomena were later made known to them, the students then determined if the exact value fell within their estimated interquartile range (= UQ – LQ, as in Fig. 10-13) without necessarily assuming a normal distribution. On the average, only about 35% of the exact values were found to fall within their respective interquartile ranges. One would expect that if the estimates had fairly reflected the true degree of uncertainty, the average should have been closer to 50%! The conclusion is that estimators in general tend to think they know more than they really do. The moral is that variability estimates generally need to be spread out sufficiently to counteract human tendencies to compress the range of uncertainty.

Problem 10-18 is a very abbreviated version of some of the questions asked of the Harvard MBA students. It might interest you to see how close the interquartile range for your estimate comes to including the correct answer for the respective questions. (This problem also makes an entertaining in-class experience for students.)

10.10 Responses to Risk and Uncertainty

Although it is generally agreed that risks and uncertainties exist in the conduct of economic analyses, there does not exist complete agreement on how much information is to be incorporated in the analysis. Some prefer to rely heavily on the recommendations obtained from analytical models; others choose to rely solely on intuition and judgment.

There is great appeal in being able to specify quantitatively, completely, and accurately an "optimum" course of action in the face of risk and uncertainty. However, what is optimum to one individual might be very unsatisfactory to another. Hence, two approaches have emerged in performing economic analyses that explicitly incorporate risk and uncertainty conditions in the analysis.

One approach is to develop a *descriptive* model that *describes* the economic performance of an individual investment alternative. As an illustration, one might descriptively model the present worths of each of several alternatives. No recommendation would be forthcoming *from the model.* Rather, the decision maker would be furnished descriptive information concerning each alternative; the final choice among the alternatives would require a separate action.

The second approach is to develop a *prescriptive* or *normative* model that includes an objective function to be maximized or minimized. The output from the model *prescribes* the course of action to be taken. In previous analyses under assumed certainty the objective was implicitly stated as the maximization of, say, present worth.

Chapters 11 and 12 are concerned with *descriptive* modeling of cash flow profiles in the face of risk and uncertainty. Chapters 13 and 14 present *prescriptive* modeling approaches that can be used to evaluate economic investment alternatives under risk and uncertainty conditions. It will be found that the analysis in Chapters 11 and 12 can be useful not only when the formal analysis is to terminate with descriptive modeling but also in providing the required inputs for the prescriptive models.

PROBLEMS

10-1. What distinction between risk and uncertainty do you think is most useful? Why?

10-2. Which of the eight causes of risk and uncertainty listed in this chapter may be said to be generally within the control (power to affect) of the economic analyst or the people from whom he or she obtains estimates?

10-3. Which of the three listed ways for a firm to change or influence degree of uncertainty is generally within the control of the economic analyst?

10-4. Under what circumstances might it be reasonable to start the analysis at the point immediately following "test point 2" in the sequence of steps shown in Fig. 10-8?

10-5. In Fig. 10-8, are there any points at which the step "Drop out projects that are clearly not contenders" might reasonably be added or deleted to make it a more reasonable representation of ideal general practice? Explain your reasoning.

10-6. What are the pros and cons of descriptive modeling versus prescriptive modeling? Which do you prefer? Why?

10-7. Classify each of the following models as descriptive or prescriptive and justify your classifications:

    a. Linear programming model

    b. Replacement model

    c. Present worth model

    d. Dynamic programming model

    e. Inventory (EOQ) model

    f. Queueing model

    g. Simulation model

    h. Quality control model

    i. Reliability model

    j. Warehouse location model

    k. Rate of return model

    l. Forecasting model

10-8. a. If you were faced with the following decision problem under risk, which action would you choose? Why?

	S_1 (0.1)	S_2 (0.7)	S_3 (0.2)
A_1	−$40	$30	$20
A_2	− 20	20	40
A_3	0	20	10

    b. Answer part (a) assuming all payoffs are in thousands of dollars.

    c. Answer part (a) assuming the decision problem is for a large, wealthy corporation and all payoffs are in tens of thousands of dollars. Further assume that it is unacceptable in that corporate environment to recommend a project with a potential payoff loss greater than, say, $250,000.

10-9. a. If you were faced with the following personal decision problem under risk, which would you choose? Why?

	S_1 (0.4)	S_2 (0.6)
A_1	$110	−$20
A_2	− 40	85
A_3	0	50

    b. Answer part (a) assuming all the payoffs are in hundreds of thousands of dollars.

10-10. If the decision problems in Problems 10-8 and 10-9 had been decisions under uncertainty, would your preferences change? Why or why not?

10-11. Estimate the expected total remaining life of a given car, either one you happen to own or some other car with which you are familiar, in terms of the following:

    a. Single best estimate

    b. Subjective continuous probability distribution (cumulative and probability density function)

    c. Beta II distribution (calculate the approximate mean and variance)

    d. Normal distribution (calculate the approximate mean and variance)

10-12. Rework Problem 10-11 except change the variable to be estimated to the number of basketball (or football) games to be won next year by the team of your choice. Of course, since the variable is discrete, the Beta II and normal distributions are only approximations.

10-13. Repair costs have been recorded as follows:

$2,250	$4,000	$ 750	$1,250
1,000	500	1,000	750
500	250	250	2,750
750	1,500	1,750	250
500	1,750	1,500	1,000

On the basis of the data above:

a. Find the median value.

b. Find the upper and lower quartile values.

c. Estimate the probability that a given repair will cost $1,000 or more.

d. Show the data in graph form with "Cost to repair in $" as the horizontal axis and "Estimated probability that repair cost will not exceed $ _____" as the vertical axis.

10-14. A heat exchanger is being installed as part of a plant modernization program. It costs $10,000, including installation, and is expected to reduce the overall plant operating cost by $2,500 per year. Estimates of the useful life of the heat exchanger range from an optimistic 12 years to a pessimistic 4 years. The most likely value is 5 years.

a. Using the Beta II approximation formula, determine the before-tax rate of return at the expected life.

b. At what life would the investment just be recovered (i.e., the payoff period)?

10-15. Estimated maintenance expenses are very uncertain, but it is thought that there is a 50% chance they will be less than $15,000 and more than $9,000. Assuming the estimate can best be described by a normal distribution, what is

a. The estimated mean and standard deviation?

b. The probability the expense will be more than $12,000?

c. The probability the expense will be more than $16,000? *Hint:* Use Appendix E and the relation

$$S = \frac{\text{Upper limit cost} - \text{Expected (mean) cost}}{\text{Standard deviation}}.$$

10-16. Using the Beta II distribution approximation formula and the fact that for independent elements (variables) being added, the expected total = the sum of the expected elements and the variance of the total = the sum of the variances of the independent elements, find the expected total cost and variance of the total cost for the following (all numbers are in hundreds of dollars):

Cost element	Optimistic cost	Most likely cost	Pessimistic cost
1. Direct labor	$79	$95	$95
2. Direct material	60	66	67
3. Indirect expenses	93	93	96

10-17. Estimate the age of your instructor (or someone else whose exact age is unknown to you) in terms of:

    a. Most likely value.

    b. Interquartile range (UQ and LQ). Assuming a normal distribution for your estimate of the age, determine the mean and standard deviation. Should your mean correspond to your answer to (a)? What does a relatively low versus high value indicate?

10-18. For each of the following physical phenomena, treat it as a variable (to you) and estimate its value in terms of an interquartile range (UQ and LQ). Allow yourself less than half a minute to estimate each, thus eliminating the possibility of "looking up" the value (outcome). Your UQ and LQ should reflect a range within which you think there is a 50% probability that the value (outcome) will be contained. For your interest in determining your (and perhaps a group's) variable estimating tendencies, the correct answers (values) for parts (a) through (f) are given at the end of the problems in Chapter 11.

    a. Area of the United States (in millions of square miles).

    b. Weight of the human heart (in ounces, not including blood).

    c. Maximum speed of Boeing 747 "jumbo jet" (in mi/hr).

    d. Elevation of Mt. McKinley in Alaska (in ft above sea level).

    e. Number of muscles in the human body (not including one's brain!).

    f. Proportion of the world's surface covered by sea (in %).

10-19. Your estimate of the cost of your new Lincoln two years from now is as follows:

Lower quartile $32,000

Upper quartile $44,000

If you assume a normal distribution:

    a. What is the estimated mean and standard deviation?

    b. What is the probability that your Lincoln will cost at least $35,000?

    c. What is the probability that your Lincoln will cost less than $30,000?

11
Sensitivity Analysis

11.1 Introduction

There are a number of procedures for *describing* analytically the effects of risk and uncertainty on capital projects. Such procedures are generally categorized as *sensitivity* or *risk analyses.* In this chapter sensitivity analysis procedures are described. Risk analysis is considered in Chapter 12.

Sensitivity analyses are performed when conditions of uncertainty exist for one or more parameters. The objectives of a sensitivity analysis are to provide the decision maker with information concerning (1) the behavior of the measure of economic effectiveness due to errors in estimating various values of the parameters and (2) the potential for reversals in the preferences for economic investment alternatives. The term "sensitivity analysis" is derived from the desire to measure the sensitivity of a decision to changes in the values of one or more parameters.

11.2 One-at-a-Time Procedure and Break-even Analysis

One-at-a-time procedures consider the sensitivity of the measure of economic effectiveness (e.g., AW) caused by changes in a single parameter. A popular form of this type of sensitivity analysis is called *break-even analysis,* which is useful in situations where there is uncertainty regarding a single element (or parameter) in an engineering economy study. The break-even point for an element in the analysis of a single project is defined as that value of the element at which the project is marginally acceptable (barely justified). Example 11-1 illustrates break-even analysis and the more general one-at-a-time procedure of sensitivity analysis.

Example 11-1

An investment project is being considered by a firm. The following most likely (or expected) estimates have been provided:

Parameter	Estimate
Investment:	$10,000
Project life:	5 yr
Salvage value:	$2,000
Annual receipts:	$5,000
Annual disbursements:	$2,200
MARR = 8%	
AW(8%) = $636	

(a) Suppose that for the single project for which most likely estimates are shown above, the parameter for which there is particular uncertainty is the life of the project; hence, it is desired to find the minimum project life at which the project will barely be justified (i.e., the break-even life).

(b) A sensitivity analysis is to be performed for those parameters whose estimated values are most uncertain. In this case the effect of up to $\pm 100\%$ changes in project life, MARR, and annual disbursements on the AW are to be considered one at a time.

Solution

(a) A break-even analysis on project life results in this equation:

$$\$5,000 - \$2,200 - (\$10,000 - \$2,000)(A/P, 8\%, N) - \$2,000(8\%) = 0,$$

$$(A/P, 8\%, N) = \frac{-\$2,640}{-\$8,000} = 0.330.$$

By interpolating in the interest tables, we find that $N = 3.7$ years. Thus, a project life of more than 3.7 years results in a positive AW and an acceptable project.

(b) The approach to be used in the one-at-a-time procedure depends on whether or not the investment will be repeated within the planning horizon. If the investment will be repeated a sufficient number of times that either an indefinite planning horizon or a least-common-multiple-of-lives planning horizon is appropriate, then it is easier to perform an AW analysis than, say, a PW analysis. However, if the investment will not be repeated, then a present worth analysis may be easier to perform. Both approaches will be used to illustrate the differences in the results.

Annual worth approach:

$$AW = -\$10,000(A/P, i\%, N) + \$5,000 - D + \$2,000(A/F, i\%, N),$$

where D denotes the annual disbursement. The results of a one-at-a-time sensitivity analysis are depicted graphically in Fig. 11-1. The abscissa expresses the changes in i, N, and D as a percentage of the estimated values.

Present worth approach:

$$PW = -\$10,000 + \$5,000(P/A, i\%, N) - D(P/A, i\%, N)$$
$$+ \$2,000(P/F, i\%, N).$$

The results of a one-at-a-time sensitivity analysis are depicted in Fig. 11-2. It appears, from the $\pm 20\%$ error regions, that both the AW and the PW are equally

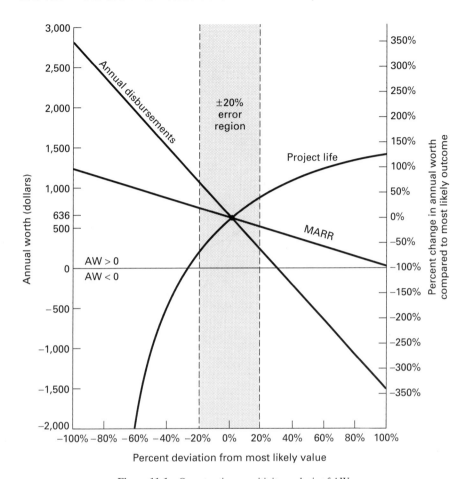

Figure 11-1   One-at-a-time sensitivity analysis of AW.

sensitive to changes in the project life and annual receipts. Both AW and PW are relatively insensitive to changes in the MARR. Furthermore, the project will be profitable (AW > 0 and PW > 0) so long as either the project life is at least 74% of the estimated value, the MARR is no more than twice the estimated value, or the annual disbursements do not increase by more than approximately 28%. The primary difference in the AW and PW analyses is the effect of project life. With AW the effect of changes in project life decreases at a faster rate, i.e., levels off sooner, than with PW. In the ± 20% error region, however, the sensitivity of project life is relatively the same for both measures of economic effectiveness.   ■

Computer spreadsheets are extremely useful when conducting sensitivity studies. A sample spreadsheet for Example 11-1 is given ahead; the cell formulas are presented first, followed by a table of PW(8%) for a large number of percentage deviations in life, MARR, and annual disbursements. Notice that the plot of results shown in Fig. 11-2 is similar to changes in AW(8%) shown in Fig. 11-1. However, Fig. 11-2 better illustrates the nonlinearity of PW(8%) to changes in the MARR.

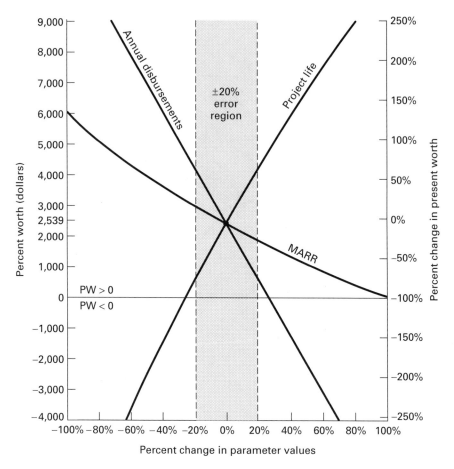

Figure 11-2 One-at-a-time sensitivity analysis of PW.

Example 11-2

Four alternative incinerators are to be compared on the basis of the following estimates:

	Incinerator			
	A	B	C	D
First cost:	$3,000	$3,800	$4,500	$5,000
Life:	10 yr	10 yr	10 yr	10 yr
Salvage value:	$0	$0	$0	$0
Annual operating disbursements:	$1,800	$1,770	$1,470	$1,320
MARR = 10%				

It is desired to determine the sensitivity of the preferred alternative to either estimation errors or changes in the annual operating disbursements.

Sample spread sheet for examle 11-1

R/C	A	B	C	D
1	Nominal Values			
2				
3	Investment:		INV	
4	Salvage Value:		SV	
5	Annual Receipts:		AR	
6	Annual Disbursements:		AD	
7	Project Life (yrs):		N	
8	MARR:		I	
9				
10				Annual
11	% Change	Life	MARR	Disbursements
12				
13	-100%	-INV+SV/(1+i)^(N*(1+A13))-PV(i,N*(1+A13),AR-AD)	-INV+SV/((1+i*(1+A13))^N-PV(i*(1+A13),N,AR-AD)	-INV+SV/(1+i)^N-PV(1,N,AR-AD*(1+A13))
14	-90%	-INV+SV/(1+i)^(N*(1+A14))-PV(i,N*(1+A14),AR-AD)	-INV+SV/((1+i*(1+A14))^N-PV(i*(1+A14),N,AR-AD)	-INV+SV/(1+i)^N-PV(1,N,AR-AD*(1+A14))
15	-80%	-INV+SV/(1+i)^(N*(1+A15))-PV(i,N*(1+A15),AR-AD)	-INV+SV/((1+i*(1+A15))^N-PV(i*(1+A15),N,AR-AD)	-INV+SV/(1+i)^N-PV(1,N,AR-AD*(1+A15))
16	-70%	-INV+SV/(1+i)^(N*(1+A16))-PV(i,N*(1+A16),AR-AD)	-INV+SV/((1+i*(1+A16))^N-PV(i*(1+A16),N,AR-AD)	-INV+SV/(1+i)^N-PV(1,N,AR-AD*(1+A16))
17	-60%	-INV+SV/(1+i)^(N*(1+A17))-PV(i,N*(1+A17),AR-AD)	-INV+SV/((1+i*(1+A17))^N-PV(i*(1+A17),N,AR-AD)	-INV+SV/(1+i)^N-PV(1,N,AR-AD*(1+A17))
18	-50%	-INV+SV/(1+i)^(N*(1+A18))-PV(i,N*(1+A18),AR-AD)	-INV+SV/((1+i*(1+A18))^N-PV(i*(1+A18),N,AR-AD)	-INV+SV/(1+i)^N-PV(1,N,AR-AD*(1+A18))
19	-40%	-INV+SV/(1+i)^(N*(1+A19))-PV(i,N*(1+A19),AR-AD)	-INV+SV/((1+i*(1+A19))^N-PV(i*(1+A19),N,AR-AD)	-INV+SV/(1+i)^N-PV(1,N,AR-AD*(1+A19))
20	-30%	-INV+SV/(1+i)^(N*(1+A20))-PV(i,N*(1+A20),AR-AD)	-INV+SV/((1+i*(1+A20))^N-PV(i*(1+A20),N,AR-AD)	-INV+SV/(1+i)^N-PV(1,N,AR-AD*(1+A20))
21	-20%	-INV+SV/(1+i)^(N*(1+A21))-PV(i,N*(1+A21),AR-AD)	-INV+SV/((1+i*(1+A21))^N-PV(i*(1+A21),N,AR-AD)	-INV+SV/(1+i)^N-PV(1,N,AR-AD*(1+A21))
22	-10%	-INV+SV/(1+i)^(N*(1+A22))-PV(i,N*(1+A22),AR-AD)	-INV+SV/((1+i*(1+A22))^N-PV(i*(1+A22),N,AR-AD)	-INV+SV/(1+i)^N-PV(1,N,AR-AD*(1+A22))
23	0%	-INV+SV/(1+i)^(N*(1+A23))-PV(i,N*(1+A23),AR-AD)	-INV+SV/((1+i*(1+A23))^N-PV(i*(1+A23),N,AR-AD)	-INV+SV/(1+i)^N-PV(1,N,AR-AD*(1+A23))
24	10%	-INV+SV/(1+i)^(N*(1+A24))-PV(i,N*(1+A24),AR-AD)	-INV+SV/((1+i*(1+A24))^N-PV(i*(1+A24),N,AR-AD)	-INV+SV/(1+i)^N-PV(1,N,AR-AD*(1+A24))
25	20%	-INV+SV/(1+i)^(N*(1+A25))-PV(i,N*(1+A25),AR-AD)	-INV+SV/((1+i*(1+A25))^N-PV(i*(1+A25),N,AR-AD)	-INV+SV/(1+i)^N-PV(1,N,AR-AD*(1+A25))
26	30%	-INV+SV/(1+i)^(N*(1+A26))-PV(i,N*(1+A26),AR-AD)	-INV+SV/((1+i*(1+A26))^N-PV(i*(1+A26),N,AR-AD)	-INV+SV/(1+i)^N-PV(1,N,AR-AD*(1+A26))
27	40%	-INV+SV/(1+i)^(N*(1+A27))-PV(i,N*(1+A27),AR-AD)	-INV+SV/((1+i*(1+A27))^N-PV(i*(1+A27),N,AR-AD)	-INV+SV/(1+i)^N-PV(1,N,AR-AD*(1+A27))
28	50%	-INV+SV/(1+i)^(N*(1+A28))-PV(i,N*(1+A28),AR-AD)	-INV+SV/((1+i*(1+A28))^N-PV(i*(1+A28),N,AR-AD)	-INV+SV/(1+i)^N-PV(1,N,AR-AD*(1+A28))
29	60%	-INV+SV/(1+i)^(N*(1+A29))-PV(i,N*(1+A29),AR-AD)	-INV+SV/((1+i*(1+A29))^N-PV(i*(1+A29),N,AR-AD)	-INV+SV/(1+i)^N-PV(1,N,AR-AD*(1+A29))
30	70%	-INV+SV/(1+i)^(N*(1+A30))-PV(i,N*(1+A30),AR-AD)	-INV+SV/((1+i*(1+A30))^N-PV(i*(1+A30),N,AR-AD)	-INV+SV/(1+i)^N-PV(1,N,AR-AD*(1+A30))
31	80%	-INV+SV/(1+i)^(N*(1+A31))-PV(i,N*(1+A31),AR-AD)	-INV+SV/((1+i*(1+A31))^N-PV(i*(1+A31),N,AR-AD)	-INV+SV/(1+i)^N-PV(1,N,AR-AD*(1+A31))
32	90%	-INV+SV/(1+i)^(N*(1+A32))-PV(i,N*(1+A32),AR-AD)	-INV+SV/((1+i*(1+A32))^N-PV(i*(1+A32),N,AR-AD)	-INV+SV/(1+i)^N-PV(1,N,AR-AD*(1+A32))
33	100%	-INV+SV/(1+i)^(N*(1+A33))-PV(i,N*(1+A33),AR-AD)	-INV+SV/((1+i*(1+A33))^N-PV(i*(1+A33),N,AR-AD)	-INV+SV/(1+i)^N-PV(1,N,AR-AD*(1+A33))

R/C	A	B	C	D
1	Nominal Values			
2				
3	Investment:		$10,000	
4	Salvage Value:		$2,000	
5	Annual Receipts:		$5,000	
6	Annual Disbursements:		$2,200	
7	Project Life (yrs):		5	
8	MARR:		8%	
9				
10				Annual
11	% Change	Life	MARR	Disbursements
12				
13	-100%	($8,000)	$6,000	$11,325
14	-90%	($6,754)	$5,592	$10,446
15	-80%	($5,556)	$5,200	$9,568
16	-70%	($4,402)	$4,822	$8,690
17	-60%	($3,292)	$4,459	$7,811
18	-50%	($2,224)	$4,109	$6,933
19	-40%	($1,196)	$3,772	$6,054
20	-30%	($208)	$3,447	$5,176
21	-20%	$744	$3,134	$4,298
22	-10%	$1,660	$2,832	$3,419
23	0%	$2,541	$2,541	$2,541
24	10%	$3,389	$2,260	$1,662
25	20%	$4,204	$1,988	$784
26	30%	$4,989	$1,726	($94)
27	40%	$5,745	$1,473	($973)
28	50%	$6,472	$1,228	($1,851)
29	60%	$7,171	$992	($2,730)
30	70%	$7,844	$763	($3,608)
31	80%	$8,492	$542	($4,486)
32	90%	$9,115	$328	($5,365)
33	100%	$9,715	$120	($6,243)

Solution   The equivalent uniform annual cost is given in Fig. 11-3 for various percent changes in the annual operating disbursements. The choice will be between alternatives A and D because one or the other will have the minimum annual cost over the entire range. Setting AC(A) = AC(D) and solving for the percent change in annual disbursements gives

$$\$3,000(A/P, 10\%, 10) + \$1,800(1 + x) = \$5,000(A/P, 10\%, 10)$$
$$+ \$1,320(1 + x),$$
$$\$2,000(A/P, 10\%, 10) = \$480(1 + x),$$
$$\$325.50 = \$480(1 + x),$$
$$x = -0.322 \text{ or } -32.2\%.$$

Thus, so long as the annual disbursements are at least 67.8% of the estimated value, then alternative D will be preferred. On the basis of the analysis, it might be concluded that the decision is relatively insensitive to changes in annual disbursements. Furthermore, this example illustrates that the break-even point for an element in the comparison of two or more alternatives is the value of the element at which the alternatives are equally desirable.   ■

A related sensitivity test that is often quite valuable is to determine the relative (or absolute) change in one or more elements (parameters) which will just reverse the

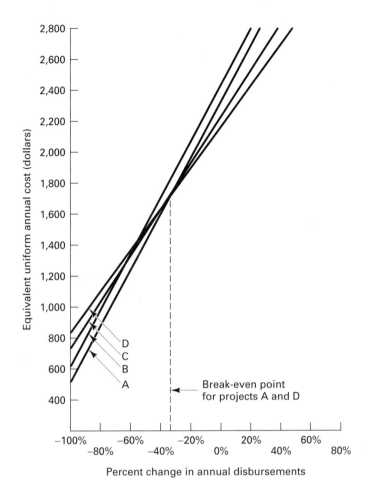

Figure 11-3   Sensitivity analysis of changes in operating disbursements for Example 11-2.

decision. Applied to the investment alternative example (Example 11-1), this means the relative change that will decrease the net AW by $636 so that it reaches $0. Table 11-1 shows this and emphasizes that decision reversal is most sensitive to relative changes in annual savings and least sensitive to relative changes in the salvage value. To further illustrate the computations, let y = estimated outcome for the investment at which AW(8%) = $0:

$$-Y(A/P, 8\%, 5) + \$2,000(A/F, 8\%, 5) + \$5,000 - \$2,200 = 0.$$

From this equation, we find $y \simeq \$12,540$, which is 25% (rounded) more than the $10,000 most likely estimate.

For examples of tabular and graphical displays of sensitivity analyses applied to comparison of alternatives, the interested reader is referred to Appendix 11-A.

TABLE 11-1 Example of Sensitivity to Decision Reversal

	Most likely estimate	To reverse decision (decrease AW to $0)		
		Estimate outcome	Change amount	Change amount as % of most likely
Investment:	$10,000	−$12,540	+$2,540	+ 25%
Life:	5 yr	3.7 yr	−1.3 yr	− 26%
Salvage value:	$2,000	−$1,740	−$3,740	−187%
Annual savings:	$5,000	$4,363	− $637	− 13%
Annual disbursements:	$2,200	$2,837	+ $637	+ 29%
Minimum attractive rate of return:	8%	16.2%	+8.2%	+103%

11.3 Multiparameter Procedures

The preceding discussion concentrated on the sensitivity of the measure of merit to changes in a single parameter one at a time. Such an approach allows the relative sensitivity of parameters to be identified. However, it overlooks the possibility of interaction among parameters. Additionally, since estimation errors will generally occur in more than one parameter, it is important to examine the sensitivity of the measure of effectiveness to multiple parameters.

Two approaches will be described in the consideration of multiparameter sensitivity analysis. First, the approach used in the one-at-a-time analysis will be extended to multiple parameters; second, an optimistic–pessimistic approach will be described.

11.3.1 Sensitivity Surface Approach

In the one-at-a-time analysis depicted in Figs. 11-1 and 11-2, sensitivity curves were obtained. If combinations of more than one parameter are analyzed, a sensitivity surface is required.

Example 11-3
Consider the data used to develop Figs. 11-1 and 11-2. Suppose the most critical parameters are believed to be the initial investment and the annual receipts. Perform a multiparameter sensitivity analysis involving the two parameters.

Solution   Let x denote the percent change in the initial investment, and let y denote the percent change in the annual receipts. The annual worth can be given as

$$AW = -\$10,000(1 + x)(A/P, 8\%, 5) + \$5,000(1 + y) - \$2,200$$
$$+ \$2,000(A/F, 8\%, 5)$$
$$= \$636.32 - \$2,504.60x + \$5,000y.$$

The investment will be profitable so long as AW ≥ 0 or

$$y \geq -0.127264 + 0.50092x.$$

Plotting the inequality relationship yields the two regions depicted in Fig. 11-4. The favorable region (AW > 0) is dominant. If errors in estimating the values of the investment and annual receipts were statistically independent and uniformly distributed over the interval ±10%, then

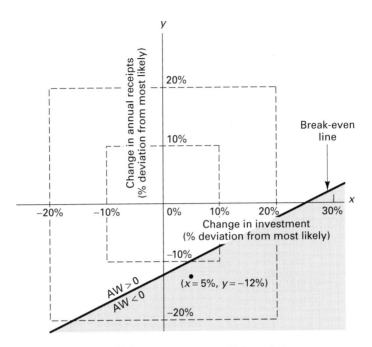

Figure 11-4 Two-parameter sensitivity analysis.

the probability of AW < 0 would be given by the ratio of the shaded area below the break-even line which is contained within the ±10 error region to the total area in the error region. However, it is not generally the case that estimation errors are either statistically independent or uniformly distributed. Furthermore, the incorporation of probabilistic considerations in sensitivity analyses lies in the domain of risk analysis, to be considered in chapter 12. If it is anticipated that ±20% estimation errors will be made, then there is greater concern for the profitability of the investment. For example, the combination of a 5% increase in the initial investment and a 12% reduction in annual receipts would result in a negative annual worth. ■

Example 11-4

Suppose that in the previous example the life of the investment were also a critical parameter. Perform a multiparameter sensitivity analysis for the three parameters.

   Solution   It is difficult to develop a three-dimensional representation of the sensitivity surface. But it is possible to gain an insight into the sensitivity of annual worth to errors in estimating the three parameters by plotting a family of break-even lines for each possible value of project life. Letting AW(N) denote the annual worth as a function of N, the following results are obtained:

$$\text{AW}(N) = -\$10,000(1 + x)(A/P, 8\%, N) + \$5,000(1 + y) - \$2,200$$
$$+ 2,000(A/F, 8\%, N) \geq 0,$$
$$\text{AW}(2) = -\$1,846.62 - \$5,607.70x + \$5,000y \geq 0,$$
$$y \geq 0.369324 + 1.12154x,$$
$$\text{AW}(3) = -\$464.24 - \$3,880.30x + \$5,000y \geq 0,$$
$$y \geq 0.092848 + 0.77606x,$$

$$AW(4) = \$224.64 - \$3,019.20x + \$5,000y \geq 0,$$
$$y \geq -0.044928 + 0.60384x,$$
$$AW(5) = \$636.32 - \$2,504.60x + \$5,000y \geq 0,$$
$$y \geq -0.127264 + 0.50092x,$$
$$AW(6) = \$909.44 - \$2,163.20x + \$5,000y \geq 0,$$
$$y \geq -0.181888 + 0.43264x,$$
$$AW(7) = \$1,121.15 - \$1,920.70x + \$5,000y \geq 0,$$
$$y \geq -0.22423 + 0.38414x.$$

The family of break-even lines is given in Fig. 11-5. For $N = 4$, 5, 6 (a $\pm 20\%$ range) the profitability of the investment continues to appear promising. However, for $N = 4$, there is very little opportunity for error in estimating the investment required and the annual receipts. For example, if $N = 4$, then with a 10% increase in the investment there must be at least a 1.55% increase in annual receipts for the annual worth to be positive. ■

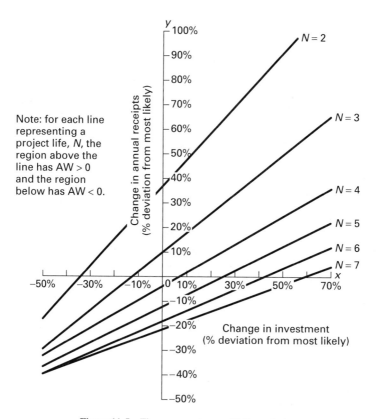

Figure 11-5   Three-parameter sensitivity analysis.

11.3.2 Optimistic–Pessimistic Approach — *For Project*

The optimistic–pessimistic approach, as its name implies, involves changing estimates of one or more elements (parameters) in a favorable outcome (optimistic) direction and in an unfavorable outcome (pessimistic) direction to determine the effect of these various changes on the economic study result.

In using this method, it is desirable for the estimator to adopt a guideline philosophy of "how optimistic" and "how pessimistic" in making the estimates. One convenient way to do this is to adopt a probabilistic statement such as: "An optimistic estimate will mean a value of the element that we would expect to be bettered or exceeded in outcome no more than, say, 5% of the time, while a pessimistic estimate is a value of the element that we would expect to be worse than the final outcome no more than, say, 5% of the time." If, for example, the 5% were changed to 25%, the optimistic and pessimistic estimates would correspond to quartile estimates described in the latter part of Chapter 10. Generally, though, 1% to 10% "tail" or "outside of" probabilities are used. Figure 11-6 illustrates this.

Example 11-5

For Example 11-1, suppose the optimistic, pessimistic, and most likely estimates are as follows:

	Estimation condition		
	Optimistic	Most likely	Pessimistic
Investment:	$10,000	$10,000	$10,000
Life:	7 yr	5 yr	4 yr
Salvage value:	$2,000	$2,000	$2,000
Annual receipts:	$6,000	$5,000	$4,500
Annual disbursements:	$2,200	$2,200	$2,400
Interest rate:	8%	8%	8%

Solution   Using the annual worth method:

$$\text{AW(optimistic)} = -\$10,000(A/P, 8\%, 7) + \$2,000(A/F, 8\%, 7) + \$3,800$$
$$= \$2,103,$$
$$\text{AW(most likely)} = -\$10,000(A/P, 8\%, 5) + \$2,000(A/F, 8\%, 5) + \$2,800$$
$$= \$636,$$
$$\text{AW(pessimistic)} = -\$10,000(A/P, 8\%, 4) + \$2,000(A/F, 8\%, 4) + \$2,100$$
$$= -\$475.$$

(11-1)

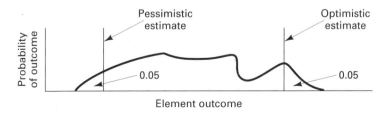

Figure 11-6   Illustration of estimates for optimistic–pessimistic approach.

TABLE 11-2 Calculated Results for All Combinations
of Estimating Conditions in Example 11-6

					Net annual worth					
		Annual disb.—O			Annual disb.—M			Annual disb.—P		
Life		O	M	P	O	M	P	O	M	P
Annual	O	$2,103*	$1,636*	$1,235	$2,103*	$1,636*	$1,235	$1,903*	$1,436	$1,035
receipts	M	1,103	636	235	1,103	636	235	903	436	35
	P	603	136	− 275	603	136	− 275	403	− 64	− 475

Key: O is optimistic outcome.   *Indicates net AW > $1,500
M is most likely outcome.
P is pessimistic outcome.

These example calculations are based on the assumption that the elements will all equal the pessimistic estimates, all equal the most likely estimates, or all equal the optimistic estimates. Of course, one can investigate the effect on calculated results when various elements equal the optimistic, most likely, or pessimistic estimates in various combinations. When this is done, it is usually helpful to summarize the results in tabular form.   ■

Example 11-6

For Example 11-5, note that estimates for the three conditions differed only for the project life, annual receipts, and annual disbursements. Table 11-2 shows the calculated results for all combinations of estimating conditions—optimistic (O), most likely (M), and pessimistic (P)—for only the three elements (parameters). Displays such as Table 11-2 can be made more easily informative by the use of graphical symbols. For example, all outcomes greater than $1,500 are marked for emphasis in Table 11-2. Figure 11-7 shows results in histogram bar form with each outcome rounded to the nearest $100. Such rounding can aid decision makers especially when showing results in a table. Other devices such as color coding, shading, etc., can be very useful for communicating in terms of tables and graphs.   ■

11.4 Summary

Several popular descriptive techniques were presented in this chapter for exploring the sensitivity of estimates in capital investment evaluations. Specifically, single-parameter and multiparameter sensitivity analyses were illustrated. In addition, break-even analysis was covered as a subset of one-at-a-time sensitivity analysis. Illustration of tabular and graphical approaches to sensitivity analysis for comparison of alternatives has been provided as an appendix to this chapter.

PROBLEMS

11-1. A certain potential investment project is critical to a firm. The following are "best" or "most likely" estimates:

Investment:	$100,000
Life:	10 yr
Salvage value:	$20,000
Net annual cash flow:	$30,000
Minimum required rate of return:	10%

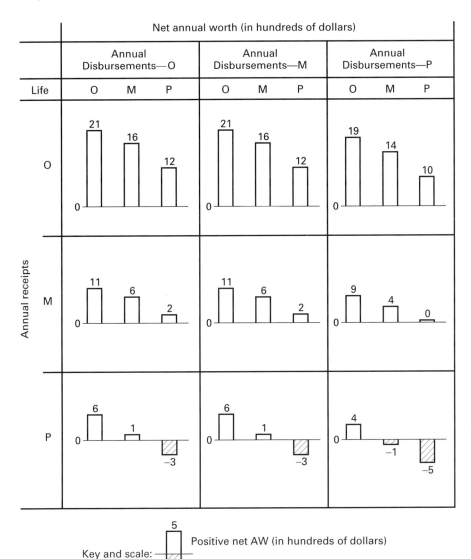

Figure 11-7   Example 11-6 results shown in histogram form with rounding to nearest $100.

It is desired to show the sensitivity of a measure of merit (net annual worth) to variation, over a range of ±50% of the expected values, in the following elements: (a) life, (b) net annual cash flow, (c) interest rate. Graph the results. To which element is the decision most sensitive?

11-2. Suppose that for a certain potential investment project the optimistic–pessimistic estimates are as follows:

	Optimistic	Most likely	Pessimistic
Investment:	$90,000	$100,000	$120,000
Life:	12 yr	10 yr	6 yr
Salvage value:	$30,000	$20,000	0
Net annual cash flow:	$35,000	$30,000	$20,000
Minimum required rate of return:	10%	10%	10%

a. What is the net annual worth for each of the three estimation conditions?

b. It is thought that the most critical elements are life and net annual cash flow. Develop a table showing the net annual worth for all combinations of estimates for those two elements assuming all other elements to be at their "most likely" values. Also, show results with histogram bars for entries in your table.

11-3. Two pumps, A and B, are being considered for a given drainage need. Both pumps operate at a rated output of 8 kW (10.7 hp) but differ in initial cost and electrical efficiency. Electricity costs 5¢ per kWh, and the minimum before-tax rate of return is 15%. Below are relevant data.

	Pump A	Pump B
Cost installed:	$3,500	$4,500
Expected life to termination at zero market value:	10 yr	15 yr
Maintenance cost per 1,000 hr of operation:	$50	$30
Efficiency:	60%	80%

a. The critical variable hardest to estimate is the number of hours of operation per year. Determine the break-even point for this variable.

b. If the operating time is greater than the break-even point in part (a), which pump is better?

c. Plot the total annual costs of each pump as a function of hours of operation.

11-4. A new all-season hotel will require land costing $300,000 and a structure costing $500,000. In addition, fixtures will cost $150,000, and working capital of 30 days' gross income at 100% capacity will be required. While the land is nondepreciable, the investment in fixtures should be recovered in 8 years, and the investment in the structure should be recovered in 25 years.

When the hotel is operating at 100% capacity, the gross income will be $1,400 per day. Fixed operating expenses, exclusive of depreciation and interest, will amount to $120,000 per year. Operating expenses, which vary linearly in proportion to the level of operation, are $80,000 per year at 100% capacity.

a. At what percentage of capacity must the hotel operate to earn a before-tax minimum attractive rate of return of 20%?

b. Plot the revenue and total cost (including cost of capital) as a function of percentage of capacity. (*Note:* The point at which revenue and the total of all costs, including capital, are equal is commonly called the *unhealthy point.*)

11-5. It is desired to determine the most economical thickness of insulation for a large cold-storage room. Insulation is expected to cost $150 per 1,000 sq ft of wall area per in. of thickness installed and to require annual property taxes and insurance of 5% of first cost. It is expected to have $0 net salvage value after a 20-year life. The following are estimates of the heat loss per 1,000 sq ft of wall area for several thicknesses:

Insulation, in.	Heat loss, Btu per hr
3	4,400
4	3,400
5	2,800
6	2,400
7	2,000
8	1,800

The cost of heat removal is estimated at $0.02 per 1,000 Btu per hr. The minimum required yield on investment is 20%. Assuming continuous operation throughout the year, analyze the sensitivity of the optimal thickness to errors in estimating the cost of heat removal. Use the AW technique (a computer spreadsheet should be considered here).

11-6. An industrial machine costing $10,000 will produce net cash savings of $4,000 per year. The machine has a five-year economic life but must be returned to the factory for major repairs after three years of operation. These repairs cost $5,000. The company's cost of capital is approximately 10%. What rate of return will be earned on purchase of this machine? Analyze the sensitivity of the internal rate of return to ±$2,000 changes in the repair cost.

11-7. The following alternatives are available to fill a given need that is expected to exist indefinitely. Each is expected to have $0 salvage value at the end of each life cycle.

	Plan A	Plan B	Plan C
First cost:	$2,000	$6,000	$12,000
Life cycle:	6 yr	3 yr	4 yr
Annual disbursements:	$3,500	$1,000	$400

a. Analyze the sensitivity of the preferred plan due to ±30% errors in estimating the annual disbursements. Use a MARR of 10%.

b. Analyze the sensitivity of the preferred plan due to ±50% errors in estimating the MARR (i.e., the MARR will vary from 5% to 15%).

11-8. An improved facility costing $50,000 has been proposed. Construction time will be 2 years with expenditures of $20,000 the first year and $30,000 the second year. Cash flows are as follows:

Year	Savings
−1	−$20,000
0	−30,000
1	10,000
2	14,000
3	18,000
4	22,000
5	26,000

The facility will not be required after 5 years and will have a market value of $5,000. Analyze the sensitivity of annual worth due to errors in estimating both the savings in the first year and the magnitude of the gradient amount. Use a table to show results of ±50% changes in both variables. The MARR is 10% per year.

11-9. It is desired to determine the optimal height for a proposed building that is expected to last 40 years and then be demolished at zero salvage. The following are pertinent data:

	Number of floors			
	2	3	4	5
Building first cost:	$200,000	$250,000	$320,000	$400,000
Annual revenue:	40,000	60,000	85,000	100,000
Annual cash disbursements:	15,000	25,000	25,000	45,000

In addition to the building first cost, the land requires an investment of $50,000 and is expected to retain that value throughout the life period. Analyze the sensitivity of the decision due to changes in estimates of the MARR between 10%, 15%, and 20%. Use the PW method and ignore income taxes.

11-10. There are five alternative machines to do a given job. Each is expected to have a market value of 100% of the investment amount at the end of its life of 4 years. The firm's minimum attractive rate of return is 12%. Analyze the sensitivity of the best choice due to the market value ranging from 0% to 100% of the initial investment.

	Machine				
	A	B	C	D	E
Investment:	$10,000	$14,000	$21,000	$27,000	$34,000
Net cash flow per yr:	1,100	1,800	2,800	3,400	4,450

11-11. Best estimates of the parameters for an investment are given below. It is expected that the investment will be repeated indefinitely.

Initial investment:	$15,000
Net annual receipt:	$2,500
Project life:	10 yr
Salvage value:	$0
MARR = 15%	

a. Perform a one-at-a-time sensitivity analysis to help determine the most critical parameter(s). One way to do this is to calculate "sensitivity to decision reversal" as in Table 11-1.

b. Perform a multiparameter sensitivity analysis using the sensitivity surface approach. Base the analysis on the initial investment and the net annual receipts.

11-12. An office building is considering converting from a coal burning furnace to one that burns either fuel oil or natural gas. The cost of converting to fuel oil is estimated to be $80,000 initially; annual operating costs are estimated to be $4,000 less than that experienced using the coal furnace. Approximately 140,000 Btus are produced per gallon of fuel oil; fuel oil is anticipated to cost $1.10 per gallon.

The cost of converting to natural gas is estimated to be $60,000 initially; additionally, annual operating and maintenance costs are estimated to be $6,000 less than that for the coal-burning furnace. Approximately 1,000 Btus are produced per cubic foot of natural gas; it is estimated natural gas will cost $0.02 per cu ft.

A planning horizon of twenty years is to be used. Zero salvage values and a 10% MARR are appropriate. Perform a sensitivity analysis for the annual Btu requirement for the heating system. (*Hint:* First calculate the break-even number of Btus (in thousands). Then determine AWs if Btu requirement is ±30% of the break-even amount.)

11-13. A company is considering investing $10,000 in a heat exchanger. The heat exchanger will last 5 years, at which time it will be sold for $2,000. The maintenance cost at the end of the first year is estimated to be $1,000 and then will increase by $500 per year over its life. As an alternative, the company may lease the equipment for $3,500 per year, including maintenance. The MARR (before tax) is 5%.

Perform a multiparameter sensitivity analysis for this problem, where x = change in initial investment of the heat exchanger and y = change in annual leasing expense. Draw a diagram to illustrate your answer.

11-14. Black Diamond Coal Company is considering an investment in a piece of property (land including mineral rights). The property will cost $24 million. It is estimated that 40 million tons of run-of-mine coal are recoverable from the property. The run-of-mine coal will be cleaned by a neighboring coal cleaning plant. The cleaning recovery is 70%, and the cleaning costs will be $5 per ton of *cleaned* coal.

The project requires a $60 million investment in mine development and equipment and will deplete the coal reserves at a rate of 2 million tons of cleaned coal per year. The investment in mine development and equipment will be capitalized and recovered by the straight-line depreciation method over a 14-year period. The depletion allowance is 10% of net before-tax income, and depletion serves to reduce taxable income (just as depreciation does). The selling price of cleaned coal is $38 per ton. Working capital is estimated at $5 million for the project. Working capital and one-half the purchase price of the property are salvageable at the end of the project's life. The operating costs are estimated at $18.50 per ton of cleaned coal. The income tax is 50% of taxable income (mining companies are subject to more taxes than a manufacturing company). Assume all investments occur at time zero and all cash flows are end-of-year cash flows. The MARR for Black Diamond Coal Company is 15%.

a. Determine the annual worth of the project.

b. How sensitive is the annual worth to plus and minus 10% changes in the average operating cost?

11-15. A single-stage centrifugal blower is to be purchased for a manufacturing plant. Suppliers have been consulted and the choice has been narrowed down to two new units of modern design, both made by the same company and both having the same rated capacity and pressure. Both are driven at 3600 rpm by identical 200-hp electric motors.

One blower has a guaranteed efficiency of 72% at full load and is offered installed for $5,000. A second blower is more expensive because of aerodynamic refinement, which gives it a guaranteed efficiency of 75% at full load.

Except for those differences in efficiency and price, the units are equally desirable in all respects such as durability, maintenance, ease of operation, and quietness. In both cases plots of efficiency versus amount of air handled are flat in the vicinity of full rated load. The application is such that whenever the blower is running it will be at full load.

Ignore income tax differences that may arise between the two blowers. Assume both blowers have negligible salvage values. The firm's MARR is 10% per year.

a. Develop a formula for calculating how much the manufacturer could afford to pay for the more efficient unit. (*Hint:* You need to specify important variables and use them in your formula.)

b. Sketch a graph, for your response to (a), in which the affordable amount is a function of the efficiency of the second (more expensive) blower.

11-16. (*Note:* This problem is based on Appendix 11-A.) The following alternatives will fill a given need. "Repeatability" can be assumed and MARR = 10%.

	Plan	
	A	B
Investment:	$2,000	$6,000
Life:	6 yr	3 yr
Annual disbursements:	$3,500	$1,000
Salvage value:	$0	$4,000

a. Calculate the "assumed-certain" equivalent annual cost for each.

b. Set up a "Sensitivity to Decision Reversal" table (recognizing the decision is reversed when $AW(A) \geq AW(B)$), and show calculations for
(i) life for project A only;
(ii) annual disbursements for project A only.

c. Set up a sensitivity surface equation and graph for showing the following proportion changes.

x = proportion change in project A investment only

y = proportion change in project A annual disbursements only

Show the area (domain) in which project A is preferred to project B.

11-17. Consider these two alternatives for solid-waste removal:

Alternative A: Build a Solid-Waste Processing Facility

Capital cost:	$108 million in 1996 (commercial operation starts in 1996)
Expected life of facility:	20 years
Annual operating costs:	$3.46 million (expressed in 1996 dollars)
Estimated salvage value:	40% of initial capital cost at all times

Alternative B: Contract with Vendors for Solid Waste Disposal After Intermediate Recovery

Capital cost:	$17 million in 1996 (this is for *intermediate* recovery from the solid waste stream)
Expected contract period:	20 years
Annual operating costs:	$2.10 million (in 1996 dollars)
Repairs to intermediate recovery system every 5 years:	$3.0 million (in 1996 dollars)
Annual fee to vendors:	$10.3 million (in 1996 dollars)
Estimated salvage value at all times:	$0

Related Data

MACRS class life:	15 years (see Table 6-4)
Study period:	20 years
Effective income tax rate:	40%
Company MARR (after tax):	10.0%
Inflation rate:	0% (ignore inflation)

a. How much more expensive (in terms of capital cost only) could alternative B be in order to break even with alternative A?

b. How sensitive is the after-tax PW of alternative B to cotermination of both alternatives at the end of *year 10?*

c. Is the initial decision to adopt alternative B in (a) reversed if our company's annual operating costs for alternative B only ($2.10 million per year) unexpectedly double? Explain why (or why not).

11-18. Refer to Problem 8-13. Based on the data provided, the preferred choice is the *challenger.* Consider the sensitivity of this selection to changes in the effective income tax rate, and further assume that long-term depreciation recapture (losses or gains) is taxed at one-half the effective income tax rate. Try to develop a computer spreadsheet as part of your solution.

[Answers to Problem 10-18: (a) 3.62 million square miles; (b) 10.6 ounces; (c) 640 miles per hour; (d) 20,320 ft; (e) 639 muscles; and (f) 70.2%.]

APPENDIX 11-A

Tabular and Graphical Illustrations of Sensitivity Analysis for the Comparison of Alternatives*

In this appendix we illustrate sensitivity analysis with a series of examples involving simple graphical and tabular techniques applied to a two-alternative case. Most of these examples could be extended to the comparison of three or more alternatives; and, of course, measures of merit other than net AW could be used.

One might conclude that use of all the types of tabular and graphical techniques to be illustrated would be "overkill" for comparison of a given set of alternatives. However, it is very important to use visual displays to communicate effectively the results of an engineering economy study. Consequently, one can select from among the techniques shown (or other variations one might create) according to which are judged to be most useful/practicable for decision-making purposes.

Assumed-Certain Example

Table 11-1-A is a typical assumed-certain or "expected" (or most likely) set of estimates for example alternatives A and B. The net annual worth (net AW) measure of merit for each alternative is shown for a before-tax minimum attractive rate of return (MARR) of 20%. As an example of calculations, the net AW for alternative B at $i = 20\%$ can be determined as

$$-[(\$200,000 - \$40,000)(A/P, 20\%, 15) + \$40,000(20\%)] + \$72,000 - \$20,000$$
$$= \$+9,775.$$

The bracketed term is, of course, the capital recovery cost for alternative B. The net AW for alternative B is not only positive, thus indicating an acceptable project, but it is $3,550 greater than the $+6,225 net AW for alternative A. Based on economic considerations alone, alternative B would be the recommended choice assuming certainty.

TABLE 11-1-A Example "Expected" Estimates with Calculated Net Annual Worths

Variable	Alternative A	Alternative B
Investment:	+$150,000	$200,000
Life:	10 yr	15 yr
Salvage value:	$0	$40,000
Annual receipts or savings:	$87,000	$72,000
Annual disbursements:	$45,000	$20,000
Net AW at $i = 20\%$:	$+6,225	$+9,775

*Adapted from J. R. Canada, and W. G. Sullivan, *Economic and Multiattribute Evaluation of Advanced Manufacturing Systems* (Englewood Cliffs, NJ: Prentice-Hall, 1989). Reprinted by permission of the publisher.

It should be noted that the analyses of mutually exclusive alternatives having different economic lives as shown herein all use the common "repeatability" assumptions.

Exploration of Sensitivity to Interest Rates

Any parameter (variable) could be varied by ±20%, for example, to establish a rank ordering of impact (significance on the chosen measure of merit). Here we suppose that the interest rate (MARR) is a major contributor to the uncertainty of project economic desirability. Figure 11-1-A shows plots of the net AWs for both alternatives as a function of the interest rate varying from 15% to 30%. It is instructive to note that the two curves intersect at an interest rate, $i = 26\%$. When $i < 26\%$, alternative B has a higher net AW and is preferred, but alternative A is preferred for $i > 26\%$. For all examples presented in the remainder of this appendix, the MARR is held constant at 20%.

Optimistic–Pessimistic Estimates

The optimistic and pessimistic estimates shown in Table 11-2-A happen to be ±10% of expected estimates in Table 11-1-A for alternative A and ±20% of expected estimates for alternative B merely for convenience in illustration. It is reasonable that

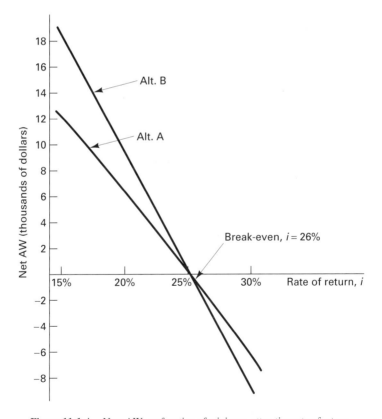

Figure 11-1-A   New AWs as function of minimum attractive rate of return.

TABLE 11-2-A   Example Optimistic and Pessimistic Estimates[a] with Calculated Net AWs

	Optimistic estimates		Pessimistic estimates	
	Alternative A	Alternative B	Alternative A	Alternative B
Investment:	$135,000	$160,000	$165,000	$240,000
Life:	11 yr	18 yr	9 yr	12 yr
Salvage value:	$0	$40,000	$0	$40,000
Annual receipts or savings:	$95,700	$86,400	$78,300	$57,600
Annual disbursements:	$40,500	$16,000	$49,500	$24,000
Net AW at $i = 20\%$:	$+24,002	$+37,464	$-12,137	$-19,460
AW(B) − AW(A)		$+13,462		$-7,323

[a] For this example, the optimistic and pessimistic estimates happen to be ±10% of expected estimates for alternative A and ±20% of expected estimates for alternative B. (*Exception:* Salvage value does not vary.)

for given "tail probabilities," some of the estimates will vary widely and some little or not at all in a favorable or unfavorable direction.

The results of Table 11-2-A are useful if one thinks that outcomes for the two alternatives will be perfectly correlated. That is, if both sets of estimates take on their "optimistic" values, B is better than A by a net AW of $13,462. On the other hand, if both sets assume their "pessimistic" values, then alternative A is better than alternative B by a net AW of $7,323.

Without perfectly correlated outcomes, there are $3^2 = 9$ combinations of estimates to consider. Each combination requires a separate set of calculations to compute the measure of merit. Probability statements regarding the final measure of merit cannot be formulated unless much more information is given.

Table 11-3-A allows one to explore the effects of different combinations of optimistic and pessimistic outcomes on the difference between the two alternatives. The diagonal that is highlighted in Table 11-3-A reflects the results obtained from Tables 11-1-A and 11-2-A. As an example, suppose that one were concerned with the

TABLE 11-3-A   Difference in Net Annual Worth of Alternative B Compared to Alternative A [AW(B) − AW(A)] for All Combinations of Optimistic, Expected, and Pessimistic Estimate Sets[a]

	Alternative A		
Alternative B	Optimistic (AW = $+24,002)	Expected (AW = $+6,225)	Pessimistic (AW = $-12,137)
Optimistic (AW = $+37,464)	$+13,462	$+31,239	$+49,601
Expected (AW = $+9,775)	−14,227	+ 3,550	−21,912
Pessimistic (AW = $-19,460)	−43,462	−25,685	− 7,323

[a] Positive-valued entries favor alternative B; negative-valued entries favor alternative A.

difference between the alternatives if alternative A results in its "optimistic" outcome and alternative B results in its "expected" outcome. For this case alternative A would be better than alternative B by a net AW of $14,227. As another example, one might be interested in the *extremes* that alternative B could be better than alternative A by a net AW of $49,601, or the reverse could be true by a net AW of $43,462.

Sensitivity Tables

Table 11-4-A shows "sensitivity to decision reversal" caused by changes in expected estimates for alternative A, then alternative B, and then the difference between the alternatives. For each variable the amount and the percent change that would just make the AWs for the two alternatives equal are shown. This shows, for example, that relatively low percentage changes in receipts or disbursements or investment required for either alternative would reverse the decision. Thus, one could say that the decision is relatively sensitive to changes in those three variables. Furthermore, one might reasonably conclude that the decision is only moderately sensitive to changes in the lives and quite insensitive to changes in the salvage values.

Sensitivity Graphs

Figure 11-2-A is a useful type of graph for exploring sensitivity to changes in two variables at a time. In this case the net AW for both alternatives is shown as a function of the net annual cash flows (i.e., annual receipts minus annual disbursements) varying over a range on the *x*-axis and for various assumed lives. All other variables are held constant at their "expected" values. Thus, one can choose particular combinations of outcomes for the variables plotted in Fig. 11-2-A to observe how the net AWs compare for alternatives A and B. For instance, a 25% increase in net annual cash flows and a life of 15 years for alternative A will cause roughly the same net AW as a 25% increase in net annual cash flows and a life of 10 years for alternative B.

TABLE 11-4-A Sensitivity to Decision Reversal (Based on Expected Estimates)

a. Alternative A

	Expected estimate	To reverse decision (Net AW same as for alternative B)		
		Estimated outcome	Change amount	Change as % of expected
Investment:	$150,000	$135,115	$-14,885	-10.0
Life:	10 yr	14.7 yr	+4.7 yr	+47.0
Salvage value:	$0	$92,208	+$92,208	—
Annual receipts				
or savings:	$87,000	$90,550	$+3,550	+ 4.1
Annual				
disbursements:	$45,000	$41,450	$-3,550	- 7.9

b. Alternative B

	Expected estimate	To reverse decision (Net AW same as for alternative A)		
		Estimated outcome	Change amount	Change as % of expected
Investment:	$200,000	$216,601	+$16,601	+ 8.3
Life:	15 yr	10.3 yr	-4.7 yr	- 31.3
Salvage value:	$40,000	$-215,468	$-255,468	-639
Annual receipts				
or savings:	$72,000	$68,450	$-3,550	- 4.9
Annual				
disbursements:	$20,000	$23,550	$+3,550	+ 17.8+

c. Difference (Alt. B − Alt. A), or, (Alt. A → Alt. B)

	Expected estimate	To reverse decision (Equal net AWs)		
		Estimated outcome	Change amount	Change as % of expected
Investment:	$50,000	Meaningful only if lives same		
Life:	5 yr	Meaningful only if lives same		
Salvage value:	$40,000	Meaningful only if lives same		
Annual receipts				
or savings:	-$15,000	-$18,550	$-3,550	23.7
Annual				
disbursements:	-$25,000	-$28,550	$-3,550	14.2
MARR ($i\%$)	Both alts. 20%	25.9%	+5.9%	29.5

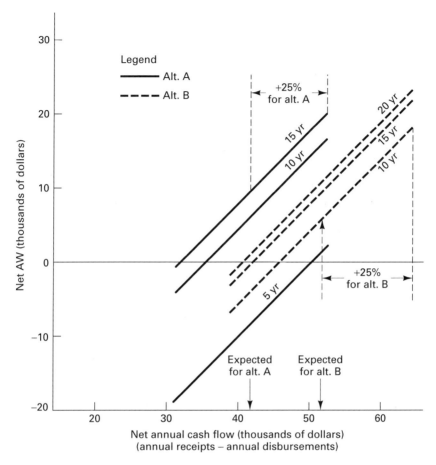

Figure 11-2-A   Sensitivity of net annual worths to net annual cash flows varying ±25% of expected and project lives varying ± 5 years from expected for alternatives A and B.

12
Analytical and Simulation Approaches to Risk Analysis

12.1 Introduction

As was noted in Chapter 11, *sensitivity analyses* are performed under *conditions of uncertainty* for an investment in order to *describe* the effects of the measure of effectiveness due to either estimation errors or changes in the values of one or more parameters. When *conditions of risk* are present, the process is referred to as *risk analysis.*

The term "risk analysis" has different interpretations among various government agencies and units, as well as nongovernment organizations. However, there is a growing acceptance that risk analysis involves the development of the probability distribution of the measure of effectiveness. Furthermore, the *risk* associated with an investment alternative is generally either given as the probability of an unfavorable value for the measure of effectiveness or determined by the variance of the measure of effectiveness.

This chapter deals with risk as a probabilistic phenomenon. In this regard analytical approaches to risk analysis are first discussed, followed by the inclusion of Monte Carlo methods for simulating risky economic situations.

12.2 Analytical Methods of Risk Analysis

Typical parameters for which conditions of risk can reasonably be expected to exist include the initial investment, yearly operating and maintenance expenses, salvage values, the life of an investment, the planning horizon, and the minimum attractive rate of return. The parameters can be statistically independent, correlated with time, and/or correlated with each other. Analytical methods of performing risk analyses are described in this section.

12.2.1 Random Cash Flows

In order to determine analytically the probability distribution for the measure of effectiveness, a number of simplifying assumptions are normally made. The simplest situation is one involving a known number of random and statistically independent cash flows. As an example, suppose the random variable A_j denotes the net cash flow occurring at the end of period j, $j = 0, 1, \ldots, N$. Hence, the present worth is given by

$$PW = \sum_{j=0}^{N} (1 + i)^{-j} A_j. \tag{12-1}$$

Since the expected value of a sum of random variables equals the sum of the expected values of the random variables, then the expected present worth is given by[1]

$$E[PW] = \sum_{j=0}^{N} (1 + i)^{-j} E[A_j], \tag{12-2}$$

where $E[\cdot]$ denotes the expected value. Furthermore, since the A_j's are statistically independent, then the variance of present worth is given by[2]

$$V[PW] = \sum_{j=0}^{N} (1 + i)^{-2j} V[A_j], \tag{12-3}$$

where $V[\cdot]$ denotes the variance, and $\sigma[\cdot] = \sqrt{V[\cdot]}$ will denote the standard deviation.

The central limit theorem, from probability theory, establishes that the sum of independently distributed random variables tends to be normally distributed as the number of terms in the summation increases.[3] Hence, as N increases, the PW tends to be normally distributed with a mean value of $E[PW]$ and a variance of $V[PW]$.

Example 12-1

Consider the following data, where the MARR = 10%:

j	$E[A_j]$	$V[A_j]$
0	−$10,000	1×10^6
1–10	$1,800	4×10^4

Determine the probability of a negative present worth.

---

[1] Recall that the expected value of a constant times a random variable equals the constant times the expected value of the random variable.

[2] Recall that the variance of a constant times a random variable equals the constant squared times the variance of the random variable.

[3] An additional condition is that the random variables have finite moments. The condition is usually met in risk analysis.

Solution

$$E[PW] = -\$10,000 + \sum_{j=1}^{10} (1.10)^{-j} \$1,800$$

$$= -\$10,000 + \$1,800(P/A, 10\%, 10)$$

$$= \$1,059.20,$$

$$V[PW] = 1 \times 10^6 + (4 \times 10^4) \sum_{j=1}^{10} (1.10)^{-2j},$$

which reduces to[4]

$$V[PW] = 1 \times 10^6 + (4 \times 10^4)(P/A, 10\%, 20)/2.10$$

$$= 116.217 \times 10^4.$$

The probability of PW ≤ 0 is obtained as follows:

$$\Pr(PW \le 0) = \Pr\left(S < \frac{0 - E[PW]}{\sqrt{V[PW]}} \right)$$

$$= \Pr\left(S \le \frac{0 - 1,059.20}{\sqrt{116.217 \times 10^4}} \right) \tag{12-4}$$

$$= \Pr\left(S \le -\frac{1,059.20}{1,078.04} \right)$$

$$= \Pr(S \le -0.982)$$

$$= 0.163,$$

where S is the standard normal deviate for which a table of probabilities is given in Appendix E.■

A slightly more complex situation involves a set of correlated cash flows. In particular, if the A_j's are not statistically independent, then the variance calculation is modified as follows:

$$V[PW] = \sum_{j=0}^{N} V[A_j](1 + i)^{-2j} + 2\sum_{j=0}^{N-1} \sum_{k=j+1}^{N} \text{Cov}[A_j, A_k](1 + i)^{-(j+k)}, \tag{12-5}$$

where $\text{Cov}[A_j, A_k]$ is the covariance between A_j and A_k.[5] Alternately, the variance can be given as

$$V[PW] = \sum_{j=0}^{N} V[A_j](1 + i)^{-2j} + 2\sum_{j=0}^{N-1} \sum_{k=j+1}^{N} \rho_{jk}\sigma[A_j]\sigma[A_k](1 + i)^{-(j+k)} \tag{12-6}$$

where ρ_{jk} is the correlation coefficient between A_j and A_k.

---

[4] The relation $\sum_{j=1}^{N} (1 + i)^{-2j} = (P/A, i\%, 2N)/(2 + i)$ is used.

[5] Recall that $\text{Cov}[X, Y] = E[XY] - E[X]E[Y]$.

If all A_j and A_k are perfectly correlated such that $\rho_{jk} = +1$, then

$$V[\text{PW}] = \left\{ \sum_{j=0}^{N} \sigma[A_j](1 + i)^{-j} \right\}^2 . \tag{12-7}$$

When dealing with correlated random variables there is greater uncertainty that a normal approximation is reasonable as the distribution for PW. However, such an assumption is often made when a large number of random variables are involved and not all are correlated. In the absence of suitable conditions for the use of a normal approximation, Monte Carlo simulation is recommended.

In performing risk analyses involving correlated cash flows, Hillier[6] argues that it is probably unrealistic to expect that accurate estimates of covariances can be obtained. Consequently, it is suggested that the net cash flow in any year be separated into those components of cash flow that one can reasonably expect to be independent from year to year and those that are correlated over time. Specifically, Hillier assumes that the net cash flow in year j can be represented by

$$A_j = X_j + Y_{j1} + Y_{j2} + \cdots + Y_{jm}, \tag{12-8}$$

where X_j denotes the component of cash flow that is statistically independent for all values of j, and Y_{jk} denotes the component of cash flow that is *perfectly* correlated with $Y_{0k}, Y_{1k}, \ldots, Y_{Nk}$. Notice that $Y_{j1}, Y_{j2}, \ldots, Y_{jm}$ are statistically independent. Thus, it is assumed that Y_{jk} is perfectly correlated over j and perfectly uncorrelated over k.

The resulting expected value of present worth and variance of present worth can be given as

$$E[\text{PW}] = \sum_{j=0}^{N} E[X_j](1 + i)^{-j} + \sum_{j=0}^{N} \sum_{k=1}^{m} E[Y_{jk}](1 + i)^{-j} \tag{12-9}$$

and

$$V[\text{PW}] = \sum_{j=0}^{N} V[X_j](1 + i)^{-2j} + \sum_{k=1}^{m} \left\{ \sum_{j=0}^{N} \sigma[Y_{jk}](1 + i)^{-j} \right\}^2 . \tag{12-10}$$

Example 12-2

A business is considering purchasing a numerically controlled milling machine. A firm quote indicates the machine can be purchased and installed for $150,000. A planning horizon of ten years is used in the analysis. The salvage value is anticipated to be normally distributed with a mean of $25,000 and a standard deviation of $2,500. Labor costs are expected to be perfectly correlated; maintenance costs are also expected to be perfectly correlated; and other annual operating costs are anticipated to be statistically independent. The expected values and standard deviations are estimated as follows:

---

[6] F. S. Hillier, "The Derivation of Probabilistic Information for the Evaluation of Risky Investments," *Management Science* 9, no. 3 (1963):443–457.

$$E[\text{labor } j] \qquad = -\$10,000 - (j-1)\$1,000,$$
$$\sigma[\text{labor } j] \qquad = 0.10E[\text{labor } j],$$
$$E[\text{maintenance } j] = -\$1,000 - (j-1)\$500,$$
$$\sigma[\text{maintenance } j] = 0.20E[\text{maintenance } j],$$
$$E[\text{other } j] \qquad = -\$3,000,$$
$$\sigma[\text{other } j] \qquad = \$500.$$

Using a MARR of 20%, it is desired to determine $E[\text{PW}]$ and $\sigma[\text{PW}]$.

Solution

$$
\begin{aligned}
E[\text{PW}] = \ &-\$150,000 + \$25,000(P/F, 20\%, 10) \\
&- [\$10,000 + 1,000(A/G, 20\%, 10)](P/A, 20\%, 10) \\
&- [\$1,000 + \$500(A/G, 20\%, 10)](P/A, 20\%, 10) \\
&- \$3,000(P/A, 20\%, 10) \\
= \ &-\$227,507
\end{aligned}
$$

$$
V[\text{PW}] = (500)^2 \sum_{j=1}^{10} (1.20)^{-2j} + (2,500)^2 (1.20)^{-20}
$$

$$
+ \left\{ \sum_{j=1}^{10} [1,000 + 100(j-1)](1.20)^{-j} \right\}^2 + \left\{ \sum_{j=1}^{10} [200 + 100(j-1)](1.20)^{-j} \right\}^2
$$

$$
= \frac{250,000(P/A, 20\%, 20)}{2.20} + 6,250,000(P/F, 20\%, 20)
$$

$$
+ \{[1,000 + 100(A/G, 20\%, 10)](P/A, 20\%, 10)\}^2
$$

$$
+ \{[200 + 100(A/G, 20\%, 10)](P/A, 20\%, 10)\}^2
$$

$$
\doteq 35.25 \times 10^6
$$

Therefore, $\sigma[\text{PW}] = \$5,937$. ∎

Example 12-3

The firm described in the previous example is considering as an alternative a semiautomated milling machine that will cost $80,000 initially and have a normally distributed salvage value in 10 years with a mean of $15,000 and a standard deviation of $1,000. As with the numerically controlled alternative, yearly labor costs and yearly maintenance costs are perfectly correlated; other annual operating costs are anticipated to be statistically independent. The following expected values and standard deviations have been estimated:

$$E[\text{labor } j] \qquad = -\$20,000 - (j-1)\$2,000,$$
$$\sigma[\text{labor } j] \qquad = 0.15E[\text{labor } j],$$
$$E[\text{maintenance } j] = -\$500 - (j-1)\$400,$$
$$\sigma[\text{maintenance } j] = 0.20E[\text{maintenance } j],$$
$$E[\text{other } j] \qquad = -\$6,000,$$
$$\sigma[\text{other } j] \qquad = \$1,000.$$

Using a MARR of 20%, it is desired to determine the probability that the numerically controlled milling machine is the preferred alternative. The present worth of each alternative is assumed to be normally distributed.

Solution

$$E[PW] = -\$80,000 + \$15,000(P/F, 20\%, 10)$$
$$- [\$20,000 + 2,000(A/G, 20\%, 10)](P/A, 20\%, 10)$$
$$- [\$500 + \$400(A/G, 20\%, 10)](P/A, 20\%, 10) - \$6,000(P/A, 20\%, 10)$$
$$= -\$219,552$$

$$V[PW] = \frac{(1,000)^2 (P/A, 20\%, 20)}{2.20} + (1,000)^2 (P/F, 20\%, 20)$$
$$+ \{[3,000 + 300(A/G, 20\%, 10)](P/A, 20\%, 10)\}^2$$
$$+ \{[100 + 80(A/G, 20\%, 10)](P/A, 20\%, 10)\}^2$$
$$\doteq 274.51 \times 10^6$$

Therefore, $\sigma[PW] = \$16,568$.

Letting PW_1 denote the present worth of the numerically controlled milling machine and PW_2 denote the present worth of the semiautomatic machine, then it is desired to determine $Pr(PW_1 > PW_2)$ or $Pr(PW_1 - PW_2 > 0)$. Letting $Y = PW_1 - PW_2$, if PW_1 and PW_2 are normally distributed, then Y will be normally distributed with an expected value of $E[Y] = E[PW_1] - E[PW_2]$ and a variance of $V[Y] = V[PW_1] + V[PW_2]$. Therefore,

$$E[Y] = -\$227,507 + \$219,552 = -\$7,955,$$
$$V[Y] = (35.25 \times 10^6) + (274.51 \times 10^6) = 309.76 \times 10^6,$$

and

$$Pr(Y > 0) = Pr\left(S > \frac{0 - (-\$7,955)}{\sqrt{309.76 \times 10^6}} \right) = Pr(S > 0.452),$$

or, using Appendix E,

Pr(numerically controlled milling machine is the most economical) = 0.3257.   ■

Example 12-4

In the previous example, suppose the machine is to be used to manufacture parts for a 10-year production contract yielding a guaranteed annual income of \$57,000. What is the probability that the firm will not make at least a 20% internal rate of return on the investment for each alternative?

Solution   So long as there do not exist multiple roots in determining the internal rate of return, then

$$Pr(i^* < MARR) = Pr(PW < 0 \mid MARR).$$

Therefore, for the numerically controlled milling machine,

$$Pr(i^* < 20\%) = Pr[\$57,500(P/A, 20\%, 10) + PW_1 < 0]$$
$$= Pr(PW_1 < -\$241,040)$$
$$= Pr\left(S < \frac{-\$13,533}{\sqrt{35.25 \times 10^6}} \right) = Pr(S < -2.297) = 0.0113.$$

For the semiautomatic milling machine,

The numerically controlled milling machine has the lower expected present worth. However, because of its smaller variance, it has a smaller probability of failing to earn at least a 20% internal rate of return. Thus, there is a 1.13% risk of that type associated with the numerically controlled machine, as opposed to a 9.73% risk for the semiautomatic machine. ∎

12.2.2 Random Project Life

Next, we consider the case in which the project life is a random variable, N, defined over the positive integers, i.e., $N = 1, 2, \ldots$. In this case, the present worth is given by

$$\text{PW} = A_0 + \sum_{j=1}^{N} (1 + i)^{-j} A_j. \tag{12-11}$$

If the A_j's are statistically independent random variables, then

$$E[\text{PW} \mid N] = E[A_0] + \sum_{j=1}^{N} (1 + i)^{-j} E[A_j] \tag{12-12}$$

and

$$E[\text{PW}] = E[A_0] + \sum_{N=1}^{\infty} \left(\sum_{j=1}^{N} (1 + i)^{-j} E[A_j] \right) \cdot P(N), \tag{12-13}$$

where $P(N)$ is the probability mass function for N.

The variance of PW is obtained by first determining $E[\text{PW}^2]$, where

$$E[\text{PW}^2 \mid N] = E\{A_0 + \sum_{j=1}^{N} (1 + i)^{-j} A_j\}^2$$

$$= E\left\{ A_0^2 + \sum_{j=1}^{N} (1 + i)^{-2j} A_j^2 \right\} + E\left\{ 2 \sum_{j=0}^{N-1} \sum_{k=j+1}^{N} (1 + i)^{-j-k} A_j A_k \right\}. \tag{12-14}$$

If the A_j's are statistically independent, then

$$E[\text{PW}^2 \mid N] = E[A_0^2] + \sum_{j=1}^{N} (1 + i)^{-2j} E[A_j^2]$$

$$+ 2 \sum_{j=0}^{N-1} \sum_{k=j+1}^{N} (1 + i)^{-j-k} E[A_j] E[A_k].$$

However, since $E[X^2] = V[X] + E^2[X]$, then

$$E[\text{PW}^2 \mid N] = (V[A_0] + E^2[A_0]) + \sum_{j=1}^{N} (V[A_j] + E^2[A_j])(P/F, i\%, 2j)$$

$$+ 2\sum_{j=0}^{N-1} \sum_{k=j+1}^{N} E[A_j]E[A_k](P/F, i\%, j+k). \tag{12-15}$$

Therefore,

$$E[\text{PW}^2] = (V[A_0] + E^2[A_0]) + \sum_{N=1}^{\infty} \left\{ \sum_{j=1}^{N} (V[A_j] + E^2[A_j])(P/F, i\%, 2j) \right.$$

$$\left. + 2\sum_{j=1}^{N-1} \sum_{k=j+1}^{N} E[A_j]E[A_k](P/F, i\%, j+k) \right\} P(N). \tag{12-16}$$

Given the values of $E[\text{PW}]$ and $E[\text{PW}^2]$, the variance of present worth is obtained from the relation

$$V[\text{PW}] = E[\text{PW}^2] - [E[\text{PW}]]^2. \tag{12-17}$$

Example 12-5

Consider an investment having the following expected values and variances for the statistically independent cash flows:

j	$E[A_j]$	$V[A_j]$
0	−$10,000	1×10^6
1	2,000	2×10^4
2	3,000	3×10^4
3	4,000	4×10^4
4	5,000	5×10^4
5	6,000	6×10^4

The life of the investment is a random variable with the following probability distribution:

N	$P(N)$
3	0.25
4	0.50
5	0.25

By using a MARR of 10%, determine the expected value of present worth and variance of present worth.

Solution   To determine the expected present worth, the following relation is applied:

$$E[\text{PW}] = E[\text{PW} \mid N = 3](0.25) + E[\text{PW} \mid N = 4](0.50)$$

$$+ E[\text{PW} \mid N = 5](0.25),$$

where

$$E[\text{PW} \mid N = 3] = -\$10,000 + [\$2,000 + \$1,000(A/G, 10\%, 3)] \cdot$$
$$(P/A, 10\%, 3) = -\$2,688.22$$
$$E[\text{PW} \mid N = 4] = -\$10,000 + [\$2,000 + \$1,000(A/G, 10\%, 4)] \cdot$$
$$(P/A, 10\%, 4) = \$714.60$$
$$E[\text{PW} \mid N = 5] = -\$10,000 + [\$2,000 + \$1,000(A/G, 10\%, 5)] \cdot$$
$$(P/A, 10\%, 5) = \$4,443.71$$

Therefore,

$$E[\text{PW}] = -\$2,688.22(0.25) + \$714.60(0.50) + \$4,443.71(0.25)$$
$$= \$796.17.$$

To determine the variance of present worth, the value of $E[\text{PW}^2]$ is first determined using Eq. 12-16.

$$E[\text{PW}^2] = (V[A_0] + E^2[A_0]) + \sum_{N=3}^{5} \left\{ \sum_{j=1}^{N} (V[A_j] + E^2 A_j])(P/F, 10\%, 2j) \right.$$
$$\left. + 2 \sum_{j=1}^{N-1} \sum_{k=j+1}^{N} E[A_j]E[A_k](P/F, 10\%, j + k) \right\} P(N)$$

The complete solution for this problem is left to the interested student. The answer is $E[\text{PW}^2]$ $= 224 \times 10^6$. Thus, using Eq. 12-17,

$$V[\text{PW}] = (224 \times 10^6) - (796.17)^2 = 223.366 \times 10^6.$$

The central limit theorem for the *random* sum of independent random variables asserts that the distribution of PW can be approximated by the normal distribution under rather general conditions. Assuming PW is normally distributed and using Appendix E yields

$$\Pr(\text{PW} < 0) = \Pr\left(S < \frac{0 - 796.17}{\sqrt{223.366 \times 10^6}} \right) = \Pr(S < -0.053) = 0.4789. \qquad \blacksquare$$

Example 12-6
An individual is considering an investment of $10,000 in a project that will return $3,000 per year for the duration of the project. The life of the investment is not known with certainty. The following probability distribution is felt to be reasonable for the life of the investment:

N	$P(N)$
1	0.10
2	0.15
3	0.20
4	0.25
5	0.15
6	0.10
7	0.05

It is desired to determine the probability of a positive present worth, the expected present worth, and the variance of present worth. A MARR of 15% is to be used.

Solution   Since the cash flows are deterministic, the probability of a positive present worth is given by

$$Pr(PW > 0) = Pr[-\$10,000 + \$3,000(P/A, 15\%, N) > 0]$$
$$= Pr[(P/A, 15\%, N) > 3.33].$$

From the interest tables it is seen that the $(P/A, 15\%, N)$ factor is greater than 3.33 for values of N greater than or equal to 5. Therefore,

$$Pr(PW > 0) = Pr(N \geq 5) = 0.15 + 0.10 + 0.05 = 0.30.$$

The expected present worth is given by

$$E[PW] = -\$10,000 + 3,000 \sum_{N=1}^{7} (P/A, 15\%, N)P(N)$$
$$= -\$10,000 + \$3,000[0.870(0.10) + 1.626(0.15)$$
$$+ 2.283(0.20) + 2.855(0.25) + 3.352(0.15)$$
$$+ 3.784(0.10) + 4.160(0.05)]$$
$$= -\$10,000 + \$3,000(2.59045) = -\$2,228.65.$$

The variance of present worth is obtained as follows:

$$E[PW^2] = \sum_{N=1}^{7} [-\$10,000 + \$3,000(P/A, 15\%, N)]^2 P(N)$$
$$= (-\$7,390)^2(0.10) + (-\$5,122)^2(0.15) + (-\$3,151)^2(0.20)$$
$$+ (-\$1,435)^2(0.25) + (\$3,136)^2(0.15) + (\$1,352)^2(0.10)$$
$$+ (\$2,480)^2(0.05) = 13.86 \times 10^6.$$

Thus,

$$V[PW] = (13.86 \times 10^6) - (-2,228.65)^2$$
$$= 8.90 \times 10^6. \qquad \blacksquare$$

Example 12-7
In the previous example suppose that each annual receipt will be identical but that the size of the first annual receipt is a random variable, A, with the following probability distribution:

A	$P(A)$
$2,000	0.20
3,000	0.50
4,000	0.30

Determine the probability of a profitable investment, the expected present worth, and the variance of present worth.

Solution   The probability of a positive present worth can be obtained as follows:

$$Pr(PW > 0) = Pr(PW > 0 \mid A = \$2,000)(0.20)$$
$$+ Pr(PW > 0 \mid A = \$3,000)(0.50)$$
$$+ Pr(PW > 0 \mid A = \$4,000)(0.30)$$

or

$$Pr(PW > 0) = Pr[(P/A, 15\%, N) > 5.00 \mid A = \$2,000](0.20)$$
$$+ Pr[(P/A, 15\%, N) > 3.33 \mid A = \$3,000](0.50)$$
$$+ Pr[(P/A, 15\%, N) > 2.50 \mid A = \$4,000](0.30),$$

which reduces to

$$Pr(PW > 0) = Pr(N \geq 10)(0.20) + Pr(N \geq 5)(0.50) + Pr(N \geq 4)(0.30)$$
$$= 0(0.20) + 0.30(0.50) + 0.55(0.30)$$
$$= 0.315.$$

The expected present worth can be obtained in a similar manner, namely,

$$E[PW] = \sum_A \sum_N E[PW \mid N, A]P(N)P(A)$$

$$= \sum_A E[PW \mid A]P(A). \tag{12-18}$$

From previous calculations it is seen that

$$E[PW \mid A = \$2,000] = -\$10,000 + \$2,000(2.59045) = -\$4,819.10,$$
$$E[PW \mid A = \$3,000] = -\$10,000 + \$3,000(2.59045) = -\$2,228.65,$$
$$E[PW \mid A = \$4,000] = -\$10,000 + \$4,000(2.59045) = \$361.80.$$

Therefore,

$$E[PW] = -\$4,819.10(0.20) - \$2,228.65(0.50) + \$361.80(0.30)$$
$$= -\$1,969.61.$$

The variance of present worth is obtained as follows:

$$E[PW^2] = \left\{ \sum_{N=1}^{7} [-\$10,000 + \$2,000(P/A, 15\%, N)]^2 \, P(N) \right\}(0.20)$$

$$+ \left\{ \sum_{N=1}^{7} [-\$10,000 + \$3,000(P/A, 15\%, N)]^2 \, P(N) \right\}(0.50)$$

$$+ \left\{ \sum_{N=1}^{7} [-\$10,000 + \$4,000(P/A, 15\%, N)]^2 \, P(N) \right\}(0.30).$$

The interested reader can solve the above to find that $E[PW^2] = 16.23 \times 10^6$, and that $V[PW] = (16.23 \times 10^6) - (1,969.61)^2 = 12.35 \times 10^6.$ ■

12.3 Monte Carlo Simulation of Risk

The Monte Carlo simulation technique is an especially useful means of analyzing situations involving risk to obtain approximate answers when a physical experiment or the use of analytical approaches is either too burdensome or not feasible. It has enjoyed widespread acceptance in practice because of the analytical power it makes possible without the necessity for complex mathematics. It is especially adaptable to computation by digital computers. Indeed, computer languages have been developed especially to facilitate Monte Carlo simulation.

The technique is sometimes descriptively called the *method of statistical trials.* It involves, first, the random selection of an outcome for each variable (element) of interest, the combining of these outcomes with any fixed amounts, and calculation if necessary to obtain one trial outcome in terms of the desired answer (measure of merit). Done repeatedly, this will result in enough trial outcomes to obtain a sufficiently close approximation of the mean, variance, distribution shape, or other characteristic of the desired answer. Figure 12-1 schematically shows this process applied to investment project analysis.

The key requisite of the Monte Carlo technique is that the outcomes of all variables of interest be *randomly* selected, i.e., that the probability of selection of all possible outcomes be in exact accord with their respective probability distributions. This is accomplished through the use of tables of random numbers and relating these numbers to the distributions of the variables. *Random numbers* are numbers that have been generated in such a way that there is an equal probability of any number appearing each time, regardless of what sequence is experienced at any prior time. Appendix C contains one page of these numbers. The following simple example will demonstrate the Monte Carlo technique.

Example 12-8

As an illustration of Monte Carlo simulation applied to one variable or element, suppose the annual net cash flow for a project is estimated to have the distribution shown in Table 12-1.

This random simulation can be accomplished through tabular methods by assigning random numbers to each outcome in proportion to the probability of each outcome. Because two-digit probabilities are given in this case, sets of only two random digits are needed, which are shown in Table 12-2.

Now one can generate net cash flow outcomes by picking random numbers[7] and determining the net cash flow that corresponds to each according to the list in Table 12-2. Table 12-3 lists ten two-digit random numbers taken arbitrarily from a table of random numbers, such as that in Appendix C, together with the corresponding net cash flows, taken from Table 12-2.   n

It may be of interest to note that the mean net cash flow based on the simulated outcomes from Example 12-8 is $175,000/10 = $17,500. This compares with a mean of $17,250 for the known distribution shown in Table 12-1. Results for ten simulated

---

[7] *Note:* The random numbers should be taken from the table in a way to assure randomness or non-repetitiveness by randomly selecting a point to begin in the table and randomly selecting the direction of movement within the table (such as up, down, to the right, etc.).

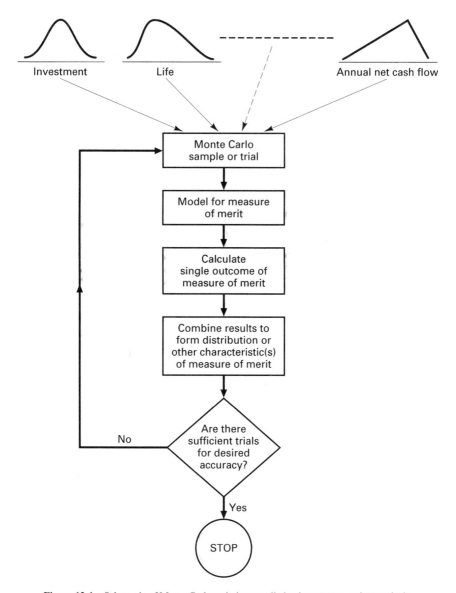

Figure 12-1   Schematic of Monte Carlo technique applied to investment project analysis.

outcomes would not always turn out this close. However, in general, the larger the number of Monte Carlo trials, the closer the approximation to the desired answer(s).

For Monte Carlo simulations in which a computer is not used, it is sometimes helpful to use a graph of the distribution function (cumulative frequency function) instead of a table matching random numbers to the various outcomes. Figure 12-2 contains a graph of the distribution function for the example in Table 12-1. Once a

TABLE 12-1   Example Frequency
Distribution for Net Annual
Cash Flow

Net cash flow	P(Net cash flow)
$10,000	0.10
15,000	0.50
20,000	0.25
25,000	0.15

TABLE 12-2   Assignment of
Random Numbers for Example
in Table 12-1

Net cash flow	Random numbers
$10,000	00–09
15,000	10–59
20,000	60–84
25,000	85–99

TABLE 12-3   Generation of Outcomes
for Example in Table 12-1

Random number	Net cash flow outcome
47	$15,000
91	25,000
02	10,000
88	25,000
81	20,000
74	20,000
24	15,000
05	10,000
51	15,000
74	20,000

graph such as that in Fig. 12-2 has been constructed, outcomes are generated as follows: a random number table such as that in Appendix C is used to obtain random values that correspond to the ordinate scale (vertical axis) with the decimal removed. For
each random number, a horizontal line is drawn until it meets the curve. Then a vertical line is dropped to the abscissa (horizontal axis), and the outcome is thus determined. The dotted line in Fig. 12-2 illustrates the generation of a sample cash flow
outcome.

   The above example is for a discrete outcome distribution, but it should be noted
that the same principle applies for continuous distributions. For continuous distributions, the tabular method is usually impractical, but the graphical method is readily
acceptable as is shown in the next section.

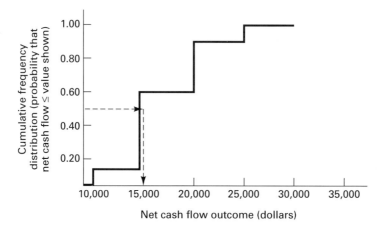

Figure 12-2   Sample cumulative frequency distribution for net cash flow.

12.3.1 Generation of Random Normal Values

It is quite common for random phenomena to possess a normal distribution and for element outcomes to be estimated as normally distributed. The Monte Carlo technique can be conveniently used for simulation of random outcomes in such cases.

The basic quantity needed to generate randomly distributed normal outcomes is called a *random normal deviate,* or random normal number. A random normal deviate is merely a random number of standard deviations from the mean of a standard normal distribution. Random normal deviates can be obtained directly from a graph of the cumulative standard normal distribution. Such a graph is shown in Fig. 12-3.

For a normal distribution, the probability of an occurrence near the mean is greater than the probability of an occurrence farther from the mean. This is reflected in Fig. 12-3, for the relative frequency of occurrence at each outcome value is proportional to the slope of the cumulative frequency curve.

To obtain random normal deviates, a table of random numbers is used to select numbers between 0.000 and 0.999 on the ordinate scale of the cumulative frequency distribution. (*Note:* More or less than three decimal places can be used as desired for accuracy.) For each random number a horizontal and vertical line can be drawn to find the corresponding random normal deviate. This is shown for two example random numbers in Fig. 12-3, and the results are summarized below.

Random number	Random normal deviate
405	−0.24
877	1.16

Tables of random normal deviates can be generated by a procedure such as that just given. Such a table is presented in Appendix D. Tables of random normal deviates save much effort, for they enable us to generate a Monte Carlo sample from a normal

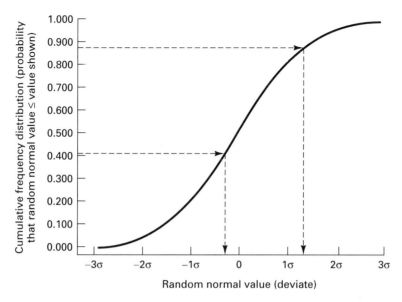

Figure 12-3   Cumulative standard normal frequency distribution for generation of random normal deviates.

distribution merely by using this relation: outcome value = mean + (RND × standard deviation), where RND denotes the random normal deviate. As an example, suppose a project has a mean life of eight years and a standard deviation of two years. Generated lives using the above random normal deviates, for example, can then be calculated as

$$8 - 0.24(2) = 7.52 \text{ year}$$

and

$$8 + 1.16(2) = 10.32 \text{ year.}$$

12.3.2 Generation of Uniformly Distributed Values[8]

Whenever the cumulative distribution function of a random variable can be expressed mathematically, random outcomes of that variable can be generated from random numbers by direct mathematical substitution. An example is the following development of a mathematical model for the generation of uniformly distributed values.

A uniform continuous distribution with a minimum value a and a maximum value b has a density function and cumulative frequency distribution as shown in Fig. 12-4. For this distribution, the mean equals $(a + b)/2$, the variance equals $(b - a)^2/12$, and the range equals $(b - a)$.

---

[8] Methods of generating random variates from a variety of different probability distributions are given by R. E. Shannon, *Systems Simulation: The Art and Science* (Englewood Cliffs, NJ: Prentice-Hall, Inc., 1975).

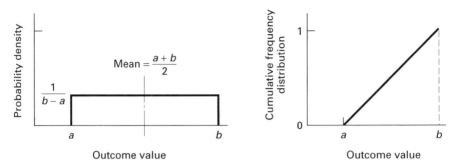

Figure 12-4   Density and distribution function for uniform continuous distribution.

To illustrate the generation of outcomes according to this distribution, let RN denote a random number, and let RN_m denote the highest-valued random number. By similar triangles, it can be seen that

$$\text{outcome value} = a + \frac{RN}{RN_m}(b - a)$$

$$= a + (\text{RN expressed as a decimal})(b - a). \qquad (12\text{-}19)$$

An equivalent statement is

$$\text{outcome value} = \frac{a + b}{2} - \frac{(b - a)}{2} + \frac{RN}{RN_m}(b - a). \qquad (12\text{-}20)$$

If an element or variable is uniformly distributed with a mean of 8 and a range of 6, random outcomes can be generated using the relation

$$8 - \frac{6}{2} + \frac{RN}{RN_m}(6) = 5 + (\text{RN expressed as a decimal})(6).$$

Example 12-9

Example of the use of Monte Carlo simulation for nonindependent elements: One of the valuable features of the Monte Carlo technique is that it provides an analysis tool for cases in which elements are not independent and thus are difficult or impossible to manipulate so as to obtain the desired answers analytically. For example, suppose that the life of a project is described by some distribution whose mean is a function of the annual cash flow of the project. Further, suppose the annual cash flow itself is described by some distribution and that it is desired to determine the distribution of the PW (present worth) of those cash flows over the life of the project. The following illustrates the use of the Monte Carlo technique for this type of situation.

The annual net cash flow of a project is estimated to be normally distributed with a mean of $10,000 and a standard deviation of $2,000. The life of the project is estimated to be uniformly distributed with a mean of 0.0005 of the annual net cash flow (rounded to the nearest integer year) and a range (difference between maximum and minimum life) of 6 years. Table 12-4 demonstrates the use of the Monte Carlo technique for ten trials to obtain an estimate of the mean of the PW of the cash flows, at an interest rate of 10%.

The estimated mean PW from the limited number of trials in Table 12-4 is $374,800/10 = $37,480. Repeated trials would doubtless result in a more accurate answer. It is worthy of note that the exact answer for this situation is not the same as $10,000(P/A, 10\%, 5) = $37,910$, where $10,000 is the mean cash flow and 5 years is the mean life.   ■

TABLE 12-4 Monte Carlo Example with Nonindependent Elements

Random normal deviate (RND)	Annual net cash flow (ANCF) [$10,000 + RND($2,000)]	Three random numbers (RN)	Project life N $\left[0.0005(\text{ANCF}) -3 + \dfrac{\text{RN}}{999}(6)\right]$	Project life N to nearest integer	$(P/A, 10\%, N)$	PW of cash flows [ANCF × $(P/A, 10\%, N)$]
0.944	$ 8,112	443	3.65	4	3.170	$25,700
−1.140	7,720	511	3.64	4	3.170	24,500
1.353	12,706	549	6.54	7	4.868	62,000
0.466	10,912	169	3.48	3	2.487	27,100
0.732	11,464	656	6.65	7	4.868	56,000
−1.853	6,394	955	5.84	6	4.355	27,800
−0.411	9,188	783	6.29	6	4.355	40,100
0.488	10,976	572	5.92	6	4.355	48,000
−0.351	9,298	842	6.75	7	4.868	45,400
−1.336	7,328	372	2.89	3	2.487	18,200
						$\Sigma = \overline{\$374,800}$

Example 12-10

Example of Monte Carlo technique applied to economic analysis for a single project: To illustrate the use of the Monte Carlo technique to calculate the measure of merit for a single project for which the element outcomes are estimated as variables, consider a case in which the estimates are as follows:

Investment:	Normally distributed with mean of $100,000 and standard deviation of $5,000
Life:	Uniformly distributed with minimum of 4 yr and maximum of 16 yr (rounded to nearest integer)
Market value (MV):	$10,000 (single outcome)
Annual net cash flow:	$14,000 with probability 0.4
	$16,000 with probability 0.4
	$20,000 with probability 0.2

It is assumed that all elements subject to variation vary independently of one another, and it is desired to obtain a good estimate of the distribution characteristics for the AW using an interest rate of 10%. For purposes of the illustration, only five repetitions of the Monte Carlo simulation will be made. However, perhaps several thousand repetitions would be needed to obtain sufficiently accurate AW distribution information. Table 12-5 shows the calculations.

The estimate of the AW based on the very limited sample as calculated in Table 12-5 is −$18,000/5 = −$3,600.

An estimate of the standard deviation of the AW can be obtained from the relation

$$\sigma[AW] = \sqrt{\frac{\sum_{i=1}^{k}(AW_i - E[AW])^2}{k - 1}} \qquad (12\text{-}21)$$

$$= \sqrt{\frac{\$58,100,000}{4}} = \$3,850.$$

This estimate should also come closer to the true standard deviation with increasing numbers of Monte Carlo trials.   n

Example 12-11

Example of Monte Carlo technique applied to economic comparison of two independent projects: Suppose two competing projects, A and B, have the following estimated distributions of AW and it is desired to estimate the distribution of the difference in AW between the projects using Monte Carlo simulation. Assume that the outcomes are independent of each other.

Project A		Project B
AW	P(AW)	
−$ 5,000	0.10	AW is normally distributed
10,000	0.30	with mean of $25,000 and
20,000	0.50	standard deviation of $10,000
30,000	0.10	

TABLE 12-5 Monte Carlo Example for a Single Project

Random normal deviate (RND)	Investment, P [$100,000 + RND($5,000)]	Three random numbers (RN)	Project life, N $\left[4 + \dfrac{RN}{1,000}(16-4)\right]$	Project life N to nearest integer	One random number	Annual receipts, A $\begin{bmatrix}\$14,000 \text{ for } 0\text{–}3\\16,000 \text{ for } 4\text{–}7\\20,000 \text{ for } 8\text{–}9\end{bmatrix}$	AW $[-\{(P-MV)(A/P,10\%,N)\\+MV(10\%)\}+A]$
0.30	$101,500	693	4 + 8.32	12	2	$14,000	−$ 432
−0.92	95,400	192	4 + 2.30	6	5	16,000	− 4,700
0.13	100,650	092	4 + 1.10	5	1	14,000	− 9,000
−0.16	99,200	490	4 + 5.87	10	4	16,000	+ 500
0.54	102,700	314	4 + 3.77	8	1	14,000	− 4,400
							$\Sigma = -\$18,032$

TABLE 12-6   Monte Carlo Comparison of Two Independent Projects

One random number	AW for project A $\begin{bmatrix} -\$5,000 \text{ for } 0 \\ 10,000 \text{ for } 1\text{--}3 \\ 20,000 \text{ for } 4\text{--}8 \\ 30,000 \text{ for } 9 \end{bmatrix}$	Random normal deviate (RND)	AW for project B [= $25,000 + RND($10,000)]	Difference in AW for two projects
4	$20,000	0.636	$31,360	$11,360
1	10,000	−0.179	22,210	12,210
4	20,000	−2.546	− 460	− 20,460
6	20,000	0.457	29,570	9,570
				Σ = $12,680

Sample calculations to obtain the desired answers are shown in Table 12-6. The mean difference in AW based on the limited simulation in Table 12-6 is $12,680/4 = $3,170. To reemphasize, a much larger number of Monte Carlo trials than illustrated is needed before meaningful estimates of the distribution of the difference in the AW for the two projects can be made.

Example 12-12

Example of Monte Carlo technique applied to economic comparison of two projects with correlated elements: Two competing projects have the following outcome distribution characteristics:

	Project	
	A	B
Investment (normally distributed)		
Mean:	$50,000	$60,000
Standard deviation:	$20,000	$5,000
Life (uniformly distributed and rounded to the nearest year)		
Minimum:	3 yr	2 yr
Maximum:	7 yr	12 yr

It is thought that the investment outcomes are completely correlated ($\rho = +1$) for the two projects; i.e., when the investment for one project occurs high, the investment for the second project occurs correspondingly high, etc. On the other hand, the lives for the two projects are thought to be independently distributed. Table 12-7 demonstrates the use of the Monte Carlo technique for generating the distribution of the difference in the capital recovery costs for the two projects. For purposes of the illustration, zero market value and 10% interest are assumed and only three trials are shown. Note that the investment amounts for the two projects are generated using the same random normal deviates, thus reflecting complete correlation. On the other hand, the lives for the two projects are generated using independent random digits, reflecting the independence of life outcomes. ■

For typical practical problems, hundreds or even thousands of Monte Carlo trials are required in order to reduce sampling variation to a sufficiently low level so that the desired answers possess the level of accuracy thought necessary. This is often too laborious a task to accomplish by hand, but it can be done very efficiently with digital computers.

TABLE 12-7   Monte Carlo Comparison of Projects with Correlated Elements

Random normal deviate (RND)	Investment (P) for A [$50,000 + RND($20,000)]	Investment (P) for B [$60,000 + RND($5,000)]	Two random numbers	Life for A $\left[3 + \dfrac{RN}{99}(4)\right]$ (rounded)
0.178	$53,560	$60,890	95	8
−0.507	49,860	57,460	04	4
0.362	57,240	61,810	08	4

Two random numbers	Life for B $\left[2 + \dfrac{RN}{99}(10)\right]$ (rounded)	Capital recovery cost for A $[P(A/P, 10\%, N)]$	Capital recovery cost for B $[P(A/P, 10\%, N)]$	Difference in capital recovery cost [project B − project A]
16	4	$10,000	$19,150	$ 9,150
98	12	15,550	8,400	− 7,150
00	2	18,030	35,000	16,970

12.4 Method for Determining Required Number of Monte Carlo Trials and Limitations

An easy method for determining the approximate number of Monte Carlo trials required to obtain sufficiently accurate answers is to keep a running tally (plot) on the average answer(s) of interest for increasing numbers of trials and to judge the number of trials at which those answer(s) have become stable enough to be within the accuracy required.

It is to be expected that the average outcome(s) will dampen or stabilize with increasing numbers of trials. This phenomenon is illustrated in Fig. 12-5 by the wavy line. Figure 12-5 also shows how a given typical permissible range of error can result in an indicated approximate number of trials required.

The Monte Carlo technique possesses real limitations that should be recognized. As for any analysis technique, the results can be no more accurate than the model and estimates used. The technique also inherently possesses the same problems

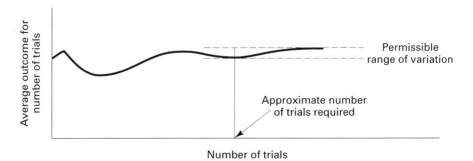

Figure 12-5   Number of trials.

of statistical variation and the need for experimental design that is encountered in direct physical experimentation. Finally, it should be emphasized that a sufficiently large number of Monte Carlo trials must be performed to reduce sampling variation (range of error) to a level that is tolerable in view of the accuracy needed and economically justified.

PROBLEMS

12-1. An individual is considering an investment of $10,000 in a venture. It is expected that annual receipts of $2,000 will occur over a 10-year period. However, the annual receipts are not guaranteed to occur at the expected level. The standard deviation for annual receipts is estimated to be $200 each year. A MARR of 10% is to be used to determine the expected value and standard deviation of present worth assuming (a) annual receipts and statistically independent and (b) annual receipts are perfectly correlated.

12-2. A firm must purchase a new 20-hp electric motor. Motor A sells for $800 and has an efficiency rating of 90%; motor B sells for $500 and has a rating of 80%. The cost of electricity is $0.06/kWh. An 8-year study period is used; equal market values are assumed for the two motors. A MARR of 20% is to be used. The annual usage of the motor is uncertain; the following probabilities have been estimated:

Annual usage	Probability
1,000 hr	0.20
2,000 hr	0.40
3,000 hr	0.30
4,000 hr	0.10

Assuming that annual usage is statistically independent, determine the probability that motor A is preferred to motor B. (*Note:* 1 hp = 0.746 kW; motors are rated by output; and input = output/efficiency.) Use the PW method.

12-3. A manufacturing plant in Wisconsin can contract snow removal at a cost of $300/day. Alternatively, a snow removal machine can be purchased for $25,000; it is estimated to have a useful life of ten years and a zero market value at that time. Annual costs for operating and maintaining the equipment are estimated to be $10,000. A MARR of 10% is used by the plant. The number of days per year the snow removal equipment is required is a random variable with the following probability distribution:

No. days/yr x	Probability $P(x)$
20	0.10
40	0.20
60	0.30
80	0.30
100	0.10

Snow removal requirements in one year are independent of the requirements in any other year.

a. Determine the mean and variance for PW and AW of the savings resulting from the purchase of the machine (i.e., cost of contracting – cost of purchase).

b. Assuming normally distributed PW, what is the probability of a positive equivalent savings by purchasing the machine?

12-4. An aluminum extrusion plant manufactures a particular product at a variable cost of $X/unit. The fixed cost associated with manufacturing the product equals $20,000/yr. The selling price for the product is $0.40/unit. The variable cost has the following estimated probability distribution:

x	P(x)
0.07	0.05
0.08	0.10
0.09	0.20
0.10	0.30
0.11	0.20
0.12	0.10
0.13	0.05

Develop the probability distribution for the break-even value of annual sales volume. What is the expected value?

12-5. An initial investment of $9,000 results in annual receipts of $2,500 until the project is terminated. The probability distribution for the life of the project is estimated to be as follows:

N	P(N)
4	0.10
5	0.25
6	0.45
7	0.15
8	0.05

Use a MARR of 15% to determine the probability the investment is profitable (i.e., its PW ≥ 0).

12-6. In Problem 12-5, suppose the MARR is a random variable with the following probability distribution:

i	P(i)
10%	0.30
15%	0.40
20%	0.30

What is the probability the investment is profitable?

12-7. In Problem 12-5, suppose the annual receipt is identical each year of the project, but let the first such receipt be distributed as follows:

R	P(R)
$2,000	0.25
2,500	0.50
3,000	0.25

What is the probability that the investment is profitable?

12-8. Two mutually exclusive investment alternatives are being considered and one of them must be selected.

Alternative A requires an initial investment of $13,000 in equipment; annual operating and maintenance costs are anticipated to be normally distributed, with a mean of $5,000 and a standard deviation of $500; the terminal market value at the end of the eight-year planning horizon is anticipated to be normally distributed with a mean of $2,000 and a standard deviation of $800.

Alternative B requires end-of-year annual expenditures over the eight-year planning horizon, with the annual expenditure being normally distributed with a mean of $7,500 and a standard deviation of $750. Using a MARR of 15% what is the probability that alternative A is the most economic alternative (i.e., the least costly)?

12-9. An investment of $15,000 is contemplated for a machine. The machine will be used for N years and will be disposed of at a market value of MV. Maintenance and operating costs for the machine are estimated to increase by $500/yr, with the first year's cost being $1,500. Using the probability distributions given below, determine the expected value and variance for equivalent uniform annual cost given a MARR of 12%.

		MV		
N	$1,000	$2,000	$3,000	$4,000
6	0.00	0.05	0.10	0.15
8	0.05	0.10	0.15	0.05
10	0.15	0.15	0.05	0.00

12-10. The market value of a prospective asset is a random variable that depends on the life of the asset according to the following table:

		Probability of market value			
Life	$5,000	$10,000	$15,000	$20,000	$25,000
2	—	—	0.20	0.50	0.30
4	—	0.20	0.50	0.20	0.10
6	0.30	0.30	0.30	0.10	—
8	0.30	0.50	0.20	—	—

It is thought that each of the asset lives is equally likely to occur. If the investment in the asset is $50,000 and the interest rate is 15%, show how one can obtain a distribution of the capital recovery cost for the asset by setting up a table and generating five trial outcomes.

12-11. The estimated element outcomes for a key project are as follows:

Investment:	Normally distributed with a mean of $1,000,000 and a variance of 16,000,000
Life:	5 yr with probability 0.2
	7 yr with probability 0.7
	9 yr with probability 0.1
Net annual cash flow:	Uniformly distributed between $120,000 and $340,000 per yr
Market value:	$0

All element outcomes are independent of each other. Demonstrate how to obtain a distribution of the AW by generating five outcomes. From this, obtain estimates of the mean and variance of the AW. Assume a MARR of 10%.

12-12. A certain project is expected to require an investment of $100,000 and to have a life that can be best described by a uniform distribution with a minimum of 5 years and a maximum of 15 years. The market value is expected to be $40,000 if the life is less than 8 years, $20,000 if the life is 8 to 12 years, and $15,000 if the life is 13 or more years. Show how to build a distribution of the capital recovery cost for this project by generating five outcomes using the Monte Carlo technique. Use an interest rate of 10% and round the project lives to the nearest whole year. From your results estimate the expected value and variance of the distribution of capital recovery cost.

12-13. Project X is expected to require an investment of $40,000 and to have a life that is normally distributed with a mean of 5 years and a standard deviation of 1 year (rounded to the nearest integer). The market value is expected to vary according to the relationship $8,000 - $1,000 \times$ (life in years).

Project Y is expected to require an investment of $50,000 and to have a life that is uniformly distributed between 5 and 15 years (rounded to the nearest integer). The market value is expected to be nil regardless of life. Show how to build a distribution of the difference in capital recovery costs for the two projects (X minus Y) by generating three outcomes using the Monte Carlo technique. Interest is 15% per year.

12-14. Records for a certain inventory item type indicate that the number of days lead time required for replenishment can be expected to have the following probability distribution:

Lead time (days)	Probability
1	0.25
2	0.50
3	0.25

Demand during any day of lead time is a normally distributed random variable independent of the number of days lead time and with a mean of 4 units and a standard deviation of 1 unit of the item. (Round up demand to nearest whole number.)

The opportunity cost of a stock-out (failing to have a unit when demanded) is estimated at $70 per unit. Inventory holding costs (cost of carrying excess inventory to guard against stock-outs) average $100 per unit per year. Show how to use the Monte Carlo technique to determine the most economical size of safety stock for the item, where safety stock is defined as the difference between the number of units in inventory when the item is reordered and the expected demand during the lead time period. Assume that on average there are five reorder periods per year and the cost of carrying safety stock equals the holding cost per unit per year multiplied by the size of the safety stock. Illustrate by simulating five reorder periods with safety stocks of 0 units and then 4 units.

12-15. Two investment alternatives, A and B, are under consideration; one must be selected. Alternative A requires an initial investment of $15,000 in equipment; annual operating and maintenance costs are anticipated to be normally distributed, with a mean of $6,000 and a standard deviation of $600; the terminal market value at the end of the 10-year study period is anticipated to be normally distributed with a mean of $2,000 and a standard deviation of $500. Alternative B requires end-of-year annual expenditures over the study period, with the annual expenditure being normally distributed with a mean of $9,000 and a standard deviation of $900. Use a MARR of 10%. What is the probability that alternative A is the preferred alternative? Answer this question using a closed-form analytical approach.

12-16. Use the Monte Carlo simulation technique to obtain ten pairs of present worth values for the alternatives described in Problem 12-15. What percentage of the present worth combinations favors alternative A?

12-17. A certain portable machine has a first cost of $25,000 (known with certainty) and no market value. Direct labor is $15,000/yr with 40% probability and $17,000/yr with 60% probability. Manufacturer's records show that the number of breakdowns per year has the following probability function:

Breakdowns (x)	$P(x)$
0	0.368
1	0.368
2	0.184
3	0.061
4	0.019

The cost of each breakdown is assumed to be normally distributed with $\mu = \$200$, $\sigma = \$40$. When two breakdowns occur, for example, you will need to generate two separate cost outcomes (not one, multiplied by two). By using Monte Carlo simulation, develop five trials of total annual equivalent operating expenses. If you desire a 15% return on your money, what *average* annual income would be required from the machine to make it an attractive investment? The life of the machine is five years.

13
Decision Criteria and Methods for Risk and Uncertainty

13.1 Introduction

This chapter will be devoted first to numerous criteria for aiding in making decisions for what are classically called *risk problems,* i.e., problems in which probabilities of various possible outcomes can be estimated. These criteria are

1. dominance,
2. aspiration level,
3. most probable future,
4. expected value,
5. expectation-variance,
6. certain monetary equivalence,
7. expected utility.

The final section illustrates the use of several decision rules and principles that can be used for *uncertainty problems,* i.e., when the probabilities of various possible outcomes cannot be estimated.

13.2 General Model for Risk Problems

Table 10-1 showed a general formulation for decision problems in which there are various possible outcomes (called *states of nature*) for which probabilities can be estimated.

A typical problem involving five alternatives and four states of nature together with associated probabilities is shown in Table 13-1. The numbers in the body of the

TABLE 13-1 Example Problem with Probabilities Known—
Payoffs in $M of Net PW

	State of nature (probability of state)			
Alternatives	S_1 (0.5)	S_2 (0.1)	S_3 (0.1)	S_4 (0.3)
I	3	−1	1	1
II	4	0	−4	6
III	5	−2	0	2
IV	2	−2	0	0
V	5	−4	−1	0

table can be thought of as returns or profits in $M (where $M denotes thousands of dollars) of net PW, such that a negative number is a cost.

The following sections describe and illustrate several criteria for decision making using the example in Table 13-1. Most of the criteria apply to classical risk problems in which probabilities of various outcomes can be estimated. However, the first two criteria or principles can be applied to decision problems even when the probabilities are not known.

13.3 Dominance Criterion or Elimination Check

The first step in making a decision when the results for all alternatives and states of nature can be quantified is to eliminate from consideration any alternatives that are clearly not to be performed regardless of the state of nature that occurs. If the result for any alternative X is better than the result for some other alternative Y, for all possible states of nature, then alternative X is said to *dominate* alternative Y, and thus Y can be dropped from further consideration.

Example 13-1
Given the problem depicted in Table 13-1, check for dominance and take appropriate action.

Solution   By systematic visual inspection, one can determine that alternative I dominates IV and that alternative III dominates V. (*Note:* Actually, III and V are equally good for state of nature S_1, but V is never better than III.) Thus, alternatives IV and V can both be eliminated from further consideration on the grounds that no rational decision maker would choose either alternative. ∎

The example problem in Table 13-1 can now be reduced to the problem shown in Table 13-2.

13.4 Aspiration-Level Criterion

The aspiration-level criterion involves selecting some level of aspiration and then choosing so as to maximize (or minimize) the probability of achieving this level. An aspiration level is simply some level of achievement (like profit) the decision maker desires to attain or some level of negative results (like cost) to be avoided.

TABLE 13-2 Example Problem in Table 13-1 After
Dominated Alternatives Have Been Eliminated—
Payoffs in $M of Net PW

Alternatives	State of nature (probability of state)			
	S_1 (0.5)	S_2 (0.1)	S_3 (0.1)	S_4 (0.3)
I	3	−1	1	1
II	4	0	−4	6
III	5	−2	0	2

Example 13-2

Given the decision problem in Table 13-2, determine which would be the best alternative if the decision maker has each of the following aspiration levels: (a) possible result of at least 5; (b) possible negative result (loss) no worse than −1.

Solution

(a)   Alternatives II and III have possible results of 5 or greater. The probabilities of these results are 0.3 and 0.5, respectively. Hence, alternative III would be the choice.

(b)   Only alternative I has a possible result that is no more negative than −1. Thus, alternative I would be the choice.   ■

It is commonly thought that some form of the aspiration-level criterion is the most widely used of all principles in management decision making. The following are cases in which use of aspiration levels makes intuitive good sense.

1. When it is costly or too time consuming to determine all the reasonable alternatives and their prospective results, one may choose to search for alternatives only until an alternative is found that gives a reasonable probability of achieving the aspiration level.

2. Occasionally a given alternative is available for only a limited time and action must be taken before information on all the reasonable alternatives and their prospective results can be developed. For example, equipment at a particular price may be available only if an agreement to purchase is made within a matter of hours or days.

3. Sometimes it is difficult or impossible to evaluate the results for each alternative, but it may be possible to determine which alternatives do meet the aspiration level of the decision maker. In this case, a reasonable criterion is to choose the alternative that maximizes the probability of achieving the aspiration level.

13.5 Most Probable Future Criterion

The most probable future criterion suggests that as the decision maker considers the various possible outcomes in a decision, he or she overlooks all except the most probable one and acts as though it were certain.

Many decisions are based on this principle, since, in fact, only the most probable future is seriously considered (thus making the problem virtually one of "assumed certainty").

Example 13-3

Given the decision problem in Table 13-2, determine which would be the best alternative using the most probable future criterion.

Solution   The most probable future is state of nature S_1 (probability $= 0.5$). The results for S_1 range from 3 for alternative I to 5 for alternative III. Of these, alternative III has the best result and would thus be the choice.   ■

13.6 Expected Value Criterion

Using the expectation principle, and thereby choosing to optimize the expected payoff or cost (expressed in equivalent terms), simplifies a decision situation by weighting all dollar payoffs or costs by their probabilities. The criterion is often known as the *expected monetary value*, or *EMV*. As long as the dollar consequences of possible outcomes for each alternative are not very large in the eyes of the decision maker, the expectation principle can be expected to be consistent with a decision maker's behavior.

✓   The general formula for finding the expected outcome (value) of a variable x for any alternative A_i having k discrete outcomes is

$$E[x] = \sum_{j=1}^{k} x_j \cdot P(x_j), \qquad (13\text{-}1)$$

where

$$E[x] = \text{expected value of } x,$$

$$x_j = j\text{th outcome of } x,$$

$$P(x_j) = \text{probability of } x_j \text{ occurring.}$$

If the notation for the general model in Table 13-1 is used, this can be expressed as

$$E[A_i] = \sum_{j=1}^{k} R_{ij} \times P(S_j). \qquad (13\text{-}2)$$

Example 13-4

Given the same decision problem in Table 13-3, show which alternative is best by the expected value criterion.

   Solution

Alternative, A_i	$E[A_i]$
I	$3(0.5) - 1(0.1) + 1(0.1) + 1(0.3) = 1.8$
II	$4(0.5) + 0(0.1) - 4(0.1) + 6(0.3) = 3.4$
III	$5(0.5) - 2(0.1) + 0(0.1) + 2(0.3) = 2.9$

Thus, alternative II, having the highest expected result, is best.   ■

13.7 Expectation-Variance Criterion

The *expectation-variance criterion* or procedure, sometimes called the *certainty equivalence method,* involves reducing the economic desirability of a project into a single measure that includes consideration of the expected outcome as well as variation of that outcome. One simple example is

$$Q = E[x] - A \cdot \sigma[x], \tag{13-3}$$

where

$$Q = \text{expectation-variance measure,}$$
$$E[x] = \text{mean or expected monetary outcome,}$$
$$\sigma[x] = \text{standard deviation of monetary outcome,}$$
$$A = \text{coefficient of risk aversion.}[1]$$

The variance of a variable x for any alternative having k discrete outcomes is

$$V[x] = \sum_{j=1}^{k} (x_j - E[x])^2 P(x_j), \tag{13-4}$$

where $V(x)$ = variance of x, and all other symbols were defined with Eq. 13-1. A more convenient form for calculating $V(x)$, which corresponds to Eq. 12–17, is

$$V[x] = \sum_{j=1}^{k} x_j^2 P(x_j) - (E[x])^2. \tag{13-5}$$

Example 13-5
Given the decision problem in Table 13-2, determine which alternative is best using the criterion in Eq. 13-3, with $A = 0.7$.

   Solution   Using Eq. 13-5 gives these results:

Alternative, A_i	$V[A_i]$
I	$(3)^2(0.5) + (-1)^2(0.1) + (1)^2(0.1) + (1)^2(0.3) - (1.8)^2 = 1.76$
II	$(4)^2(0.5) + (0)^2(0.1) + (-4)^2(0.1) + (6)^2(0.3) - (3.4)^2 = 8.84$
III	$(5)^2(0.5) + (-2)^2(0.1) + (0)^2(0.1) + (2)^2(0.3) - (2.9)^2 = 5.69$

---

[1] Donald Farrar (*The Investment Decision Under Uncertainty,* Englewood Cliffs, NJ: Prentice-Hall, Inc., 1962) and others have shown that as long as there is a diminishing marginal utility of money, the correspondence between a firm's coefficient of risk aversion and its utility function of monetary outcome is

$$A = -\frac{U''(E[x])}{2}.$$

   That is, the coefficient of risk aversion is equal to the negative of one-half of the second derivative of the utility function evaluated at the expected monetary outcome.

Using Eq. 13-3 and calculated results for $E[x]$ from the above example of the expected value criterion gives:

Alternative	$Q = E[x] - 0.7\sqrt{V[x]}$
I	$1.8 - 0.7\sqrt{1.76} = 0.87$
II	$3.4 - 0.7\sqrt{8.84} = 1.32$
III	$2.9 - 0.7\sqrt{5.69} = 1.23$

Thus, alternative II is highest and therefore the best by this particular expectation-variance criterion.   ∎

There are innumerable other expection-variance criteria that can be applied depending on the risk preferences and sophistication of the decision maker and his or her analyst. For example, Cramer and Smith, in a classic article,[2] recognize that the desirability of an investment project is a function of not only the expected value and variance but also of the investment amount in the individual project. Hence, they developed an evaluation model of the form

$$Q = E[x] - A\sigma[x]^a I^b, \tag{13-6}$$

where

$$Q = \text{certainty equivalence or expectation-variance measure,}$$
$$E[x] = \text{expected monetary outcome,}$$
$$A = \text{coefficient of risk aversion,}$$
$$\sigma[x] = \text{standard deviation of monetary outcome,}$$
$$I = \text{project investment amount,}$$
$$a \text{ and } b = \text{constants.}$$

Cramer and Smith further show detailed examples of how one can empirically obtain all the constants for the use of the model. The following is a simple application.

Example 13-6

Suppose the outcomes and investments required for two competing projects are estimated as follows:

Alternative	Outcomes in net AW		Investment required
	$E[AW]$	$\sigma[AW]$	
A	$10,000	$25,000	$22,500
B	8,000	4,000	40,000

It is desired to show which alternative would be preferred if each of the following criteria is used:

(a) expected value,

[2] R. H. Cramer and B. E. Smith, "Decision Models for the Selection of Research Projects," *The Engineering Economist* 9, no. 2 (Winter 1964).

(b)   expectation-variance, using Eq. 13-3 with a high coefficient of risk aversion, A, for both projects;

(c)   expectation-variance, using Eq. 13-6 with $A = 0.40$ for project A and 0.75 for project B and constants $a = b = 0.50$.

Solution

(a)   Project A is better; $E(\text{AW})_A > E(\text{AW})_B$.

(b)   Project B would be better, intuitively. (*Note:* This is true if the coefficient of risk aversion, A, for both projects were anything higher than 0.095.)

$$A = \frac{\$10,000 - \$8,000}{\$25,000 - \$4,000} = 0.095$$

(c)   Project A:

$$Q = \$10,000 - 0.4(\$25,000)^{0.5}(\$22,500)^{0.5} = \$520.$$

Project B:

$$Q = \$8,000 - 0.75(\$4,000)^{0.5}(\$40,000)^{0.5} = \$5,000.$$

Thus, on the basis of the above calculations, project B is better.   ■

13.8 Certain Monetary Equivalence Criterion

An offshoot of the expectation-variance criterion is to determine subjectively the *certain monetary equivalence* of any set of results for any alternative. The *certain monetary equivalent,* or *CME,* is merely the monetary amount for certain at which the decision maker would be indifferent between that amount and various possible monetary outcomes. This concept is very useful in practice and can be applied to situations involving gains (payoffs) or losses (costs). While it can be most meaningfully applied to risk situations in which various payoffs or costs and their respective probabilities are known, it can also be used in situations involving uncertainty regarding payoffs/costs, or probabilities, or both. Below are several examples.

Example 13-7
Suppose a decision maker is faced with either a possible loss of $100,000 with probability 0.01 or no loss with probability 0.99. He desires to decide what is the maximum amount he would be willing to pay in order to avoid the risk of loss.

    Solution   The desired quantity is a certain monetary equivalent. It is quite subjective, depending on the decision maker's risk preferences, particularly considering the consequences of the possible monetary outcomes in relation to the total assets at his disposal. As a guide, one might calculate the expected monetary value, EMV, as follows:

$$\text{EMV} = -\$100,000(0.01) + \$0(0.99) = -\$1,000.$$

If the possible $100,000 loss poses little threat in the eyes of the decision maker, so that he is called *risk-neutral* in this situation, he could reasonably designate the EMV as his CME. However, most decision makers are at least somewhat *risk-averse* and thus will be willing to pay a certain amount of more than $1,000—say, $5,000—in order to avoid the chance of loss of $100,000. This illustrates why most people are willing to purchase liability insurance even though the known cost of the policy is higher than the expected losses to be covered by the policy. On the other hand, occasionally there are decision makers who are *risk-seeking* in nature and would pay only something less than $1,000 for certain in order to avoid the risk of loss. A possible, but not necessarily rational, extreme is a decision maker who enjoys the risk of loss of $100,000 so much (for example, he may like to boast about it to his friends) that he is unwilling to pay any amount to avoid the risk. ■

Example 13-8

Suppose a decision maker is again confronted with the example problem in Table 13-2 and desires to designate her CME for each alternative so as to choose the best.

Solution   Again, the specification of CMEs is very subjective, reflecting the decision maker's relative weighting of the consequences of the various possible gains and losses. As a starting point, one might consider the EMVs for alternatives I, II, and III, which were previously calculated to be 1.8, 3.4, and 2.9, respectively. After considering this and the range of possible outcomes involved, a particular decision maker might choose CMEs of, say, 1.0, 2.0, and 2.5, respectively. In this case, alternative III, with the highest CME, would be preferred.

The CME criterion can also be used for situations in which probabilities are neither known nor estimable. In such cases, any CMEs determined are even more subjective than otherwise; nevertheless, they can express one's "gut feelings" about the uncertainties and risk preferences involved. ■

Example 13-9

Suppose a decision maker has a prospective project that is estimated to bring the following possible equivalent returns (for which the respective probabilities are thought to be too nebulous to estimate):

$$\text{Return in \$M:} \quad -50, \text{ or } 150, \text{ or } 250.$$

It is desired to demonstrate the determination of his CME for the project; i.e., what single lump amount would he accept as being just as valuable to him as the variable return?

Solution   This requires very subjective judgment based on the decision maker's intuitive feelings about the desirability (or nondesirability) of the possible gains or losses and their respective likelihoods. If he very much abhors the possible loss of $50M, he might be willing to *pay* something like $10M or $25M (CME = −$10M or −$25M) or more to avoid the risk. On the other hand, if he is strongly attracted to the possible gain of $150M or $250M, and if he thinks the probabilities of them are quite high, he might specify a CME approaching the $250M gain.■

13.9 Expected Utility Criterion

The *expected utility criterion* or method has particular usefulness for analyzing projects in which the potential gain or loss is of significant size compared to the total funds available to the firm. More specifically, if the marginal utility or desirability of each dollar potentially to be gained or lost is not a constant, the utility of dollars rather than just the amount of dollars is relevant, and it may then be worthwhile to use the expected utility method rather than the probabilistic monetary method.

The expected utility method consists of determining the *cardinal utility*—e.g., relative degree of usefulness or desirability to the decision maker—of each of the possible outcomes of a project or group of projects on some numerical scale and then calculating the expected value of the utility to use as the measure of merit.

The application of this method is based on the premise that it is possible to measure the attitudes of an individual or decision maker toward risk. If the decision maker is consistent with herself, then a relation between monetary gain or loss and the utility or relative desirability of that gain or loss can be obtained through the decision maker's answers to a series of questions and resultant computations as explained ahead.

13.9.1 Steps in Deriving Utility-of-Money Function

1. Select two possible monetary outcomes within the range of interest. For example, say you pick $0 and $10,000.

2. Assign arbitrary utility indices to these monetary outcomes, the only restriction being that the index for the higher monetary outcome be higher than the index for the lower monetary outcome. For example, say you assign an index of 1 to a $0 outcome and 20 to a $10,000 outcome.

3. The utility value of other monetary outcomes can be found by having the decision maker answer questions based on the following relation: Given any three monetary amounts, $X < $Y < $Z, and known utility values for any two of these amounts, the utility of the third amount can be found by the equation:

$$U[\$Y] = P \times U[\$X] + (1 - P) \times U[\$Z], \qquad (13\text{-}7)$$

where

$$P = \text{probability,}$$
$$U[X] = \text{utility of } \$X, \text{etc.}$$

   (a) *To obtain utility values for monetary amounts within any two amounts,* $X *and* $Z, for which utility values have been assigned or calculated, ask questions such as, "What monetary amount for certain, $Y, would you desire just as highly as a *P*% chance of $X and a (1 − *P*%) chance of $Z? (*Note:* It is generally thought that it is easiest for decision makers to think in terms of $P = 0.5$, though P can be any value between 0 and 1.)

   For example, say you let $P = 50\%$, and suppose the decision maker decides he would desire $3,000 for certain just as much as a 50% chance of $0 outcome and a 50% chance of $10,000 outcome. The utility of $3,000 can then be calculated as

$$U[\$3,000] = 0.5 \times U[\$0] + 0.5 \times U[\$10,000]$$
$$= 0.5 \times 1 + 0.5 \cdot 20$$
$$= 10.5.$$

   (b) To obtain utility values for monetary outcomes less than or greater than those for which utility values have been assigned or calculated, ask questions such as, "What relative chances of monetary outcomes of $X versus $Z would be just as desirable as a certain monetary outcome of $Y?" (*Note:* $X or $Z is the amount for which the utility value is to be determined.)

   For example, suppose it is desired to find the utility of $20,000 given the utility values obtained above for $0, $3,000, and $10,000.

Suppose the question posed is, "What relative chances of monetary outcomes of $3,000 versus $20,000 would be just as desirable as a certain outcome of $10,000?" Suppose further that the considered answer by the decision maker is 40% chance of $3,000 and 60% chance of $20,000. The utility value of $20,000 can then be calculated as

$$0.4 \times U[\$3,000] + 0.6 \times U[\$20,000] = U[\$10,000],$$
$$0.4 \times 10.5 + 0.6 \times U[\$20,000] = 20,$$
$$U[\$20,000] = 26.3.$$

4. Questions and computations in step 3 can be continued as long as utility values are needed. These can, in turn, be graphed to show utility values for the entire range of monetary outcomes of interest. A graph based on the above values is shown in Fig. 13-1.

In carrying out the utility derivation procedure, inconsistencies in the decision maker's replies may be discovered (e.g., two or more utility values calculated for the same monetary outcome or an extremely jagged utility-of-money function). If this happens, it becomes necessary to re-question to obtain judgments that are internally consistent.

The use of expected utility value as a decision criterion has a real advantage over the expected monetary value such as expected annual worth or expected present worth. Procedures based on expected monetary values virtually overlook the severe consequences of widely varying possible outcomes and merely take a weighted aver-

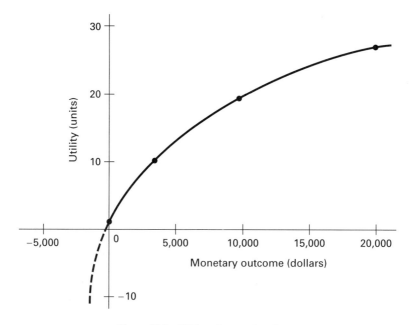

Figure 13-1   Utility-of-money function.

age of all outcomes. The expected utility procedure overcomes this objection by incorporating these variance influences directly into the computations. A large loss may be assigned a large negative utility by the individual, or he may assign a very great positive utility to a large increment in wealth, thus automatically bringing variance influences into the calculated results. This is demonstrated in the following example.

Example 13-10

As an example of calculation of expected utility, suppose the decision maker having the utility-of-money function in Fig. 13-1 is faced with a project that is expected to have monetary outcomes according to the following probabilities:

Monetary outcome (in net PW)	Probability
$20M	0.05
10M	0.15
0M	0.30
−2M	0.50

It is desired to compare calculated results using expected utility and expected monetary value or outcome.

   Solution   The necessary calculations are shown in Table 13-3. Thus, the expected utility method indicates an unfavorable project (expected utility is less than 1.0 when the utility of a monetary outcome of $0 is 1.0). In contrast, the expected monetary outcome (EMV) indicates a favorable project (expected net PW is greater than $0).   ∎

   Again, it should be emphasized that the expected utility method is useful for analyzing projects in which the potential gain or loss is of significant size compared to the monetary resources of the individual or firm for which the analysis is made. If a graph of the utility-of-money function can be closely approximated by a straight line over the range between the maximum and minimum monetary returns under consideration, then the expected monetary value can be used in place of the expected utility without significant error.

TABLE 13-3   Calculation of Expected Monetary Outcome (EMV) and Expected Utility

Monetary outcome (in net PW)	Probability of monetary outcome	Monetary outcome × probability	Utility of outcome	Utility × probability
$20M	0.05	$1.0M	26.3	1.31
10M	0.15	1.5M	20.0	3.00
0M	0.30	0M	1.0	0.30
−2M	0.50	−1.0M	−10.0	−5.00
		Expected net PW: Σ = $1.5M		Expected utility: Σ = −0.39

13.9.2 The Use of Utility Measurements

The method of assigning utilities to outcomes can be quite useful for gaining understanding of the rationale behind decisions made in situations involving risk. Indeed, if a decision maker specifies a utility-of-money function and if the economic analyst can predict monetary outcomes of individual projects that the decision maker accepts or believes, then the analyst can specify the project acceptance or rejection choices that would presumably turn out to be the same as those of the decision maker (neglecting nonmonetary factors). Thus, the problem of the analyst in dealing with decision under risk would be "solved." That is, the analyst could provide the manager with recommendations consistent with the manager's own thinking, thus allowing the the manager to address other problems.

Several limitations to use of the expected utility method should be recognized. First, it is often time consuming, and it is difficult to obtain a consistent utility-of-money function for an individual or organization. Second, responses for determining utility-of-money functions may well change over time; indeed, they may change even from day to day because of changes in the mood or temporary outlook of the person being questioned. Finally, a utility function for a particular set of alternative projects is not necessarily valid for another set of alternatives. Many intangible considerations taint the choice of any specific weighting. A decision maker might indicate a utility function that clearly shows a conservative approach in her attitudes toward corporate actions, but she might have an entirely different set of attitudes for investing with her own personal finances, such as in the stock market.

Under some conditions, it is expedient to employ methods that retain the concepts of utility functions without having to enumerate the full range of utilities and continuing through formal expected utility calculations. These informal uses can serve well to solidify subjective evaluations of risk situations.

13.9.3 Multicriteria Utility Models

Significant developments in the theory and practice of using multicriteria (e.g., multiple objectives, attributes, or factors) utility models have been summarized in a definitive work by Keeney and Raiffa.[3]

In general, the utility, $U(x) = U(x_1, x_2, \ldots, x_n)$, of any combination of outcomes $(x_1, x_2, \ldots, x_n)$ for n criteria $(X_1, X_2, \ldots, X_n)$ can be expressed as either (a) an additive or (b) a multiplicative function of the individual criteria utility functions $U_1(x_1)$, $U_2(x_2), \ldots, U_n(x_n)$, provided that each pair of criteria is

1. preferentially independent of its complement (i.e., the preference order of consequences for any pair of criteria does not depend on the levels at which all other criteria are held); and

2. utility independent of its complement (i.e., the conditional preference order for lotteries involving only changes in the levels of any pair of criteria does not depend on the levels at which all other criteria are held).

---

[3] R. L. Keeney and H. Raiffa, *Decisions with Multiple Objectives: Preferences and Value Tradeoffs* (New York: John Wiley & Sons, Inc., 1976).

The mechanics of building additive and multiplicative utility functions is beyond the scope of this book. The interested reader is referred to Keeney and Raiffa.

13.10 Expectation and Variance Criteria Applied to Investment Projects

The following examples show how expectation and variance criteria may be calculated for a project in which one or more of the variables (elements) are thought to vary according to independent discrete probabilities. The same type of analysis could be used for comparison of two or more alternatives.

Example 13-11

A single project is estimated to have a variable life and other element outcomes as follows:

Investment:	$10,000
Life:	3 yr, with probability = 0.3
	5 yr, with probability = 0.4
	7 yr, with probability = 0.3
Salvage value:	$2,000
Annual receipts:	$5,000
Annual disbursements:	$2,200

Find the expected annual worth and the standard deviation of the annual worth if the MARR is 8%.

Solution

For life = 3 yr, net AW is

$$\$5,000 - \$2,200 - [(\$10,000 - \$2,000)(A/P, 8\%, 3) + \$2,000(8\%)] = -\$460.$$

For life = 5 yr, net AW is

$$\$5,000 - \$2,200 - [(\$10,000 - \$2,000)(A/P, 8\%, 5) + \$2,000(8\%)] = \$630.$$

For life = 7 yr, net AW is

$$\$5,000 - \$2,200 - [(\$10,000 - \$2,000)(A/P, 8\%, 7) + \$2,000(8\%)] = \$1,110.$$

$$E[\text{AW}] = \sum_{\text{yr}} \text{AW} \times P(\text{AW})$$

$$= -\$460(0.3) + \$630(0.4) + \$1,110(0.3) = \$446$$

$$V[\text{AW}] = \sum_{\text{yr}} (\text{AW})^2 \times P(\text{AW}) - (E[\text{AW}])^2$$

$$= (-\$460)^2 \times 0.3 + (\$630)^2 \times 0.4$$

$$+ (\$1,110)^2 \times 0.3 - (\$446)^2$$

$$= \$401,000$$

$$\sigma[\text{AW}] = \sqrt{\$401,000} = \$631$$

Note that the expected net AW at an expected life of 5 years, which is $446, is less than the net AW at the assumed-certain life of 5 years, which is $630. Variation of project life can have a very marked effect on the results of an economic evaluation. In general, the greater the life variation, the higher the expected capital recovery cost based on that variation compared to the capital recovery cost at the assumed-certain life equal to the expected life. ■

Example 13-12

Assume the same conditions as in Example 13-11 except that annual receipts is also a random variable and is $7,000 with a probability of 0.33 or $4,000 with a probability of 0.67. Further, assume that the variation of project life occurs independently of variation of annual receipts. Show the net AW for all possible occurrences, and compute the expected net AW.

	Project life		
Annual receipts	3 yr $(P = 0.3)$	5 yr $(P = 0.4)$	7 yr $(P = 0.3)$
$7,000 $(P = 0.33)$	$1,540	$2,630	$3,110 ⎤
$4,000 $(P = 0.67)$	−1,460	−370	110 ⎦ Net AW

Solution

$$E[AW] = \sum_{R} \sum_{N} (AW \mid N, R)P(N)P(R),$$

where R denotes annual receipts and N denotes project life. Hence,

$$E[AW] = \$1,540(0.3)(0.33) + \$2,630(0.4)0.33)$$
$$+ \$3,110(0.3)(0.33) - \$1,460(0.3)(0.67)$$
$$- \$370(0.4)(0.67) + \$110(0.3)(0.67)$$
$$= \$437. \quad ■$$

In view of the risk and uncertainty regarding numerous variables or elements as typically found in economic analyses, it is reasonable that the measure of merit or desirability for one or more projects can be expressed as one or more continuous distributions. As an illustration, if the net annual worths for projects A and B in Example 13-6 (using the expectation-variance criterion) were distributed normally, the situation could be depicted as in Fig. 13-2. It can be seen in Fig. 13-2 that the probability of a loss (negative net AW) for project A is much higher than for project B; hence, project B might be chosen. If the respective distributions are extremely skewed rather than normally distributed, the indicated decision may differ. For example, suppose that project A is skewed to the right and project B is skewed to the left as shown in Fig. 13-3. On the basis of these conditions, project B might no longer be considered the more desirable.

If a probabilistic monetary model involves simple mathematical functions for the individual elements considered, then it sometimes can be mathematically manipulated so as to obtain directly the desired parameters or characteristics of the measure of merit. Chapter 12 described how Monte Carlo simulation can be utilized to obtain approximations for models of virtually unlimited complexity.

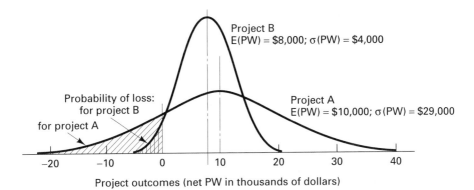

Figure 13-2   Outcome data for alternative projects assuming normal distributions.

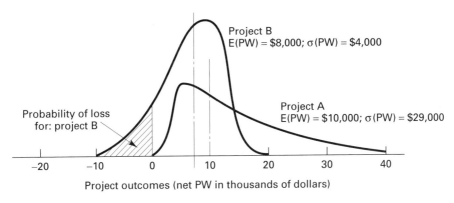

Figure 13-3   Same alternative projects as in Fig. 13-2, except with skewed distributions.

13.11 Miscellaneous Decision Rules for   *when probability is not known*
Complete Uncertainty

In this section we will describe some decision rules or principles for choosing from among alternatives in situations in which there is complete uncertainty about certain probabilities. These decision rules apply to situations in which there are a number of alternatives (*courses of action*) and a number of possible outcomes (*states of nature*), and in which the result (*effect*) of each alternative on each possible outcome is known but the probability of occurrence of each possible outcome is not known.

The most difficult aspect of using these decision rules is deciding which one to use for making a decision. In effect, these decision rules reflect various degrees of optimism or pessimism and should be chosen according to which reflect certain management views involving intuition and appropriateness for a particular situation. The greatest defense for the use of any of these rules is that their use will promote explicitness and consistency in decision making under complete uncertainty.

TABLE 13-4 Example Problem Involving Complete
Uncertainty Payoffs—Net PW ($M)

Alternatives	State of nature			
	S_1	S_2	S_3	S_4
I	3	−1	1	1
II	4	0	−4	6
III	5	−2	0	2

A representation of a typical problem is given by the matrix in Table 13-4.

Note that this is the same as the problem in Table 13-2, except that now probabilities are not known. The following sections explain and illustrate each of several decision rules for this type of problem.

13.11.1 Maximin or Minimax Rule

✓ The *maximin rule* suggests that the decision maker determine the minimum profit (payoff) associated with each alternative and then select the alternative that maximizes the minimum profit. Similarly, in the case of costs, the *minimax rule* suggests that the decision maker determine the maximum cost associated with each alternative and then select the alternative that minimizes the maximum cost. These decision rules are conservative and pessimistic, for they direct attention to the worst outcome and then make the worst outcome as desirable as possible.

Example 13-13

Given the payoffs for each of three alternatives and for each of four possible states of nature (chance occurrences) in Table 13-4, determine which alternative would maximize the minimum possible payoff.

    Solution The minimum possible payoff for alternative I is −1, for alternative II is −4, for alternative III is −2. Hence, alternative I would be chosen as maximizing these minimum payoffs. ∎

13.11.2 Maximax or Minimin Rule

The maximax or minimin rules are direct opposites of their counterparts discussed above and thus reflect extreme optimism. The *maximax rule* suggests that the decision maker determine the maximum profit associated with each alternative and then select the alternative that maximizes the maximum profit. Similarly, in the case of costs, the *minimin rule* indicates that the decision maker should determine the minimum cost associated with each alternative and then select the alternative that minimizes the minimum cost.

Example 13-14

Given the same payoff matrix as in Table 13-4, determine which alternative would maximize the maximum payoff.

    Solution The maximum possible payoff for alternative I is 3. Similarly, for II the maximum payoff is 6, and for III it is 5. The highest of these is 6, which occurs with alternative II; so alternative II is the maximax choice. ∎

13.11.3 Laplace Principle or Rule

The Laplace rule simply assumes that all possible outcomes are equally likely and that one can choose on the basis of expected outcomes as calculated using equal probabilities for all outcomes. There is a common tendency toward this assumption in situations where there is no evidence to the contrary, but the assumption (and, there-fore, the rule) is of highly questionable merit.

Example 13-15

Given the same payoff matrix as in Table 13-4, determine which alternative is best using the Laplace rule.

Solution

$$E[\text{alt. I}]: \qquad 3 \times \tfrac{1}{4} - 1 \times \tfrac{1}{4} + 1 \times \tfrac{1}{4} + 1 \times \tfrac{1}{4} = 1.00$$

$$E[\text{alt. II}]: \qquad 4 \times \tfrac{1}{4} + 0 \times \tfrac{1}{4} - 4 \times \tfrac{1}{4} + 6 \times \tfrac{1}{4} = 1.50$$

$$E[\text{alt. III}]: \qquad 5 \times \tfrac{1}{4} - 2 \times \tfrac{1}{4} + 0 \times \tfrac{1}{4} + 2 \times \tfrac{1}{4} = 1.25$$

Thus, alternative II, giving the highest expected payoff, is best. ∎

13.11.4 Hurwicz Principle or Rule

The Hurwicz rule is intended to reflect any degree of moderation between extreme optimism and extreme pessimism that the decision maker may wish to choose. The rule may be stated explicitly as

Select an index of optimism, a, such that $0 \leq a \leq 1$. For each alternative, compute the weighted outcome: $a \times$ (value of profit or cost if most favorable outcome occurs) + $(1 - a) \times$ (value of profit or cost if least favorable outcome occurs). Choose the alternative that optimizes the weighted outcome.

A practical difficulty of the Hurwicz rule is that it is difficult for the decision maker to determine a proper value for a, the weighting factor. The Hurwicz rule also lacks several of the desirable properties of a good decision rule, and it can even lead to results that are obviously counter to one's intuition.

Example 13-16

Given the same payoff matrix as in Table 13-4, calculate which alternative would be best, using the Hurwicz rule, for an index of optimism of 0.75. Also graph the calculated payoff for each alternative over the entire range of the index of optimism.

Solution

$$\text{alt. I:} \qquad 0.75(3) + 0.25(-1) = 2.0$$

$$\text{alt. II:} \qquad 0.75(6) + 0.25(-4) = 3.5$$

$$\text{alt. III:} \qquad 0.75(5) + 0.25(-2) = 3.25$$

Thus, alternative II, giving the highest payoff, is best. The graph is shown in Fig. 13-4. ∎

13.11.4.1 Minimax Regret Rule.

The minimax regret rule, proposed by L. J. Savage, is similar to the minimax and maximin rules but is intended to counter some of the ultraconservative results given by those rules. This rule suggests that the decision maker examine the maximum possible *regret* (loss because of not having

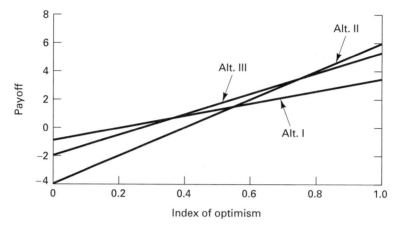

Figure 13-4   Graphed payoffs over range of optimism indices.

chosen the best alternative for each possible outcome) associated with each alternative and then select the alternative that minimizes the maximum regret.

Example 13-17

Given the same payoff matrix as in Table 13-4, show which alternative would be chosen on the basis of minimizing the maximum regret. Develop a regret matrix to obtain a solution.

Solution   The result is shown in Table 13-5.

TABLE 13-5   Regret Matrix for Problem of Table 13-4

Alternatives	State of nature				Maximum of states
	S_1	S_2	S_3	S_4	
I	2(= 5 − 3)	1	0	5	5
II	1(= 5 − 4)	0	5	0	5
III	0(= 5 − 5)	2	1	4	4 (min. of all max.)

Thus, it can be seen that the worst (highest) regret for alternative I is 5, for alternative II is 5, and for alternative III is 4. The minimum of these maximum regrets is 4 for alternative III, and thus alternative III is the choice.   ∎

PROBLEMS

13-1. Given the following matrix of net PWs (in $M) for four alternatives and various possible business conditions for which probabilities can be estimated.

Alternatives	Business condition (and probability)		
	Excellent (0.3)	Fair (0.5)	Poor (0.2)
I	30	25	−15
II	15	15	15
III	35	30	−5
IV	45	10	−5

Show what can be determined, or which alternative is preferred, using each of the following criteria:

a. does any alternative dominate all the other alternatives? Is any alternative dominated by any other alternative? Which alternative(s) is (are) left for consideration after checking for dominance?

b. if the aspiration level is to minimize the chance of a loss and maximize the chance to make at least 28

c. most probable future

d. expected value

e. expectation-variance, using the function in Eq. 13-3 where $\sigma[x]$ is approximated as (maximum outcome − minimum outcome)/5 and where $A = 1.0$

f. certain monetary equivalence, using your own subjective assessment assuming you are the top decision maker for a firm with $100 million in assets

13-2. Suppose you were faced with the same matrix of net PWs as in Problem 13-1 but you do not have probability estimates for the various business conditions. Show which alternative should be chosen using each of the following decision rules or principles:

a. maximin rule

b. maximax rule

c. Laplace principle

d. Hurwicz principle, with 0.5 optimism

e. minimax regret rule

13-3. Given the following matrix of equivalent annual costs (in $M) for three alternatives and various possible conditions.

	State of nature (and probability)		
Alternatives	Kind (0.3)	Erratic (0.6)	Perverse (0.1)
Alpha	20	28	40
Bravo	25	25	25
Charlie	11	27	45

Show what can be determined, or which alternative is preferred, using each of the following criteria:

a. is any alternative dominated?

b. if the aspiration level is to have a possible cost no greater than 26

c. the most probable future

d. expected value

e. certain monetary equivalence, using your own subjective assessment assuming you are a manager who is risk-neutral and thus you base your decisions on long-run averages

13-4. Suppose you are faced with the same matrix of equivalent annual costs as in Problem 13-3 except that you feel the probability estimates are invalid and cannot be used. Show which alternative should be chosen using each of the following decision rules or principles:

a. minimax rule

b. minimin rule

    c. Laplace principle

    d. Hurwicz principle, with 0.8 index of optimism

    e. minimax regret rule

    f. minimize variability, or range (i.e., max. minus min.)

13-5. Given the matrix of costs below, for various mutually exclusive alternatives, show which is best using the following decision rules or principles:

    a. minimax rule

    b. minimin rule

    c. Laplace principle

    d. Hurwicz principle, with $\frac{2}{3}$ optimism

    e. minimax regret rule

	State of nature			
Alternative	A	B	C	D
I	18	18	10	14
II	14	14	14	14
III	5	26	10	14
IV	14	22	10	10
V	10	12	12	10

13-6. Suppose the flip of a fair coin will determine whether you gain $\$X$ or lose $\$X$. What certain amount would you personally be willing to pay (or accept), instead of the random outcome, if $\$X$ is

    a. $0.10

    b. $1.00

    c. $10

    d. $100

    e. $1,000

    f. $10,000

13-7. Suppose a business opportunity has a 0.25 chance of making a PW of $\$X$ and a 0.75 chance of making $0. For what certain amount would you be just willing to sell the opportunity if the money is for you personally and $\$X$ is

    a. $1,000

    b. $10,000

    c. $100,000

    d. $1,000,000

13-8. Answer Problem 13-7 if the money belongs to a large corporation for which you are the decision maker. Are you more risk-averse if making such decisions for a corporation or for yourself personally?

13-9. Entrepreneur Y has a utility index of 108 for $11,000 and 75 for $0. He is indifferent between a 0.5 chance at $11,000 plus a 0.5 chance at a $20,000 loss and a certainty of $0. What is his utility index for a loss of $20,000?

13-10. Entrepreneur Z has a utility index of 10 for $18,750, 6 for $11,200, and zero for $0. What probability combination of $0 and $18,750 would make her indifferent to $11,200 for certain?

13-11. Two economists, Alfred M. Dismal and J. Maynard Science, are arguing about the relative merits of their respective decision rules. Dismal says he always takes the act with the greatest expected monetary value; Science says she always takes the act with the greatest expected utility, and her utility function for money is $U = 10 + 0.2M$, where M is the monetary payoff. For decisions involving monetary payoffs, who will make the better choices?

13-12. You have a date for the economic analysis ball; the admission is $20, which you do not have. On the day of the dance your psychology instructor offers you either $16 for certain or a 50–50 chance at nothing or $24. Which choice would you make, assuming you had no other sources of funds or credit. Why? If the utility of $16 is 20, and the utility of $0 is zero, what does this imply about the utility of $20?

13-13. Develop a utility function for yourself for the monetary outcomes of −$100,000, −$10,000, +$10,000, +$40,000, and +$200,000. Start with the following monetary outcomes and arbitrarily assigned units:

Monetary outcome	Utility units
$ 1,000	10
15,000	30

Write the questions you ask yourself, and show your calculations. Finally, plot the results with monetary outcome on the *x*-axis.

13-14. Suppose that the utility-of-money function of a decision maker is described as utility = ln(monetary outcome in thousands of dollars) between the monetary outcome limits of $100 and $1,000,000. The monetary outcomes and associated probabilities for two competing projects are as follows:

Project	Monetary outcome (gain)	Probability
A	$ 1,000	0.33
	10,000	0.33
	19,000	0.33
B	$ 3,000	0.3
	10,000	0.4
	12,000	0.1
	13,000	0.2

Show which project is preferable by (a) the expected monetary method and by (b) the expected utility method. (*Note:* ln means logarithm to the base *e*.)

13-15. A certain project requires an investment of $10,000 and is expected to have net annual receipts minus disbursements of $2,800. The salvage value as a function of life, together with associated probabilities, is

Life	Salvage value	Probability
3	$4,000	0.25
5	2,000	0.50
7	0	0.25

Find the expected net AW and the standard deviation of net AW if the MARR is 8%.

13-16. Work Problem 13-15 with the change that the net annual receipts minus disbursements is a random variable independent of the life and is estimated to be $1,800 with probability 0.2, $2,800 with probability 0.6, and $3,800 with probability 0.2.

13-17. Project Stochastic is estimated to require an investment of $25,000, have a life of 5 years and $0 salvage value, and have an annual net cash flow of $5,000 with 30% probability, $10,000 with 50% probability, and $12,000 with 20% probability. If the MARR is 15%, calculate the expected value and variance of the net AW for project Stochastic.

13-18. Project Variate is estimated to require an investment of $25,000 and have an annual net cash flow of $16,000 and a $0 salvage value. The life for project Variate is estimated to be 1 year with 10% probability, 5 years with 50% probability, and 10 years with 40% probability. If the MARR is 15%, calculate the expected value and variance of the net AW.

13-19. Plot a frequency histogram for the projects in Problems 13-17 and 13-18, distinguishing between the two by shaded coding. Which project would probably be thought more desirable if the decision maker were (a) conservative, thus not prone to take risks, or (b) a maximizer of expectations, regardless of risk.

13-20. Suppose that the expectation-variance decision function for a given project is equal to the net AW minus a constant times the standard deviation of the net AW.

    a. For the projects in Problems 13-17 and 13-18, determine which appears to be the more desirable if the constant coefficient is 0.6.

    b. At what value of the constant coefficient are the two projects equally desirable?

13-21. The mean and standard deviation of the rate of return for project X are estimated to be 15% and 5%, respectively. Similarly, the mean and standard deviation of the rate of return for a competing project Y are estimated to be 25% and 18%, respectively.

    a. If the expectation-variance function for the decision maker is the expected rate of return minus 0.1 times the variance of the rate of return (in integer amounts), show which project would be more desirable.

    b. For the function in part (a), at what value of the coefficient applied to the variance would the projects be considered equally desirable?

13-22. Suppose that in Problem 13-15 the interest on capital is 4% with probability 0.5 and 12% with probability 0.5, and suppose that interest varies independently of the life of the project.

    a. Calculate the expected net AW.

    b. Plot histograms of outcomes for Problem 13-15 and for this problem to compare variability.

13-23. A specific project requires an investment of $100,000 and is expected to have a salvage value of $20,000. It is thought equally likely that the life will turn out to be either 6, 10, or 12 years. The net annual cash inflow is twice as likely to be $30,000 as either $35,000 or $18,000. If the minimum required rate of return is 10%:

    a. Develop a table showing the net AW for all combinations of the two variables.

    b. Assuming that the variable outcomes are independent, calculate $E[AW]$ and $V[AW]$.

14
Decision Tree Analysis

14.1 Introduction

Decision trees, also commonly called *decision flow networks* and *decision diagrams,* are powerful means for depicting and facilitating the analysis of important problems, especially those that involve sequential decisions and variable outcomes over time. Decision trees have great usefulness in practice because they make it possible to look at a large complicated problem in terms of a series of smaller simpler problems and they enable objective analysis and decision making that includes explicit consideration of the risk and effect of the future.

The name *decision tree* is descriptive of the appearance of a graphical portrayal, for it shows branches for each possible alternative for a given decision and branches for each possible outcome (event) that can result from each alternative. Such networks reduce abstract thinking to a logical visual pattern of cause and effect. When costs and returns are associated with each branch and probabilities are estimated for each possible outcome, then analysis of the decision tree can clarify choices and risks.

14.2 A Deterministic Decision Tree

The most basic form of decision tree occurs when each alternative can be assumed to result in a single outcome—that is, when certainty is assumed. The replacement problem in Fig. 14-1 illustrates this. The problem as shown reflects that the decision on whether to replace the old machine with the new machine is not just a one-time decision, but rather one that recurs periodically. That is, if the decision is made to keep the old machine at decision point 1, then later, at decision point 2, a choice again has to

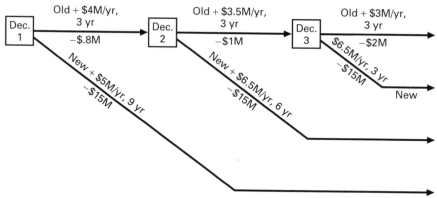

Figure 14-1 Deterministic replacement example.

be made. Similarly, if the old machine is chosen at decision point 2, then a choice again has to be made at decision point 3. For each alternative, the cash inflow is shown above the arrow and the cash investment opportunity cost is shown below the arrow.

For this problem, one is concerned initially with which alternative to choose at decision point 1. But an intelligent choice at decision point 1 should take into account the later alternatives and decisions that stem from it. Hence, the correct procedure in analyzing this type of problem is to *start* at the most distant decision point, determine the best alternative and quantitative result of that alternative, and then "roll back" to each successive decision point, repeating the procedure until finally the choice at the initial or present decision point is determined. By this procedure, one can make a present decision that directly takes into account the alternatives and expected decisions of the future.[1]

For simplicity in this example, timing of the monetary outcomes will first be neglected, which means that a dollar has the same value regardless of the year in which it occurs. Table 14-1 shows the necessary computations and decisions. Note that the monetary outcome of the best alternative at decision point 3 ($7.0M for the "Old")

Table 14-1 Monetary Outcomes and Decisions at Each Point—
Deterministic Replacement Example of Fig. 14-1

Decision point	Alternative	Monetary outcome		Choice
3	Old	$3M(3) - $2M	= $ 7.0M	Old
	New	$6.5M(3) - $15M	= $ 4.5M	
2	Old	$7M + $3.5M(3) - $1M	= $16.5M	
	New	$6.5M(6) - $15M	= $24.0M	New
1	Old	$24M + $4M(3) - $0.8M	= $35.2M	Old
	New	$5M(9) - $15M	= $30.0M	

[1] This procedure is a special (and simple) case of dynamic programming (DP). For a discussion of the use of DP in replacement studies, the reader is referred to C. S. Park and G. P. Sharpe-Bette, *Advanced Engineering Economics* (New York: John Wiley & Sons, 1990).

becomes part of the outcome for the "Old" alternative at decision point 2. Similarly, the best alternative at decision point 2 ($24.0M for the "New") becomes part of the outcome for the "Old" alternative at decision point 1.

By following the computations in Table 14-1, one can see that the answer is to keep the "Old" now and plan to replace it with the "New" at the end of 3 years. But this does not mean that the old machine should necessarily be kept for a full 3 years and then a new machine bought without question. Conditions may change at any time, thus necessitating a fresh analysis—probably a decision tree analysis—based on estimates that are reasonable in light of conditions at that later time.

14.3 A Deterministic Decision Tree Considering Timing

For decision tree analyses, which involve working from the most distant decision point to the nearest decision point, the easiest way to take into account the timing of money is to use the present worth approach and thus discount all monetary outcomes to the decision points in question. To demonstrate, Table 14-2 shows computations for the same replacement problem of Fig. 14-1 using an interest rate of 25% per year.

Note from Table 14-2 that when taking into account the effect of timing by calculating present worths at each decision point, the indicated choice is not only to keep the "Old" at decision point 1, but also to keep the "Old" at decision points 2 and 3 as well. This result is not surprising since the high interest rate tends to favor the alternatives with lower initial investments, and it also tends to place less weight on long-term returns.

14.4 Consideration of Random Outcomes

The deterministic replacement example of Fig. 14-1 did not include one of the most powerful elements in the use of decision trees: the formal consideration of variable outcomes to which probabilities of occurrence can be assigned. Suppose that for each

Table 14-2   Decisions at Each Point with Interest = 25% per yr
for Deterministic Replacement Example of Fig. 14-1

Decision point	Alternative	PW of monetary outcome		Choice
3	Old	$3M(P/A,3) − $2M$ $3M(1.95) − $2M	= $3.85M	Old
	New	$6.5M(P/A,3) − $15M$ $6.5M(1.95) − $15M	= −$2.33M	
2	Old	$3.85(P/F,3) + $3.5M(P/A,3) − $1M$ $3.85(0.512) + $3.5M(1.95) − $1M	= $7.79M	Old
	New	$6.5M(P/A,6) − $15M$ $6.5M(2.95) − $15M	= $4.18M	
1	Old	$7.79M(P/F,3) + $4M(P/A,3) − $0.8M$ $7.79M(0.512) + $4M(1.95) − $0.8M	= 10.98M	Old
	New	$5.0M(P/A,9) − $15M$ $5.0M(3.46) − $15M	= $2.30M	

alternative there are two possible monetary outcomes, depending on whether the
demand is "high" or "low." In such a case, the decision tree problem of Fig. 14-1
would appear as in Fig. 14-2. Note that for each alternative in Fig. 14-2 there is shown
a circle from which are drawn arrows to represent each possible chance event or state
of nature which can result, such as demand being either "high" (H) or "low" (L).

In order to solve this problem—that is, to determine the best alternative for each
decision point, etc.—it is necessary to determine the outcome (usually expressed in
monetary units) and the probability of occurrence for each possible chance event.
Then the criterion (measure of merit) for choice (usually expected PW of monetary
outcomes) can be decided and the solution computed by the same procedure as
before; that is, criterion outcomes and decisions are determined for the most distant
decision points first, and then the procedure is successively repeated, moving back in
time until the decision for decision point 1 is determined.

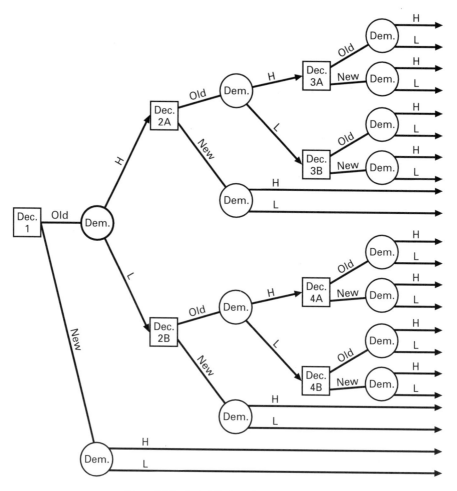

Figure 14-2 Probabilistic replacement example.

14.5 A Classical Decision Tree Problem

The following is a brief description of a classical problem for which decision trees are very useful for analysis and solution.[2]

An oil wildcatter must decide whether to drill or not to drill at a given site before his option expires. He is uncertain whether the hole will turn out to be dry, wet, or a gusher. The net payoffs (in present worths) for each state are $-70,000, $50,000, and $200,000, respectively. The initially estimated probabilities that each state will occur are 0.5, 0.3, and 0.2, respectively. Figure 14-3 is a decision flow network depicting this simple situation. Table 14-3 shows calculations to determine that the best choice for the wildcatter is to drill based on an expected monetary value of $20,000 versus $0 if he does not drill. Nevertheless, this may not be a clear-cut decision because of the risk of a $70,000 loss and because the wildcatter might reduce the risk by obtaining further information.

Suppose it is possible for the wildcatter to take seismic soundings at a cost of $1,000. The soundings will disclose whether the terrain below has no structure (outcome NS), an open structure (outcome OS), or a closed structure (outcome CS).

Instead of using Bayesian methods for revision of probabilities (to be discussed later), let us assume that the probabilities of the various possible well outcomes given the various seismic sounding outcomes are as shown in Fig. 14-4, which is a flow diagram for the entire problem.[3] The solution of the problem using the EMV criterion is shown in Table 14-4. It should be noted that the alternative "seismic soundings" is now best with an expected monetary outcome of $31,550.

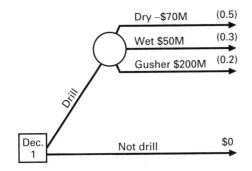

Figure 14-3 Oil wildcatter problem diagram, before consideration of seismic soundings.

Table 14-3 Expected Monetary Calculations for the Oil Wildcatter Problem Before Consideration of Seismic Soundings

Drill:	$-\$70,000(0.5) + \$50,000(0.3) + \$200,000(0.2) = \$20,000$
Not drill:	$= \$0$

[2] H. Raiffa, *Decision Analysis: Introductory Lectures on Choices Under Uncertainty* (Reading, MA: Addison-Wesley, 1968).

[3] Problem 14-11 at end of this chapter is a statement of the sampling or added-study probabilities which, when combined with the prior probabilities in Fig. 14-3, result in the posterior probabilities in Fig. 14-4.

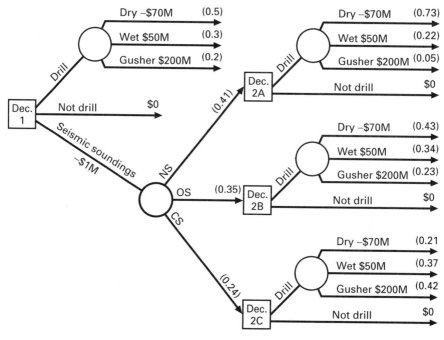

Figure 14-4   Oil wildcatter problem with seismic soundings taken into consideration.

Table 14-4   Expected Monetary Calculations for the Oil Wildcatter Problem with Consideration of Seismic Soundings

Decision point	Alternative	Expected monetary value		Choice
2A	Drill	−$70M(0.73) + $50M(0.22) + $200M(0.05) =	−$30M	
	Not drill		$0	Not drill
2B	Drill	−$70M(0.43) + $50M(0.34) + $200M(0.23) =	$32.9M	Drill
	Not drill		$0	
2C	Drill	−$70M(0.21) + $50M(0.37) + $200M(0.42) =	$87.5M	Drill
	Not drill		$0	
1	Drill		$20M	
	Not drill		$0	
	Seismic soundings	$0(0.41) + $32.9M(0.35) + $87.5M(0.24) − $1M	= $31.55M	Seismic soundings

14.6 Constructing a Decision Tree

Now that decision trees (diagrams) have been introduced and the mechanics of using the diagrams to arrive at an initial decision have been illustrated, the steps involved can be summarized as follows:

✓

1. Identify the points of decision and alternatives available at each point.
2. Identify the points of uncertainty and the type or range of possible outcomes at each point.
3. Estimate the values needed to make the analysis, especially the probabilities of different outcomes and the costs/returns for various outcomes and alternative actions.
4. Analyze the alternatives, starting with the most distant decision point(s) and working back, to choose the best initial decision.

The preceding example used the expected monetary value (EMV) as the decision criterion. However, if outcomes can be expressed in terms of utility units, then the decision maker can use the expected utility as a decision criterion. Alternatively, the decision maker may be willing to express his certain monetary equivalent (CME) for each chance outcome node and use that as his decision criterion.

Because a decision diagram can quickly become discouragingly, if not unmanageably, large, it is generally best to start out by structuring a problem simply by considering only major alternatives and outcomes in order to get an initial understanding or "feel" for the problem. Then one can develop more information on alternatives and outcomes that seem sufficiently important to affect the final decision until one is satisfied that the study is sufficiently complete (in view of the nature and importance of the problem and the time and study resources available).

The proper diagramming of a decision problem is, in itself, generally very useful to the understanding of the problem, and it is essential to correct subsequent analysis.

The placement of decision points and chance outcome nodes from the initial decision point to the base of any later decision point should give a correct representation of the information that will and will not be available when the decision maker actually has to make the choice represented by the decision point in question. The decision tree diagram should show the following:

1. all initial or immediate alternatives among which the decision maker wishes to choose;
2. all uncertain outcomes and future alternatives that the decision maker wishes to consider because they may directly affect the consequences of initial alternatives;
3. all uncertain outcomes that the decision maker wishes to consider because they may provide information that can affect her future choices among alternatives and hence indirectly affect the consequences of initial alternatives.

It should also be noted that the alternatives at any decision point and the outcomes at any outcome node must be

1. mutually exclusive, i.e., no more than one can possibly be chosen, and
2. collectively exhaustive, i.e., one must be chosen or something must occur if the decision point or outcome node is reached.

14.7 Use of Bayesian Statistics to Evaluate the Worth of Further Investigative Study

One alternative that frequently exists in an investment decision problem is further research or investigation before deciding on the investment. This means making an intensive objective study, hopefully by a fresh group of people. It may involve such aspects as undertaking additional research and development study, making a new analysis of market demand, or possibly studying anew future operating costs for particular alternatives.

The concepts of Bayesian statistics provide a means for utilizing subsequent information to modify estimates of probabilities and also a means for estimating the value of further economic investigative study. Chapter 15 provides additional discussion of Bayesian statistical approaches to decision making.

To illustrate how the worth of sample information is obtained, consider the one-stage decision situation shown in Fig. 14-5, in which each alternative has two possible chance outcomes: "high" or "low" demand. It is estimated that each outcome is equally likely to occur, and the monetary result expressed as PW is shown above the arrow for each outcome. Again, the amount of investment for each alternative is shown below the respective lines. Based on these amounts, the calculation of the expected monetary outcome (net PW) is shown in Table 14-5, which indicates that the "Old" should be chosen.

To demonstrate the use of Bayesian statistics, suppose that one is considering the advisability of undertaking an independent intensive investigation before deciding upon the "Old" versus the "New." Suppose also that this further study would cost $0.1M. In order to use the Bayesian approach, it is necessary for management to assess the conditional probabilities that the intensive investigation will yield certain results. These probabilities reflect explicit measures of management's confidence in the ability of the investigation to predict the outcome. Sample assessments are shown

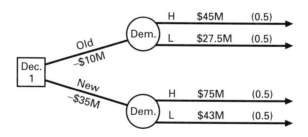

Figure 14-5   One-stage replacement problem.

Table 14-5   Expected Monetary Values for
Problem in Fig. 14-5

Old:	$45M(0.5) + $27.5M(0.5) − $10M = $26.25M
New:	$75M(0.5) + $43M(0.5) − $35M = $24.0M

Table 14-6 Management's Assessment of Confidence
in Investigation Results

$P(h \mid H) = 0.70$
$P(h \mid D) = 0.20$
$P(d \mid H) = 0.30$
$P(d \mid D) = 0.80$

Key:	Investigation-predicted demand	Actual demand
	h = High	H = High
	d = Low	D = Low

in Table 14-6. As an explanation, $P(h/H)$ means the probability that the predicted demand is "high," given that the actual demand will turn out to be "high."

Appendix 14-A contains a formal statement of Bayes's theorem as well as a tabular format for ease of calculations in the discrete outcome case. Tables 14-7 and 14-8 use this format for revision of probabilities based on the data in Table 14-6 and the prior probabilities of 0.5 that the demand will be high and 0.5 that the demand will be low.

The probabilities calculated in Tables 14-7 and 14-8 can now be used to assess the alternative of further investigation. Figure 14-6 shows a decision tree diagram for this alternative as well as the two original alternatives. Note the demand probabilities entered on the branches according to whether the investigation indicates "high" (H) or "low" (D) demand.

Table 14-7 Computation of Posterior Probabilities Given That
Investigation-Predicted Demand Is High (h)

(1)	(2)	(3)	(4) = (2)(3)	(5) = (4)/Σ(4)
State (actual demand)	Prior probability P(state)	Confidence assessment $P(h \mid$ state)	Joint probability	Posterior probability P(state $\mid h$)
H	0.5	0.70	0.35	0.78
D	0.5	0.20	0.10	0.22
			$\Sigma = \overline{0.45}$	

Table 14-8 Computation of Posterior Probabilities Given That
Investigation-Predicted Demand Is Low (d)

(1)	(2)	(3)	(4) = (2)(3)	(5) = (4)/Σ(4)
State (actual demand)	Prior probability P(state)	Confidence assessment $P(d \mid$ state)	Joint probability	Posterior probability P(state $\mid d$)
H	0.5	0.30	0.15	0.27
D	0.5	0.80	0.40	0.73
			$\Sigma = \overline{0.55}$	

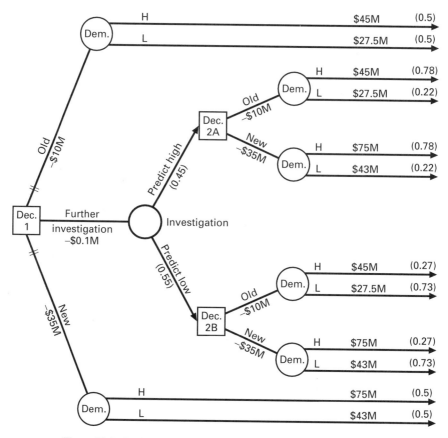

Figure 14-6   Replacement problem with alternative of further investigation.

The expected outcome for the alternative of further investigation can now be calculated. This is done by the standard decision tree principle of determining the decision at the most distant points and working back. This is shown in Table 14-9. It is worthy to note that the 0.45 and 0.55 probabilities that investigation-predicted demand will be "high" and "low," respectively, are obtained from the total in column (4) of the Bayesian revision calculations shown in Tables 14-7 and 14-8.

Table 14-9   Expected Monetary Outcome for Replacement Problem of Fig. 14-6

Decision point	Alternative	Expected monetary outcome		Choice
2a	Old	$45M(0.78) + $27.5M(0.22) − $10M	= $31.15M	
	New	$75M(0.78) + $43M(0.22) − $35M	= $32.96M	New
2b	Old	$45M(0.27) + $27.5M(0.73) − $10M	= $22.23M	Old
	New	$75M(0.27) + $43M(0.73) − $35M	= $16.64M	
1	Further investigation	$32.96M(0.45) + $22.23M(0.55) − $0.1M = $26.96M		Further investigation
	Keep old	(from Table 14-5): $26.25M		
	New	(from Table 14-5): $24.00M		

Thus, from Table 14-9, it can be seen that the alternative of further investigation, with an expected return of $26.96M, is the best present course of action by a slight margin. While the figures used here do not reflect much advantage to the further investigation, the advantage potentially can be great.

It is often thought useful to show results of calculations and choices between alternatives directly on the decision diagrams. For example, the replacement problem in Fig. 14-6 (and the calculated results and choices in Table 14-9) might be shown as in Fig. 14-7. The numbers in small boxes next to each outcome node represent the expected value (or other indicator of desirability) of outcomes beyond that point. The "double slash" marks for all alternatives except one emanating from each decision point indicate alternatives that would *not* be chosen. The number in the small box next to each decision point indicates the value of the best alternative at each point.

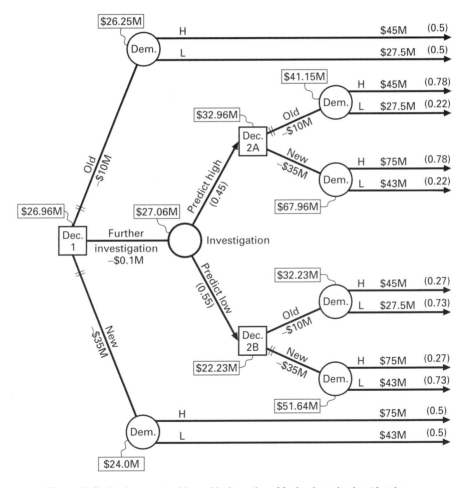

Figure 14-7 Replacement problem with alternative of further investigation (showing useful notation conventions).

14.8 Alternate Method of Analysis

Some analysts and decision makers prefer to show the criterion values (like expected monetary value) at the end of each possible path through the tree. (*Note:* If the monies along the path are at significantly different points in time, the criterion values should be expressed in terms of equivalent worths, such as PW or AW.) Then, the "roll back" technique can be used to determine the optimal choice at each decision point and obtain the same initial decision as when using the previous method. For example, the problem in Fig. 14-6 and Fig. 14-7 could be shown as in Fig. 14-8, which also includes the "notation conventions" explained in the last section. The criterion values at the end of each branch are placed in oval boxes to distinguish them from the other outcome and investment values emanating from each chance node and decision point. As an example, the criterion value for the third path from the top, $34.9M, in Fig. 14-8 is

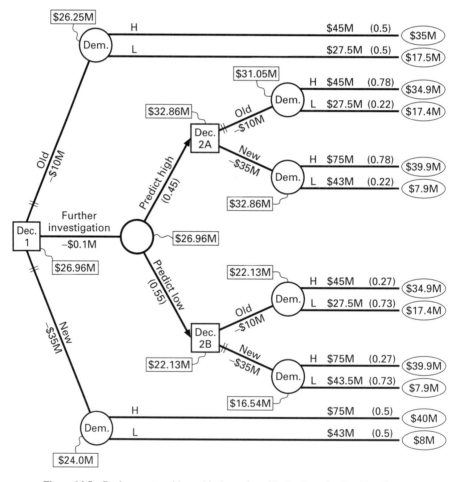

Figure 14-8 Replacement problem with alternative of further investigation (showing outcomes at end of each branch and subsequent analysis).

Table 14-10   Expected Monetary Outcomes for Replacement Problem of Figure 14-8

Decision point	Alternative	Expected monetary outcome		Choice
2a	Old	$34.9M(0.78) + $17.4M(0.22)	= $31.05M	
	New	$39.9M(0.78) + $7.9M(0.22)	= $32.86M	New
2b	Old	$34.9M(0.27) + $17.4M(0.73)	= $22.13M	Old
	New	$39.9M(0.27) + $7.9M(0.73)	= $16.54M	
1	Further investigation	$32.86M(0.45) + $22.13M(0.55) = $26.96M		Further investigation
	Old	35M(0.5) + 17.5M(0.5) = $26.25M		
	New	40M(0.5) + 8M(0.5) = $24.00M		

obtained by adding −$0.1M, −10M, and $45M for that path. Calculations of expected monetary outcomes in Fig. 14-8 are shown in Table 14-10.

14.9 Examples of Decision Tree Applications

The decision tree technique can be useful in a very wide range of decision situations. To give some idea of the breadth of potential applications, two examples follow.

14.9.1 Small Versus Large Asset

Figure 14-9 shows a situation in which a firm is initially faced with the decision between a small machine and a large machine for a use in which demand for the machine is uncertain but subject to probabilistic estimates. Further, if the firm should

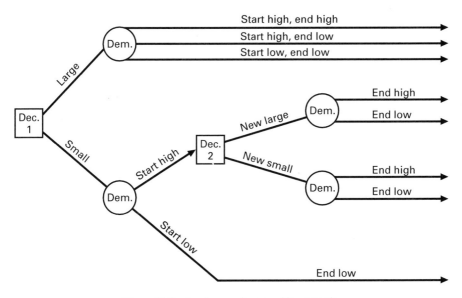

Figure 14-9   Small versus large machine example.

invest in a small machine now, it has the future choice of whether to invest in another small machine according to the anticipated demand at the time of that future decision.

14.9.2 Facilities Modernization

Figure 14-10 shows a situation in which a firm is faced with the decision of whether to invest in major automation of the plant's facilities. The new equipment is supposed to result in reduced labor cost, but its technical performance (perf.) is critical and subject to variation. Also, the monetary outcome is influenced strongly by the total market demand for the product and by whether or not competitors (compet.) also automate. The diagram shows two decision stages, but, of course, further stages can be enumerated if that is thought desirable.

14.10 Advantages and Disadvantages of Decision Tree Analysis

The systematic approach of decision tree analysis has its merits and demerits. Indeed, what is a pro to some analysts and decision makers may well be a con to others. The following is a synthesis of often claimed advantages:

1. *Makes uncertainty explicit.* The uncertainty the analyst feels about estimates or projects is recognized and incorporated in the analysis.

2. *Promotes more reasoned estimating procedures.* Requiring that estimates be given as probability distributions rather than as single values, and requiring that these estimates be broken into elements, forces more attention on the estimating.

3. *Encourages consideration of whole problem.* The systematic approach forces the analyst or decision maker to come to quantitative grips with the interactions between various facets of his or her problem.

4. *Helps communication.* It facilitates the provision of inputs in an unambiguous quantitative manner from experts and analysts as needed and it provides these results to the decision maker in a clear manner.

5. *Helps determine need for data and study.* The systematic examination of the value of information in a decision context helps suggest the gathering and compilation of data from new sources.

6. *Stimulates generation of new alternatives.* Detailed decision analysis helps the decision maker and his or her staff to think hard about new, viable alternative actions.

7. *Helps "sell" decision.* A hard, thorough analysis can be used to emphasize that a decision has not been made frivolously and to rally support for the decision.

8. *Provides framework contingency planning.* Decision analysis not only results in an initial decision but it can be used as a basis for continuous reevaluation of a decision problem that has a distant time horizon.

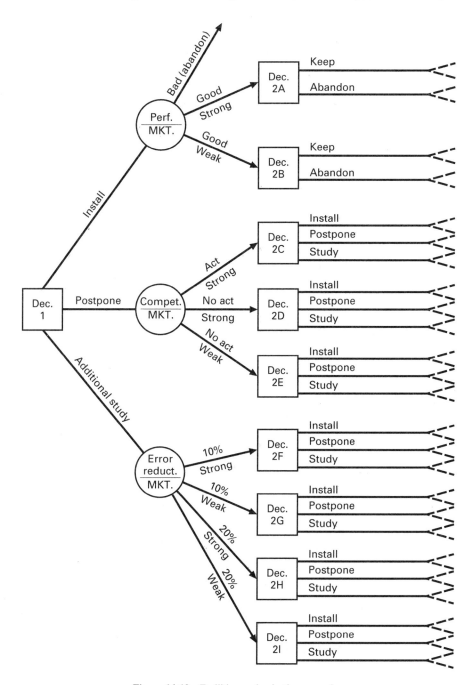

Figure 14-10   Facilities modernization example.

The following is a synthesis of frequently claimed disadvantages:

1. *Tends to exclude consideration of intangibles.* Because a well-done decision analysis is thorough, there is a tendency to place too great a reliance on the quantitative results.

2. *The clear basic questions are often the most difficult.* Often the decision maker would rather take refuge in the fuzziness and complexity of real-life situations than reveal preferences in a number of broken-down or starkly simple decision situations.

3. *Requires expert articulation of the thought process.* A decision maker may be a great synthesizer of interconnected considerations through subconscious thinking, but he may be unable to give a verbal description of his thought process, thus making it appear that he is much more restricted in the complexity of his analysis than is actually true.

4. *Decision analysts tend to lack compassion.* Persons who elect to get into formalized, systematic analysis are so prone to attach numbers to everything that they tend to exclude many human and artistic qualities, thereby inhibiting creativity.

14.11 Summary

The decision tree approach may appear to be complex, but it needs to be no more complex than the decision situation involved. Any investment problem can be examined at many levels of detail. A major difficulty in setting up a decision tree analysis is to strike the appropriate level. In general, the appropriate level allows decision makers to consider major future alternatives commensurate with the consequences of those alternatives without becoming so concerned with detail and refinement that the key factors are obscured.

Decision tree methodology is a basis for investment analysis, evaluation, and decision and is a means of making explicit the process that should be at least intuitively present in good investment decision making. Use of this methodology will help force a consideration of alternatives, define problems for further investigation, and clarify for the decision maker the nature of the risks she faces and the estimates she must make.

PROBLEMS

14-1. Because of shifting rock formations, a community will be in danger of the collapse of an upstream dam for a year starting now. A permanent replacement dam has been started, but it will take a year to complete. If there is a dam collapse, it will destroy the town, but no deaths would be expected because there is an efficient warning system.

One alternative is to make temporary repairs on the existing dam and to construct temporary levees. Such repairs would greatly decrease the probability of the collapse of the dam. If the collapse should occur, the levees might or might not hold the water back from the town.

The decision maker in this problem has decided to call in experts to give opinions. Because the collapse of the old dam depends in part on the underlying geological features of the area, the experts can better assess the likelihood of dam collapse if they conduct some expensive geological tests.

Diagram the decision maker's problem in the form of a decision tree.

14-2. The president of the High Point Carolina Company, Bob Foscue, must decide quickly whether or not to lease a large manufacturing area that has become available adjacent to the firm's present facilities. He is convinced that if he does not lease the extra space now, it will not be feasible for him to move or otherwise expand his plant capacity for at least two years. If he does not lease the space and runs out of production capacity during the period, he may be forced next year to make a difficult decision that could be critical to the future of his small business. This decision will be between (a) failure to fill all customer orders by rationing limited supplies among existing accounts and (b) using outside contract furniture builders to produce for demand in excess of the firm's capacity. The consequences of choice (a) would be a significant slowing of the firm's aggressive growth momentum. However, choice (b) has an excellent chance of working out well except for two potentially fatal dangers: the contract suppliers might develop dangerous competition with the firm's designs or they might fail to meet quality and shipping quantity requirements.

Foscue believes that if he does not lease the additional space, the question of whether or not he will run out of production capacity and be forced to choose between (a) and (b) depends on two key factors: (1) retailer demand for his furniture during the coming year's buying season and (2) his own decision on whether or not to continue an existing merchandising arrangement with a large direct-mail catalog firm. The direct merchandising arrangement provides an outlet for considerable sales volume, but at modest prices. Foscue is very sure that the merchandising firm will not offer materially improved prices or terms. His own final decision about renewal of the contract will depend on his overall evaluation of supply and demand factors at the time. His current decision about the lease, by limiting the amount of furniture he can supply without going to outside contractors, may have some impact on his subsequent decision on the renewal of the contract. However, Foscue does not consider it feasible or worthwhile to lay out all of the important developments occurring prior to the time of renegotiation. He is fairly confident, however, that if he does not renew the contract, then total demand will be low enough so that he can meet it fully without either expanding beyond his present manufacturing space or relying upon contract producers.

Draw a decision tree that would be appropriate for a first model of Foscue's problem.

14-3. A purchasing manager is faced with deciding whether or not to stock a large supply of metal. The uncertain variable is the future price of the metal. The following are present worths of consequences and prior probabilities for the various perceived outcomes:

Future price	P(future price)	PW if do stock	PW if do not stock
High	0.3	$100,000	$0
Medium	0.5	– 10,000	0
Low	0.2	–100,000	0

For $6,000 it is possible to hire a consulting firm that would be able to make a fairly accurate forecast in terms of whether the price will go up or down as follows:

If the future is going to be:	Then the probabilities the consultant will predict the price will go up or down are as follows:	
	Up	Down
High	0.9	0.1
Medium	0.4	0.6
Low	0.8	0.2

a. Diagram the problem in the form of a decision tree.

b. Determine what would be the better alternative using the EMV criterion.

14-4. The Norva Company has already spent $80,000 developing a new electronic gauge and is now considering whether or not to market it. Tooling for production would cost $50,000. If the gauge is produced and marketed, the company estimates that there is only one chance in four that the gauge would be successful. If successful, the net cash inflows would be $100,000 per year for 8 years. If not successful, the net cash outflows would be $30,000 per year for 2 years, after which time the venture would be terminated. The MARR is 20% per year.

a. Draw a decision tree and determine the better alternative using the EMV criterion based on present worths.

b. Suppose the market research group can make a market survey that with probability 0.8 will *predict* a success if the gauge will turn out to be a success and with probability 0.9 will *predict* failure if the gauge will turn out to be unsuccessful. Should the survey be undertaken first? What is the expected value of the survey to the company?

14-5. Suppose, given the alternatives in the small versus large machine example in Fig. 14-9, the demands are assumed to be random variables with present worths of outcomes as follows:

At Decision Point 1:

If "Large," normal distribution with:

$$\text{Expected outcome} = \$500M$$
$$\text{Standard deviation} = \$200M$$

If "Small," discrete distribution with:

$$P(\text{Start high}) = 0.70$$
$$P(\text{Start low}) = 0.30, \text{ with outcome}$$
$$\text{"End low" having uniform distribution between}$$
$$\$150M \text{ and } \$450M$$

At Decision Point 2:

If "New large," normal distribution with:

$$\text{Expected outcome} = \$650M$$
$$\text{Standard deviation} = \$250M$$

If "New small," normal distribution with:

$$\text{Expected outcome} = \$550M$$
$$\text{Standard deviation} = \$150M$$

Demonstrate the use of Monte Carlo simulation (see Chapter 12) for developing data to approximate the distribution of the *difference* between the present worths of the "Large" and "Small" alternatives at decision point 1. Set up a table to show your random numbers and random deviates and the subsequent calculations, and demonstrate by generating five full outcomes for the desired distribution.

14-6. Given the following two-stage decision situation shown in Fig. 14-11, determine which is the best initial decision. Use the expected PW method and a MARR of 12%. To give the problem a physical context, the following letter symbols have been employed for each alternative:

>   BSW—Build small warehouse
>   RLW—Rent large warehouse
>   BA—Build addition
>   NC—No change

14-7. A firm must decide between purchasing an automatic machine that costs $50,000 and will last 10 years and have $0 salvage value or purchasing a manual machine that costs $20,000 and will last 5 years and have $0 salvage value. If the manual machine is purchased initially, after 5 years a decision will have to be made between a manual machine having the same characteristics affecting cost as the first manual machine and a semiautomatic machine costing $40,000 that would have a $20,000 salvage value after 5 years of life. The annual operating costs for each of the machines is as follows: automatic, $10,000; manual, $14,000; semiautomatic, $11,000.

  a. Graphically construct a decision tree to represent this situation.

  b. Determine which decision would be made at each point using the PW method and a MARR of 10%.

  c. At what interest rate would the decision between the manual and semiautomatic machine be reversed?

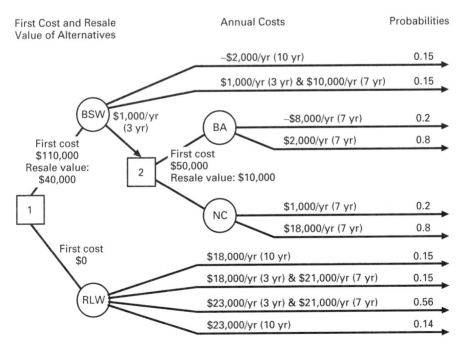

Figure 14-11   Two-stage decision situation for Problem 14-6.

14-8. Suppose one is faced with the same alternatives and dollar outcome consequences as in the replacement problem depicted in Fig. 14-6. However, the initial estimates of probability of demand are: high, 0.6; low, 0.4. Furthermore, management's assessment of confidence in further investigation results, using the notation in Table 14-6, are

$$P(h \mid H) = 0.80,$$
$$P(h \mid D) = 0.40,$$
$$P(d \mid H) = 0.20,$$
$$P(d \mid D) = 0.60.$$

Calculate the choice at each decision point to determine the best initial decision. How close is the initial decision with these revised probabilities to the initial decision for the original problem depicted in Fig. 14-6?

14-9. Figure 14-12 is a decision tree portrayal of a building lease versus buy problem with input data supplied. Investment requirements are shown as negative numbers; probabilities

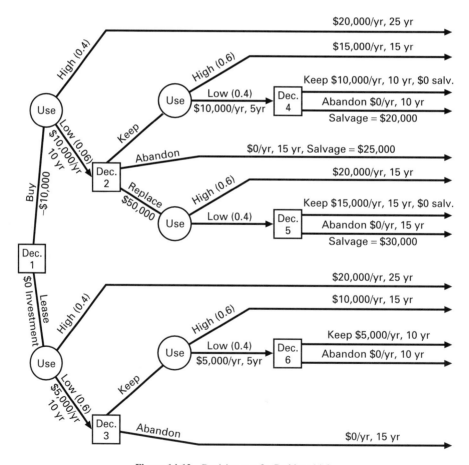

Figure 14-12 Decision tree for Problem 14-9.

associated with each outcome are shown in parentheses. The annual cash savings and duration of those savings are shown together at each relevant outcome. Salvage values in the cases of abandonments are assumed to occur at the end of the 25-year study period. Determine the best decision using the expected net PW method with a MARR of 0%.

14-10. Figure 14-13 is a simplified portrayal of the relevant factors for deciding whether to start an applied research project. Determine the answer, assuming that the decision points are each one year apart and the MARR is 20%. Use the expected net PW method.

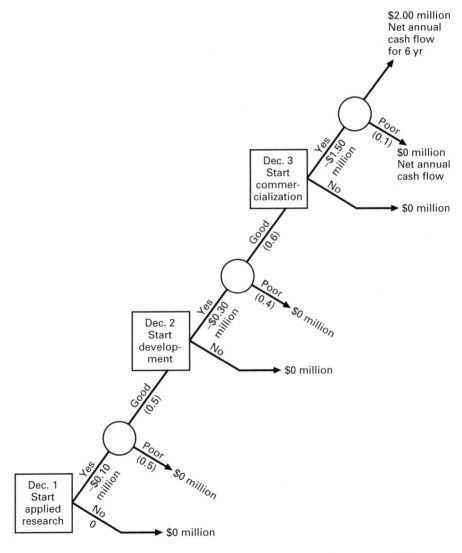

Figure 14-13   Simplified tree showing only relevant information, for Problem 14-10.

14-11. Suppose, for the oil wildcatter problem for which prior probabilities are shown in Fig. 14-3, the probabilities of the possible outcomes for the added study (seismic soundings) are as follows:

Given that the well state will turn out to be:	Then the probability that the seismic soundings will indicate ____ structure is:		
	No	Open	Closed
Dry	0.6	0.3	0.1
Wet	0.3	0.4	0.3
Soaking	0.1	0.4	0.5

Given the structure indicated by the seismic sounding, use Bayes's theorem to calculate the posterior probabilities regarding what the well state will turn out to be. Check your results against the probabilities shown in Fig. 14-4.

APPENDIX 14-A

Bayes's Theorem and Tabular Format for Calculation of Posterior Probabilities

In general, if there are n mutually exclusive, exhaustive possible outcomes S_1, S_2, S_i, S_n, and the results of additional study, such as sampling or further investigation, are X, such that X is discrete and $P(X) \neq 0$, and prior probabilities $P(S_i)$ have been established, then Bayes's theorem for the discrete case can be written as

$$P(S_i \mid X) = \frac{P(X \mid S_i)P(S_i)}{\sum_i P(X \mid S_i)P(S_i)}. \tag{14-A-1}$$

The posterior probability $P(S_i \mid X)$ is the probability of outcome S_i given that additional study resulted in X. The probability of X and S_i occurring, $P(X \mid S_i)P(S_i)$, is the "joint" probability of X and S_i, or $P(X, S_i)$. The sum of all the joint probabilities is equal to the probability of X. Therefore, Eq. 14-A-1 can be written as

$$P(S_i \mid X) = \frac{P(X \mid S_i)P(S_i)}{P(X)}. \tag{14-A-2}$$

A format for application is presented in Table 14-A-1. The columns are as follows:

COLUMN

(1) S_i: the potential states of nature

(2) $P(S_i)$: the estimated prior probability of S_i (*Note:* This column sums to unity.)

Table 14–A–1 Format for Applying Bayes's Theorem in Discrete Outcome Cases

(1) State	(2) Prior probability	(3) Probability of sample outcome, or confidence assessment, X	(4) = (2)(3) Joint probability	(5) = (4)/Σ(4) Posterior probability $P(S_i \mid X)$
S_1	$P(S_1)$	$P(X \mid S_1)$	$P(X \mid S_1)P(S_1)$	$P(X \mid S_1)P(S_1)/P(X)$
S_2	$P(S_2)$	$P(X \mid S_2)$	$P(X \mid S_2)P(S_2)$	$P(X \mid S_2)P(S_2)/P(X)$
$\vdots$	$\vdots$	$\vdots$	$\vdots$	$\vdots$
S_i	$P(S_i)$	$P(X \mid S_i)$	$P(X \mid S_i)P(S_i)$	$P(X \mid S_i)P(S_i)/P(X)$
$\vdots$	$\vdots$	$\vdots$	$\vdots$	$\vdots$
S_n	$P(S_n)$	$P(X \mid S_n)$	$P(X \mid S_n)P(S_n)$	$P(X \mid S_n)P(S_n)/P(X)$
	$\sum_{i=1}^{n} P(S_i) = 1.0$		$\sum_{i=1}^{n} P(X \mid S_i)P(S_i) = P(X)$	$\sum_{i=1}^{n} P(S_i \mid X) = 1.0$

(3) $P(X \mid S_i)$: the conditional probability of getting sample or added study results X, given that S_i is the true state

(4) $P(X \mid S_i)P(S_i)$: the point probability of getting X and S_i; the summation of this column is $P(X)$, which is the probability that the sample or added study results in outcome X

(5) $P(S_i \mid X)$: the posterior probability of S_i given that sample outcome resulted in X; numerically, the ith entry is equal to the ith entry of column (4) divided by the sum of column (4) (*Note:* This column sums to unity.)

15

Statistical Decision Techniques

15.1 Introduction

Numerous statistical techniques to aid in decision making have been developed in the last two decades. Some of the more powerful of these techniques are included in the body of knowledge called *statistical decision theory*. Statistical decision theory is commonly characterized as the mathematical analysis of decision making when, although the state of the world is not known, further information can be obtained by experimentation.

Statistical decision theory also often involves the use of *subjective probabilities* to express the decision maker's degree of belief in the possible outcomes. A practice that is commonly associated with statistical decision theory is the use of Bayesian statistics, which was initially referred to in Chapter 14.

15.2 Bayesian Statistics

Bayesian statistics is characterized by the adjustment of "prior" probabilities for an unknown parameter or factor to more reliable "posterior" probabilities based on the results of sample evidence or evidence from further study. Bayes's theorem, usually employed in this adjustment, is developed with applications below.

Bayes's theorem is an extension of joint and conditional probability theory. The probability of two events occurring is given by probability theory as

$$P(A, B) = P(A|B)P(B), \tag{15-1}$$

where

$$P(A, B) = \text{probability of the two events } A \text{ and } B \text{ occurring together,}$$

$$P(A|B) = \text{probability of } A \text{ occurring given that } B \text{ has occurred,}$$

$$P(B) = \text{probability of } B \text{ occurring.}$$

It can quickly be reasoned that the probability of drawing the ace of spades from a standard deck of 52 cards is 1 in 52. As an example of the application of Eq. 15-1, this probability can also be computed as follows:

$$P(\text{ace, spade}) = P(\text{ace | spade})P(\text{spade})$$
$$= (1 \text{ ace/13 spades})(13 \text{ spades/52 cards})$$
$$= (1/13)(13/52) = 1/52.$$

Note that the condition of getting the ace of spades is equivalent to drawing the spade ace. This leads to a second axiom:

$$P(A, B) = P(B, A).$$

Since $P(B, A) = P(B|A) \cdot P(A)$, then

$$P(A|B)P(B) = P(B|A)P(A). \tag{15-2}$$

Bayes's theorem is derived from Eq. 15-2. Provided that $P(B)$ and $P(A) \neq 0$, then

$$P(A|B) = \frac{P(B|A)P(A)}{P(B)},$$

$$P(B|A) = \frac{P(A|B)P(B)}{P(A)}. \tag{15-3}$$

Thus, the probability of an ace, given a spade, can be computed in an indirect way using Eq. 15-3 as

$$P(\text{ace | spade}) = \frac{P(\text{spade | ace})P(\text{ace})}{P(\text{spade})}$$
$$= \frac{(1 \text{ spade/4 aces})(4 \text{ aces/52 cards})}{(13 \text{ spades/52 cards})}$$
$$= \frac{(1/4)(1/13)}{(1/4)} = 1/13.$$

In general, if there are n mutually exclusive, exhaustive possible outcomes S_1, $S_2, \ldots, S_n$, and the results of additional study, such as sampling or further investigation, X, such that X is discrete and $P(X) \neq 0$, and prior probabilities $P(S_i)$ have been established, then Bayes's theorem for the discrete case can be written as

$$P(S_i | X) = \frac{P(X|S_i)P(S_i)}{P(X)}. \tag{15-4}$$

The posterior probability $P(S_i | x)$ is the probability of outcome S_i given that additional study resulted in X. The probability of S and S_i occurring, $P(X | S_i)P(S_i)$, is the "joint" probability of X and S_i. The sum of all the joint probabilities is equal to the probability of X. Therefore, Eq. 15-4 can be written as

$$P(S_i | X) = \frac{P(X|S_i)P(S_i)}{\sum_i P(X|S_i)P(S_i)}. \tag{15-5}$$

A format for application was presented earlier in Appendix 14-A. As a side note of interest, when X has a continuous density function $f_1(x)$ and S has a continuous density function $f_2(s)$, such that all conditional density functions are continuous, then the continuous equivalent of Bayes's theorem states that

$$f_2(S|X) = \frac{f_1(X|S)f_2(S)}{\int_s f_1(X|S)f_2(S)dS}.$$ (15-6)

Examples in the remainder of this chapter will be devoted to discrete outcome cases.

Example 15-1

Example involving sampling to revise probabilities for production process: Let us consider the following application: A production process requires that equipment be set up for a fixed run of 200 units. If the setup is good, defects occur with a probability of 0.05 and in a random fashion. A bad setup occurs randomly with a probability of 0.2 and then the random defect rate is 0.25. Letting S_1 represent a good setup, S_2 a bad setup, and X the event that a sample of one is found to be defective, it is desired to calculate the posterior probabilities that the setup is good or bad given that X has occurred. Table 15-1 shows the necessary calculations, where $P(S_1)$ equals 0.8 and $P(S_2)$ equals 0.2.

TABLE 15-1 Posterior Probability Calculation for Production Process

S_i	$P(S_i)$	$P(X\|S_i)$	$P(X\|S_i)P(S_i)$	$P(S_i\|X)$
S_1	0.80	0.05	0.04	4/9 = 0.44
S_2	0.20	0.25	0.05	5/9 = 0.56
	$\sum_i = \overline{1.00}$		$\sum_i = P(X) = \overline{0.09}$	$\sum_i = \overline{1.00}$

Thus, the prior probability that the setup is good, 0.8, is revised to a posterior probability of 0.44 based on the evidence that a sample unit is defective.   ∎

Example 15-2

Example of use of Bayes's theorem for discrete outcome investment analysis: As a further example, consider an investment project with a return (expressed in net PW) of $6,000 if event S_1 occurs and –$4,000 if event S_2 occurs. The prior probability estimates are 0.4 for S_1 and 0.6 for S_2. Thus the expected return, denoted $E[R]$, is

$$E[R] = 0.4(\$6,000) + 0.6(-\$4,000) = 0.$$

The alternative of not investing also has an expected return of zero, for there would be no gain or loss. Additional study will result in either X_1, which indicates a net PW of $6,000, or X_2, which indicates a net present worth loss of $4,000. If S_1 will occur, then X_1 will be indicated with a probability of 0.8. Similarly, if S_2 will occur, X_2 will be indicated with a probability of 0.6. The problem is summarized as follows:

$$P(S_1) = 0.4, \qquad P(X_1|S_1) = 0.8,$$
$$P(S_2) = 0.6, \qquad P(X_2|S_1) = 1.0 - P(X_1|S_1) = 0.2,$$
$$E[S_1] = \$6,000, \qquad P(X_2|S_2) = 0.6,$$
$$E[S_2] = -\$4,000, \qquad P(X_1|S_2) = 1.0 - P(X_2|S_2) = 0.4.$$

TABLE 15-2   Computation of Posterior Probabilities Given X_1

S_i	$P(S_i)$	$P(X_1 \mid S_i)$	$P(X_1 \mid S_i)P(S_i)$	$P(S_i \mid X_1)$
S_1	0.4	0.8	0.32	0.32/0.56 = 0.57
S_2	0.6	0.4	0.24	0.24/0.56 = 0.43
	$\sum_i = 1.0$		$\sum_i = P(X_1) = 0.56$	$\sum_i = 1.00$

The posterior probabilities resulting from the additional study can now be computed from the above information. New expected returns then can be computed, and the decision to invest or not invest in the project can be determined as a function of the additional study outcome X.

Table 15-2 shows the computation of posterior probabilities if the added study results in X_1. When X_1 occurs, the probability of S_1 is revised from the "prior" 0.4 to the "posterior" 0.57. The expected return given X_1, denoted $E[R \mid X_1]$, can then be computed as

$$E[R \mid X_1] = 0.57(\$6,000) + 0.43(-\$4,000) = \$1,714.$$

Hence, if X_1 occurs, the project should be undertaken to obtain the positive expected net PW return.

The computation of posterior probabilities if the additional study results in X_2 is presented in Table 15-3. The occurrence of X_2 results in a posterior probability of S_1 of 0.18, and the expected return given X_2, denoted $E[R \mid X_2]$, is computed to be

$$E[R \mid X_2] = 0.18(\$6,000) + 0.82(-\$4,000) = -\$2,182.$$

Since $E[R \mid X_2]$ is negative, the decision would be not to invest in the project if X_2 occurs, thus resulting in an expected return of zero.

Considering the decision rule—to invest if additional study results in X_1, and to reject the project if X_2 occurs—the overall expected return is now positive and is calculated to be

$$E[R] = \begin{cases} \$0 & \text{if } X = X_2, \\ \$1,714 & \text{if } X = X_1. \end{cases}$$

From Tables 15-2 and 15-3, $P(X_1) = 0.56$ and $P(X_2) = 0.44$. Thus, with sampling or additional study,

$$E[R] = E[R \mid X_1]P(X_1) + E[R \mid X_2]P(X_2)$$
$$= (\$1,714)0.56 + (\$0)0.44 = \$960.$$

In general, the overall expected return, given additional study or sample information resulting in X_j, is

$$E[R \mid SI] = \sum_j \max(E[A_i] \mid X_j)P(X_j), \tag{15-7}$$

where $E[R \mid SI]$ is the expected return given sample information.

The change in the expected value from \$0 to \$960 is often called the *expected value of sample information* (EVSI). Expressed symbolically,

$$EVSI = E[R \mid SI] - E[R]. \tag{15-8}$$

■

TABLE 15-3   Computation of Posterior Probabilities Given X_2

S_i	$P(S_i)$	$P(X_2 \mid S_i)$	$P(X_2 \mid S_i)P(S_i)$	$P(S_i \mid X_2)$
S_1	0.4	0.2	0.08	0.18
S_2	0.6	0.6	0.36	0.82
	$\sum_i = 1.0$		$\sum_i = P(X_2) = 0.44$	$\sum_i = 1.00$

15.3 Expected Value of Perfect Information

The *expected value of perfect information* (EVPI) is the maximum possible EVSI and is the maximum expected loss due to imperfect information as to what will be the state of nature in a situation involving risk. Interpreted another way, the expected value of perfect information is the amount that could be gained, on the average, if the future regarding a particular decision situation became perfectly predictable and decisions changed to the optimal choice(s) based on the new known conditions. Another term synonymous with the expected value of perfect information is the *expected opportunity loss* (EOL).

Figure 15-1 shows the steps for computation of the EVPI for the usual situation of multiple alternatives and discrete outcomes. The discrete outcome investment project examined earlier will be used as an example of the application of EVPI. In this case, the alternatives are to invest or not to invest. If perfect information were available, the project would be accepted and have a net present value of $6,000 when S_1 is to occur or it would be rejected when S_2 is to occur. The expected present worth, given perfect information (certainty), denoted $E[R \mid PI]$, is thus $6,000 with an expected frequency of occurrence of 0.4, or it is $0, occurring with a probability of 0.6. Overall,

$$E[R \mid PI] = 0.4(\$6,000) + 0.6(\$0) = \$2,400.$$

This reasoning process corresponds to the right-hand side of Fig. 15-1.

Prior to sampling or added study, the expected return for the investment project was computed to be $0. This corresponds to the left-hand side of Fig. 15-1. Thus the EVPI is $2,400 – $0, or $2,400. This is a measure of the maximum possible expected to be gained by sampling for further information. It should be noted from the previous section that the $960 EVSI is expected to be gained (out of the $2,400 EVPI) by sampling one unit.

In general, EVPI (EOL) can be expressed as

$$\text{EVPI} = \text{EOL} = E[R \mid PI] - E[R]. \tag{15-9}$$

The general formula for computation of expected return under certainty is

$$E[R \mid PI] = \sum_i P(S_i) \cdot \max[\text{return}(A_1, \dots, A_j, \dots, A_m \mid S_i)], \tag{15-10}$$

where
$$E[R \mid PI] = \text{expected return given perfect information,}$$
$$P(S_i) = \text{probability of the } i\text{th outcome,}$$
$$A_j = \text{the } j\text{th alternative,}$$
$$\text{return}(A_j \mid S_i) = \text{value of alternative } j \text{ given that } S_i \text{ outcome occurs,}$$
$$\max[\text{return}(A_1, \dots, A_j, \dots, A_m \mid S_i)] = \text{decision rule that the value of the the returns, given that } S_i \text{ occurs, is the maximum return over all } m \text{ alternatives.}$$

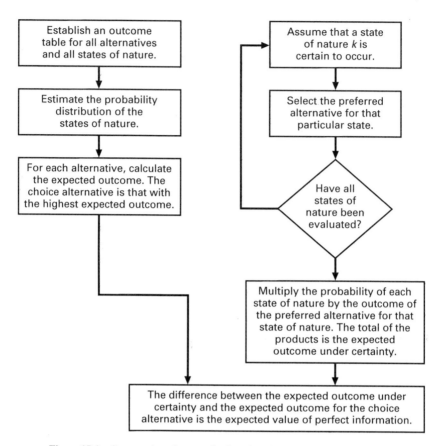

Figure 15-1   Computation of expected value of perfect information for multiple alternatives and discrete outcomes.

Example 15-3

Additional example of computation of EOL *through use of opportunity loss concept:* The *expected opportunity loss* (EOL) concept will be further demonstrated with the multiple-alternative, discrete outcome example given in Table 15-4. In this example, the returns (expressed in net PW) depend on whether business conditions are good, average, or poor. The probabilities of these three possible business conditions are given as 0.25, 0.50, and 0.25, respectively. The expected return for each alternative can be calculated as:

alternative I: $300M(0.25) + $200M(0.50) - $80M(0.25) = $155M,

alternative II: $400M(0.25) + $200M(0.50) - $200M(0.25) = $150M,

alternative III: $100M(0.25) + $240M(0.50) + $0M(0.25) = $145M.

Using the maximum expected return decision criterion, alternative I would be chosen. The *opportunity loss* (OL) for each business condition can be derived by the following rationale:

TABLE 15-4 Outcomes in Net PW for Three Alternatives
Given Discrete Business Conditions

	Business condition		
	Good $P(\text{good}) = 0.25$	Average $P(\text{average}) = 0.50$	Poor $P(\text{poor}) = 0.20$
Alternative I	$300M	$200M	−$ 80M
Alternative II	400M	200M	−200M
Alternative III	100M	240M	0M

If good conditions prevail, alternative II would be preferred, and having chosen I incorrectly represents an OL of $400M − $300M = $100M. Similarly, if average business conditions occur, failure to select the optimal alternative III rather than alternative I represents an OL of $240M − $200M = $40M. If poor conditions resulted, alternative III would be the best choice at $0M. Choice of alternative I instead would cause an OL of $0M − (−$80M) = $80M. Thus, for the above problem EOL is computed to be

$$\text{EOL} = (\$100M)0.25 + (\$40M)0.50 + (\$80M)0.25 = \$65M.$$

In general, EOL can be computed as:

$$\text{EOL} = \sum_i (\text{OL} \mid S_i)P(S_i). \qquad (15\text{-}11)$$

This implies that the expected return for the three alternatives, given perfect information, would be $65M greater than the expected return with no added information. To confirm this implication, the expected return with perfect information should be equal to the expected return with no added information ($155M) plus the EOL ($65M), or $220M. The computation of $E[R \mid PI]$ is shown in Table 15-5 to be $220M.

TABLE 15-5 Expected Return with Perfect Information

	Business condition (S_i)		
	Good	Average	Poor
Best alternative $\mid S_i$	II	III	III
$P(S_i)$	0.25	0.50	0.25
Outcome for best alternative, $E(R \mid PI)_i$	$400M	$240M	$0M

$E(R \mid PI) = \sum_i E(R \mid PI)_i P(S_i) = \$400M(0.25) + \$240M(0.50) + \$0M(0.25) = \$220M$

■

Example 15-4

As an example of evaluation of sample information schemes, suppose the Seyab Company is expanding its product line and a decision has to be made as to the size of additional facilities. The alternatives are to build a large plant (A_1), a modest plant (A_2), or to lease facilities as required (A_3). The possible outcomes in the strategy period are that demand will be high (S_1), good (S_2), fair (S_3), or low (S_4). Table 15-6 summarizes the net PW for each alternative for each of the four outcomes and also the best assessment by management of the (prior) probabilities of each possible demand outcome. From the data in Table 15-6, one can calculate the expected outcomes for each alternative as

$$E[A_1] = 0.2(\$18 \text{ million}) + 0.45(\$12 \text{ million}) + 0.25(\$6 \text{ million})$$
$$+ 0.1(-\$12 \text{ million}) = \$9.30 \text{ million},$$
$$E[A_2] = 0.2(\$12 \text{ million}) + 0.45(\$12 \text{ million}) + 0.25(\$9 \text{ million})$$
$$+ 0.1(-\$6 \text{ million}) = \$9.45 \text{ million},$$
$$E[A_3] = 0.2(\$13 \text{ million}) + 0.45(\$10 \text{ million}) + 0.25(\$6 \text{ million})$$
$$+ 0.1(-\$1 \text{ million}) = \$8.50 \text{ million}.$$

Thus, based on the prior probabilities and expected outcomes, the indicated decision would be to build a modest plant (A_2) for an expected net PW of \$9.45 million.

Suppose that a scheme for further consideration is that a market consultant be called in for further study. It is judged that the consultant will predict one of three possible ranges of demand outcome: good to high (X_1), fair to good (X_2), or poor to fair (X_3). The performance of the consultant based on management's subjective belief (or history) is given in Table 15-7. The interpretation of Table 15-7 is that if the demand is going to turn out to be high, the consultant will predict good–high with a probability of 0.6, i.e., $P(X_1 \mid S_1) = 0.6$; or he will predict fair–good with a probability of 0.3, i.e., $P(X_2 \mid S_1) = 0.3$; or he will predict poor–fair with a probability of 0.1, i.e., $P(X_3 \mid S_1) = 0.1$. Similarly, $P(X_1 \mid S_2) = 0.3$, $P(X_2 \mid S_2) = 0.5$, etc.

The expected value of the project, using the sample information (further study results) from the consultant can be computed by determining what the optimal expected return will be for each possible prediction (X_j) by the probability of each prediction occurring. Tables 15-8, 15-9, and 15-10 present the computations of posterior probabilities given that X_1, X_2, or X_3 occurs, respectively.

TABLE 15-6 Seyab Company Returns (in Net PW)

	Demand and probability of demand			
Alternative	High, S_1 $P(S_1) = 0.2$ (\$ million)	Good, S_2 $P(S_2) = 0.45$ (\$ million)	Fair, S_3 $P(S_3) = 0.25$ (\$ million)	Fair, S_4 $P(S_4) = 0.1$ (\$ million)
Large plant (A_1)	\$18	\$12	\$6	−\$12
Modest plant (A_2)	12	12	9	−6
Lease (A_3)	13	10	6	−1

TABLE 15-7 Seyab Company—Indications of
Confidence in Consultant's Study

	... the probabilities the consultant will predict _____ are		
Given the demand will turn out to be _____ ...	Good–high X_1	Fair–good X_2	Poor–fair X_3
High (S_1)	0.6	0.3	0.1
Good (S_2)	0.3	0.5	0.2
Fair (S_3)	0.1	0.3	0.6
Poor (S_4)	0.1	0.1	0.8

TABLE 15-8   Seyab Company—Computation of Posterior Probabilities Given That the Consultant Predicts X_1

S_i	$P(S_i)$	$P(X_1 \mid S_i)$	$P(S_i)P(X_1 \mid S_i)$	$P(S_i \mid X_1)$
S_1	0.20	0.60	0.120	0.414
S_2	0.45	0.30	0.135	0.466
S_3	0.25	0.10	0.025	0.086
S_4	0.10	0.10	0.010	0.034
			$P(X_1) = 0.290$	

TABLE 15-9   Seyab Company—Computation of Posterior Probabilities Given That the Consultant Predicts X_2

S_i	$P(S_i)$	$P(X_2 \mid S_i)$	$P(S_i)P(X_2 \mid S_i)$	$P(S_i \mid X_2)$
S_1	0.20	0.30	0.060	0.162
S_2	0.45	0.50	0.225	0.608
S_3	0.25	0.30	0.075	0.203
S_4	0.10	0.10	0.010	0.027
			$P(X_2) = 0.370$	

TABLE 15-10   Seyab Company—Computation of Posterior Probabilities Given That the Consultant Predicts X_3

S_i	$P(S_i)$	$P(X_3 \mid S_i)$	$P(S_i)P(X_3 \mid S_i)$	$P(S_i \mid X_3)$
S_1	0.20	0.10	0.020	0.059
S_2	0.45	0.20	0.090	0.265
S_3	0.25	0.60	0.150	0.441
S_4	0.10	0.80	0.080	0.235
			$P(X_3) = 0.340$	

If X_1 is predicted, then the posterior probabilities of S_1, S_2, S_3, and S_4 are given in Table 15-8 as 0.414, 0.466, 0.086, and 0.034, respectively. The expected value of the alternatives can be then calculated as

$$E[A_1] = 0.414(\$18 \text{ million}) + 0.466(\$12 \text{ million}) + 0.086(\$6 \text{ million})$$
$$+ \ 0.034(-\$12 \text{ million}) = \$13.15 \text{ million},$$

$$E[A_2] = 0.414(\$12 \text{ million}) + 0.466(\$12 \text{ million}) + 0.086(\$9 \text{ million})$$
$$+ \ 0.031(-\$6 \text{ million}) = \$11.13 \text{ million},$$

$$E[A_3] = 0.414(\$13 \text{ million}) + 0.466(\$10 \text{ million}) + 0.086(\$6 \text{ million})$$
$$+ \ 0.034(-\$1 \text{ million}) = \$10.52 \text{ million}.$$

Therefore, given X_1 is predicted, A_1 (large plant) would be the choice with an expected net PW of \$13.15 million.

Similarly if X_2 is predicted, then the posterior probabilities of S_1, S_2, S_3, and S_4 are calculated in Table 15-9 as 0.162, 0.608, 0.203, and 0.027, respectively. The expected return of each alternative can be similarly computed as \$11.11 million for A_1, \$10.91 million for A_2, and \$9.38 million for A_3. Thus, alternative A_1 is again preferred, but with an expected return of \$11.11 million.

Finally, if X_3 is predicted, the posterior probabilities of S_1, S_2, S_3, S_4 are calculated in Table 15-10 as 0.059, 0.265, 0.441, and 0.235, respectively. The expected returns in this instance are \$4.07 million for A_1, \$6.45 million for A_2, and \$5.83 million for A_3. Hence, A_2 is the optimal alternative with an expected return of \$6.45 million.

It should be noted from the summations in the fourth columns of Tables 15-8, 15-9, and 15-10 that the probabilities that X_1, X_2, and X_3 will be predicted are, respectively, 0.29, 0.37, and 0.34.

The expected return given sample information $E[\text{R} \mid \text{SI}]$ can now be computed using the general relationship

$$E[\text{R} \mid \text{SI}] = \sum_j \max (E[A_i \mid X_j] P(X_j)). \qquad (15\text{-}12)$$

Thus, $E[\text{R} \mid \text{SI}] = (\$13.15 \text{ million})(0.29) + (\$11.11 \text{ million})(0.37) + (\$6.45 \text{ million})(0.34) =$ \$10.12 million. To determine the EVSI, one need only compute

$$\text{EVSI} = E[\text{R} \mid \text{SI}] - E[\text{R}]$$
$$= \$10.12 \text{ million} - \$9.45 \text{ million} = \$0.67 \text{ million}.$$

This result indicates that management, based on the expected monetary principle, should be willing to pay up to \$0.67 million for the consultant's services in performing the added study if no other added study alternatives are available.

Suppose that the Seyab Company, rather than hiring a consultant, could follow the scheme of engaging its own staff in a fresh intensive study to predict outcomes. To evaluate the expected value of this new (sampling) information, management of the firm will again need to determine conditional probabilities that this type of added study will result in certain predictions given that the demand will turn out either high, good, fair, or poor. Table 15-11 shows the expected performance that reflects management's confidence in the added study. In this case, the staff will predict four possible outcomes corresponding to the actual expected demand—high, good, fair, or poor.

The posterior probabilities of demand turning out to be either S_1, S_2, S_3, or S_4 for each possible staff study prediction X_1, X_2, X_3, or X_4 can be calculated as shown in previous examples. Table 15-12 summarizes the results of these calculations. The expected returns for each alternative and each staff study prediction are summarized in Table 15-13.

Table 15-14 indicates the expected return of the best alternative and also probabilities of staff prediction for X_1, X_2, X_3, and X_4, respectively. The expected returns come directly from Table 15-13, but the probabilities are the result of the posterior probability calculations not shown but for which the results were summarized in Table 15-12. The right-hand column of Table 15-14 shows the computation of the expected return given the sample (added study)

TABLE 15-11   Seyab Company—Indications of Confidence in Staff Study

Given the demand will turn out to be _____ ...	... the probabilities the staff will predict _____ are			
	High X_1	Good X_2	Fair X_3	Poor X_4
High (S_1)	0.7	0.2	0.1	0.0
Good (S_2)	0.3	0.5	0.1	0.1
Fair (S_3)	0.1	0.3	0.4	0.2
Poor (S_4)	0.0	0.2	0.3	0.5

TABLE 15-12 Seyab Company—Summarization of
Calculated Posterior Probabilities for Staff Study

S_i	Posterior probability			
	$P(S_i \mid X_1)$	$P(S_i \mid X_2)$	$P(S_i \mid X_3)$	$P(S_i \mid X_4)$
S_1	0.467	0.111	0.103	0.000
S_2	0.450	0.625	0.231	0.310
S_3	0.083	0.208	0.513	0.345
S_4	0.000	0.056	0.153	0.345

TABLE 15-13 Seyab Company—Summarization of
Expected Returns for Staff Study

Alternative	Prediction ($ million)			
	X_1	X_2	X_3	X_4
$E(A_1)$	14.30	10.07	5.87	1.65
$E(A_2)$	11.75	10.37	7.71	4.76
$E(A_3)$	11.07	8.89	6.51	4.83

TABLE 15-14 Seyab Company—Calculation of
Expected Return for Staff Study

Staff prediction (X_j)	Expected return of best alternative ($ million)	$P(X_j)$	Expected return $\times P(X_j)$ ($ million)
X_1	14.30	0.300	4.29
X_2	10.37	0.360	3.73
X_3	7.71	0.195	1.50
X_4	4.83	0.145	0.71
			$\sum_j = \$10.23$ million

information to be $10.23 million. From this the expected value of sample information can be calculated as $10.23 million − $9.45 million = $0.78 million.

The difference between the EVSI and the cost of additional study is called the *expected net value of sample information,* denoted ENVSI. Symbolically,

$$\text{ENVSI} = \text{EVSI} - \text{cost of sample information,} \qquad (15\text{-}13)$$

which is a more valuable decision criterion than EVSI alone. Suppose that for this example problem it is thought that the consultant would cost $0.50 million for the added study and that if the firm's staff conducted the study it would cost $0.65 million. In this event, the two alternatives for the added study can be compared through the following calculation.

Consultant:

$$\text{ENVSI} = \$0.67 \text{ million} - \$0.50 \text{ million} = \$0.17 \text{ million}$$

Staff:

$$\text{ENVSI} = \$0.78 \text{ million} - \$0.65 \text{ million} = \$0.13 \text{ million}$$

Thus, the expected net value of the sample information (added study) is slightly greater for the consultant alternative, and the consultant should thus be chosen on the basis of expectations.■

PROBLEMS

15-1. The Quick Key Lock Company produces two types of locks and two models of each type. The following table summarizes the probabilities of demand for each type and model:

	Type	
Model	Cartridge (C)	Bolt (B)
Standard (S)	0.5	0.1
Pick-proof (P)	0.2	0.2

  a. What is $P(S)$? $P(P)$? $P(C)$? $P(B)$?

  b. What is the conditional probability of a lock being type B given that the lock is model P?

  c. What is $P(B|S)$?

  d. What is the joint probability of the bolt type being the pick-proof model?

  e. If one knew that a lock drawn randomly was the bolt type, what is the probability that that particular lock is the standard model?

15-2. The Parcel Delivery Service has analyzed costs and longevity of its trucks. The length of service varies from $2\frac{1}{2}$ to $4\frac{1}{2}$ years with rare exception. The average cost per mile ranges from $0.105 to $0.165. The following matrix summarizes experience with 1,000 vehicles in which the ages and costs are shown discretely for simplicity.

	Cost per mile	
Length of service	$0.12	$0.15
3 yr	120	280
4 yr	270	330

  a. What is the prior probability of the cost of $0.12 per mile? Of $0.15 per mile? Of 3 years of service? Of 4 years?

  b. What are the following conditional probabilities: $P(\$0.12 \mid 3 \text{ yr})$? $P(\$0.15 \mid 3 \text{ yr})$? $P(\$0.12 \mid 4 \text{ yr})$? $P(\$0.15 \mid 4 \text{ yr})$?

  c. A truck at retirement is found to be in the $0.12 per mile category. What is the posterior probability that it is 4 years old?

15-3. The Carolina Clay Company manufactures brick. A grading process occurs after each firing. A sample is drawn and a quality test is made. Grades A, B, and C have a 0.95, 0.75, and 0.50 proportion passing the quality test, respectively. Historically, the distribution of grade A, B, and C lots has been 32%, 44%, and 24%, respectively.

	Pass	Fail	Total
Grade A			
Grade B			
Grade C			
Total			

a. One brick is randomly drawn from each of 1,000 random lots. Fill in the matrix with the expected results.

b. What is the probability of pass? Fail?

c. A lot of unknown grade was sampled once. The brick passed the test. What are the posterior probabilities that the brick was drawn from a grade A lot? Grade B lot? Grade C lot?

15-4. The returns (in PW) for three investment alternatives are summarized below:

	Business condition		
Alternative	Good $(P = 0.30)$	Fair $(P = 0.60)$	Poor $(P = 0.10)$
A_1	$120,000	$60,000	-$100,000
A_2	90,000	70,000	-40,000
A_3	-30,000	50,000	90,000

a. Based on the decision rule to maximize expected return, which alternative is best?

b. Which alternative(s) would be chosen if it were known with certainty that business conditions would be good, fair, or poor, respectively?

c. What would be the expected return if perfect information were available?

d. What is the expected value of perfect information?

e. Construct an opportunity loss matrix.

f. On the basis of the decision rule to minimize EOL, which alternative should be chosen?

15-5. In Problem 15-4, additional information conditional upon sample outcomes X_1 and X_2 is obtainable. The prior and posterior probabilities are presented below:

	Business conditions		
Probability	Good	Fair	Poor
Prior probability	0.30	0.60	0.10
Posterior probability if X_1 occurs	0.40	0.50	0.10
Posterior probability if X_2 occurs	0.15	0.65	0.20

The probabilities of X_1 or X_2 occurring are equally likely:

$$[P(X_1) = P(X_2) = 0.5].$$

a. Compute the $E[R]$ for each alternative if X_1 occurs; if X_2 occurs.

b. What is the $E[R]$ if sample information is obtained?

c. What is the value of the sample information?

d. What is the expected net gain if the sample information costs $5,000?

PART THREE
Additional Topics in Capital Investment Decision Analysis

16
Mathematical Programming for Capital Budgeting

16.1 Introduction

A number of mathematical programming algorithms have been used to assist management in developing capital budgets. Some of the commonly used mathematical programming approaches are outlined in this chapter. The emphasis of this chapter is on the formulation of capital budgeting problems rather than on the details of the algorithms. Consequently, in a number of instances, heuristic approaches are used to obtain solutions to the example problems.

16.2 Indivisible, Independent Investment Opportunities

In order to illustrate the use of mathematical programming in capital investment decision analyses, we will consider a situation in which m new independent, indivisible[1] investment *opportunities* are available. Investment opportunity i has a worth of p_i (p_i can be positive or negative), an initial investment of c_i, and annual operating and maintenance costs of a_i. There exists a capital budget limitation of $\$C$ for new investments; similarly, a limitation of $\$A$ exists on total annual operating and maintenance costs for new investments. It is desired to select the set of investment opportunities that maximizes present worth subject to the budgetary limitations.

Letting x_i be defined to be 0 if opportunity i is not selected for investment and letting x_i be defined to be 1 if opportunity i is selected for investment, the following mathematical programming formulation of the investment decision problem for a single time period is obtained:

[1] An investment opportunity is indivisible if it cannot be broken up into parts; either it is undertaken as a whole or it is not undertaken at all.

maximize $p_1x_1 + p_2x_2 + \ldots + p_mx_m,$

subject to $c_1x_1 + c_2x_2 + \ldots + c_mx_m \leq C$

                  $a_1x_1 + a_2x_2 + \ldots + a_mx_m \leq A,$

                  $x_1 = (0, 1), \qquad i = 1, \ldots, m,$

where p_i = worth of opportunity i,

          c_i = initial investment of opportunity i,

          a_i = total annual operating and maintenance costs,

          x_i = a binary decision variable for opportunity i
                (either 0 or 1).

The optimum set of investment opportunities can be determined by solving the zero–one integer linear programming problem formulated above. Branch and bound, implicit enumeration, and dynamic programming are examples of solution procedures that can be used to solve the budget allocation problem. A number of heuristic approaches also exist for determining "good," if not optimum, solutions to the problem.

With m investment opportunities there exist 2^m combinations of the m binary decision variables. As depicted in Fig. 16-1, for $m = 5$ there exist 32 combinations to be considered. Even though a number of the combinations might be infeasible because of the constraints that exist, for large firms faced with numerous investment opportunities the number of feasible combinations can still be very large. Hence, it is desirable to use mathematical programming algorithms to achieve a systematic and objective selection process; total enumeration of all possible combinations is obviously not a feasible solution procedure.

16.3 Indivisible, Dependent Investment Opportunities

The previous discussion considered budgetary constraints in a situation involving independent investment opportunities. In a number of instances investment opportunities are *dependent*. As an illustration, two or more investment opportunities might be *mutually exclusive*. Similarly, the selection of one particular opportunity might be *contingent* on the selection of one or more other opportunities.

If opportunities j and k are mutually exclusive, then the following constraint can be added to the mathematical programming formulation: $x_j + x_k \leq 1$. Similarly, if the selection of opportunity e is contingent on the selection of either opportunity f or g, then the following constraint applies: $x_e \leq x_f + x_g$ or $x_e - x_f - x_g \leq 0$. (As before, the x's are zero–one decision variables.)

Example 16-1

A firm is considering two different computing systems (1 and 2) and three different software packages (1, 2, and 3). Software packages 1 and 2 can be used only on computing system 1; software package 3 can be used on either computing system. Alternatively, the firm can develop its own software. In this instance the following investment opportunities are defined to meet the objective of minimizing the cost of computing.

1. Purchase computing system 1.
2. Purchase computing system 2.

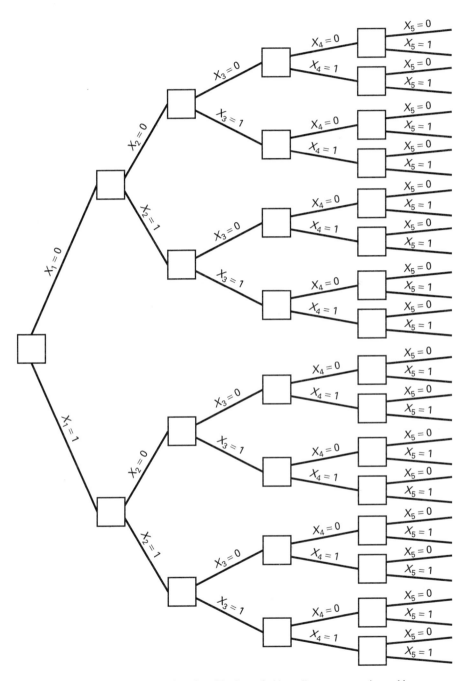

Figure 16-1   Tree representation of combinations of a binary linear programming problem with five investment opportunities.

3. Purchase software package 1.
4. Purchase software package 2.
5. Purchase software package 3.
6. Prepare own software package.

The following constraints apply for this situation:

$$
\begin{aligned}
x_1 + x_2 & \le 1, \\
x_3 + x_4 + x_5 + x_6 & \le 1, \\
-x_1 \quad\quad x_3 + x_4 & \le 0, \\
-x_1 - x_2 \quad\quad x_5 + x_6 & \le 0,
\end{aligned}
$$

where $x_i = 0, 1$ and $1 \le i \le 6$.

The first two constraints indicate that the two computing systems and four software options are mutually exclusive investment opportunities, respectively. The remaining two constraints express the contingency relationships between software packages and computing systems.

One approach that can be used when dependencies exist is to solve the problem first assuming all investment opportunities are independent. If the solution satisfies the dependency constraints, then an optimum solution has been obtained. On the other hand, if one or more of the dependency constraints are violated, than a branch and bound approach can be used. ■

Example 16-2

As a second example, consider a three-period capital budgeting problem. Estimates of cash flows are as follows:

Investment Opportunity		Net cash flow ($M), end of year[a]:				Net PW ($M) at 12%[b]
		0	1	2	3	
A1		−225	150 (60)	150 (70)	150 (70)	+135.3
A2	Mutually exclusive	−290	200 (180)	180 (80)	160 (80)	+146.0
A3		−370	210 (290)	200 (170)	200 (170)	+119.3
B1		−600	100 (100)	400 (200)	500 (300)	+164.1
B2	Independent	−1,200	50 (250)	600 (400)	600 (400)	+151.9
C1	Mutually	−160	70 (80)	70 (50)	70 (50)	+8.1
C2	exclusive and dependent on acceptance of	−200	90 (65)	80 (65)	60 (65)	−13.1
C3	A1 or A2	−225	90 (100)	95 (60)	100 (70)	+2.3

[a] Estimates in parentheses are annual operating expenses (which have already been subtracted in determination of net cash flows).

[b] For example, net PW for A1 $= -\$225,000 + \$150,000(P/A, 12\%, 3) = +\$135,300$.

The MARR is 12% and the ceiling on investment funds available is $1,200,000. In addition, there is a constraint on operating funds for support of the alternative selected, and it is $400,000 in year 1. From these constraints on funds outlays and the interrelationships among opportunities indicated above, we shall formulate this situation in terms of a linear integer programming problem.

First, the net present worth of each investment opportunity at 12% is calculated. The objective function then becomes

$$\text{maximize net PW} = 135.3X_{A1} + 146.0X_{A2} + 119.3X_{A3} + 164.1X_{B1}$$
$$+ 151.9X_{B2} + 8.1X_{C1} - 13.1X_{C2} + 2.3X_{C3}.$$

The budget constraints are the following.

Investment funds constraint:

$$225X_{A1} + 290X_{A2} + 370X_{A3} + 600X_{B1} + 1{,}200X_{B2}$$
$$+ 160X_{C1} + 200X_{C2} + 225X_{C3} \le 1{,}200$$

First year's operating cost constraint:

$$60X_{A1} + 180X_{A2} + 290X_{A3} + 100X_{B1} + 250X_{B2}$$
$$+ 80X_{C1} + 65X_{C2} + 100X_{C3} \le 400$$

Interrelationships among the investment opportunities give rise to these constraints on the problem.

$X_{A1} + X_{A2} + X_{A3} \le 1$	A1, A2, A3 are mutually exclusive
$X_{B1} \le 1$	B1, B2 are independent
$X_{B2} \le 1$	
$X_{C1} + X_{C2} + X_{C3} \le X_{A1} + X_{A2}$	accounts for dependence of C1, C2, C3 on A1 *or* A2

Finally, if all decision variables are required to be either 0 (not in optimal solution) or 1 (included in optimal solution), the last constraint on the problem would be written as

$$X_j = 0, 1 \quad \text{for } j = \text{A1, A2, A3, B1, B2, C1, C2, C3.}$$

As one can see, a fairly simple problem such as this one would require an inordinate amount of time to solve by listing and evaluating all mutually exclusive combinations. Consequently, it is recommended that a suitable computer program be utilized to obtain solutions for all but the most simple capital budgeting problems. ■

16.4 Independent Collections of Mutually Exclusive Opportunities

Next we consider a capital budgeting problem involving indivisible investment opportunities consisting of independent collections of mutually exclusive investment opportunities. A diagram that depicts this situation is given below.

Source i $(i = 1,2,\dots,m)$	i_1			i_2			$\dots$	i_m		

| Opportunity j $(j = 1,2,\dots,n_i)$ | 1 | 2 | $\dots$ | n_1 | 1 | 2 | $\dots$ | n_2 | 1 | 2 | $\dots$ | n_m |

Sources are independent and opportunities are mutually exclusive.

To motivate the discussion, let us consider a firm that has m sources of investment opportunities, with source i providing n_i mutually exclusive investment opportunities for consideration. A single budgetary constraint limits the total amount invested. Extending the notation introduced to date, we see that the budget allocation problem is formulated as follows:

$$\text{maximize} \quad \sum_{i=1}^{m} \sum_{j=1}^{n_i} p_{ij} x_{ij},$$

$$\text{subject to} \quad \sum_{i=1}^{m} \sum_{j=1}^{n_i} c_{ij} x_{ij} \leq C,$$

$$\sum_{j=1}^{n_i} x_{ij} \leq 1, \qquad i = 1, \dots, m,$$

$$x_{ij} = (0, \ 1) \text{ for all } i, j,$$

where m = number of sources of investment opportunities,

n_i = number of mutually exclusive investment opportunties available from source i,

p_{ij} = present worth of investment opportunity j from source i,

c_{ij} = initial investment required for investment opportunity j from source i,

$x_{ij} = \begin{cases} 1, & \text{if investment opportunity } j \text{ from source } i \text{ is selected} \\ 0, & \text{otherwise,} \end{cases}$

C = budget limit.

The first constraint expresses the budget limitation; the remaining generalized upper-bound constraints express the mutually exclusive conditions on all opportunities from the same source.

The budget allocation problem is a generalized upper-bound variation of the classic knapsack problem, studied extensively in the mathematical programming literature.[2] Exact solution procedures exist for such a formulation and are computationally feasible for several hundred x_{ij}.

---

[2] H. A. Taha, *Operations Research—An Introduction*, 4th ed. (New York: Macmillan Publishing Company, 1987), Chapter 9.

From a historical perspective, to solve this particular capital budgeting problem, Lorie and Savage[3] suggested the following heuristic approach:

1. Let α be a nonnegative multiplier; let $f_{ij}(\alpha) = p_{ij} - \alpha c_{ij}$; and let j^* be the index j having the maximum $f_{ij}(\alpha)$ value for each i; if $\max_j f_{ij}(\alpha) \leq 0$, then j^* is undefined.

2. Find the smallest value of α such that

$$\sum_{i=1}^{m} c_{ij*} \leq C,$$

where $c_{ij*} = 0$ when j^* is undefined. The investment opportunities having investments c_{ij*} are to be funded.

In the case in which all opportunities are independent, the *Lorie-Savage problem* is formulated as

$$\text{maximize } \sum_{i=1}^{m} p_i x_i,$$

$$\text{subject to } \sum_{i=1}^{m} c_i x_i \leq C,$$

$$x_i = (0, 1).$$

Applying the Lorie-Savage procedure for independent collections of mutually exclusive opportunities is equivalent to ranking the opportunities in decreasing order of $p_i \div c_i$ and funding opportunities in order until further funding would exceed the budget limit. If the last opportunity funded is opportunity k, then the quantity $p_i - \alpha c_i \geq 0$ for funded opportunities; for those opportunities not funded, $p_i - \alpha c_i < 0$, where

$$\alpha \in \left[\frac{p_{k+1}}{c_{k+1}}, \frac{p_k}{c_k} \right].$$

Example 16-3

A firm has available \$275,000 for allocation among three plants. Plant A has available three mutually exclusive investment opportunities; plant B has available a single investment opportunity; and plant C has available two mutually exclusive investment opportunities. The net present worths for the investment opportunities are given in Table 16-1, along with the required initial investments. Determine the budget allocation using the Lorie-Savage procedure.

TABLE 16-1   Data for the Lorie-Savage Example

i	j	Net present worth	Initial investment
1	1	\$60,000	\$110,000
	2	66,000	150,000
	3	68,000	215,000
2	1	9,000	50,000
3	1	37,000	120,000
	2	38,000	165,000

[3] J. H. Lorie and L. J. Savage, "Three Problems in Capital Rationing," *Journal of Business* 28, no. 4 (1955):229–239.

1. Let $\alpha = 0$ ($C = 275$).

i	j	p_{ij}	c_{ij}	$f_{ij}(\alpha)$	$\max_j f_{ij}(\alpha)$	j^*	c_{ij}^*
1	1	60	110	60	—	—	—
	2	66	150	66	—	—	—
	3	68	215	68	68	3	215
2	1	9	50	9	9	1	50
3	1	37	120	37	—	—	—
	2	38	165	$\dfrac{38}{279}$	38	2	$\dfrac{165}{430 > 275}$

2. Let $\alpha = 0.10$.

i	j	p_{ij}	c_{ij}	$f_{ij}(\alpha)$	$\max_j f_{ij}(\alpha)$	j^*	c_{ij}^*
1	1	60	110	49.0	—	—	—
	2	66	150	51.0	51	2	150
	3	68	215	46.5	—	—	—
2	1	9	50	4.0	4	1	50
3	1	37	120	25.0	25	1	120
	2	38	165	21.5	—	—	$\overline{320 > 275}$

3. Let $\alpha = 0.15$.

i	j	p_{ij}	c_{ij}	$f_{ij}(\alpha)$	$\max_j f_{ij}(\alpha)$	j^*	c_{ij}^*
1	1	60	110	43.50	—	—	—
	2	66	150	43.50	43.50	1,2	110
	3	68	215	35.75	—	—	—
2	1	9	50	1.50	1.50	1	50
3	1	37	120	19.00	19.00	1	120
	2	38	165	13.25	—	—	$\overline{280 > 275}$

4. Let $\alpha = 0.20$.

i	j	p_{ij}	c_{ij}	$f_{ij}(\alpha)$	$\max_j f_{ij}(\alpha)$	j^*	c_{ij}^*
1	1	60	110	38	38	1	110
	2	66	150	36	—	—	—
	3	68	215	25	—	—	—
2	1	9	50	−1	−1	Undefined	—
3	1	37	120	13	13	1	120
	2	38	165	5	—	—	$\overline{230 < 275}$

The recommended budget allocation is to undertake opportunity 1 ($j = 1$) at plant A ($i = 1$) and opportunity 1($j = 1$) at plant C ($i = 3$) for a present worth of 97. It should be noted that the Lorie-Savage procedure did not yield an optimum solution; in this instance, investments in opportunity 2 at plant A and opportunity 1 at plant C will yield a present worth of 103 and satisfy the budget constraint. (As an exercise, the curious reader is asked to solve this example using the branch and bound method.) ∎

16.5 Divisible Investment Opportunities

In some cases, investment opportunities are divisible. Namely, it is possible to undertake some portion of the investment instead of being forced to "do all" or "nothing." A number of such investments can be formulated as linear programming problems using continuous decision variables; others can be formulated as integer linear programming problems.

Example 16-4

A major industrial firm has $60 million available for the coming year to be allocated among three processing plants. Because of personnel levels and ongoing projects at the plants, it is necessary that at least $6 million be allocated to plant 1, $16 million be allocated to plant 2, and $10 million be allocated to plant 3. Because of the production facilities available at plant 3, no more than $34 million can be utilized without major new capital expansion; such expansion cannot be undertaken at this time. A number of investment opportunities exist at the various plants. Each plant has submitted budget requests in which the opportunities are grouped into categories by anticipated rate of return. For simplicity, the rate of return is expressed as a percentage of investment. Upper limits have been placed on the investment in each category. The data for the budgeting problem are given in Table 16-2.

A linear programming formulation of the problem is given as follows:

$$\text{maximize} \quad 0.16x_1 + 0.12x_2 + 0.14x_3 + 0.20x_4$$
$$+ \; 0.12x_5 + 0.10x_6 + 0.16x_7 + 0.18x_8$$

$$\text{subject to} \quad x_1 + x_2 + x_3 + x_4 + x_5 + x_6 + x_7 + x_8 \leq 60$$
$$x_1 + x_2 + x_3 \geq 6$$
$$x_4 + x_5 \geq 16$$
$$x_6 + x_7 + x_8 \geq 10$$
$$x_6 + x_7 + x_8 \leq 34$$

TABLE 16-2   Data for the Linear Programming Example

Plant	Budget category	Rate of return	Maximum investment
1	1	16%	$12 million
	2	12%	10 million
	3	14%	18 million
2	4	20%	12 million
	5	12%	6 million
3	6	10%	14 million
	7	16%	20 million
	8	18%	8 million

$$0 \le x_1 \le 12, \qquad 0 \le x_2 \le 10, \qquad 0 \le x_3 \le 18,$$
$$0 \le x_4 \le 12, \qquad 0 \le x_5 \le 6, \qquad 0 \le x_6 \le 14,$$
$$0 \le x_7 \le 20, \qquad 0 \le x_8 \le 8,$$

where x_j represents the amount (in millions) to be invested in budget category j.

An optimum solution to the linear programming problem can be obtained by using the simplex method. The optimum solution will be found to be $x_1 = 12, x_2 = 0, x_3 = 4, x_4 = 12, x_5 = 4$, $x_6 = 0, x_7 = 20$, and $x_8 = 8$. Hence, $16 million will be allocated to plant 1, $16 million will be allocated to plant 2, and $28 million will be allocated to plant 3. ∎

If it is assumed that an individual investment opportunity is divisible such that fractional opportunities or projects can be pursued, then a number of the binary linear programming formulations considered previously can be solved by replacing the binary requirement with the restriction $0 \le x_i \le 1$.

Weingartner[4] proves that the number of fractional-valued x_i's will be no greater than the number of inequality constraints. Hence, if the first Lorie-Savage example problem is solved as a linear programming problem, at most one fractional x_i will occur.

16.6 The MARR Controversy

The controversy over the establishment of the minimum attractive rate of return has been alluded to in earlier chapters. However, because of the intensity of the issue as it relates to the mathematical formulation of capital budgeting problems, it deserves additional consideration. The essential question is what the value of the MARR should be in present worth calculations for capital budget determinations when a constraint exists on capital available for investment.

The issue was first dealt with explicitly by Baumol and Quandt;[5] they concluded that a meaningful discount rate cannot be determined simultaneously with capital budgeting. Lusztig and Schwab[6] later contended otherwise. Bernhard[7] studied the issue and came to the following conclusion:

> As concluded by Baumol and Quandt, in programming models of the Lorie-Savage type, internal procurement of economically meaningful values of the discount rate, k, is not permissible. We have seen here . . . that the more recently proposed procedure of Mao, and of Lusztig and Schwab, does not, in general, give meaningful results and hence does not counter the earlier finding of Baumol and Quandt.

[4] H. M. Weingartner, *Mathematical Programming and the Analysis of Capital Budgeting Problems* (Chicago: Markham Publishing Co., 1967).

[5] W. J. Baumol and R.E. Quandt, "An Expected Gain-Confidence Limit Criterion for Portfolio Selection," *Management Science* 10, no. 1 (1963):174–182.

[6] P. Lusztig and B. Schwab, "A Note on the Application of Linear Programming to Capital Budgeting," *Journal of Financial and Quantitative Analysis* 3, no. 4 (1968):426–431.

[7] R. H. Bernhard, "Some Problems in the Use of Discount Rate for Constrained Capital Budgeting," *AIIE Transactions* 3, no. 3(1971):180–184.

In developing the mathematical programming formulations, an assumption was made that the MARR is based on the opportunity cost of capital. Unfortunately, the opportunity cost will depend on the collection of investments accepted (and rejected!). Bernhard recommends that the manager's utility function for reinvestment funds made available over time be used to determine the budget allocation. Independently, Atkins and Ashton[8] considered the Lorie-Savage formulation allowing fractional investments and concluded, "Our analysis, while questioning the logic of the Baumol and Quandt paper, does not question the correctness of their conclusion that a capital budgeting model which ignores the owner's consumption preference over time is of no practical importance."

Although the MARR controversy is an important issue, from a pragmatic point of view the objective in using mathematical programming formulations is to obtain budget allocations that are better than would otherwise be obtained. Final decisions will generally be different from the output from the mathematical model because of the nonquantifiables, multiple criteria, and changing conditions. Consequently, in an operating setting, it is felt that the MARR used by the firm in evaluating mutually exclusive investment alternatives will yield results that "satisfice," if not optimize!

16.7 Capital Rationing

The capital rationing problem considered in Chapter 9, can be formulated as a special type of linear programming problem: the transportation problem. Recall that a number of sources of investment funds may be available, along with a number of investment opportunities. Each dollar borrowed from source j costs c_j; each dollar invested in investment opportunity i returns r_i. The dollar amount available from source j is denoted b_j; the maximum amount that can be invested in opportunity i is denoted a_i. The net return resulting from borrowing a dollar from source j and investing it in opportunity i is denoted $c_{ij} = r_i - c_j$. The amount borrowed from source j and invested in opportunity i is denoted x_{ij}.

In order to ensure that money will not be invested unless it is profitable to do so, a dummy source of funds is defined with

$$b_{n+1} = \sum_{i=1}^{m} a_i \quad \text{and} \quad c_{i,n+1} = 0.$$

Similarly, in order to ensure that money is not borrowed unless it is profitable to do so, a dummy investment opportunity is defined with

$$a_{m+1} = \sum_{j=1}^{n} b_j \quad \text{and} \quad c_{m+1,j} = 0.$$

The capital rationing problem is formulated as follows:

---

[8] D. R. Atkins and D. J. Ashton, "Discount Rates in Capital Budgeting: A Reexamination of the Baumol and Quandt Paradox," *The Engineering Economist* 21, no. 3 (1976):159–172.

$$\text{maximize} \sum_{i=1}^{m+1} \sum_{j=1}^{n+1} c_{ij} x_{ij},$$

$$\text{subject to} \sum_{i=1}^{m+1} x_{ij} = b_j, \quad j = 1, \ldots, n+1,$$

$$\sum_{j=1}^{n+1} x_{ij} = a_i, \quad i = 1, \ldots, m+1,$$

$$x_{ij} \geq 0 \quad \text{for all } i, j.$$

The capital rationing problem is formulated using the well-known transportation problem formulation. It is intuitive to note that the optimum solution to the capital rationing problem is based on this principle: *Invest the cheapest money in the opportunity with the greatest return; continue until the marginal cost of securing a dollar is equal to or greater than the marginal return obtained.*

Example 16-5

Consider a situation in which three sources of funds are available and five investment opportunities exist. Suppose $c_1 = 0.10$, $c_2 = 0.12$, $c_3 = 0.15$, $r_1 = 0.30$, $r_2 = 0.20$, $r_3 = 0.14$, $r_4 = 0.11$, and $r_5 = 0.08$; similarly, $b_1 = 100$, $b_2 = 50$, $b_3 = 20$, $a_1 = 15$, $a_2 = 25$, $a_3 = 40$, $a_4 = 65$, and $a_5 = 70$. Since

$$\sum_{i=1}^{5} a_i = 215 \quad \text{and} \quad \sum_{j=1}^{3} b_j = 170,$$

a "dummy" source is defined with $b_4 = 215$ and $c_4 = 0$; also a "dummy" investment opportunity is defined with $a_6 = 170$ and $r_6 = 0$.

As given in Fig. 16-2, the optimum solution is to borrow 100 from source 1; 15 is invested in opportunity 1, 25 is invested in opportunity 2, 40 is invested in opportunity 3, and 20

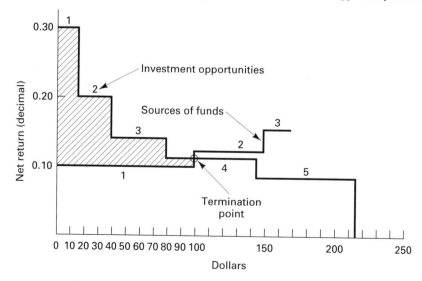

Figure 16-2   Graphical representation of capital rationing solution.

is invested in opportunity 4. The solution is obtained by allocating as much as possible to the cells having the greatest c_{ij} values.

Solving this capital rationing problem as a transportation problem gives the solution illustrated in Fig. 16-3. Opportunities 1, 2, and 3 are completely funded from source 1; 20 is borrowed from source 1 and is invested in opportunity 4. ∎

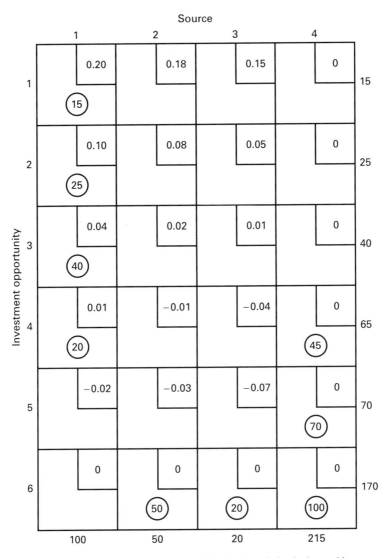

Figure 16-3 Transportation tableau solution for the capital rationing problem.

16.8 Portfolio Selection

A class of investment problems that has received considerable attention among researchers is that referred to by Markowitz[9] as the portfolio selection problem. The portfolio problem involves the selection of an optimal set of investments (the portfolio) based on the expected yields, the risks associated with the investments, and the utility of the investor.

Markowitz used the variance of the return obtained as a surrogate for risk. He assumed the objective of the investor to be the maximization of expected utility, where expected utility can be expressed as a function of the expected value and variance of return.

If y_i is defined as the *percentage* of the available funds invested in investment opportunity i, and if R_i is a random variable denoting the return obtained from opportunity i, then the expected value and variance of return are given as

$$E[R] = \sum_{i=1}^{m} y_i E[R_i],$$

$$V[R] = \sum_{i=1}^{m} \sum_{j=1}^{m} y_i y_j \, \text{Cov}\,[R_i, R_j],$$

where $\text{Cov}[R_i, R_j]$ is the covariance of R_i and R_j and $\text{Cov}[R_i, R_i] = V[R_i]$.

Multiple values of the y_i can yield the same variance; similarly, the same expected return can be obtained for multiple combinations of the y_i (portfolios). The risk-averse investor will want to select one of the portfolios that yields either minimum variance ($\mathcal{V}$) for a given expected return ($\mathcal{E}$) or maximum $\mathcal{E}$ for a given $\mathcal{V}$. Such portfolios are referred to as *mean-variance ($\mathcal{E}$-$\mathcal{V}$) efficient portfolios.*

As depicted in Fig. 16-4 (by the bold line segment), a number of $\mathcal{E}$-$\mathcal{V}$ efficient portfolios can exist. If the utility function can be represented in such a way that the expected utility is given as $E[U] = f(\mathcal{E}, \mathcal{V})$ then some $\mathcal{E}$-$\mathcal{V}$ efficient portfolio will maximize $E[U]$. In Figure 16-4 iso-utility curves are shown with $E[U_1] > E[U_2] > E[U_3]$. The optimum portfolio is portfolio Y, as shown.

If it is assumed that the investor's utility function is at least twice differentiable, the utility function may be expanded by Taylor's series about its mean to yield

$$U(R) = U(\mu) + U'(\mu)[R - \mu] + \frac{U''(\mu)}{2}[R - \mu]^2 + \ldots,$$

where $\mu = E[R]$ and $U'(\mu)$ is the first derivative of U evaluated at μ. The expected utility is given by

$$E[U(R)] = \mu + \frac{U''(\mu)}{2}\sigma^2 + \ldots,$$

where $\sigma^2 = V[R]$. Letting $\theta = -U''(\mu)/2$, which was defined in Chapter 13 as the coefficient of risk aversion, then dropping all terms beyond those given above yields

$$E[U(R)] \doteq \mu - \theta\sigma^2.$$

[9] H. M. Markowitz, *Portfolio Selection: Efficient Diversification of Investments* (New York: John Wiley & Sons, Inc., 1959).

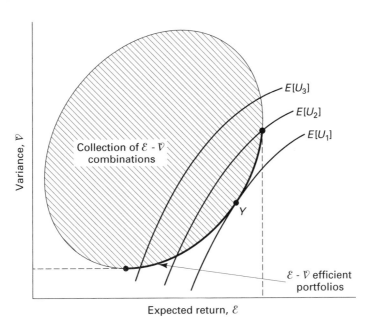

Figure 16-4   Mean-variance ($\mathcal{E}$-$\mathcal{V}$) efficient portfolios.

If the investor is risk-averse, $\theta > 0$.

The above result indicates the expected utility can be approximated by a linear function of $\mathcal{E}$ and $\mathcal{V}$. Hence, the portfolio problem can be formulated as a quadratic programming problem.

$$\text{maximize } \sum_{i=1}^{m} y_i E[R_i] - \theta \sum_{i=1}^{m} \sum_{j=1}^{m} \text{Cov } [R_i, R_j] y_i, y_j$$

$$\text{subject to } \sum_{i=1}^{m} y_i \leq 1,$$

$$y_i \geq 0.$$

A number of different objective functions have been proposed in the literature, including the maximization of $\mu - \theta\sigma$. As noted in Chapter 13, empirical results by Cramer and Smith[10] tend to support the use of the standard deviation instead of the variance in approximating the utility function.

Among the several criteria that may be used in selecting the portfolio are the following:[11]

1. expected value criterion
   maximize $E[R]$

---

[10] R. H. Cramer and B. E. Smith, "Decision Models for the Selection of Research Projects," *The Engineering Economist* 9, no. 2 (1964):1–20.

[11] The constraints $\sum_{i=1}^{m} y_i \leq 1$ and $y_i \geq 0$ are required for each formulation.

2. portfolio criterion
   minimize $V[R]$,
   subject to $E[R] \geq \alpha$

3. aspiration level criterion
   maximize $P[R \geq \beta]$

4. fractile criterion
   maximize z,
   subject to $P[R \geq z] \geq \gamma$

5. chance constrained criterion
   maximize $E[R]$,
   subject to $P[R \geq \delta] \geq \varepsilon$

where $E[\cdot]$, $V[\cdot]$, and $P[\cdot]$ denote the expected value, variance, and probability operations, respectively, and α, β, γ, δ, and ε are appropriately defined constants.

Example 16-6

Consider a simplified portfolio situation involving only two investment opportunities with the following formulation:

$$\text{maximize } z = 10y_1 + 12y_2 - 4y_1^2 - 6y_1y_2 - 5y_2^2,$$
$$\text{subject to} \quad y_1 + y_2 \leq 1,$$
$$y_1 \geq 0, \quad i = 1, 2.$$

Taking the partial derivatives of z with respect to y_1 and y_2, setting the results equal to zero, and solving produces the following:

$$\left.\begin{array}{l} \dfrac{\partial z}{\partial y_1} = 10 - 8y_1 - 6y_2 = 0 \\[2mm] \dfrac{\partial z}{\partial y_2} = 12 - 6y_1 - 10y_2 = 0 \end{array}\right\} \; y_1 = \frac{7}{11} \;\text{ and }\; y_2 = \frac{9}{11}.$$

Since

$$y_1 + y_2 = \frac{16}{11} > 1,$$

the constraint will be active and $y_1 + y_2 = 1$. Thus, $y_2 = 1 - y_1$, and the objective function becomes

$$z = 10y_1 + 12(1 - y_1) - 4y_1^2 - 6y_1(1 - y_1) - 5(1 - y_1)^2$$

or

$$z = 7 + 2y_1 - 3y_1^2.$$

By differentiating z with respect to y_i, setting the result equal to zero, and solving we find $y_1 = \frac{1}{3}$; hence, $y_2 = \frac{2}{3}$.

16.9 Goal Programming

The concept of goal programming was developed by Charnes and Cooper.[12] A number of applications and extensions of the technique are described by Ignizio[13] and Lee.[14]

The goal-programming approach involves the specification of several desired levels of attainment (goals) and the establishment of priorities for the goals. Generally, the priorities are expressed as either ordinal or cardinal priorities. In the case of ordinal priorities, the most important goal is satisfied first, followed by the satisfaction of as many other goals as possible, in rank order. When a cardinal scale is used, it is typically assumed that "goal attainment" is additive and an aggregate, linear objective function is developed.

Goal programming was developed initially in conjunction with linear programming formulations. More recently, the concept has been applied in conjunction with both integer programming and nonlinear programming.

In terms of capital investment decision making, it is generally the case that managers have multiple goals. Although the number of reported applications of the use of goal programming in capital budgeting is small, the approach seems promising. For this reason, the goal programming approach is illustrated with the following example.

Example 16-7

Consider a production manger who must schedule the production of two products on three machines. The unit profits, machining times, and production capacities of the equipment are summarized below (Product $1 = x$ and Product $2 = y$).

	Machines			
Product	1	2	3	
1	0.2 hr	0.4 hr	0.3 hr	Time/unit
2	0.4 hr	0.1 hr	0.3 hr	Time/unit
Capacity	40 hr	40 hr	40 hr	Time/week

Product	Maximum demand	Profit
1	80/wk	$5/unit
2	80/wk	$20/unit

The linear programming formulation of the production scheduling problem is given as follows:

$$\text{maximize } 5x + 20y,$$
$$\text{subject to } 0.2x + 0.4y \leq 40,$$
$$0.4x + 0.1y \leq 40,$$
$$0.3x + 0.3y \leq 40,$$
$$x \leq 80,$$
$$y \leq 80,$$
$$x \geq 0, \ y \geq 0.$$

[12] A. Charnes and W. W. Cooper, *Management Models and Industrial Applications of Linear Programming* (New York: John Wiley & Sons, Inc., 1961).

[13] J. P. Ignizio, *Goal Programming and Extensions* (Lexington, MA: D.C. Heath & Company, 1976).

[14] S. M. Lee, *Goal Programming for Decision Analysis* (Philadelphia, PA: Auerbach Publishers, Inc., 1972).

As depicted in Fig. 16-5, the optimum solution is $x = 40$ and $y = 80$. The unused machining capacities for the machines are 0, 16, and 4 hours per week, respectively. Hence, machine 2 will be only 60% utilized.

The manager is concerned not only with the utilization of the machines but also with the profits realized for the firm. He also recognizes that the market forecasts of 80 units per week for the products could be in error.

The manager would prefer to have the three machines equally loaded, but he does not want to sacrifice profits unduly. The maximum profit condition ($x = 40$, $y = 80$) would result in an $1,800 profit. He feels that a satisfactory minimum profit level would be $1,500.

The following alternative formulation is being considered:

$$\text{minimize } (d_1^-, d_2^- + d_3^- + d_4^-),$$

$$\text{subject to} \quad 5x + 20y + d_1^- - d_1^+ = \$1,500,$$

$$0.2x + 0.4y + d_2^- \quad = \quad 40,$$

$$0.4x + 0.1y + d_3^- \quad = \quad 40,$$

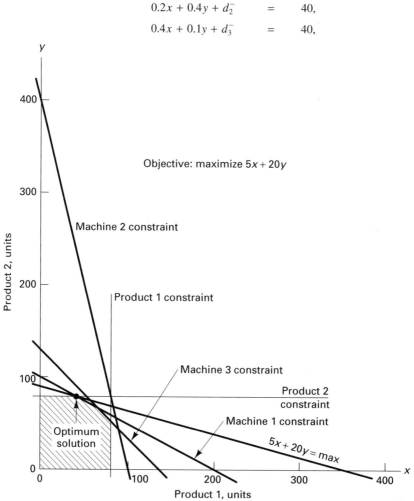

Figure 16-5   Linear programming formulation of production scheduling problem.

$$0.3x + 0.3y + d_4^- = 40,$$
$$x \leq 80,$$
$$y \leq 80,$$
$$x, y, d_j^-, d_j^+ \leq 0.$$

The slack variable, d_j^-, and surplus variable, d_j^+, measure the deviation from the right-hand side of the constraints. (Since the upper limit of 40 hours' production capacity cannot be exceeded, the surplus variables are omitted for $j = 2, 3, 4$.) The objective function is expressed using an ordinal scaling of the profit goal (make at least a $1,500 profit) and the machine balance goal (minimize the sum of unused capacities).[15] To solve the problem, the set of optimum solutions is determined using the following objective: minimize d_1^-. The resulting set of solutions is depicted in Fig. 16-6, for $d_1^- = 0$.

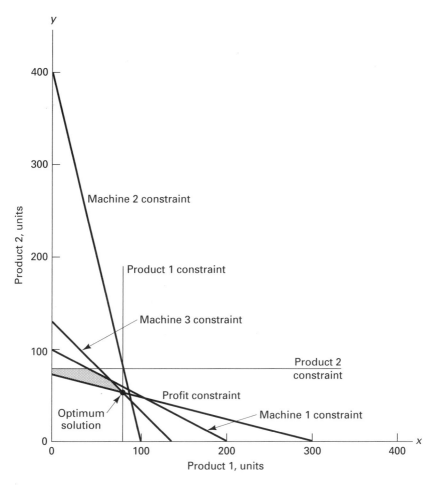

Figure 16-6   Goal programming formulation of production scheduling problem.

[15] An alternate expression of the machine balance goal will be considered subsequently.

Next, the following linear programming problem is solved:

$$\text{minimize } d_2^- + d_3^- + d_4^-,$$

$$\begin{aligned}
\text{subject to } \quad 0.2x + 0.4y + d_2^- &= 40, \\
0.4x + 0.1y + d_3^- &= 40, \\
0.3x + 0.3y + d_4^- &= 40, \\
5x + 20y - d_1^+ &= \$1,500, \\
x &\leq 80, \\
y &\leq 80, \\
x, y, d_j^-, d_1^+ &\geq 0.
\end{aligned}$$

The optimum solution is $x = 700/9$, $y = 500/9$, $d_2^- = 20/9$, $d_3^- = 10/3$, $d_4^- = 0$, and $d_1^+ = 0$. Hence, the profit obtained is \$1,500, with machine utilizations of 94.44%, 91.67%, and 100.00%, respectively.

An alternative expression of the machine balance goal is to minimize the maximum difference in the d_j^- values, i.e.,

$$\text{minimize maximum } (\mid d_2^- - d_3^- \mid, \mid d_2^- - d_4^- \mid, \mid d_3^- - d_4^- \mid),$$

$$\begin{aligned}
\text{subject to } \quad 0.2x + 0.4y + d_2^- &= 40, \\
0.4x + 0.1y + d_3^- &= 40, \\
0.3x + 0.3y + d_4^- &= 40, \\
5x + 20y - d_1^+ &= \$1,500, \\
x &\leq 80, \\
y &\leq 80, \\
x, y, d_j^-, d_1^+ &\geq 0.
\end{aligned}$$

Letting z denote the maximum absolute deviation among the slack variables, the formulation becomes

$$\text{minimize } z,$$

$$\begin{aligned}
\text{subject to } \quad z &\geq \mid d_2^- - d_3^- \mid, \\
z &\geq \mid d_2^- - d_4^- \mid, \\
z &\geq \mid d_3^- - d_4^- \mid, \\
0.2x + 0.4y + d_2^- &= 40, \\
0.4x + 0.1y + d_3^- &= 40, \\
0.3x + 0.3y + d_4^- &= 40, \\
5x + 20y - d_1^+ &= \$1,500, \\
x &\leq 80, \\
y &\leq 80, \\
x, y, d_j^-, d_1^+ &\geq 0.
\end{aligned}$$

The first three constraints ensure that the minimum value of z will equal the maximum absolute deviation among the slack variables. A constraint of the form

$$|u - v| \leq w$$

can be written as two linear constraints,

$$u - v \leq w,$$
$$u - v \geq -w.$$

Therefore, the optimization problem can be written as

minimize z,

$$
\begin{aligned}
\text{subject to } d_2^- - d_3^- - z &\leq 0, \\
-d_2^- + d_3^- - z &\leq 0, \\
d_2^- - d_4^- - z &\leq 0, \\
-d_2^- + d_4^- - z &\leq 0, \\
d_3^- - d_4^- - z &\leq 0, \\
-d_3^- + d_4^- - z &\leq 0, \\
0.2x + 0.4y + d_2^- &= 40, \\
0.4x + 0.1y + d_3^- &= 40, \\
0.3x + 0.3y + d_4^- &= 40, \\
5x + 20y - d_1^+ &= 1{,}500, \\
x &\leq 80, \\
y &\leq 80.
\end{aligned}
$$

A linear programming solution to the above problem yields the following values for the decision variables: $x = 700/9$, $y = 500/9$, $d_1^+ = 0$, $d_2^- = 20/9$, $d_3^- = 10/3$, $d_4^- = 0$, and $z = 10/3$. Hence, a profit of \$1,500 is obtained, along with machine utilizations of 94.44%, 91.67%, and 100.00%, respectively. For this example the same solution is obtained for the goal programming problem using (1) a minimax objective and (2) an objective of minimizing the sum of the deviations. ∎

Goal programming problems are special cases of multicriteria optimization problems; considerable research has been performed on the general class of multicriteria or (multiattribute) optimization problems. Although multicriteria optimization offers considerable potential for improved decision making in capital budgeting, further development is required before it will receive widespread application. Based on the research performed to date, it appears that interactive approaches involving the decision maker will be used. For a discussion of interactive approaches and multicriteria optimization, see Geoffrion, Dyer, and Feinberg.[16] Chapter 19 contains several popular techniques for consideration of multiple objectives and criteria, and Chapter 20 is devoted entirely to the analytic hierarchy process.

[16] A. M. Geoffrion, J. S. Dyer, and A. Feinberg, "An Interactive Approach for Multi-Criterion Optimization, with an Application to the Operation of an Academic Department," *Management Science* 19, no. 4 (1972):357–368.

16.10 Summary

Applications of mathematical programming to capital budgeting problems have yet to be widely experienced in U.S. industry. There are several reasons for this, including (1) the separation of the decision maker from the analysis, (2) the difficulty in quantifying strategic and risky considerations involved with capital budgeting situations, and (3) the focus on optimization that seems to supersede unrealistically the need for "satisficing" in actual decision making. Thus, a brighter future for mathematical programming usage in capital budgeting may result when analysts assist in understanding the structure and process of this decision-making activity in the complex and often ambiguous environment that characterizes industry and government. We anticipate that better communication and a problem-solving orientation based on achieving "near-optimal" solutions will go far toward making mathematical programming an important part of capital budgeting practice in the United States.

REFERENCES

Baumol, W. J., and R. E. Quandt, "Investment and Discount Rates Under Capital Rationing—A Programming Approach," *Economic Journal* 75, no. 298 (June 1965):317–329.

BERNHARD, R. H., "Mathematical Programming Models for Capital Budgeting—A Survey, Generalization, and Critique," *Journal of Financial and Quantitative Analysis* 4, no. 2 (1969):111–158.

CHARNES, A., W. W. COOPER, AND M. H. MILLER, "Application of Linear Programming to Financial Budgeting and the Costing of Funds," *Journal of Business* (Jan. 1959):20–46.

FREELAND, J. R., AND M. J. ROSENBLATT, "An Analysis of Linear Programming Formulations for the Captial Rationing Problem," *The Engineering Economist* 24, no. 1 (Fall 1978):49–61.

HAWKINS, C. A., AND R. A. ADAMS, "A Goal Programming Model for Capital Budgeting," *Financial Management* (Spring 1974):52–57.

HAYES, J. W., "Discount Rates in Linear Programming Formulations of the Capital Budgeting Problem," *The Engineering Economist* 29, no. 2 (Winter 1984):123–126.

IGNIZIO, J. P., *Linear Programming in Single and Multiple Objective Systems,* Englewood Cliffs, NJ: Prentice-Hall, 1982.

LUSZTIG, P., AND B. SCHWAB, "A Note on the Application of Linear Programming to Capital Budgeting," *Journal of Financial and Quantitative Analysis* 3, no. 5 (Dec. 1968):427–431.

MYERS, S. C., "Interactions of Corporate Financing and Investment Decisions—Implications for Capital Budgeting," *Journal of Finance* 29 (March 1974):1–26.

WEINGARTNER, H. M., "Capital Budgeting of Interrelated Projects: Survey and Synthesis," *Management Science* 12 (March 1966):485–516.

WHITMORE, G. A., AND L. R. AMEY, "Capital Budgeting Under Rationing: Comments on the Lusztig and Schwab Procedure," *Journal of Financial and Quantitative Analysis* 8, no. 1 (Jan. 1973):127–135.

PROBLEMS

16-1. A firm is considering the development of several new products. The products under consideration are listed below; products in each group are mutually exclusive.

Group	Product	Development cost	Annual net cash income
	A1	$ 500,000	$ 90,000
A	A2	650,000	110,000
	A3	700,000	115,000
B	B1	600,000	105,000
	B2	675,000	112,000
C	C1	800,000	150,000
	C2	$1,000,000	175,000

At most one product from each group will be selected. The firm has a minimum attractive rate of return of 10% and a budget limitation on development costs of $2,100,000. The life of all products is assumed to be ten years, with no salvage value. Set up this situation as a linear programming problem.

16-2. Use a computer program (e.g., LINDO) to solve the following capital budgeting formulation:

$$\text{maximize } 18x_1 + 20x_2 + 15x_3 + 22x_4,$$
$$\text{subject to } 4x_1 + 6x_2 + 5x_3 + 6x_4 \le 15,$$
$$5x_1 + 3x_2 + 6x_3 + 4x_4 \le 12,$$
$$10x_1 + 5x_2 + 4x_3 + 12x_4 \le 30,$$
$$x_j = (0, 1,) \quad j = 1, \ldots, 4.$$

16-3. In Problem 16-2 suppose investment opportunities 1 and 3 are mutually exclusive. Suppose investment opportunity 2 is contingent on the selection of opportunity 4. Solve the problem using a computer program (e.g., LINDO).

16-4. A firm is confronted with ten investment opportunities. The investment opportunities will require commitments of investment funds for up to three years. The present worths and investment requirements (for each of the first three years) are given below. Opportunity 2 is contingent on the selection of opportunity 7; opportunity 3 is contingent on the selection of both opportunities 1 and 4; and opportunity 6 is contingent on the selection of either mutually exclusive opportunity 8, 9, or 10. Determine the budget allocation that maximizes present worth using a computer program (e.g., LINDO).

Investment opportunity (i)	Present worth (p_i)	Year 1 (c_{1i})	Year 2 (c_{2i})	Year 3 (c_{3i})
1	20	24	16	6
2	16	12	10	2
3	11	8	6	6
4	4	6	4	7
5	4	1	6	9
6	18	18	18	20
7	7	13	8	0
8	19	14	8	12
9	24	16	20	24
10	4	4	6	8
Budget limit =		50	30	30

16-5. A firm has $300,000 available to be allocated among four production plants. Plant 1 has proposed three mutually exclusive projects; plant 2 has available two mutually exclusive projects; plant 3 has proposed three mutually exclusive investment opportunities; and plant 4 has available two mutually exclusive projects. The relevant data are given below. Use the Lorie-Savage procedure to obtain an allocation, and compare the result with that obtained using an exact solution procedure.

Investment source i	Mutually exclusive investment opportunity (by source)	Initial investment required	Present worth
1	1	$100,000	$25,000
	2	125,000	20,000
	3	175,000	22,000
2	1	80,000	20,000
	2	100,000	15,000
3	1	50,000	10,000
	2	75,000	15,000
	3	90,000	12,000
4	1	40,000	10,000
	2	60,000	12,000
Budget limitation =		$300,000	

16-6. A firm has available six sources of investment funds and ten independent investment opportunities. The interest costs of the investment capital and the rates of return for the investment opportunities are given below. Assuming divisible investment opportunities, determine the optimum borrowing and investment program.

Source of funds	Interest cost	Maximum investment funds available
1	8%	$ 50,000
2	9	80,000
3	10	150,000
4	12	250,000
5	15	500,000
6	18	1,000,000

Investment opportunity	Rate of return	Maximum amount of investment
1	5%	$ 30,000
2	8	80,000
3	20	50,000
4	15	80,000
5	12	100,000
6	10	200,000
7	25	50,000
8	10	75,000
9	12	40,000
10	7	60,000

16-7. In Problem 16-6, if the investment opportunities are not divisible, how would you solve the problem? Develop a mathematical formulation of the capital rationing problem for indivisible investment opportunities when borrowed funds are to be used. What are the potential pitfalls of using a criterion involving rates of return instead of present worths?

16-8. An individual has identified two investment opportunities. Every $100 invested in opportunity 1 yields a present worth of $20; in opportunity 2 a present worth of $24 is anticipated for every $100 invested. A maximum of $10,000 is available for investment. The individual wishes to hedge against risk by diversifying her investment; it is required that the amount invested in opportunity 1 be from 40% to 60% of the total. Formulate the investment problem as a linear programming problem, and determine the optimum investment program.

16-9. In Problem 16-8, suppose a goal-programming approach is to be taken. The individual has a primary objective of investing the $10,000 in such a way that a present worth of at least $2,230 is obtained. A secondary goal is to minimize the difference in the amount invested in each opportunity. Determine the investment program that would result.

16-10. In Problem 16-8, determine the effect on the optimum solution resulting from the presence of a third investment that has a present worth of $22 per $100 investment. It will be required that from $30,000 to $40,000 be invested in each opportunity rather than from 40% to 60% to be invested in opportunity 1.

16-11. The Frosty Fir Company wants to best utilize the wood resources in one of its forest regions. Within this region, there is a sawmill and a plywood mill; thus timber can be converted to lumber or plywood.

Producing a marketable mix of 1,000 board feet of lumber products requires 1,000 board feet of spruce and 3,000 board feet of Douglas fir. Producing 1,000 ft^2 of plywood requires 2,000 board feet of spruce and 4,000 board feet of Douglas fir. This region has available 32,000 board feet of spruce and 72,000 board feet of Douglas fir.

Sales commitments require that at least 4,000 board feet of lumber and 12,000 square feet of plywood be produced during the planning period. The profit contributions are $40 per 1,000 board feet of lumber products and $60 per 1,000 ft^2 of plywood. Let L be the amount (in 1,000 board feet) of lumber produced and P be the amount (in 1,000 ft^2) of plywood produced.

Formulate the problem as a goal programming model based on the following priority setup:

Priority 1. Meet the sales demand.
Priority 2. Maximize the total profit, up to $9,000.
Priority 3. Produce under all resource limitations.

16-12. Three alternatives are being considered for a proposed AS/RS application. Their cash flow estimates are shown below. A and B are mutually exclusive while C is a high-speed palletizer that represents an optional add-on feature to alternative A. Investment funds are limited to $5 million. Another constraint on this project is the engineering personnel needed to design and implement the solution. No more than 10,000 labor hours of engineering time can be committed to the project. Set up a 0–1 integer programming formulation of this capital budgeting problem.

	Alternative		
	A	B	C
Initial investment ($ millions)	4.0	4.5	1.0
Labor requirement (hrs.)	7,000	9,000	3,000
After-tax annual savings, yrs. 1–4 ($ millions)	1.3	2.2	0.9
PW (10%)	0.12×10^6	2.47×10^6	1.85×10^6

→ Identifies areas where
   cost reduction can
      or should be made

→ puts costs on activities
   that create costs

→ leads to
   competitive adv

Disad —
→ costly to track
→ must review annually
   to see if justified

17
Activity-Based Costing [1]

17.1 Introduction

Rightfully or wrongfully so, poor capital budgeting practices have been blamed, in part, for the competitive decline of manufacturing in the United States.[2] Capital budgeting, as discussed in Chapter 9, is fraught with many complexities including *estimation* of future cash flows associated with proposed capital expenditures.[3] Closely connected to estimation, the topic of this chapter is *activity-based costing (ABC)*. ABC is important to improved capital budgeting practices because it potentially ✓ offers an improved means for producing more accurate estimates of cash flows that truly reflect activities undertaken to create value. With more accurate cost and revenue estimates, allocation of scarce capital can be accomplished such that future wealth of an enterprise is maximized. This chapter is intended to describe ABC in the context of engineering economy analyses as well as to provide an overview in terms of its development and usage.

Recently, some companies have adopted the tenets of ABC in their cost management systems. However, in most cases an implementation decision was a survival reaction driven by increased competition—not as an action justified by its intrinsic good sense. Cooper[4] gives several examples of such implementations: Tektronix's

---

[1] This chapter was written with Dr. Jerome P. Lavelle, Department of Industrial and Manufacturing Systems Engineering, Kansas State University, Manhattan, KS.

[2] R. H. Hayes and D. A. Garvin, "Managing as if Tomorrow Mattered," *Harvard Business Review* (May–June 1982).

[3] R. Aggarwal, *Capital Budgeting Under Uncertainty* (Englewood Cliffs, NJ: Prentice-Hall, 1993).

[4] R. Cooper, "Cost Classification in Unit-Based and Activity-Based Manufacturing Cost Systems," *Journal of Cost Management* (Fall 1990).

Portable Instruments Division developed its new cost system after facing intense Japanese competition; John Deere Component Works developed its new cost system when their market position changed from sole supplier to active competitor; and Siemens Electric Motor Works developed their new cost system after facing intense competition from Eastern European companies. More and more companies are beginning to recognize the value-added information that an ABC system provides.

17.2 What Is Activity-Based Costing?

Activity-based costing (ABC) is a methodology for enterprise-wise management of business costs. It focuses on detailing costs and assigning those costs to the items that cause them to occur. The underlying philosophy of ABC is stated by Liggett, as:

> Certain *activities* are carried out in the manufacture of *products*. Those *activities* consume a firm's *resources,* thereby creating *costs.* The *products,* in turn, consume *activities.* By determining the amount of resource (and the resulting *cost*) consumed by an *activity* and the amount of an *activity* consumed in manufacturing a *product,* it is possible to directly trace manufacturing *costs* to *products*.[5]

Thus, one can see that *activities* are at the heart of an ABC system—but what are activities? Activities to produce a marketable product have classically been divided into two areas: direct and indirect (or overhead) activities. Direct activities are actions that accomplish a task *directly* associated with the final product or service produced. Examples of direct activities include component insertion, painting, drilling, backboard wiring, thread tapping, and wave solder. All of these have a direct impact on the product being produced, and without them the physical product would not be the same. Indirect activities are actions that are necessary to provide a good or service to a customer but do not *directly* affect the product or service produced. Rather, their effect is *indirect.* Examples of indirect activities might include material procurement, kitting, issuing a purchase order, material handling, machine setup, maintenance, and inspection. In addition, *overhead* items such as marketing, sales, utility expenses, machine depreciation, engineering and technical support, and customer warranties can also be classified as indirect activities. As you can see, all of these items, both direct and indirect, are intimately necessary in delivering the product to the customer. Thus, the concept of an activity in ABC includes activities that are direct and indirect in nature.

It is important to determine the cost of each activity to provide for a tracing of the true cost components in some set of tasks. Example 17-1 shows the calculation of activity costs in a telephone customer-order processing work cell.

Example 17-1

Five workers in a telephone-order business cell process customer orders for one product line of the firm. The workers carry out two basic tasks, categorized as *telephone* and *non-telephone tasks* by the company. The *telephone tasks* include taking product orders, logging customer

---

[5] H. R. Liggett, J. Trevino, and J. P. Lavelle, "Activity-Based Cost Management Systems in an Advanced Manufacturing Environment," *Economic and Financial Justification of Advanced Manufacturing Technologies,* H. R. Parsaei, ed. (New York: Elsevier Science Publishers, 1992).

complaints and feedback, and providing on-line product inquiry assistance. These tasks are performed 75%, 10%, and 15% of the time, respectively, when a worker is engaged in a *telephone task*. The *non-telephone tasks* include invoicing and billing orders, order shipping preparation, and inventory updating and maintenance. These non-telephone tasks are performed 35%, 25%, and 40% of the time, respectively, when the worker is engaged in such *non-telephone tasks*. The five workers include two workers dedicated to 100% telephone tasks, one worker who is 50% telephone and 50% non-telephone tasks, and two workers who are split 10% telephone and 90% non-telephone tasks. Assuming a 2,000 hour work-year and a labor-benefits rate of $15.00 per hour, the annual costs of the telephone-order business cell are calculated as follows:

One full-time employee annual costs = (2,000 hrs/yr) × (15$/hrs) = $30,000/year.

Telephone Tasks:

Taking product orders:

$30,000 × 2.7 employees × 75% of time in task   = $60, 750 per year

Logging customer complaints/feedback:

$30,000 × 2.7 employees × 10% of time in task   = $8, 100 per year

Providing on-line product inquiry assistance:

$30,000 × 2.7 employees × 15% of time in task   = $12, 150 per year

Subtotal = $81,000 per year

Non-Telephone Tasks:

Invoicing/billing orders:

$30,000 × 2.3 employees × 35% of time in task = $24, 150 per year

Order shipping preparation:

$30,000 × 2.3 employees × 25% of time in task = $17, 250 per year

Inventory updating/maintenance:

$30,000 × 2.3 employees × 40% of time in task = $27, 600 per year

Subtotal = $69,000 per year   ■

The concept of *resources* is also important in an ABC system. Resources are assets available in the enterprise for use in direct and indirect activities. Examples include production equipment, labor, tooling, computer time, and floor space. Associated with the use of these resources is a *cost*. When the resource is used, a cost can be assigned to the activity (direct or indirect) that caused that cost. In this way costs, and the activities that *drive* those costs, can be identified and managed. The principle of activity-based costing systems, then, involves identifying activities that consume resources that have costs associated with them.

17.3 Traditional Versus Activity-Based Costing Systems

In this chapter, traditional costing systems are referred to as absorption-based accounting systems. This designation refers to the manner in which overhead and indirect expenses are assigned to cost centers. As described later, this assignment is sometimes done on an arbitrary basis and results in cost centers often *absorbing* costs that they do not directly cause. Absorption-based accounting is thus distinguished from cause-and-effect–based (ABC) accounting systems.

Activity-based costing systems are different from traditional costing in three important aspects. First, they depart from the way in which manufacturing overhead costs are managed. Second, they focus on process costs and not product costs—providing the basis for more accurate cost reduction. Last, they provide a more detailed cost structure than that found in typical traditional (absorption-based) costing systems. These differences allow for more accurate and timely management of both direct and indirect costs.

17.3.1 Overhead Allocation

Traditional cost accounting systems allocate overhead (indirect) costs in terms of volume-based production indicators like direct labor hours, material cost, and direct labor cost. Of these, many firms choose direct labor hours as their basis for allocation. The effect of allocating overhead costs in this manner is to spread an overhead cost, such as utility expense or process engineering support, in direct proportion to the volume-based indicator for each costing cell in the enterprise. This means that centers with low direct labor content are allocated (or charged) a lower proportion of overhead costs than those with higher direct labor content. This allocation is made regardless of the fact that these cost cells may be consuming a disproportionately higher percentage of those overhead resources. Activity costing, on the other hand, traces the costs back to the activities that consume resources and cause costs to occur. This provides a framework for accurately identifying sources for cost reduction and puts the burden of cost justification on those activities that are indeed creating the costs.

Consider the example of allocating overhead to two production costing centers, one a high-volume dedicated assembly line (with a mix of automation and manual operations, and few product changeovers) and the other a lower-volume flexible production cell (with heavy automation, little direct labor content, and small lot sizes). Because of the fixed nature of the high-volume dedicated line, it will require fewer support costs than will the flexible assembly cell. Table 17-1 lists items that the low-volume line (small lot size, etc.) will require more of due to its flexible, low-volume nature.

With traditional costing systems, high-volume products often subsidize lower-volume products because lower-volume products are assigned less than their share of overhead and indirect costs. In the case of the high-volume dedicated assembly line, a traditional accounting system approach would allocate more overhead cost to it, in comparison to the flexible assembly cell, based on the fact that there is more direct labor content in the operations in that costing center. Allocation is made in this manner even though the low-volume products (those manufactured on the flexible cell) typically cause a disproportionate share of the overhead, indirect, and support expenses.

TABLE 17-1   Additional Resources Consumed by
Low-Volume Flexible Assembly Cell

•Material pick orders	•Manufacturing engineering support	•Routing sheets
•Manufacturing assembly aids	•Machine program downloads	•Material handling requests
•Setup time	•Shipping invoices	•Work orders
•Scheduling, receiving, warehousing, testing, and maintenance resources		

Manufacturers recognize this and tend to price the undercosted low-volume products at a premium in order to account for the "spreading" of overhead in the costing system. This causes two effects. First, the reporting system will identify these products as having the highest market margins, and thus sales and marketing forces will target them to increase their commissions (usually at the expense of the more profitable high-volume products). Second, prices on the overcosted higher-volume products tend to be higher than necessary. This leaves a manufacturer vulnerable to other firms, who have either a less diverse product offering or a costing system that leads to accurate costing, to offer the same product at a lower cost. Table 17-2 illustrates several examples of allocation bases for traditional versus activity-based costing systems.

TABLE 17-2   Cost Basis for Traditional Versus Activity-Based Costing

Overhead cost item	Traditional costing	ABC costing
Utility expenses	Direct labor hours	Metered utility expense
Maintenance activities	Direct labor hours	Maintenance hours billed
Order entry	Direct labor hours	No. of orders processed
Receiving activities	Direct labor hours	No. of receipts processed
Engineering support	Direct labor hours	No. of hours billed
Design change orders	Direct labor hours	No. of orders processed
Inspection activities	Direct labor hours	No. of items inspected
Production setups	Direct labor hours	No. of setups required

17.3.2 An Example

To illustrate several ABC concepts, let's consider the JPL Company, a maker of pagers for the telecommunications industry. Currently JPL is manufacturing two models of pagers with an equal annual production volume of 50,000 units each. Historically, JPL has used a traditional (absorption-based) accounting system for product costing. As such, the company uses direct labor hours as the allocation base for all its manufacturing overhead costs. The total manufacturing overhead costs as well as the total direct material and direct labor costs of both products for the current year are given in Table 17-3.

Recently the company installed an activity-based costing system and subsequently identified six main activities to be responsible for most of the manufacturing overhead costs. The overhead costs were distributed to these six activities through activity accounting. Also, six corresponding cost drivers and their budget consequences for the current year were established. The activities, their costs drivers, resultant overhead costs, and budgeted rates are provided in Table 17-4.

TABLE 17-3   Total Manufacturing Costs for
Models A and B (current year)

Category	Model A	Model B
Direct material cost:	$2,800,600	$1,500,000
Direct labor cost:	$ 350,000	$ 250,000
Direct labor hours:	35,000	25,000
Total manufacturing overhead costs:	$12,000,000	

TABLE 17-4 Activity-Based Costing Data for JPL Example

Activity	Cost driver	Costs	Budgeted rate
Production:	No. of machine hours	$8,000,000	200,000 hours
Engineering:	No. of engineering change orders	1,000,000	40,000 orders
Material handling:	No. of material moves	1,000,000	60,000 moves
Receiving:	No. of batches	800,000	500 batches
Quality assurance:	No. of inspections	800,000	20,000 inspections
Packing and shipping:	No. of products	400,000	100,000 products

TABLE 17-5 Cost Driver Rates for Models A and B

Activity	Model A	Model B
Production:	50,000 machine-hours	150,000 machine-hours
Engineering:	15,000 orders	25,000 orders
Material handling:	20,000 moves	40,000 moves
Receiving:	150 batches	350 batches
Quality assurance:	6,000 inspections	14,000 inspections
Pack and ship:	50,000 products	50,000 products

Besides establishing the data in Table 17-3, the ABC system also measured the levels of each activity (or cost driver rates) required by the two products. These data are given in Table 17-5.

For post-audit purposes, the company ran a cost comparison between the two costing systems. To arrive at product costs for models A and B, the following two sections show the computations involved in the traditional (absorption-based) costing system and ABC system, respectively.

17.3.2.1 Cost Comparison Using the Traditional Costing Method

1. Obtain the overall manufacturing overhead cost rate:

cost rate = (total overhead costs) ÷ (total direct labor hours)

= $12,000,000/60,000 hours

= $200 per direct labor hour

2. Compute total cost per unit for each model:

Activity	Model A	Model B
Direct material cost:	$ 2,800,600	$1,500,000
Direct labor cost:	$ 350,000	$ 250,000
Allocated overhead:	(35,000 hr × $200)	(25,000 hr × $200)
	= $ 7,000,000	= $5,000,000
Total costs:	$10,150,600	$6,750,600
Cost per unit (based on 50,000 units):	$203	$135

17.3.2.2 Cost Comparison Using the Activity-Based Costing Method

1. Establish applied activity rate of the six activities:

applied activity rate = (activity cost in Table 17-4)/
(budgeted cost driver rate in Table 17-4)

2. Determine the activity cost to be traced to each pager model:

Cost traced to each pager model = Cost driver rate (from Table 17-5)
× applied activity rate (from step 1)

3. Compute total cost per unit for each model:

	Applied activity rate	Model A	Model B
Production:	$40/mach. hr.[*]	$2,000,000[†]	$6,000,000
Engineering:	$25/ECO	375,000	625,000
Material handling:	$16.67/move	333,400	666,600
Receiving:	$1,600/batch	240,000	560,000
Quality assurance:	$40/inspection	240,000	560,000
Pack and ship:	$4/product	200,000	200,000
		$3,388,400	$8,611,600
Total overhead costs:		$12,000,000	
Direct material cost:		$2,800,600	$ 1,500,000
Direct labor cost:		350,000	250,000
Total costs:		6,539,000	10,361,600
Cost per unit:		$130.78	$207.23

[*]$40/machine-hour = $8,000,000/200,000 machine hours
[†]$2,000,000 = $40/machine-hour × 50,000 machine-hours

17.3.2.3 Summary of Comparison.

The two costing systems produced different cost estimates as summarized in Table 17-6.

Using absorption-based (traditional) costing, model A costs more to make than model B. However, based on activity-based costing, model A is less expensive to make than model B. Traditional costing bases its cost estimates on the assumption that model A is responsible for more overhead costs than model B, using direct labor hours as an allocation base. However, because of differences in the number of *transactions* that affect indirect (overhead) costs, the reality is that model B actually incurs more overhead cost than model A, as demonstrated through the activity-based analysis. Consequently, this demonstrates how the traditional costing method generates distorted

TABLE 17-6 Product Cost Comparison
(Traditional Versus ABC)

	Model A	Model B
Traditional costing system	$203.00	$135.00
Activity-based costing system	130.78	207.23

product costs. Companies making decisions based on distorted costs may unknowingly price some of their products out of the market while selling others at a loss. Likewise, firms may make key decisions, based on competitors' market prices, that do not consider the true costs of manufacturing their new or revised products.

17.3.3 Process Focus

Activity-based costing systems focus on activities and the cost of those activities, rather than on products. This focus identifies real opportunities for process improvement and cost reduction. The ABC system, then, functions as a data provider to management decisions by providing feedback on resources consumed per operating period. This functionality is beyond the scope of traditional costing systems, which originally were designed primarily for inventory valuation for financial and tax statements. By focusing on activity costs, a manager can accurately track and evaluate the activities that together constitute the total product cost. Managers can directly see the effects of order entry, equipment setup, maintenance and engineering support, inspection, and many other costs. Traditional costing systems do not provide enough detail to track costs at that level, and they often make no provision for reporting those costs in a timely manner. As indicated in Table 17-7, ABC is useful in product-line management and enterprise-wide decisions.

In an ABC system costs are created by, and assigned to, one of two sources that create costs: products or customers. Either a *product* activity is causing the cost (such as material procurement) or a *customer* is causing the cost (such as the number of distribution points required to service a customer). ABC accounts for all costs and traces them back to one of these two sources. Traditional costing systems don't always account for the total cost of operations. Many times they ignore "below-the-line" expenses like R&D, distribution, customer service, and administration and tend to group these costs into a single expense pool. Many companies treat these expenses as fixed costs and make no attempt to assign them to customers, distribution channels, markets, or even products.[6]

TABLE 17-7   Strategic Decisions Supported by ABC Systems

Decision type	Examples
Product-line management	Product pricing
	Make-versus-buy decisions
	Product mix
	Facility expansion
	Employee reduction
	Off-shore expansion
Enterprise-wide	Advanced technology investment
	TQM implementation
	Life-cycle cost management
	JIT implementation
	Capital investment management

[6] M. C. O'Guin, *The Complete Guide to Activity Based Costing* (Englewood Cliffs, NJ: Prentice-Hall, 1991).

ABC allows a firm to distinguish costs that are customer-driven and to explicitly include them as part of the cost of serving specific customers. Because of this, a company may find that some customers are too expensive to service. ABC data provide a basis either to discontinue serving that customer or to negotiate changes in the producer–customer relationship. In such cases, costs may be shared or possibly transferred from the producer to the customer. Traditional costing systems do not provide a level of detail and assignment to be able to identify the costs of such situations.

As an example of the potential effect of the cost in servicing various types of customers, consider the case of the child's boardgame called CootieRays, manufactured by the FernWood Corporation. FernWood Corp. originally marketed CootieRays through several distribution channels, including direct customer orders, small specialty toy stores, distributor's mail catalogs focused on new parents, large franchise toy stores, and small and large discount stores. At first FernWood Corp. was delighted to be marketing its product through so many channels and grouped the cost of distributing CootieRays through all channels into a single cost category. However, with the hiring of a new Chief Financial Officer at FernWood Corp., each product line was asked to detail the cost of servicing its individual customers. After breaking out the costs of marketing CootieRays to such different types of customers, FernWood Corp. changed its focus and selected the marketing channels that made the most sense from a cost and marketing standpoint. Until they looked at the cost to service each type of customer, they had no idea what those accounts were costing them to maintain. As a result, FernWood Corp. decided to discontinue servicing several types of customers because of the cost burden this had on the total product profitability. Several cost items that FernWood Corp. considered when evaluating the costs of maintaining each customer included

- the number of inventory pick orders and retrieval costs from storage needed,
- the number of customer orders, invoices, and accounts receivable items created,
- the packaging costs required for customer pick order sizes,
- the cost of shipping to customers (carrier required, distance, and frequency),
- the number of ship points required to service the customer,
- the time of inventory in facility staging/storage areas for each customer,
- the cost of promotion for each customer,
- the cost of required customer support for returns and problem resolution.

17.4 ABC as a Tool for the 1990s and Beyond

Activity-based costing may be the strategic system most frequently implemented by companies in the immediate future. The framework of the ABC system allows for an accurate accounting of contemporary manufacturing and overhead costs. This supports enhanced product costing and operations management control. Johnson[7] describes activity-costing and its effect on enterprise competitiveness as follows:

[7] H. R. Johnson, "Managing Costs: An Outmoded Philosophy," *Manufacturing Engineering* (May 1989).

Managing with information from financial accounting systems impedes business performance today because traditional accounting data do not track sources of competitiveness and profitability in the global economy. Cost information, per se, does not track sources of competitive advantage such as quality, flexibility and dependability; furthermore, traditional product cost information, based on plantwide indirect allocations, is increasingly inaccurate and unreliable in today's technological environment.

To be profitable in the long run, manufacturers must stop doing two things that traditional management accounting encourages them to do. First, they must stop managing the costs of activities that will disappear if they do what it takes to be competitive in the global market. Second, they must stop basing business decisions on financial product cost information.

Manufacturers are grappling with many issues as they plan to compete over the next decade. Some of these are complex issues that strike at the very core of business strategy and include items such as investment analysis, the function and responsibility of engineering design, and the cost of quality. Let's consider each of these issues in the context of ABC and illustrate how an ABC system provides "value-added" information and functionality over the traditional cost accounting systems.

17.4.1 Capital Budgeting/Economic Justification Using ABC Data

Capital budgeting can be defined as decisions, most often made at the upper level of management, that map available funds to investment alternatives. As discussed in Chapter 9, it involves issues such as setting an appropriate MARR, identifying and evaluating sources of funding, allowing for dependencies between projects, evaluating riskiness of alternatives, and analyzing issues surrounding budget periods, communication, and the timing of projects. However, much of what "engineering" practitioners consider capital budgeting is fundamental discounted cash flow analysis—or engineering economy analyses—discussed in Chapters 1 through 9.

Estimating economic consequences is an integral part of the capital budgeting process. Cash flow costs and benefits, interest rates, investment horizons, income tax and inflation effects, and many other items are necessary in the alternative evaluation phase of "engineering" capital budgeting. Because estimated effects are so important to the capital budgeting process, an accurate accounting of true costs becomes important in providing a realistic and meaningful analysis. Effective cost management, then, is vitally important to competitive enterprises in the context of the capital budgeting process. Activity-based costing systems identify and track the "right" cost items, as well as the levels of those items, that are necessary in capital budgeting analyses.

17.4.2 An Example

ABC, because of its emphasis on accurate accounting and detailing of explicit items that affect cost, is an excellent source of accurate data in the capital budgeting process. As an example, consider the case of an evaluation for replacing a manual

TABLE 17-8 Traditional Engineering
Economy Analysis Data; Table of
Incremental Cash Flows

	Flexible assembly cell
Investment cost:	$250,000
Annual savings from labor wages and benefits:	$100,000
Annual Maintenance costs:	$ 20,000
Useful life:	8 yr
Market value:	$ 50,000
MARR% (before-tax)	20%

assembly line with an automated flexible assembly cell. Table 17-8 gives the details of the "apparent" incremental costs to be included in an engineering economy analysis of going from the manual to the flexible alternative.

A traditional before-tax analysis would result in the measures of merit given in Table 17-9. From that table it seems obvious that investment in the automated flexible assembly cell is justified regardless of the measure of merit used for decision-making purposes.

However, by considering a more complete set of costs, as can be detailed through activity-based costing, the analysis may result in a different recommendation. Consider the additional, indirect expenses not assigned to the flexible assembly cell given in Table 17-10. When these costs are included in the analysis, a reversal in the previous recommendation could result. There are nonmonetary effects that are often difficult to account for in dollars and include in such analyses, and multiattribute (multicriteria) decision models are sometimes used systematically to incorporate such effects (see Canada and Sullivan[8] for examples involving advanced manufacturing technology evaluation). However, an ABC system permits the "right" data to be developed for use in engineering economy studies.

TABLE 17-9 Before-Tax Economic
Evaluation Results; Measures of
Merit for Incremental Cash Flows

Measure of merit	Flexible assembly cell
Simple payback period	3.1 yr
Net present value	$68,590
Internal rate of return (IRR)	28.1%
External rate of return (ERR) (ε = MARR)	22.9%

[8] J. R. Canada and W. G. Sullivan, *Economic and Multi-Attribute Evaluation of Advanced Manufacturing Systems* (Englewood Cliffs, NJ: Prentice-Hall, 1989).

TABLE 17-10   Indirect Expenses Not Always Included in Engineering
Economy Analyses; Flexible Assembly Cell Example

List of some nontraditional costs

•Added administrative cost: due to tracking working capital and other expenses and particulars
•Process engineering and technician support costs: due to creating and maintaining manufacturing
shop aids, and programmable machine instructions
•Tooling and fixturing expense and maintenance costs: due to new tools/fixtures required by
automated machinery, storage of these items, and ongoing maintenance
•Training costs: due to added cost of preparing employees for operation of machinery, and instruction
in operation and safety procedures
•Added utility costs: due to increase in power needs to drive equipment; includes electric, hydraulic,
and pneumatic power sources
•Added materials handling costs: due to specialized containerization often required with advanced
automation
•Floor space opportunity costs: due to space requirements with flexible automation equipment for
operation and footprint clearances
•Indirect batch costs: many flexible cells lead to smaller lot sizes that tend to promote or cause an
increase in costs by increasing purchase orders, material movements, setups, routing sheets, material
pick orders, and work orders required

17.4.3 JPL Company Example Revisited[9]

As another example of the use of ABC data in engineering economy analyses,
refer to the earlier example involving the JPL Company. Suppose that the annual production is increased to a volume of 75,000 units for each model. Assume further that
a massive cost cutting and process improvement program has trimmed engineering
costs to $100,000 per year and reduced the number of engineering change orders
(ECOs) to 400 per year—150 ECOs for model A and 250 for model B.

Given this new information, you are asked to develop a basis for estimating the
justifiable investment in material handling equipment to reduce the number of material moves each year by 45,000 (i.e., from $1,000,000 to $250,000). Additional
annual maintenance costs for the new equipment are expected to total $100,000.
Assume straight-line depreciation over the equipment's useful life of five years and no
terminal market or book value. The after-tax MARR is 15%, and the effective (federal, state, and local combined) income tax rate is 40%. A cash flow analysis of the
proposed material handling investment follows:

EOY	BTCF	Depreciation	Taxable income	40% Income taxes	ATCF
0	$-\$P$	—	—	—	$-\$P$
1	650,000	$\$P/5$	$\$650,000 - \$P/5$	$-\$260,000 + 0.08P$	$\$390,000 + 0.08P$
2	650,000	$P/5$	$650,000 - P/5$	$- 260,000 + 0.08P$	$390,000 + 0.08P$
3	650,000	$P/5$	$650,000 - P/5$	$- 260,000 + 0.08P$	$390,000 + 0.08P$
4	650,000	$P/5$	$650,000 - P/5$	$- 260,000 + 0.08P$	$390,000 + 0.08P$
5	650,000	$P/5$	$650,000 - P/5$	$- 260,000 + 0.08P$	$390,000 + 0.08P$

[9] Various methods of cost estimating are discussed in Chapter 5. In this section ABC data are used
to estimate a justifiable investment amount.

Assumptions

MARR = 15% after tax

Effective income tax rate = 40%

Straight-line depreciation over five years

No market value and book value in year 5

Project Basis: Eliminate the bulk of material handling costs ($750,000 per annum) by investing in automated material handling equipment. Maintenance costs of $100,000 per annum are offset against the savings in material handling to give projected net savings of $650,000 annually.

It was found that a maximum investment of $1,649,074 could be justified in these circumstances, given the assumptions made.[10]

In these calculations, no value was attributed to possible improvements in product yield, which may be possible by the elimination of damage during handling.

17.4.4 Product Design Decisions Using ABC Data

In today's manufacturing firms there is an increased recognition of the effect that product design has on many downstream operations. Decisions made at design-time affect nearly every aspect of producing a product, delivering it to the customer, and servicing it over its intended life cycle. Much of the focus on design activities has been as part of the current total quality management (TQM) movement in the United States. Programs with acronyms like DF*x*, QFD, and DOE (Table 17-11) have all focused attention on the *process* of product design.

Activity-based costing data can be very important to the product designer, and some of the first ABC systems (like those at Tektronix and Hewlett-Packard) were designed specifically for this purpose. Activity-based systems identify activities that create costs—both direct and indirect—so that product designers can evaluate the effect that decisions like component selection, tolerances, surface finish requirements, and material selection have. Such decisions ultimately affect items like the number of parts required, the machine routing required, and the testing and inspection required. As an example, tighter tolerances mean more inspection resources—inspectors, time on the line, equipment requirements, training—which translates to more costs.

TABLE 17-11 Quality Programs That Focus Attention at the Design Stage

DF*x*	Design For *x*, where *x* can be any aspect of the product life cycle (e.g., manufacturability, testability, disposability, reworkability, or serviceability).
QFD	Quality Function Deployment, uses a house-of-quality matrix that aids in mapping customer requirements into product design features.
DOE	Design of Experiments, part of the philosophies developed by Genichi Taguchi,[11] which espouse a focus on design variables and a reduction in variation in those variables through the use of design-of-experiment testing during the design phase of product development.

[10] $P = (\$390,000 + 0.08P)(P/A, 15\%, 5)$

$P = \$1,206,792 + 0.2682P$

or $P = \$1,649,074$

[11] G. Taguchi, et al. *Quality Engineering in Production Systems* (New York: McGraw-Hill, 1989).

Many company-wide quality efforts focus on, and recognize the importance of, cost data on product designs. Designers who use well-structured ABC data are fully enlightened on *all* of the effects of their design decisions. Often, however, companies are led into inappropriate action by thinking, "Well, we have cost data—let's incorporate its use during product design." By managing costs using traditional volume-based indicators, companies are encouraged, in their product designs, to seek inappropriate design objectives such as eliminating (reducing) direct labor hours, machine hours, or material costs. Each of these design objectives ultimately increases total product costs by adding to the indirect and overhead cost items.

Another important contemporary design issue involves the temporal aspect of the design interval. Timeliness of product introduction to the marketplace is highly important in today's highly competitive marketplace, and many times it ultimately affects a product's success or failure. As such, design lead times need to be shortened to permit a firm to react to product and market changes. Today's design environment calls for a drive toward short cycle-times and rapid introduction of products to the marketplace. Designs with few part types require less manufacturing time and money. Fewer part types translate to fewer engineering drawings, specification sheets, tooling sheets, routing sheets, engineering shop aids, and other types of documentation. Also, fewer parts require less incoming inspection resources, storage space, manufacturing machine types, test sets, and mean reduced setup and changeover expenses at the machine level. All of these items translate into lowered indirect expenses.

Many companies are focusing on getting designs to market quicker. Hewlett-Packard's "Product Generation Team" is one example of a concerted corporate-wide effort to bring cost-effective and streamlined products to market in the shortest interval possible. Use of ABC data encourages a focus on minimizing costs and increasing efficiencies in product designs. Two primary tenets of design-for-manufacturing (DFM) guidelines are to minimize the number of components in product design and to maximize the modularity of product designs. Use of ABC data facilitates these, and many other, salutary effects on product designs.

17.5 Current Issues in the Use of ABC Systems

Activity-based costing is a philosophy of cost accounting that is not really new. Cooper[12] points out that economist John M. Clark questioned the unit-based allocation of manufacturing costs to production units as early as 1923. However, the precepts of ABC were ignored until very recently, when several important effects have increased interest in the concept. First, the direct-labor portion of products' costs has decreased to a relatively low percentage for the first time in history. Second, intense worldwide competition has forced firms, in many sectors, to make an order-of-magnitude increases of effectiveness and efficiency in their operations. Last, current technology in micro- and personal computers and information technology has greatly increased the collection, inspection, and use of data in manufacturing operations possible on large-scale and real-time bases.

---

[12] R. Cooper, "Cost Classification in Unit-Based and Activity-Based Manufacturing Cost Systems," *Journal of Cost Management* (Fall 1990).

These effects have combined to re-awaken U. S. companies to the benefits of ABC-type data. During the past ten years we have witnessed a proliferation of journal articles, commentaries, textbooks, case studies, and management consultants discussing the virtues of activity-based costing. Discussed below are a few of the current issues in the use of activity-based costing data and in activity-based costing system implementations.

17.5.1 ABC System Design Issues

The implementation of an activity-based costing system is not a trivial matter, and several key issues that must be addressed in the system design phase greatly affect any implementation. Primary among these are the intended uses, or objectives, of the system. This affects the degree of definition of cost pools, activity centers, and cost drivers. If the system's purpose is for product costing, less detail may be required compared to systems designed to assess specific problems or manage certain costs. Systems whose use is toward enterprise-wide continuous improvement systems' initiatives require even more detail. Kaplan[13] suggests that cost systems can be built to address three different problems: (1) inventory valuation, (2) operational control, and (3) individual product cost measurement. Brimson, on the other hand, lists four such potential objectives: (1) product costing, (2) managing cash and inventory, (3) cost control, and (4) decision support.[14]

So, it is imperative that the firm recognize the intended use of *its* system and design it with a clear recognition of objectives up front. Detail in definition is directly affected by the complexity of the processes within the company. Large and complex processes create a necessity for more detail in cost drivers and other structures than do simpler processes.

Researchers are currently addressing many issues related to optimal design characteristics of ABC systems for firms with different characteristics and in different industrial sectors. Characteristics like the numbers and levels of cost pools, activity centers, first- and second-stage drivers, and hierarchies are currently being addressed. Recent research examples include Beaujon and Singhal,[15] who are focused on the effect that alternate ABC system design choices have on activity decisions. Other researchers are focused on managing the degree of cost error in the system as a trade-off variable with system functional use. Future research will look into such issues as the potential uses of ABC in government, nonprofit, service, and military sectors. Although activity-based costing is not a new concept, there are many interesting research issues that are only now beginning to be addressed.

17.5.2 Hesitancy to Invest in ABC

All new investments in manufacturing and business firms require an impetus for such action. Investments in new insurance carriers, new automated production

---

[13] R. S. Kaplan, "One Cost System Isn't Enough," *Harvard Business Review* (Jan.–Feb. 1988).

[14] J. A. Brimson, *Activity Accounting: An Activity-Based Costing Approach* (New York: Wiley & Sons, Inc., 1991).

[15] J. B. Beaujon and R. S. Singhal, "Understanding the Activity Costs in an Activity-Based Cost System," *Journal of Cost Management* (Spring 1990).

equipment, new office-wide word processing software, or a new cost accounting system are examples of management decisions that affect the way business is done. Often, however, management is reluctant to commit to such investments. U. S. managers are notorious for looking for short-term solutions to problems—be they simple or complex.

An individual championing an ABC implementation may face much resistance. One common obstacle is the, "We've got a costing system now and it isn't broke, so why do you want to fix it?" attitude. As with most systemwide information systems, it will require sound data and detailed plans to convince decision makers to invest. As with other "change the way we do business" investments, an ABC implementation may require a "proving" phase as a pilot project.

In the future, more and more firms will be moving toward incorporating the power of activity-based costing as part of their business systems. In considering ABC implementations, companies should consider the wisdom of two quotes, the first attributed to Henry Ford and the second anonymous: "If you need it, and don't invest in it, then you'll pay for it without getting it," and "If you do what you've always done, then you'll get what you've always gotten." ABC strikes at the heart of many competitive issues currently facing U. S. firms, which would do well to consider the effects that investing, *or not investing,* will have on them in the future.

17.5.3 ABC System Integration Issues

Most ABC research has focused on software implementations at the PC- or network-based levels. Little has been accomplished in integrating ABC principles with other enterprise-wide integrated information systems. Formal inclusion in enterprise-level software architectures is perhaps the next frontier for ABC. This will occur, however, only after ABC gains acceptance as a system that provides valuable systemwide information. Until that time, it is envisioned that ABC systems will remain as stand-alone (i.e., unregulated) systems.

In ABC implementations, isolated systems suffer from problems of effort duplication, data errors, and update inconsistencies with other business software systems. Duplication costs, system validity, and the reaction of some employees to put little faith in data not coming from "the" company-wide business system are other problems that may affect these implementations.

17.6 PC-Based and Activity-Based Cost Accounting Software

Borden[16] recently published an excellent overview and analysis of nine of the most popular PC- or network-based activity-based accounting software programs. His analysis did not include mainframe-based software or generic-type software used to build activity costing programs. A list of the programs Borden evaluated is given in Table 17-12.

Borden's article gives a detailed description of each of the nine software products in Table 17-12 and compares each across twenty attributes, including network capabilities, number of installations, base price, and software language. Borden did

---

[16] J. P. Borden, "Activity-Based Management Software," *Journal of Cost Management* 7, no. 4 (Winter 1994).

TABLE 17-12   PC/Network-Based
ABC Software Borden Evaluated

Activity-based software	Provided by
Net Prophet	Sapling Software
EASYABC	ABC Technologies
CMS-PC	ICMS
QUOTE-A-PROFIT	Manufacturing Management Systems
CASSO	Automation Consulting
TR/ACM	Deloitte & Touche
ACTIVA	Price Waterhouse
The Profit Manager Series	KPMG Peat Marwick
ABCost Management	Coopers & Lybrand

not give a single recommendation, only a prudent statement that software selection and implementation should weigh the factors of cost, software feature set, product and user support, ease of user interface, and vendor upgrade potential.

17.7 Summary

In the late 1980s organizations, mainly manufacturing companies, heard about a new method of costing called activity-based costing (ABC). The aim of ABC is *complete traceability* of all factors that affect cost and performance. Armed with this information, an organization can more effectively make changes. Management wants more accurate information faster so it can respond more quickly to market opportunities. The difference between the product cost reported under traditional systems and ABC systems is dramatic. The ABC system reveals that a high percentage of the company's products generate losses in the long run. In general, many products identified as profitable by the traditional product costing system are found to be unprofitable by the ABC system. In many cases an organization's disappointing performance is a direct consequence of distorted cost information that contributes to management decisions to build and market unprofitable products. Now with ABC systems organizations can attain objectives such as cycle-time reduction, quality improvement, and profit maximization more easily through integration and management of value-adding activities.

PROBLEMS

17-1. Do you think that ABC systems will be implemented on a wide-scale basis by manufacturers in the United States? Why or why not?

17-2. What do you see as the major obstacles that face a company that has decided to "go for it" and implement ABC on an enterprise-wide basis?

17-3. What are the characteristics of the firm that you think has the most to gain from an ABC implementation? In what environments will ABC have the most impact?

17-4. Do you think that ABC can be a panacea, or do you see it as just another management fad that eventually will blow away and be forgotten?

17-5. In a small group (or by yourself), brainstorm and make a list of all of the monetary and nonmonetary attributes that could (should) be included in an economic analysis justification of implementing an ABC system.

17-6. Categorize each of the costs given below as either direct or indirect. Assume a traditional costing system is in place when you answer.

machine run costs	machine operator wages
insurance costs	machine depreciation
utility costs	cost of product sales force
material handling costs	support (administrative) staff salaries
engineering drawings support	cost of materials
cost to market the product	machine labor overtime expenses
cost of storage	cost of tooling and fixtures

17-7. Parent's Helpers manufactures two types of baby carriages in the upper end of the baby stroller market, the Smoothie and the LowRider. The Smoothie is a high-volume product with annual sales of 50,000 nationwide, and the LowRider realizes 10,000 annual sales in the same market. Each buggy requires 2.50 hours of direct labor for product completion. Costs for direct material for one unit of each design are given as

	Smoothie	LowRider
Direct material	$30.00	$45.00

Parent's Helpers has a total manufacturing overhead cost of $3 million per year, and the direct labor wages are $12.00 per hour for manufacture and assembly of the two products. The LowRider has a special design that requires more precision manufacturing processes; higher inspection tolerances; more engineering and drafting support; and more complex setups, tooling, and fixturing than the Smoothie. Also, it is necessary to manufacture the LowRider in small lots, which requires a larger number of production orders compared to the Smoothie.

a. Calculate the total unit manufacturing cost of each of the two buggy designs if one uses *direct labor hours* as a basis for assigning overhead costs. Use the relationships below to calculate the total unit cost to manufacture each design.

$$\text{Total unit manufacturing cost} = \text{unit direct material cost (UDMC)} +$$
$$\text{unit direct labor cost (UDC)} +$$
$$\text{unit overhead cost (UOC)},$$

where UOC = (total overhead cost)/(total direct hours per unit)

and

all annual direct labor hours = annual direct labor hours$_{\text{Smoothie}}$
+ annual direct labor hours$_{\text{LowRider}}$

b. Ms. Buggers, the CEO, just returned from an ABC seminar and directed you, her cost accounting engineer, to calculate the unit manufacturing cost for the two products based on activities as a basis for distributing overhead costs. Use the data collected and organized in the tables below to recalculate the unit manufacturing cost for the Smoothie and LowRider buggies from an ABC perspective.

c. Calculate the difference in the total unit manufacturing cost using the two methods of allocating overhead costs given in parts (a) and (b). Is the effect pronounced? Give an example of how this discrepancy may affect management decision making if selling price is set at 130% of total unit manufacturing cost.

Given: Annual Data Table

Activity center	Cost driver	Total traceable overhead cost (a)	Smoothie Annual events or transactions (b)	LowRider Annual events or transactions (c)
A. Machine setups	# of setups (msu)	$ 800,000	8,000	17,000
B. Machine-related	# of machine hours (mh)	1,450,000	80,000	20,000
C. Production orders	# of orders (ord)	100,000	1,500	2,500
D. Quality inspections	# of inspections (insp)	75,000	600	1,800
E. Engineering support	# of engineering hours (eh)	575,000	20,000	8,750
		TOTAL = $3,000,000		

Overhead Cost Calculation Table

Activity center	Traceable costs [(a) prev. page]	Total factory events or transactions (d) = (b) + (c)	Overhead rate per activity (e) = (a)/(d)	Smoothie		LowRider	
				No. of events or transactions [(b) prev. page]	Total cost of activity assigned (f) = (e) × (b)	No. of events or transactions [(c) prev. page]	Total cost of activity assigned (g) = (e) × (c)
A							
B							
C							
D							
E							

Total overhead cost assigned to product (x) = _____

Annual production of product (y) = _____

Unit overhead cost per product (x/y) = _____

18
Dealing with Inflation in Capital Investment Analysis

18.1 Introduction

Except for parts of Chapter 5 ("Introduction to Cost Estimating"), we have assumed that prices for goods and services in the marketplace are relatively unchanged over substantial periods of time, or that the effect of such changes is constant (or cancels out) among all cash flows for the alternatives under consideration. Unfortunately, these are not generally realistic assumptions. *Inflation,* which is the phenomenon of rising prices bringing about a reduction in the purchasing power of a given unit of money, is a fact of life and can significantly affect the economic comparison of alternatives.

Until the mid-1950s, the U. S. dollar was widely accepted as a fixed measure of the worth of resources. Because the purchasing power of a given sum of money is not constant when inflation is present, individuals and companies alike realize that investment opportunities must be evaluated with money treated as a variable measure of the worth of a resource.

Annual rates of inflation (often referred to as *escalation*) vary widely for different types of goods and services and over different periods of time. For example, the U. S. Government-prepared Consumer Price Index rose less than 2% per year during the 1950s, but increased to approximately 7% per year during the 1970s. It is expected that inflation averaging 3% to 4% per year will continue to be a concern in our economy throughout the 1990s. Although it appears that such inflation will extend into the long-term future, it is possible that its opposite, *deflation,* can occur as was true during the depression of the 1930s.

If all cash flows in an economic comparison of alternatives are inflating at the same rate, inflation can be disregarded in before-tax studies. In cases where all incomes and all expenses are not inflating at the same rate, inflation gives rise to differences in economic attractiveness among alternatives and should be taken into account. When the effects of inflation are not included in an engineering economy study, an erroneous choice among competing alternatives can result. That is, reversals in preference may occur by assuming that inflation affects all investment opportunities to the same extent. Consequently, the objective of maximizing shareholders' (owners') wealth is inadvertently compromised. To avoid this difficulty, this chapter addresses fundamental concepts and terminology when dealing specifically with inflation in capital investment decisions.

18.2 Actual Dollars Versus Real Dollars

Inflation describes the situation in which prices of fixed amounts of goods and services are increasing. As prices rise, the value of money, that is, its purchasing power (in real dollars, as defined below), decreases correspondingly.

Let us define two distinct kinds of dollars (or other monetary units such as pesos or francs) with which we can work in economic analyses, if done properly:

1. *Actual dollars:* the actual number of dollars as of the point in time they occur and the usual kind of dollar terms in which people think. Sometimes called *then-current dollars,* or *current dollars,* or even *inflated dollars,* they will be denoted as "A$" whenever a distinction needs to be made in this book.

2. *Real dollars:* dollars of purchasing power as of some base point in time, regardless of the point in time the actual dollars occur. Sometimes called *constant worth dollars,* or *constant dollars,* or even *uninflated dollars,* they will be denoted as "R$" whenever a distinction needs to be made in this book. If the base point in time, k, needs to be specified (it is usually the time of the study or the initial investment), it can be shown with a superscript [i.e., $R\$^{(k)}$].

Actual dollars at any time, n, can be converted into real dollars at time n, of purchasing power as of any base time k, by

$$R\$_n^{(k)} = A\$_n \left(\frac{1}{1+f}\right)^{n-k} = A\$_n (P/F, f\%, n-k). \qquad (18\text{-}1)$$

Similarly,

$$A\$_n = R\$^{(k)}(1+f)^{n-k} = R\$_n^{(k)}(F/P, f\%, n-k), \qquad (18\text{-}2)$$

where f is the average *inflation rate* per period over the $n - k$ periods.

18.3 Real Interest Rate, Combined Interest Rate, and Inflation Rate

We now define several types of rates and show how they are used:

1. *Real interest rate:* Increase in real purchasing power expressed as a percent per period, or the interest rate at which R$ outflow is equivalent to R$ inflow. It is sometimes known as *real monetary rate* or *uninflated rate* and denoted as i_r when it needs to be distinguished from i_c (below).

2. *Combined interest rate:* Increase in dollar amount to cover real interest and inflation expressed as a percent per period; is the interest rate at which A$ outflow is equivalent to A$ inflow. It is sometimes known as *actual rate* or *inflated rate* and is denoted as i_c whenever it needs to be distinguished from i_r (above).

3. *Inflation rate:* As defined previously, the increase in price of given goods or services as a percent per time period. It is denoted as f. The overall rate for an individual or organization is sometimes called the *general inflation rate.*

Because the real interest rate and the inflation rate have a multiplicative or compounding effect,

$$i_c = (1 + i_r)(1 + f) - 1, \tag{18-3}$$

$$i_c = i_r + f + (i_r \times f). \tag{18-4}$$

Also,

$$i_r = \frac{i_c - f}{1 + f}. \tag{18-5}$$

Where f is not large relative to the accuracy desired, then

$$i_c \cong i_r + f \tag{18-6}$$

and

$$i_r \cong i_c - f. \tag{18-7}$$

18.4 What Interest Rate to Use in Economy Studies

In general, the interest rate that is appropriate for time-value calculations in engineering economy studies depends on the type of cash flow estimates as follows:

Method	If cash flows are estimated in terms of:	Then the interest rate to use is:
A	Actual $, A$	Combined interest rate, i_c
B	Real $, R$	Real interest rate, i_r

The above is made intuitively consistent if one thinks in terms of method A as working with inflated (actual) dollars and interest, and method B as being applicable to uninflated (real) dollars and interest. Method A is the most natural to use because we usually think in terms of A$. Since interest paid or earned is based on A$, it is a combined interest rate, i_c. However, method B is sometimes easier to use.

18.5 Summary of Formulas for Relating Single Sum A$ and R$ over Time

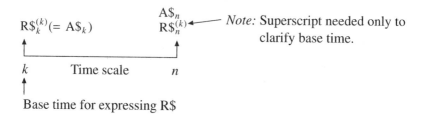

Method	Type of dollars or conversion	Moving forward in time	Moving backward in time
A	A$ (inflated $)	$A\$_n = A\$_k(F/P, i_c, n - k)$	$A\$_k = A\$_n(P/F, i_c, n - k)$
B	R$ (uninflated $)	$R\$_n^{(k)} = R\$_k^{(k)}(F/P, i_r, n - k)$	$R\$_k^{(k)} = R\$_n^{(k)}(P/F, i_r, n - k)$
		Inflating (at given time)	Deflating (at given time)
	From R$ to A$, or from A$ to R$, at a given time	$A\$_n = R\$_n^{(k)}(F/P, f, n - k)$	$R\$_n^{(k)} = A\$_n(P/F, f, n - k)$
	Using index values	$I_n = I_k(F/P, f, n - k)$	$I_k = I_n(P/F, f, n - k)$

Example 18-1

A certain expense at the end of 1997 is estimated to be $10,000. The end of 1997 is the base point for considering inflation. (Thus $10,000 = A\$_{97} = R\$_{97}^{(97)}$.) Find its equivalent worth in 2002 for the following circumstances (paralleling the formulas above), using the following rates:

$$\text{real rate, } i_r = 4\%,$$
$$\text{inflation rate, } f = 6\%.$$

Thus,

$$\text{combined rate } i_c = (1 + 0.04)(1 + 0.06) - 1$$
$$= 0.1024 = 10.24\%$$
$$\cong 10\%.$$

Solution   In 2002 (denoted 02) dollars:

$$A\$_{02} = A\$_{97}(F/P, i_c\%, 2002 - 1997)$$
$$= \$10,000(F/P, 10.24\%, 5) = \$16,282.$$

(*Note:* If 10% approximation were used, the answer would be $16,105.)
In 1997 (denoted 97) dollars:

$$R\$_{02}^{(97)} = R\$_{97}^{(97)}(F/P, i_r\%, 2002 - 1997)$$
$$= \$10,000(F/P, 4\%, 5) = \$12,167.$$

In 2002 dollars, beginning with 2002 equivalent in 1997 dollars:

$$A\$_{02} = R\$_{02}^{(97)}(F/P, f\%, 2002 - 1997)$$
$$= \$12,167(F/P, 6\%, 5) = \$16,282. \quad\blacksquare$$

18.6 Manipulating Series That Are Uniform in R$

If cash flows expressed in R$ are uniform each year (and thus the A$'s increase each year at the average rate of inflation, f), they can be conveniently converted to equivalent worth(s) at other point(s) in time using uniform series formulas at the firm's $i_r =$ real MARR.

Thus, if there is a uniform series in which each end-of-period payment, A, for n periods, is expressed in R$$^{(k)}$ (and thus is denoted $A^{(k)}$), then

$$P_0^{(k)} = (A^{(k)})(P/A, i_r, n) \tag{18-8}$$

and

$$F_n^{(k)} = (A^{(k)})(F/A, i_r, n). \tag{18-9}$$

The superscripts make clear the base point in time at which the dollars (like present or future equivalents) are expressed. Normally $k = 0$, but estimates can be converted to any base point in time at the inflation rate, f, using Eq. 18-1 or 18-2.

An equivalent formula for finding the present worth of a uniform series that is escalating at the inflation rate f for n years is

$$P_0^{(0)} = \frac{A^{(0)}(F/P, f, 1)[1 - (P/F, i_c\%, n)(F/P, f\%, n)]}{i_c - f}. \tag{18-10}$$

Note that $A^{(0)}(F/P, f, 1) = A^{(1)}$. Thus

$$P_0^{(0)} = \frac{A^{(1)}[1 - (P/F, i_c\%, n)(F/P, f\%, n)]}{i_c - f}. \tag{18-11}$$

For the special case in which $i_c = f$, so that the real monetary rate $= 0$, Eqs. 18-10 and 18-11 become

$$P_0^{(0)} = A^{(0)}(n) \tag{18-12}$$

and

$$P_0^{(0)} = A^{(1)}(P/F, f, 1)(n). \tag{18-13}$$

Example 18-2

A person who is earning $21,600 salary for (assumed end of) year 1 expects that salary to inflate (escalate) at 10%/year, which is the same as the general rate of inflation. If her real monetary rate, i_r, is 5%, then $i_c = 0.05 + 0.10 + 0.05(0.10) = 0.155$, or 15.5%. Find the present worth, $P_0^{(0)}$, for three years of salary.

(a) Use the approach of Eq. 18-8 to find $P_0^{(1)}$ and then convert to $P_0^{(0)}$.

(b) Use Eq. 18-11.

Solution

(a) $P_0^{(1)} = A^{(1)}(P/A, i_r, 3)$

$\qquad = \$21,600(P/A, 5\%, 3) = \$58,821$

$\quad P_0^{(0)} = P_0^{(1)}(P/F, f, 1)$

$\qquad = \$58,821(P/F, 10\%, 1) = \underline{\underline{\$53,474}}$

(b) $P_0^{(0)} = \dfrac{\$21,600[1 - (P/F, i_c, 3)(F/P, f, 3)]}{i_c - f}$

$\qquad = \dfrac{\$21,600[1 - (P/F, 15.5\%, 3)(F/P, 10\%, 3)]}{0.155 - 0.10}$

$\qquad = \underline{\underline{\$53,496}}$ (same as for part (a) except for round-off error) ■

18.7 Manipulating Series That Inflate (Escalate) at Rate Different from General Inflation

When a cash flow series expressed in A$ inflates or escalates at a rate different from general inflation, it will not be a uniform series when expressed in R$. The use of what might be called "differential escalation rates" can be a handy computational convenience in such a case. The following example shows more general ways to convert a series of cash flows subject to whatever rate of inflation (which might vary from year to year) into either R$ or A$, and then to manipulate them (in this case, into present worths) at the general inflation rate.

Example 18-3

Given the same individual salary situation as in Example 18-2, the only difference is that the salary will inflate (escalate) at 8%/year, which differs from the general inflation rate of 10%. To repeat, the end-of-first-year salary $A\$_1 = R\$_1^{(1)} = \$21,600$, $i_r = 5\%$, $f = 10\%$, and $i_c = 15.5\%$.

(a) Show her salary for three years expressed in R$ and in A$.

(b) Show the present worth (as of beginning of the first year) of both ways of expressing this salary.

Solution

In A$$_n$$

Year n	(1) $A\$_n = A\$_1(F/P, 8\%, n - 1)$	(2) $(P/F, 15.5\%, n)$	(3) = (1) × (2) $PW_0^{(0)}$
1	$\$21,600(F/P, 8\%, 0) = \$21,600$	0.8658	$\$18,701$
2	$21,600(F/P, 8\%, 1) = 23,328$	0.7495	17,485
3	$21,600(F/P, 8\%, 2) = 25,194$	0.6489	16,348
			$\overline{\$52,534}$

In R$$_n^{(0)}$$

Year n	$R\$_n^{(0)} = A\$_n(P/F, 10\%, n)$	$(P/F, 5\%, n)$	$PW_0^{(0)}$
1	$\$21,600(P/F, 10\%, 1) = \$19,637$	0.9524	$\$18,702$
2	$23,328(P/F, 10\%, 2) = 19,278$	0.9070	17,485
3	$25,194(P/F, 10\%, 3) = 18,928$	0.8638	16,350
			$\overline{\$52,537}$

Note from column (1) that even though her A\$ salary is going up (at 8% per year), the R\$ (purchasing power) of that salary is going down (approximately $10\% - 8\% = 2\%$ per year). The present worths of both ways of expressing the salary are the same, except for minor round-off error. ∎

Example 18-4

Mary Q. Contrary wishes to retire in the year 2020 with personal savings of \$500,000 in 1995 spending power. Assume that the expected inflation rate in the economy will average 3.5% during the next 25 years. Mary plans to invest in a 6.0% per year savings account, and her salary is expected to increase by 6.5% per year between 1995 and 2020. Assume that the first deposit occurs at the end of 1996. If Mary puts aside 10% of her salary for retirement purposes, how much will her total salary have to be in 1995 to make her retirement plan a reality?

This example demonstrates the flexibility of a spreadsheet, even in instances where all of the calculations are based on a piece of information (the 1995 salary) that we do not yet know. If we deal in actual dollars, the cash flow relationships are straightforward. A spreadsheet model for solving this problem is given in Figure 18-1. The formula in cell F10 converts the desired ending balance into actual dollars. The salary is paid at the end of the year, at which point 10% is placed in a bank account. The interest calculation is based on the cumulative deposits and interest in the account at the beginning of the year, but not on the deposit made at the end of the year. The salary is increased and the cycle repeats.

The spreadsheet model in Figure 18-1 allows us to enter the formulas for the geometric gradient representing the salary increase (column C), the 10% of the salary that goes into the savings account (column D), and the bank balance at the end of the year (column E) without knowing the 1995 salary. Some spreadsheet packages have a "solver" feature that will automatically determine the desired 1995 salary. This example illustrates an approach that is not as elegant, but is nonetheless fast and will work for software that does not have a solver feature.

The approach is to revise the 1995 salary systematically and compare the ending bank balance (copied to cell F11 for ease of viewing on the screen) with the desired year 2020 balance. To save keystrokes, the 1995 salary is broken down by powers of 10 into separate cells, in the range B4..B9. The 1995 salary is recombined with the formula in cell B11 and copied to cell C17 for use in the calculations. Starting with the highest power of 10 (cell B4), we bracket the salary that will set cells F10 and F11 equal. The results for this example are shown in Figure 18-2. ∎

18.8 Effect of Inflation on Before-Tax and After-Tax Economic Studies[1]

As illustrated previously, if subsequent benefits from an investment bring constant quantities of A\$ over time, then inflation will diminish the real value (R\$) of the future benefits and, hence, the real rate of return. If, on the other hand, all before-tax costs and benefits are changing at equal rates, then inflation has no net effect on before-tax economic analyses of alternatives. Unfortunately, however, this is not true for after-tax economic analyses.

---

[1] Section 18.8 is adapted from P. E. DeGarmo, W. G. Sullivan, and J. R. Canada, *Engineering Economy*, 7th ed. (New York: Macmillan Publishing Company, 1984). Adapted by permission of the publisher.

R/C	A	B	C	D	E	F	G
1							
2	1995 Salary	Figure Entry		Annual Salary Increase		F2	
3				Savings (% of Salary)		F3	
4	100000	B4		Savings Interest Rate		F4	
5	10000	B5		2020 Amount (R$)		F5	
6	1000	B6		Average Inflation Rate		F6	
7	100	B7					
8	10	B8					
9	1	B9					
10				Desired 2020 Amount (A$)		+F5*(1+F6)^20	
11	1995 Salary	+B4*A4+B5*A5+B6*A6+B7*A7+B8*A8+B9*A9	Balance Year 2020 (A$)		+E41*(1+$F$4)+D42		
12							
13							
14			Salary	Deposit	Bank Balance		
15					at EOY		
16		Year	(A$)	(A$)	(A$)		
17		1995	+B11				
18		1996	+C17*(1+$F$2)	+C18*($F$3)	+E17*(1+$F$4)+D18		
19		1997	+C18*(1+$F$2)	+C19*($F$3)	+E18*(1+$F$4)+D19		
20		1998	+C19*(1+$F$2)	+C20*($F$3)	+E19*(1+$F$4)+D20		
21		1999	+C20*(1+$F$2)	+C21*($F$3)	+E20*(1+$F$4)+D21		
22		2000	+C21*(1+$F$2)	+C22*($F$3)	+E21*(1+$F$4)+D22		
23		2001	+C22*(1+$F$2)	+C23*($F$3)	+E22*(1+$F$4)+D23		
24		2002	+C23*(1+$F$2)	+C24*($F$3)	+E23*(1+$F$4)+D24		
25		2003	+C24*(1+$F$2)	+C25*($F$3)	+E24*(1+$F$4)+D25		
26		2004	+C25*(1+$F$2)	+C26*($F$3)	+E25*(1+$F$4)+D26		
27		2005	+C26*(1+$F$2)	+C27*($F$3)	+E26*(1+$F$4)+D27		
28		2006	+C27*(1+$F$2)	+C28*($F$3)	+E27*(1+$F$4)+D28		
29		2007	+C28*(1+$F$2)	+C29*($F$3)	+E28*(1+$F$4)+D29		
30		2008	+C29*(1+$F$2)	+C30*($F$3)	+E29*(1+$F$4)+D30		
31		2009	+C30*(1+$F$2)	+C31*($F$3)	+E30*(1+$F$4)+D31		
32		2010	+C31*(1+$F$2)	+C32*($F$3)	+E31*(1+$F$4)+D32		
33		2011	+C32*(1+$F$2)	+C33*($F$3)	+E32*(1+$F$4)+D33		
34		2012	+C33*(1+$F$2)	+C34*($F$3)	+E33*(1+$F$4)+D34		
35		2013	+C34*(1+$F$2)	+C35*($F$3)	+E34*(1+$F$4)+D35		
36		2014	+C35*(1+$F$2)	+C36*($F$3)	+E35*(1+$F$4)+D36		
37		2015	+C36*(1+$F$2)	+C37*($F$3)	+E36*(1+$F$4)+D37		
38		2016	+C37*(1+$F$2)	+C38*($F$3)	+E37*(1+$F$4)+D38		
39		2017	+C38*(1+$F$2)	+C39*($F$3)	+E38*(1+$F$4)+D39		
40		2018	+C39*(1+$F$2)	+C40*($F$3)	+E39*(1+$F$4)+D40		
41		2019	+C40*(1+$F$2)	+C41*($F$3)	+E40*(1+$F$4)+D41		
42		2020	+C41*(1+$F$2)	+C42*($F$3)	+E41*(1+$F$4)+D42		
43							

Figure 18-1.   Sample spreadsheet for solving Example 18-4.

In general, given two projects with the same before-tax rate of return, it can be shown that inflation results in a smaller after-tax rate of return than that for a project that does not have benefits that increase with inflation. This is because even though the benefits may increase at the same rate as inflation, the depreciation charges do not increase, which results in larger income tax payments. The net result is that even though the after-tax cash flow in A$ is increased with inflation, that increase is not large

R/C	A	B	C	D	E	F	G
1							
2	1995 Salary	Figure Entry		Annual Salary Increase		6.5%	
3				Savings (% of Salary)		10.0%	
4	100000	0		Savings Interest Rate		6.0%	
5	10000	8		2020 Amount (R$)		$500,000	
6	1000	7		Average Inflation Rate		3.5%	
7	100	1					
8	10	7					
9	1	1					
10				Desired 2020 Amount (A$)		$994,894	
11	**1995 Salary**	**87171**		Balance Year 2020 (A$)		$994,895	
12							
13							
14			Salary	Deposit	Bank Balance		
15					at EOY		
16		Year	(A$)	(A$)	(A$)		
17		1995	$87,171				
18		1996	$92,837	$9,284	$9,284		
19		1997	$98,872	$9,887	$19,728		
20		1998	$105,298	$10,530	$31,441		
21		1999	$112,143	$11,214	$44,542		
22		2000	$119,432	$11,943	$59,158		
23		2001	$127,195	$12,719	$75,427		
24		2002	$135,463	$13,546	$93,499		
25		2003	$144,268	$14,427	$113,535		
26		2004	$153,645	$15,365	$135,712		
27		2005	$163,632	$16,363	$160,218		
28		2006	$174,268	$17,427	$187,258		
29		2007	$185,595	$18,560	$217,053		
30		2008	$197,659	$19,766	$249,842		
31		2009	$210,507	$21,051	$285,883		
32		2010	$224,190	$22,419	$325,455		
33		2011	$238,762	$23,876	$368,859		
34		2012	$254,282	$25,428	$416,418		
35		2013	$270,810	$27,081	$468,484		
36		2014	$288,413	$28,841	$525,435		
37		2015	$307,160	$30,716	$587,677		
38		2016	$327,125	$32,713	$655,650		
39		2017	$348,388	$34,839	$729,828		
40		2018	$371,033	$37,103	$810,721		
41		2019	$395,151	$39,515	$898,879		
42		2020	$420,835	$42,084	$994,895		
43							

Figure 18-2.   Spreadsheet results for Example 18-4.

enough to offset both the increased income taxes and inflation. It can be concluded, then, that inflation reduces the real after-tax PW (and IRR) because of the devaluation of fixed depreciation schedules (or other types of unresponsive A$ annuities such as interest payments). Such unresponsive amounts cause taxable income to increase in an *actual dollar analysis* so that income taxes increase and after-tax cash flows decrease. This observation is confirmed in Example 18-5.

Example 18-5

Your firm has decided it *must acquire* a new piece of machinery that includes the latest safety features required by OSHA. The machinery may either be (1) purchased for cash or (2) leased from another company. Mr. Williams, the president of the firm, has requested that you perform an after-tax analysis of these two means of obtaining the machinery when the following estimates and conditions are applicable.

1. The study period is five years and the estimated useful life of the machinery also is five years. Straight-line depreciation is elected with an estimated zero book value at the end of the useful life of the purchased machinery. The effective incremental income tax rate (t) is 50%. Also, a 10% investment tax credit can be taken. All recaptured depreciation (if any) is taxed at 50% of the gain.

2. The following interest rates and inflation estimates are used:
   (a)   The real after-tax MARR (i_r) is 10% per year.
   (b)   Annual inflation (f) is expected to average 8%.
   (c)   The combined after-tax MARR (i_c) is 18.8% (i.e., $i_c = 0.10 + 0.08 + 0.10(0.08) = 0.188$ or 18.8%).

3. Annual savings, operating costs, and maintenance costs, and the terminal *market value* respond to inflation. In the case of leasing the machinery, the yearly lease payment does *not* grow with inflation. When purchasing the machinery, depreciation write-offs do not respond to inflation.

4. Annual cash flow estimates:

	Machinery	
	Purchase	Lease
Savings:	$5,000	$5,000
Operating costs:	2,000	$2,000
Maintenance cost:	1,000	(included in lease contract)
Lease fee:	—	$6,000
		(payable at end of year)

All of the annual cash flows above have been estimated in real dollars.

5. Investment costs:

Purchase machine	Lease machine
Initial cost = $20,000	Deposit = $1,500
	(refundable at end of 5 years with no interest)
Market value = $1,500 at end of year 5 (in real dollars)	

6. The analysis is to be performed after taxes, and an inflation rate (f) of 8% is estimated to apply to *all* cash flows that respond to inflation.

Based on this information, should your firm purchase or lease the machinery?

Solution   An after-tax cash flow analysis is performed with actual dollar estimates in Table 18-1 for the "purchase machinery" option and in Table 18-2 for the "lease machinery" alternative. Because annual savings, operating costs, and maintenance costs inflate each year at 8%, the real dollar before-tax cash flows are converted to the corresponding actual dollar estimates in column C by using Eq. 18-1. The column C entries are then combined with depreciation write-offs in Table 18-1 and lease payments in Table 18-2 to arrive at the taxable income associated with each alternative. Notice that the firm cannot claim depreciation on the leased machinery.

After-tax cash flow in column G of Table 18-1 is determined in view of an investment tax credit in year 0 and depreciation recapture, which is taxed at 50%, in year 5. The net present worth at $i_c = 18.8\%$ is −\$7,601 for the "purchase machinery" alternative. Similarly, the present worth of column G in Table 18-2 is −\$4,395 for the leasing alternative. A recommendation should be made to lease the machinery so that present worth of cost is minimized.

It should be noted that a combined interest rate is used to discount the after-tax cash flows in Tables 18-1 and 18-2. In this regard, *the after-tax MARRs of most companies are directly stated as combined interest rates.* Furthermore, most companies make economic studies in terms of A\$ estimates because decision makers lean toward a measure of financial profitability that includes the effects of inflation.

Referring to Example 18-5, if an after-tax analysis had been performed *ignoring the effects of inflation,* the recommended course of action would have been to purchase the equipment! Thus, assuming a 0% inflation rate in this particular example would have led to an *incorrect* selection between alternatives.   ■

18.9 Deflation

The previous section concentrated on inflation because that is the dominant condition experienced in the past and expected in the future. However, we should also recognize deflation. Deflation is the opposite of inflation—a decrease in the monetary price of goods and services, which correspondingly means an *increase* in the real value or purchasing power of money. Deflation can be handled exactly comparably to inflation in economic analyses. That is, estimates can be made in terms of either R\$ or A\$, and the corresponding interest rate to use should be either the real rate, i, or the composite rate,

$$i_c = (1 + i_r)(1 - f) - 1 = i_r - f - i_r \times f, \qquad (18\text{-}14)$$

where f is the rate of deflation.

18.10 Summary

It is important to be consistent in using the correct interest rate for the type of analysis (actual or real dollars) being done. Two mistakes commonly made are as follows:

Interest rate (MARR)	Type of analysis	
	A\$	R\$
i_c	(Correct)	Mistake 1 Bias is against capital investment
i_r	Mistake 2 Bias is toward capital investment	(Correct)

Table 18-1 Purchase Equipment Alternative—Actual Dollar Analysis

Year	(A) Before-tax cash flow (R$)	(B) Adjustment $(1+f)^{year}$	(C) Before-tax cash flow (A$)	(D) Depreciation (A$)	(E) Taxable income (C) − (D)	(F) Cash flow for income taxes $-t \times (E)$	(G) After-tax cash flow (A$) (C) + (F)
0	−$20,000	1.000	−$20,000			$2,000$^a^	−$18,000
1	2,000^b	1.080	2,160	$4,000	−$1,840	920	3,080
2	2,000	1.166	2,332	4,000	− 1,668	834	3,166
3	2,000	1.260	2,520	4,000	− 1,480	740	3,260
4	2,000	1.360	2,720	4,000	− 1,280	640	3,360
5	2,000	1.469	2,938	4,000	− 1,060	531	3,469
5	1,500	1.469	2,204		2,204	− 1,102	1,102
						PW(18.8%) = −$ 7,601	

^{a} Investment tax credit = 0.10($20,000).
^{b} $5,000 (annual savings) − $2,000 − $1,000 = $2,000.

458

Table 18-2 Lease Equipment Altenative—Actual Dollar Analysis

Year	(A) Before-tax cash flow (R$)	(B) Adjustment $(1+f)^{year}$	(C) Before-tax cash flow (A$)	(D) Lease payments (A$)	(E) Taxable income (C) − (D)	(F) Cash flow for income taxes $-t \times$ (E)	(G) After-tax cash flow (A$) (C) + (F)
0	−$1,500	1.000	−$1,500				−$1,500
1	3,000[a]	1.080	3,240	$6,000	−$2,760	$1,380	− 1,380
2	3,000	1.166	3,498	6,000	− 2,502	1,251	− 1,251
3	3,000	1.260	3,780	6,000	− 2,220	1,110	− 1,110
4	3,000	1.360	4,080	6,000	− 1,920	960	− 960
5	3,000	1.469	4,407	6,000	− 1,593	796.5	− 796.5
5			1,500[b]				1,500
							PW(18.8%) = −$4,395

[a] $5,000 (annual savings) − $2,000 = $3,000.
[b] The deposit on the leased equipment, refunded with no interest.

In mistake 1 the combined interest rate (i_c), which includes an adjustment for the inflation rate (f), is used in equivalent worth calculations for cash flows estimated in real dollars. Since real dollars have constant purchasing power expressed in terms of the base time period and do not include the effect of general price inflation, we have an inconsistency. There is a tendency to develop future cash flow estimates in terms of dollars with purchasing power at the time of the study and then use the combined interest rate in the analysis. The result of mistake 1 is a bias against capital investment. The cash flow estimates in real dollars for a project are numerically lower in value than actual dollar estimates with equivalent purchasing power (assuming that $f > 0$). Additionally, the i_c value (which is greater than the i_r value that should be used) further reduces (understates) the equivalent worth of the results of a proposed capital investment.

In mistake 2 the cash flow estimates are in actual dollars, which include the effect of inflation (f), but the real interest rate (i_r) is used for equivalent worth calculations. Since the real interest rate does not include an adjustment for general price inflation, we again have an inconsistency. The effects of this mistake, opposite to those in mistake 1, result in a bias toward capital investment by overstating the equivalent worth of the future benefits.

PROBLEMS

18-1. a. Labor cost is currently $20,000 per year (expressed in year 0 dollars) and is expected to inflate at 5%/yr. Find the PW (at time 0) of the series (assumed to occur at end of each year for 15 years) if the real MARR is 10% and the combined MARR is approximately 15%.

   b. The Whackya Bank will pay you 6% for savings account deposits. The average inflation rate is expected to be 10%. What will be your real return rate (use exact formula, not approximation)?

   c. In the year 2010 you expect to donate A$ = $10,000 to your alma mater. You expect inflation to be only 5%, but your real MARR will be 15%. It is now 1996.

   (i) What will your donation be worth in year 2010 expressed in today's purchasing power (i.e., $R\$_{10}^{(96)}$)?

   (ii) What is the equivalent worth today of your future donation?

18-2. If the average inflation rate is expected to be 8% per year into the foreseeable future, how many years will it take for the dollar's purchasing power to be one-half of what it is now? (That is, the future point in time when it takes two dollars to buy what can be purchased today for one dollar.)

18-3. John and Mary Doe have computed that $3,700 deposited annually into an interest-bearing account with $i_c = 15\%$ will grow to $1 million in 40 years $(N=40$ years). If the average annual inflation rate during this 40-year period is 7%, what is the spending power equivalent *in today's dollars* of the $1 million that John and Mary will have accumulated?

18-4. a. It is desired to estimate the 1997 construction cost of a new plant for which 1992 component costs, and applicable rates of inflation, are as follows:

	1992 cost (millions)	Projected annual inflation rate (1992 through 1997) (%)
Labor:	$1	10
Building materials:	5	0
Equipment:	3	15
Total	$9	

b. Using the answer to part (a) and using a (combined) MARR of 25%, find the equivalent worth of the 1997 construction cost as of 1992 (i.e., $R\$_{92}^{(92)}$).

18-5. If you buy a lathe now, it costs $100,000. If you wait 2 years to purchase the lathe, it will cost $135,000. Suppose you decide to purchase the lathe now, reasoning that you can earn 18% per year on your $100,000 if you do not purchase the lathe. If the inflation rate (f) in the economy during the next 2 years is expected to average 12% per year, did you make the right decision?

18-6. Your company has just issued bonds with a face value of $1,000. They mature in 10 years and pay annual interest of $100. At present they are being sold for $887. If the average annual inflation rate over the next 10 years is expected to be 6%, what is the real rate of return per year on this investment?

18-7. The "actual" (then-current) costs for a utility service are expected to be as follows:

End of year	"Actual" costs
1995	$10,000
1996	10,000
1997	12,000

Assume that the average inflation rate is 8%/yr and that the real interest on money is 4%/yr, so the combined interest rate can be approximated as 12%.

a. Find the equivalent worth of the sum of all costs as of the end of 1995.

b. Express each of the three costs in "real dollars" of purchasing power as of the end of 1995.

c. Using the results of (b), find the equivalent worth of the sum of all costs as of the end of 1995. How should this answer compare to the answer to (a)?

18-8. A firm desires to determine the most economic equipment-overhauling schedule alternative to provide for service for the next nine years of operation. The firm's real minimum attractive rate of return is 8%, and the inflation rate is estimated at 7%. The following are alternatives with all costs expressed in real (constant worth) dollars.

  I. Completely overhaul for $10,000 now.
  II. A major overhaul for $7,000 now that can be expected to provide 6 years of service and then a minor overhaul costing $5,000 at the end of 6 years.
  III. A minor overhaul costing $5,000 now as well as at the end of 3 years and 6 years from now.

18-9. Joe Futile wishes to set aside money for his son's college education. His goal is to have a bank savings account containing an amount equivalent to $30,000 of today's purchasing power at the time of his son's eighteenth birthday. The estimated inflation rate is 7% per year. If the bank pays 5% compounded annually, what lump sum of money should Joe deposit in the bank account on his son's fourth birthday?

18-10. Operation and maintenance costs for alternatives A and B are estimated on different bases as follows:

Year end in which cost incurred	Alternative A in "Actual $"	Alternative B in time 0 "Real $"
1	$110,000	$100,000
2	112,000	100,000
3	114,000	100,000
4	116,000	100,000

If the average inflation rate is 10%/yr and the real interest on money is 5%/yr, show which alternative has the lower equivalent (a) present worth of costs at time 0 and (b) future worth of costs at time 4.

18-11. A large corporation's electricity bill now amounts to $400 million. During the next 10 years, electricity usage is expected to increase by 75%, and the estimated electricity bill 10 years hence has been projected to be $920 million. Assuming electricity usage and rates increase at uniform annual rates over the next ten years, what is the annual rate of inflation of electricity prices expected by this corporation?

18-12. The AZROC Corporation needs to acquire a small computer system for one of its regional sales offices. The purchase price of the system has been quoted at $50,000, and the system will reduce manual office expenses by $18,000 per year in real dollars. Historically, these manual expenses have inflated at an average rate of 8% per year, and this is expected to continue into the future. A maintenance agreement will also be contracted for, and its cost per year in actual dollars is constant at $3,000.

What is the minimum (integer-valued) life of the system such that the new computer can be economically justified? Assume that the computer's market value is zero at all times. The firm's MARR is 25% and includes an adjustment for anticipated inflation in the economy. Show all calculations.

18-13. A high school graduate has decided to invest 5% of her first-year's salary in a mutual fund. This amounts to $1,000 in the first year. She has been told that her savings should keep up with expected salary increases, so she plans to invest an *extra* 8% each year over a ten-year period. Thus, at the end of year 1 she invests $1,000; in year 2, $1,080; in year 3, $1,166.40; and so on through year 10. If the average rate of inflation is expected to be 5% over the next 10 years, and if she expects a 2% *real* return on this investment, what is the future worth of the mutual fund at the end of the tenth year?

18-14. The incremental design and installation costs of a total solar system (heating, air conditioning, hot water) in a certain Virginia home were $14,000 in 1996. The annual savings in electricity (in 1996 dollars) have been estimated at $2,500. Assume that the life of the system is 15 years.

    a. What is the internal rate of return on this investment if electricity prices do not escalate during the system's life?

    b. What average annual inflation rate on electricity would have to be experienced over the system's life to provide a combined rate of return of 25% for this investment?

18-15. A gas-fired heating unit is expected to meet an annual demand for thermal energy of 500 million Btu, and the unit is 80% efficient. Assume that each thousand cubic feet of natural gas, if burned at 100% efficiency, can deliver one million Btu. Suppose further that natural gas is now selling for $2.50 per thousand cubic feet. What is the present worth of fuel cost for this heating unit over a 12-year period if natural gas prices are expected to inflate at an average rate of 10% per year? The firm's combined MARR is 15%.

18-16. A small heat pump, including the duct system, now costs $2,500 to purchase and install. It has a useful life of 15 years and incurs annual maintenance of $100 per year in real (year 0) dollars over its useful life. A compressor replacement is required at the end of the eighth year at a cost of $500 in real dollars. The yearly cost of electricity for the heat pump is $680 based on prices at the beginning of the investor's time horizon. All costs are expected to escalate at 6%, which is the projected general inflation rate. The firm's interest rate, which includes an allowance for general inflation, is 15%. No market value is expected from the heat pump at the end of 15 years.

    a. What is the annual equivalent cost, expressed in actual dollars, of owning and operating this heat pump?

    b. What is the annual cost in year 0 (real) dollars of owning and operating the heat pump?

18-17. Your company *must* obtain some new production equipment for the next six years and is considering leasing. You have been directed to accomplish an actual dollar after-tax study of the leasing approach. The pertinent information for the study is as follows:

    Lease costs: First year, $80,000; second year, $60,000: third through sixth years, $50,000 per year. Assume that a six-year contract has been offered by the lessor that fixes these costs over the six-year period.

    Other costs (not covered in contract): $4,000 in year 0 dollars, and estimated to increase 10% each year.

    Effective income tax rate: 40%.

    a. Develop the actual dollar after-tax-cash-flow (ATCF) for the leasing alternative.

    b. If the real MARR (i_r) after taxes is 5% and the annual inflation rate (f) is 9.524%, what is the actual dollar after-tax equivalent annual cost for the leasing alternative?

18-18. A company has two different machines it can purchase to perform a specified task. Both machines will perform the same job. Machine A costs $150,000 initially while machine B (the deluxe model) costs $200,000. It has been estimated that costs will be $1,000 for machine A and $500 for machine B in the first year. Management expects these costs to increase with inflation, which is expected to average 10% per year. The company uses a 10-year study period, and its effective income tax rate is 50%. Both machines qualify as 5-year MACRS property. Which machine should the company choose?

18-19. Because of tighter safety regulations, an improved air filtration system must be installed at a plant that produces a highly corrosive chemical compound. The investment cost of the system is $260,000 in time 0 (1997) dollars. The system has a useful life of ten years and a recovery period of five years. MACRS depreciation is used, with a zero salvage value for tax purposes. However, it is expected that the MV of the system at the end of its 10-year life will be $50,000 in time 0 dollars. Costs of operating and maintaining (O&M) the system, estimated in time 0 dollars, are expected to be $6,000 per year. Annual property tax is 4% of the investment cost and *does not inflate*. Assume that the plant has a remaining life of 20 years, and that O&M costs, replacement costs, and MV inflate at 6% per year.

If the effective income tax rate is 40%, set up a table to determine the ATCF over a 20-year period. The after-tax rate of return desired on investment capital is 12% (including the effect of inflation). What is the PW of cost of this system after income taxes have been taken into account? Develop the real-dollar ATCF. (Assume that the annual general price inflation rate is 4.5% over the 20-year period.) *Suggestion:* Consider using a computer spreadsheet to solve this problem.

18-20. A certain engine lathe can be purchased for $150,000 and depreciated over 3 years to a zero salvage value with the straight-line method. This machine will produce metal parts that will produce revenues of $80,000 (time 0 dollars) per year. It is a policy of the company that annual revenues will be escalated each year to keep pace with the general inflation rate, which is expected to average 5%/yr ($f = 0.05$). Labor, materials, and utilities totaling $20,000 (time 0 dollars) per year are all expected to increase at 9% per year. The firm's effective income tax rate is 50%, and its after-tax MARR (i_c) is 26%.

Perform an actual-dollar (A$) analysis and determine the annual ATCFs of the above investment opportunity. Use a life of three years and work to the nearest dollar. What interest rate would be used for discounting purposes?

18-21. XYZ rapid prototyping (RP) software costs $20,000, lasts one year, and will be expensed. The cost of the upgrades will increase by 10% per year starting at the beginning of year 2. How much can be spent now for a RP software upgrade agreement that lasts 3 years and can be depreciated to zero over 3 years? The MARR is 20% (i_c), and the effective income tax rate (t) is 34%.

19
Multiple Attribute Decision Making[1]

19.1 Introduction

Although it is very useful to use cost or profit as a measure of desirability, many decisions between alternatives cannot be measured only in these terms. Most firms have other objectives, such as customer service, goodwill, community reputation, job satisfaction, safety, employment stability, etc. These factors, which cannot be expressed directly in cost or profit terms, often are called *intangibles, irreducibles* or *nonmonetary attributes.*

Even though a decision maker may have readily definable objectives, he or she might still have a significant problem defining the attributes (sometimes also called criteria) by which the attainment of objectives can be measured. For example, the attainment of a "safety" objective in an automobile purchase decision might be measured by such criteria as weight, maximum possible speed, and interior air bags.

The ultimate aim of the analyst, or decision maker, with respect to multiple objectives and criteria, should be to use rational methods of evaluating them so that a single measure of value may be associated with each alternative in a decision problem.

To provide perspective and motivation for the problems associated with multiple attribute decision making, consider an example involving the choice of a computer-aided design (CAD) workstation. Typical data are summarized in Table 19-1. Three vendors and "do nothing" comprise the list of feasible alternatives (choices) in the decision problem, and a total of seven attributes is judged sufficient for purposes of

[1] Some of this chapter is based on J. R. Canada and W. G. Sullivan, *Economic and Multiattribute Evaluation of Advanced Manufacturing Systems* (Englewood Cliffs, NJ: Prentice-Hall, 1989).

Table 19-1 CAD Workstation Selection Problem

	Alternative			
Attribute	Vendor A	Vendor B	Vendor C	Reference ("Do nothing")
Cost of purchasing the system	$115,000	$338,950	$32,000	$0
Reduction in design time	60%	67%	50%	0
Flexibility	Excellent	Excellent	Good	Poor
Inventory control	Excellent	Excellent	Excellent	Poor
Quality	Excellent	Excellent	Good	Fair
Market share	Excellent	Excellent	Good	Fair
Machine utilization	Excellent	Excellent	Good	Poor

discriminating among the alternatives. Aside from the question of which workstation to select, other significant questions come to mind in multiattribute decision making:

1. How are the criteria (attributes) chosen in the first place?
2. Who makes the subjective judgments regarding nonmonetary criteria such as "quality"?
3. What response is required—a partitioning of alternatives or a rank ordering of alternatives, for instance?

Several simple, though workable and credible, models for selecting among alternatives such as those in Table 19-1 are described in this chapter.

19.2 Choice of Criteria

The choice of attributes (criteria) by which to judge alternative designs, systems, products, processes, etc., is one of the most important tasks in multiple attribute decision making. (The most important task, of course, is to identify feasible alternatives from which to select.) It has been observed that the articulation of attributes for a particular decision can, in some cases, shed enough light on the problem to make the final choice obvious to all involved!

Consider again the data in Table 19-1. These general observations regarding the attributes used to discriminate among alternatives can immediately be made:

1. Each attribute distinguishes at least two alternatives—in no case should identical values for an attribute apply to all alternatives;
2. Each attribute captures a unique dimension or facet of the decision problem (i.e., attributes are independent and nonredundant);
3. All attributes, in a collective sense, are assumed to be sufficient for purposes of selecting the best alternative;
4. Differences in values assigned to each attribute are presumed to be meaningful in distinguishing among feasible alternatives.

In practice, selection of a set of attributes is usually the result of group consensus, and it is clearly a subjective process. The final list of attributes, both monetary and nonmonetary, is therefore heavily influenced by the decision problem at hand as well as an intuitive feel for which criteria will or will not pinpoint relevant differences among feasible alternatives. If too many attributes are chosen, the analysis will become unwieldy and difficult to manage. Too few attributes, on the other hand, will limit discrimination among alternatives. Again, judgment is required to decide what number is "too few" or "too many." If some attributes in the final list lack specificity and/or cannot be quantified, it will be necessary to subdivide them into lower-level attributes that can be measured.

To illustrate the above points, we might consider adding an attribute called "cost of operating and maintaining the system" in Table 19-1 to capture a vital dimension of the CAD system's life-cycle cost. The attribute "flexibility" should perhaps be subdivided into two other more specific criteria such as "ability to interface with computer-aided manufacturing equipment" (e.g., numerically controlled machine tools) and "capability to create and analyze solid-geometry representations of engineering design concepts." Finally, it might be constructive to aggregate two attributes in Table 19-1, namely, "quality" and "market share." Because there is no difference in values assigned to these two attributes across the four alternatives, they could be combined into a single attribute, perhaps named "achievement of greater market share through quality improvements."

19.3 Selection of a Measurement Scale

Identifying feasible alternatives and appropriate attributes represents a large portion of the work associated with multiple attribute decision making. The next task is to develop metrics (measured scales or descriptors) that permit various states of each attribute to be represented. For example, in Table 19-1 "dollars" was the obvious choice for the metric of purchase price. A subjective assessment of flexibility was made on a metric having five gradations that ranged from "poor" to "excellent." The gradations were poor, fair, good, very good, and excellent. In many problems, the metric is simply the scale upon which a physical measurement is made. For instance, anticipated noise pollution for various routings of an urban highway project might be a relevant attribute whose metric is "decibels."

19.4 Dimensionality of the Decision Problem

This first way of dealing with data such as that shown in Table 19-1 is called *single-dimensioned analysis*. (The dimension corresponds to the number of metrics used to represent the attributes that discriminate among alternatives.) Collapsing all information to a single dimension is popular in practice because many analysts and decision makers believe that a complex problem can be made tractable in this manner. In fact, several useful models presented later are single-dimensioned. Such models are termed *compensatory* because changes in values of a particular attribute can be offset by, or traded off against, opposing changes in another attribute.

A second basic way to process information in Table 19-1 is to retain the individuality of the attributes as the best alternative is being determined. That is, there is no attempt to collapse attributes to a common scale. This is referred to as *full-dimensioned analysis* of the multiple attribute decision problem. For example, if r^* attributes have been chosen to characterize the alternatives under consideration, the predicted values for all r^* attributes are considered in the choice.

19.5 Selected Analysis Techniques

Numerous techniques have been developed over the past many years to facilitate analysis of multiple attribute decision problems. To give the readers an introduction to the diversity and application of available methods, we will concentrate on the following, which are given in approximate order of increasing complexity and power:

1. alternative–attributes score card;
2. ordinal scaling;
3. weighting factors;
4. weighted evaluation of alternatives;
5. Brown-Gibson model.

19.5.1 Alternatives–Attributes Score Card

Graphical techniques are very powerful because they allow one to describe the multiple attribute decision problem so that decision makers can readily understand and absorb large amounts of information. Graphical techniques often do not include specific "weighting" of criteria, which may be advantageous because attribute weights are often nebulous and/or differ greatly among various decision makers.

A scorecard is a matrix of alternatives versus attributes together with numbers and/or other symbols to represent the outcome expected for each alternative with respect to each attribute. Table 19-2 is an example in which qualitative and quantitative estimates are provided for the performance of five attributes used to judge the "value" or "worth" of four alternatives for modernizing a manufacturing company. Ease of interpretation of the scorecard (to facilitate decision making) can be obtained by such devices as symbols and/or colors for "best" and "worst" alternatives for each attribute.

Figure 19-1 shows the use of shaded circles to portray visually the relative evaluations of alternatives with respect to the five criteria given in Table 19-2. As shown in the key, the evaluations for each attribute except net present worth were categorized in five ways, from "exceptional" (full shading) down to "poor" (no shading). Net present worth is shown to be "exceptional" (with full shading) for $500M, with proportionately less shading down to $0M. While the use of shaded circles does cause loss of the specific language and quantitative information given in Table 19-2, it is easy to scan for relative comparisons.

In examining the scorecard information in Table 19-2 and/or Fig. 19-1, one might be led to conclude that alternative P-4 and perhaps P-3 are definitely not as desirable as alternatives P-1 and P-2. Often a final decision can be made using this method, but it should be recognized that any relative weights/importances of the

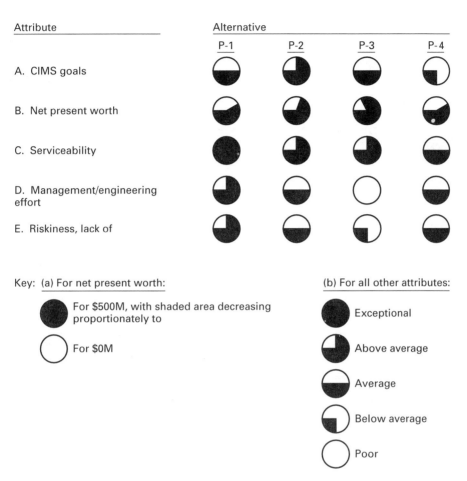

Figure 19-1.   Graphical portrayal (using shaded circles) for four alternatives and five attributes.

Table 19-2   Example Scorecard for Four Alternatives and Five Attributes

	Alternative			
Attribute	P-1	P-2	P-3	P-4
A. CIMS goals	Average	Above average	Average	Below average
B. Net present worth	$300M	$350M	$400M	$290M
C. Serviceability	<1 hr	1–1.5 hr	1–1.5 hr	1.5–2 hr
D. Management/ engineering effort	500 hr	700 hr	1,100 hr	700 hr
E. Riskiness, lack of	Above average	Average	Below average	Average

☐ Best alternative for attribute

┌┄┄┄┐ Worst alternative for attribute

attributes have not been assigned formally. *In future examples, we have elected to drop alternatives P-3 and P-4 from the analysis to simplify the ensuing discussion.*

19.5.2 Ordinal Scaling

An ordinal scaling is simply a ranking of criteria in order of preference. Before attributes are weighted (or alternatives are evaluated), it is often desired to rank them in order of decreasing preference. This might be done by presenting the decision maker with a list of attributes (or alternatives) and asking him or her to rank them in order of preference. There is, however, a procedure that may make this task easier and provides a check on the internal consistency of the value judgments obtained. This is called the *method of paired comparisons* and is illustrated below.

Consider the five attributes in Table 19-2, which we now want to rank in order. They are designated as follows:

        A. CIMS goals;

        B. net present worth;

        C. serviceability;

        D. management/engineering effort;

        E. riskiness, lack of.

The method of paired comparisons submits attributes to the decision maker two at a time for preference judgments. In general, if there are N factors, $N(N-1)2$ pairs must be judged. Assume that the results of this process are indicated by the following list of preference statements (as usual, the symbol > means "is preferred to" and = means "is equal to").

    1. $A < B$ 6. $B > D$

    2. $A > C$ 7. $B > E$

    3. $A > D$ 8. $C > D$

    4. $A > E$ 9. $C = E$

    5. $B > C$ 10. $D < E$

A good way to depict the pairwise comparisons above and then determine rankings is shown by the matrix in Table 19-3. In that matrix "P" is shown for each pair in which the row factor (attribute) is preferred to the column factor. Note that the diagonal of the matrix is empty (since a given factor cannot be preferred to itself). A good way to make sure that all pairs of factors have been considered is to recognize that there should be a P either above *or* below the diagonal for all pairs, or in case of equal preferences, an "=" is shown both above *and* below the diagonal. Note that on the right-hand side of the matrix is shown the number of times the factor in each row is preferred. Thus, for this example, it is found that the rank order of attributes is $B > A > C = E > D$.

The foregoing scheme of deducing rankings assumes transitivity of preferences. That is, if $A > C$ and $B > A$, then B must $> C$. If the number of times a given factor is preferred to another is equal for two or more factors (except in the case of ties),

Table 19-3   Illustration of Preference Comparisons[a]

	A	B	C	D	E	Number of times preferred
A	—		P	P	P	3
B	P	—	P	P	P	4
C			—	P	=	1 1/2
D				—		0
E			=	P	—	1 1/2

[a] "P" if row factor is preferred to column factor, or "=" if they are equally preferred.

there is evidence of lack of consistency (i.e., intransitivity), which suggests the need for questioning preference judgments for the attributes involved.

19.5.3 Weighting Factors

Many numerical formula methods for assigning weights exist that are easy to use but generally less defensible than direct assignment of weights based on preference comparisons among criteria. Several of these formula-based methods are described below, and example calculations are shown in Table 19-4 for the same five criteria (attributes) given in Table 19-2. The weights are expressed as percentages.

1. *Uniform or equal weights.* Given N attributes, the weight for each is

$$W_i = 1/N \times 100\%.$$

2. *Rank sum weights.* If R_i is the rank position of attribute i (with 1 the highest rank, etc.) and there are N attributes, then rank sum weights, W_i, for each attribute, may be calculated as

$$W_i = \frac{N - R_i + 1}{\displaystyle\sum_{i=1}^{N} (N - R_i + 1)} \times 100\%.$$

3. *Rank reciprocal weights.* Rank reciprocal weights, using the same notation as above, may be calculated as

$$W_i = \frac{1/R_i}{\displaystyle\sum_{i=1}^{N} (1/R_i)} \times 100\%.$$

When comparing the methods in Table 19-4, note that the rank reciprocal method gives the highest weight for the first-ranked attribute. One might choose among the weighting methods above according to which provides the closest approximation to the independently judged weight for the highest ranked attribute. If we take that independently judged weight to be the 40%, it is seen to be closest to the 44% for the rank reciprocal method.

Table 19-4 Example Calculation of Weights by Several Formulas

Attribute	Uniform $W_i = 100\%/N$	Rank sum (A) R_i^a	Rank sum (B) $N - R_i + 1$	Rank sum (C) $W_i = \dfrac{(B) \times 100\%}{\Sigma(B)}$	Rank reciprocal (D) = 1/(A) $1/R_i$	Rank reciprocal (E) $W_i = \dfrac{(D) \times 100\%}{\Sigma(D)}$
Net present worth	20	1	5	33	1	44
CIMS goals	20	2	4	27	0.5	22
Serviceability	20	3	3	20	0.33	14
Riskiness, lack of	20	4	2	13	0.25	11
Management/ engineering effort	20	5	1	7	0.20	9
	$\Sigma = 100$		$\Sigma = 15$	$\Sigma = 100$	$\Sigma = 2.28$	$\Sigma = 100$

[a] Ranks show with 1 = highest (best), etc.

19.5.4 Weighted Evaluation of Alternatives

Once weights have been assigned to attributes, the next step is to assign numerical values regarding the degree to which each alternative satisfies each attribute. This is generally a difficult judgment task using an arbitrary scale of, say, between 0 to 10 or 0 to 1,000 to reflect relative evaluations for each alternative and each attribute.

Example 19-1

Suppose that we are comparing two alternatives on the basis of how well they satisfy the five attributes having rank reciprocal weights developed in Table 19-4. The attributes, together with the subjective evaluation of how well a particular alternative meets each on the basis of a scale of 0 to 10, are shown in Table 19-5. [*Note:* These evaluation ratings roughly correspond to the graphical portrayals of alternatives P-1 and P-2 in Fig. 19-1, with a fully shaded circle representing an evaluation rating of 10, a half-shaded circle representing an evaluation rating of 5, etc.]

Once the evaluations have been made, the results can be calculated as in Table 19-6 to arrive at weighted evaluations of attributes for each alternative. Thus the summed weighted evaluation is 72.9 for alternative P-1 and 73.1 for alternative P-2, as calculated using the following equation:

weighted evaluation = (normalized attribute weight × evaluation rating)/10.

This indicates alternative P-2 is marginally better even though it happened to have lower evaluation ratings for three out of five attributes.   ■

19.5.5 Brown-Gibson Model

A noteworthy variation on the weighted evaluation procedure is the Brown-Gibson model,[2] which was first developed in 1972 and later generalized by Huang and Ghandforoush.[3] The model transforms objective (usually economic) measurement scales for alternatives under consideration into a score between 0 and 1 so that the sum of the (objective) scores totals 1. For subjective attributes, such as lack of riskiness, nominal ratings are then transformed into numerical scores, and the relative importances of the attributes are weighted so that the scores of each attribute over all alternatives and the weighted (subjective) scores over all attributes each sum to 1. Next, weights are assigned to the objective, and the subjective scores (which also sum to 1) in order to show the relative desirability for each alternative.

Table 19-5 Example Evaluation Ratings of How Well Each Alternative Satisfies Each Attribute

	Alternative	
Attribute	P-1	P-2
A. CIMS goals	7.5	9
B. Net present worth	6	7
C. Serviceability	10	7.5
D. Riskiness, lack of	8	6
E. Management/engineering effort	8	6

[2] J. Canada and W. Sullivan, *Economic and Multiattribute Evaluation of Advanced Manufacturing Systems* (Englewood Cliffs, NJ: Prentice-Hall, 1989).

[3] Huang, P. and P. Ghandforoush, "Procedures Given for Evaluating and Selecting Robots," *Industrial Engineering,* 16, No.4 (April 1984): 44-48.

Table 19-6 Calculation of Weighted Evaluations of Alternatives

Attribute	Normalized attribute weight (from table)	Alternative P-1		Alternative P-2	
		Evaluation rating	Weighted evaluation	Evaluation rating	Weighted evaluation
A. CIMS goals	22	7.5	16.5	9	19.8
B. Net present worth	44	6	26.4	7	30.8
C. Serviceability	14	10	14.0	7.5	10.5
D. Riskiness, lack of	11	8	8.8	6	6.6
E. Management/ engineering effort	9	8	7.2	6	5.4
			$\Sigma = 72.9$		$\Sigma = 73.1$

Expressed in equation form, the combined measure (weighted evaluation) for each alternative k, WE_k, is

$$WE_k = (\alpha)(OM_k) + (1 - \alpha)(SM_k),$$

where

$$OM_k = \text{objective measure for alternative } k,$$
$$SM_k = \text{subjective measure for alternative } k,$$
$$\alpha = \text{relative importance weighting for } OM_k,$$

and where OM_k, SM_k, and $\alpha \geq 0$ and ≤ 1.

We will demonstrate determination of OM_k, SM_k, and WE_k with two examples involving alternative communication systems. The examples also include the use of certain formulas or approaches for determining weights and evaluation ratings which are different than the more general methods for determining weighted evaluations shown previously.

19.5.5.1 Objective Measure.
The objective measure for each alternative k, OM_k, can be calculated by one of the following:

(a) If the measure is returns or profits for each alternative k, call it OFP_k (such as net PW), then

$$OM_k = \frac{OFP_k}{\displaystyle\sum_{k=1}^{K} OFP_k}, \qquad \text{where } K = \text{total number of alternatives.}$$

(b) If the measure is costs (e.g., equivalent annual costs), call it OFC_k:

$$OM_k = \frac{1}{OFC_k \times S}$$

where

$$S = \sum_{k=1}^{K} \frac{1}{OFC_k}.$$

Example 19-2
Alternative computer networking (communications) systems are expected to have the following total equivalent annual costs:

Alternative k	OFC_k
I	$130,000
II	150,000
III	165,000

The OM for each alternative is determined as follows:

$$S = \frac{1}{\$130,000} + \frac{1}{\$150,000} + \frac{1}{\$165,000}$$
$$= 0.000020418,$$

and

$$OM_I = \frac{1}{\$130,000 \times 0.000020418} = 0.377.$$

Similarly,

$$OM_{II} = \frac{1}{\$150,000 \times 0.000020418} = 0.326$$

$$OM_{III} = \frac{1}{\$165,000 \times 0.000020418} = \underline{0.297}$$

$$\Sigma = 1.000. \quad \blacksquare$$

19.5.5.2 Subjective Measure.   The subjective measure for each alternative k, SM_k, can be calculated for N subjective attributes, as we demonstrated earlier for weighted evaluations in Section 19.5.4.

$$SM_k = \sum_{i=1}^{N} (\text{subjective attribute weight})_i \times (\text{subjective evaluation rating})_{ik}.$$

For Example 19-3, we assume that subjective attribute weights are assigned with a 20 being the highest, and that the subjective evaluation ratings are assigned as follows:

$$\text{excellent} = 4 \qquad \text{good} = 2 \qquad \text{poor} = 0$$
$$\text{very good} = 3 \qquad \text{fair} = 1$$

Example 19-3
For the same alternative CIM systems for which objective measures were shown in Example 19-2, Table 19-7 gives raw subjective weights and evaluation ratings for each. Table 19-8 shows those numbers each divided by their respective totals, so they will add to 1. We might call these "unitized" weights and ratings. Now we calculate the SM for each:

$$SM_I = \frac{20}{60}\left(\frac{2}{7}\right) + \frac{15}{60}\left(\frac{3}{8}\right) + \frac{10}{60}\left(\frac{2}{6}\right) + \frac{10}{60}\left(\frac{1}{4}\right) + \frac{5}{60}\left(\frac{4}{10}\right) = 0.320.$$

Table 19-7 Raw Subjective Weights and Evaluation Ratings for Alternative Communications Systems

Raw weight	Subjective attribute	Raw evaluation ratings for alternative			Σ evaluation ratings
		I	II	III	
20	Quality of results	Good (2)	Excellent (4)	Fair (1)	(7)
15	Ease of use	Very good (3)	Fair (1)	Excellent (4)	(8)
10	Competitive advantage	Good (2)	Good (2)	Good (2)	(6)
10	Adaptability	Fair (1)	Poor (0)	Very good (3)	(4)
5	Expandability	Excellent (4)	Very good (3)	Very good (3)	(10)
$\sum_i = 60$					

Table 19-8 Unitized Subjective Weights and Evaluation Ratings for Alternative Communications Systems[a]

Unitized weight	Subjective attribute	Unitized evaluation ratings for alternatives			Σ evaluation ratings
		I	II	III	
20/60	Quality of results	2/7	4/7	1/7	1.0
15/60	Ease of use	3/8	1/8	4/8	1.0
10/60	Competitive advantage	2/6	2/6	2/6	1.0
		1/4	0	3/4	1.0
10/60	Adaptability	4/10	3/10	3/10	1.0
5/60	Expandability	$SM_k = 0.320$	0.302	0.378	$\sum_k SM_k = 1.0$
$\sum_i = 1.00$					

[a] Final SMs are shown at the bottom for each alternative.

Similarly (as shown on the bottom of Table 19-8),

$$SM_{II} = 0.302$$
$$SM_{III} = \underline{0.378}$$
$$\Sigma = 1.000.$$ ∎

19.5.5.3 Weighted Evaluation.
The final weighted evaluation results for each alternative k (WE_k) for any α ($0 \leq \alpha \leq 1$) can now be calculated as $WE_k = (\alpha)(OM_k) + (1 - \alpha)(SM_k)$. If $\alpha = 0.3$, the WEs for the three communication systems alternatives in Examples 19-2 and 19-3 are

$$WE_I: \quad (0.3)(0.377) + (1 - 0.3)(0.320) = 0.337,$$
$$WE_{II}: \quad (0.3)(0.326) + (1 - 0.3)(0.302) = 0.309,$$
$$WE_{III}: \quad (0.3)(0.297) + (1 - 0.3)(0.378) = \underline{0.345},$$
$$\Sigma = 1.000.$$

Thus, for $\alpha = 0.3$, alternative III is slightly better than alternative I, which is somewhat better than alternative II.

19.5.5.4 Sensitivity Analysis.   Very often it is useful to depict the effects of differing weights on the weighted evaluation as illustrated in Fig. 19-2. The vertical dashed line at $\alpha = 0.3$ helps confirm the above results for the three alternatives. The graph shows, among other things, that alternative I is preferred to alternative II for all values of α, and that alternative I becomes preferred to alternative III for all α slightly greater than 0.4. Indeed, the break-even value of α can be calculated by equating the WEs for alternatives I and III as follows:

$$(\alpha)(0.377) + (1 - \alpha)(0.320) = (\alpha)(0.297) + (1 - \alpha)(0.378),$$

$$\alpha = 0.42.$$

19.5.5.5 Pros and Cons of the Brown-Gibson Model.   The Brown-Gibson model is an enhanced weighted evaluation procedure that makes decision makers more aware of important criteria, thus yielding a more realistic decision model. Other advantages include that it is easy to use and easy to understand. Furthermore, graphical results show overall dominance and break-even points among alternatives.

Another noteworthy item is that the use of this procedure does not remove subjectivity from the evaluation process; it merely attempts to recognize and quantify the subjectivity. As a result, the subjective factors should be ranked carefully. The computation

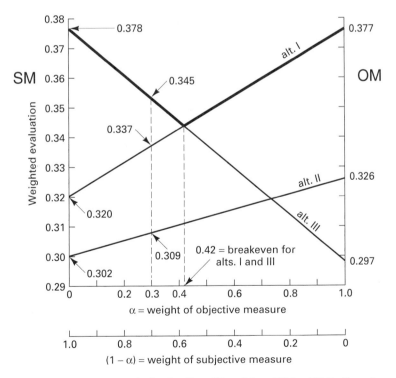

Figure 19-2.   Weighted evaluation for all possible weights of OM and SM in Examples 19-2 and 19-3.

of the objective factor weights as shown earlier can pose a problem if the decision maker is not aware that this model is only sensitive to multiplicative differences in cost. Two alternatives costing $10,000 and $15,000 would yield the same objective factor measure as two alternatives costing $100,000 and $150,000, even though the cost differentials are $5,000 and $50,000, respectively. Decision makers need to be aware of this and take it into account when making their final decision.

19.6 Summary

This chapter has outlined several simple methodologies for analyzing alternatives characterized by multiple attributes. Most such comparisons, judgments, or ratings generally are relatively easy to make. Then the decision maker can combine these into an overall evaluation for each alternative. This should make the decision maker more confident and ultimately make his or her job easier.

Chapter 20 will describe another multiple attribute methodology called the analytic hierarchy process. There are numerous other techniques, often very complex and/or restrictive, which are beyond the scope of this book.

The determination of what methodology(ies) to use depends on the nature of the decision problem and the preferences of the decision maker(s). While this determination is a very inexact science, it can be shown that the use of widely differing models (and subjective probability distributions, if applicable) will often have less effect on the probable quality of the solution than does the unintended deletion of possible outcomes, alternatives, or important attributes.

A variety of low-cost software is available for multiple attribute decision analysis techniques. The interested reader is referred to Hodge and Canada.[4]

REFERENCES

BROWN, P., AND D. GIBSON, "A Quantified Model for Facility Site Selection—Application to a Multiplant Location Problem," *AIIE Transactions* 4, no. 1 (March 1972):1–10.

CANADA, J., "Evaluation of Computer-Integrated Manufacturing Systems," *1985 Annual International Industrial Engineering Conference Proceedings,* Institute of Industrial Engineers, Norcross, GA, December 8-11, 1985, pp. 162–172.

CANADA, J., AND W. SULLIVAN, *Economic and Multiattribute Evaluation of Advanced Manufacturing Systems.* Englewood Cliffs, NJ: Prentice-Hall, 1989.

FALKNER, C., AND S. BENHAJLA, "Multi-Attribute Decision Models in the Justification of CIM Systems," *The Engineering Economist* 35, no. 2 (Winter 1990):91–114.

FRAZELLE, E., "Suggested Techniques Enable Multi-Criteria Evaluation of Material Handling Alternatives," *Industrial Engineering* 17, no. 2 (Feb. 1985):42–48.

HUANG, P., AND P. GHANDFOROUSH, "Procedures Given for Evaluating, Selecting Robots," *Industrial Engineering* 16, no. 4 (April 1984):44–48.

HUBER, R., "Justification: Barrier to Competitive Manufacturing," *Production* (Sept. 1985):46.

MACCRIMMON, K. R., "Decision Making Among Multiple Attribute Alternatives: A Survey and Consolidated Approach," Memo RM-4823-ARPA, Rand Corporation, December 1968.

[4] G. L. Hodge and J. R. Canada, "Low Cost Microcomputer Software for Non-Traditional Economic Decision Analysis," *The Engineering Economist* 35, no. 2 (Winter 1990):161–167.

MORRIS, W., *Engineering Economic Analysis.* Reston, VA: Reston Publishing, 1976.

OZERNOY, V., "A Framework for Choosing the Most Appropriate Discrete Alternative Multiple Criteria Decision-Making Method in Decision Support Systems and Expert Systems," *Toward Interactive and Intelligent Decision Support Systems,* Vol. 2. New York: Springer-Verlag, 1986.

SANDERS, G., P. GHANDFOROUSH, AND L. AUSTIN, "A Model for the Evaluation of Computer Software Packages," *Computers and Industrial Engineering* 7, no. 4 (1983):309–315.

SULLIVAN, W., "Models IE's Can Use to Include Strategic, Non-Monetary Factors in Automation Decisions," *Industrial Engineering* 18, no. 3 (March 1986):42–50.

SULLIVAN, W., AND H. LIGGETT, "A Decision Support System for Evaluating Investment in Manufacturing Local Area Networks," *Manufacturing Review* 1, no. 3 (Oct. 1988):151–157.

PROBLEMS

19-1. Select a class or type of significant decision problem in your work or personal life. Name the three or more significant attributes to be considered for that class of problem. Assume that the weighting of these attributes is nebulous, so you think it desirable to develop only a matrix of alternatives versus attributes, with entries to show how well each alternative meets each attribute in whatever measures are appropriate (such as in dollars, time, rank, and so on). Then use colors or other symbol codings to facilitate ease of understanding of the differences by the persons making the final selection among the alternatives.

19-2. Select a class or type of decision problem in your work or personal life involving multiple objectives (not necessarily the same as for Problem 19-1). Name the three to five most important objectives for that class of problem, as you see them, and do the following:

   a. Using the method of paired comparisons, rank the objectives in order of decreasing importance.

   b. Weight them and check for consistency and then normalize the weight to sum to 100.

   c. Identify two to four alternatives for the decision problem and evaluate how well each alternative meets each objective on a scale of 0 to 10. Then multiply the evaluation ratings by the objective weights and sum these for each objective to compute a weighted evaluation for each alternative.

19-3. a. Weight the relative importance of the three or four most important attributes you would consider in selecting a job. Assume that these are the only attributes you will quantitatively consider. Show how you make comparisons for internal consistency and normalize the factor weights to sum to 100.

   b. Using the attribute weights developed in (a), obtain a weighted evaluation of two alternative jobs in which you might be interested by evaluating how well each job satisfies each attribute using a scale from 0 to 10.

19-4. a. Weight the relative importance of the three to six most important attributes you would consider in selecting a personal car. Assume that these are the only attributes you will quantitatively consider. Show how you make comparisons for internal consistency and normalize the attribute weights to sum to 1,000.

   b. Using the attribute weights developed in (a), obtain a weighted evaluation of three alternative makes of cars you might consider for purchase by evaluating how well each make of car satisfies each attribute. Use a scale from 0 to 20.

19-5. Describe a significant multiattribute decision problem of existing or potential meaningful interest to you. Examples are housing, car, job, travel, and equipment. Identify at least four most relevant attributes that are as independent of each other as possible and at least two or three mutually exclusive alternatives. (*Note:* In the process of further analysis, you might well add or delete alternatives and/or attributes.)

a. Describe the problem as meaningfully as possible using two or more graphical techniques. What conclusion (choices), if any, can you make from these?

b. Rank order the attributes. Can you combine this with one or more of the graphical techniques in part (a) to make choices? If so, describe.

c. Weight the attributes by the method or formula(s) of your choice and show at least two checks and/or revisions of weights for consistency between initial judgments and preferences.

d. For each attribute show a graph of attribute outcome (*x*-axis) versus evaluation rating (*y*-axis).

e. Using parts (c) and (d), show calculations of the weighted evaluations for each alternative and display the results graphically for ease of comparison.

19-6. Two highway alignments have been proposed for access to a new manufacturing plant. Based on the data and information given below, comparison can be made of the two alignments. The better one must be chosen as the proposed highway alignment that will be the connector between an interstate highway and the proposed site. The route length for alignment A is 4.7 miles and for alignment B, 5.1 miles. The monetary costs are as follows:

Item	Alignment A	Alignment B
Land	$ 4,044,662	$ 4,390,000
Bridges	10,134,000	8,701,000
Pavement	4,112,500	4,462,500
Grade drainage	7,050,000	7,650,000
Erosion control	470,000	510,000
Clearing and grading	188,000	204,000
Total	$25,999,162	$25,917,500

Nonmonetary criteria are as follows:

	Alignment A	Alignment B
Maintenance	Moderate	High
Noise pollution	Very good	Good
Cost savings (on gas)	Excellent	Poor
Accessibility to another major roadway	U.S. Highway 41	None
Impact on wildlife	Little	Little
Relocation of residences	2	3
Road condition	Flat	Hilly

Use the Brown-Gibson model to determine which highway alignment you would recommend. Show all work.

19-7. a. Use the weighted evaluation model to make a selection of one of the three used automobiles for which some data are given below. State your assumptions regarding miles driven each year, life of the automobile (how long *you* would keep it), market

(resale) value at the end of life, interest cost, price of fuel, cost of annual maintenance, and attribute weights and other subjectively based determinations.

b. Use the data developed in part (a) and the Brown-Gibson model to make a selection.
Take as your objective measure the total annual cost of capital recovery plus the cost
of fuel. Do your answers in parts (a) and (b) agree? Explain why they should (or
should not) agree.

	Alternative		
Attribute	Domestic 1	Domestic 2	Foreign
Price	$8,400	$10,000	$9,300
Gas mileage	25 mpg	30 mpg	35 mpg
Type of fuel	Gasoline	Gasoline	Diesel
Comfort	Very good	Excellent	Excellent
Aesthetic appeal	5 out of 10	7 out of 10	9 out of 10
Passengers	4	6	4
Ease of servicing	Excellent	Very good	Good
Performance on road	Fair	Very good	Very good
Stereo system	Poor	Good	Excellent
Ease of cleaning upholstery	Excellent	Very good	Poor
Storage space	Very good	Excellent	Poor

19-8. Utilize the weighted evaluation technique to analyze the following important situation:
You need to determine which job to take, given acceptable offers from firms A, B, and
C. You first determine a set of attributes that are most important in evaluating and then
construct a decision matrix with alternatives as columns and attributes as rows, and fill
in. (Results are given in the matrix below.) Analyze the alternatives with the technique
according to your perceptions (be complete—it took years to get to this point!).

	Alternative (firm)		
Attribute	A	B	C
Starting salary	$30,000	$28,500	$33,000
Opportunity for advancement	Excellent	Excellent	Fair
Management attitude	Very good	Excellent	Good
Location	Good	Fair	Poor
Type of work	Excellent	Poor	Very good
Opportunity for continuing education	Very good	Good	Fair

19-9. The following are evaluation ratings (on a scale of 0 to 20) of how well each of two alternatives satisfies each of four attributes. Show which alternative would be best using the
weighted evaluation methodology and each of the following attribute weighting formulas. (*Note:* Normalize the summed weighted evaluations to equal 100.)

	Alternative	
Attribute	1	2
A	15	12
B	8	20
C	14	14
D	13	12

  a. Uniform.
  b. Rank sum, with $A > B > C > D$.
  c. Rank reciprocal, with $A > B > C > D$.

19-10. Volunteer alumni A and B have contributed $100M and $150M, respectively (higher is better!). However, evaluations of their leadership success (on a 1 to 5 maximum scale) are 4 and 3, respectively. Contributions (objective) are considered to be half as important as leadership success (subjective). Using the weighted evaluation method combining these measures, who should be the "Alumnus of the Year"? What part of 1.0 total "weighted brownies" does he or she score?

19-11. You have just been transferred! Your company has offered two alternatives for a new position with them, and you decide to stay with your present employer.

	Alternative	
Attribute	Dry Gulch	Rapid Creek
Salary	$40,000	$35,000
Proximity to relatives	2,000 miles	500 miles
Promotion potential	Fair	Good
Commuting time (per day)	40 minutes	90 minutes

  Use the Brown-Gibson model to select your new job, with salary being the objective measure.

19-12. Consider the data given below.

	AMS Alternative				
Attribute	A	B	C	Do nothing	Minimum acceptable
P: Purchase price	$115,000	$339,400	$32,000	$0[a]	(Open)
D: Reduced design time	60%	67%[a]	50%	0	25%
L: Impact on manufacturing lead time	Excellent[a]	Excellent	Good	Poor	Good
F: Flexibility	Excellent[a]	Very good	Good	Poor	Good
Q: Quality improvement	Very good	Excellent[a]	Fair	Poor	Good
M: Market share	Up 2%[a]	Up 2%	Up 1%	Down 2%	(Open)

[a] Best outcome.

  If $M > Q > F > P > D > L$, which alternative would you recommend using the weighted evaluation technique? Carefully develop a rationale by which nonmonetary attributes are rated in your analysis.

20
The Analytic Hierarchy Process

20.1 Introduction

The analytic hierarchy process (AHP) was developed and documented primarily by Thomas Saaty.[1] Applications of this methodology have been reported in numerous fields, such as transportation planning, portfolio selection, corporate planning, marketing, and others.

The strength of the AHP method lies in its ability to structure a complex, multiperson, multiattribute, and multiperiod problem hierarchically. Pairwise comparisons of the elements (usually, alternatives and attributes) can be established using a scale indicating the strength with which one element dominates another with respect to a higher-level element. This scaling process can then be translated into priority weights (scores) for the comparison of alternatives.

The use of the AHP to solve a decision problem consists of five stages:

1. construction of a decision hierarchy by breaking down the decision problem into a hierarchy of decision elements and identifying decision alternatives;

2. determination of the relative importance of attributes and subattributes (if any);

3. determination of the relative standing (weight) of each alternative with respect to each next higher level attribute or subattribute;

4. determination of indicator(s) of consistency in making pairwise comparisons;

5. determination of the overall priority weight (score) of each alternative.

[1] Thomas L. Saaty, *The Analytic Hierarchy Process* (New York: McGraw-Hill Book Company, 1980); Thomas L. Saaty, *Decision Making for Leaders* (Belmont, CA: Wadsworth Publishing Company, Inc., 1982).

The process of computing the priority vectors and consistency ratios is quite laborious and is best accomplished with a computer program. The AHP methodology has been incorporated in at least two software packages: Expert Choice[2] and Auto-Man.[3] Expert Choice is a generic decision problem software package, whereas Auto-Man (a public-domain package) was designed specifically for the evaluation of manufacturing alternatives. *To demonstrate manually the AHP stages, we will evaluate the automation alternatives P-1, P-2, and P-3 previously considered in Chapter 19.*

20.2 Construction of the Decision Hierarchy

The AHP begins by decomposing a complex decision problem into a hierarchy of sub-problems. We will concentrate on what Saaty calls "functional" hierarchies as applied to multiattribute decision problems. He uses the term "element" to apply to the overall objective, attributes, subattributes, sub-subattributes, and so on, and alternatives of a problem as follows:

> The top level, called the *focus,* consists of only one element—the broad, overall objective. Subsequent levels may each have several elements, although their number is very small—between 5 and 9. Because the elements in one level are to be compared with one another against a criterion in the next higher level, the elements in each level must be of the same order of magnitude.[4]

Figure 20-1 shows the standard form of the AHP hierarchy. The focus, or objective, of the decision problem is shown at the top level of the hierarchy. The second level consists of attributes considered important in achieving the overall objective. Subsequent levels are created by dividing attributes into subattributes, subattributes into sub-subattributes, and so on. The alternatives are shown in the bottom level. Figure 20-2 shows a typical four-level hierarchy applied to a career choice problem.

It is important to note that the selected attributes and subattributes should be *independent.* The mathematics of the AHP (as presented in this chapter) are based on the principle of hierarchic composition. This principle states that the elements on a single level of the hierarchy are independent and their relative importance (priority weights) does not depend on the elements at the next lower level of the hierarchy. Consider the Level II attributes shown in Fig. 20-2. According to the principle of hierarchic composition, the attributes *Money, Job security, Family life,* and *Work environment* are independent: a change in the value of one of these attributes will not change the value of any other attribute. Furthermore, the importance of these attributes with respect to career choice satisfaction is not affected by the set of alternatives under consideration.

    [2] E. H. Forman, T. L. Saaty, M. A. Selly, and R. Waldron, *Expert Choice* (McLean, VA: Decision Support Software, Inc., 1983).

    [3] Stephen F. Weber, *AutoMan: Decision Support Software for Automated Manufacturing Investments* (U. S. Department of Commerce, National Institute of Standards and Technology, NISTIR 89-4166, 1989).

    [4] Thomas L. Saaty, *Decision Making for Leaders* (Belmont, CA: Wadsworth Publishing Company, Inc., 1982), p. 28.

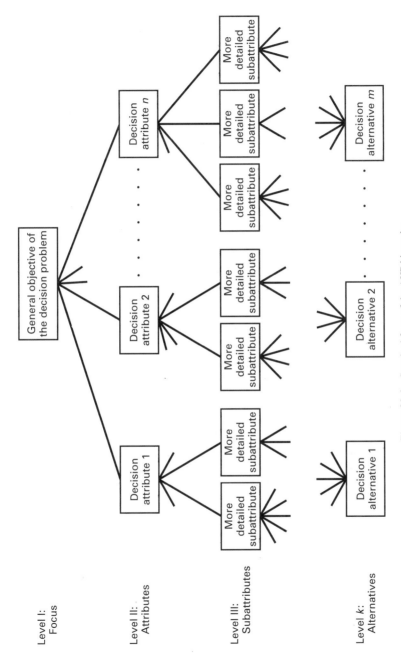

Figure 20-1 Standard form of the AHP hierarchy.

Level I:
Focus

Level II:
Attributes

Level III:
Subattributes

Level *k*:
Alternatives

General objective of
the decision problem

Decision
attribute 1

Decision
attribute 2

Decision
attribute *n*

More
detailed
subattribute

More
detailed
subattribute

More
detailed
subattribute

More
detailed
subattribute

More
detailed
subattribute

More
detailed
subattribute

Decision
alternative 1

Decision
alternative 2

Decision
alternative *m*

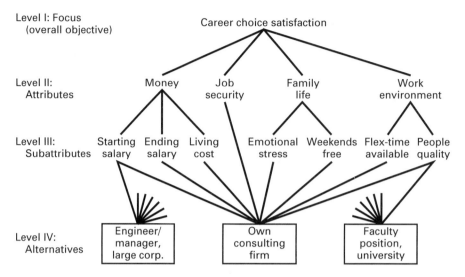

Level I: Focus
(overall objective)

Level II:
Attributes

Level III:
Subattributes

Level IV:
Alternatives

Figure 20-2 Complete hierarchy for example career choice problem. (*Note:* For
clarity, full lines indicate relationships between Level IV, and levels above it are shown
only for alternative "Own Consulting Firm." Similar and equal numbers of lines
applicable for each of the other two alternatives are shown only partially.) Source: J. R.
Canada, et al., "How to Make a Career Choice; the Use of the Analytic Hierarchy
Process," *Industrial Management* 27, no. 5 (1985).

The following guidelines from Arbel and Seidmann should be considered when
constructing hierarchies:[5]

1. The number of levels used should be chosen to represent effectively the
   problem at hand.
2. The order of the levels should reflect a logical causal relationship between
   adjacent levels.
3. The number of members in a particular level should be chosen to describe
   the level in adequate detail, but should not cause unnecessary complexity.

Let us apply the AHP method to an adaptation of the problem first introduced in
Table 19-2. We will limit our consideration to three alternatives (P-1, P-2, and P-3) for
modernizing a manufacturing company. The focus of the decision problem is selecting the "best overall automated system," and the "best" system can be defined by five
reasonably independent attributes: A. CIMS goals, B. Net present worth, C. Service-
ability, D. Management/engineering effort, and E. Riskiness, lack of. In this example,
the attributes capture the required level of detail so that subattributes are not necessary. The resulting three-level hierarchy is shown in Fig. 20-3.

---

[5] A. Arbel and A. Seidmann, "Performance Evaluation of Flexible Manufacturing Systems," *IEEE
Transactions on Systems, Man, and Cybernetics* SMC-14, no. 4 (July–Aug. 1984):606–617.

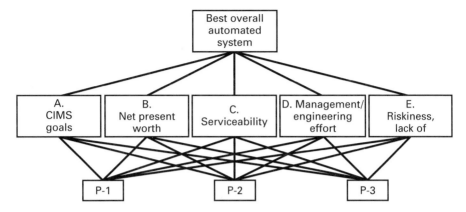

Figure 20-3   Decision hierarchy for the automation alternatives example.

20.3 Determining the Importance of Attributes and Subattributes

Once the hierarchy is established, priorities (relative importance weights) must be established for each set of elements at every level of the hierarchy. Priority data are obtained by asking various decision makers to evaluate a set of elements at one hierarchical level in a pairwise fashion regarding their relative importance with respect to an element at the next higher level of the hierarchy. This is accomplished by having decision makers respond to questions of the following form: "With respect to the best overall manufacturing system, how much more important is net present worth than lack of riskiness?" and, at the level of the alternatives, "With respect to serviceability, how strongly do you prefer P-1 to P-2?"

The response to each question takes the form of a value from one to nine and its reciprocal. The magnitude of the response indicates the strength of preference of one decision element to another. Saaty's suggested numbers to express degrees of preference between the two elements x and y are as follows:

If x is … as (than) y,	then the preference number to assign is
Equally important/preferred	1
Weakly more important/preferred	3
Strongly more important/preferred	5
Very strongly more important/preferred	7
Absolutely more important/preferred	9

Even numbers (2, 4, 6, 8) can be used to represent compromises among the preferences above. For inverse comparisons such as y to x, the reciprocal of the preference number for x to y (above) is used. The major assumptions of the AHP comparison procedure are

1. the relative weights of attributes within a level, conditional on each attribute in the immediately preceding level, are unidimensional;
2. pairwise judgments of attributes encompass all relevant aspects of importance;
3. decision makers can evaluate subproblems in an accurate and relatively consistent manner.

The results of the pairwise comparisons are placed in a matrix. Table 20-1 shows a matrix of preference numbers for the pairwise comparisons between attributes considered important to the selection of a manufacturing system. For example, Table 20-1 shows that, with respect to the overall focus, Serviceability (C) is weakly more important than Management/engineering effort (D) and is equally as important as Riskiness, lack of (E), and so on. Similar statements could be made about other rows. Note that the entries in each column are the reciprocals of the entries in the corresponding row, indicating the inverse relation of relative strength when attribute y is compared to attribute x versus x being compared to y. The only combinations that really require thoughtful judgment are those above or below the diagonal, because the "mirror-image" counterpart of each number is the reciprocal of that number and the diagonal entries are all 1 (since any given element must surely be equally preferred with that element). Note that the right-hand side of Table 20-1 shows the same paired comparison results both in decimal form and summed by columns to facilitate later calculations.

When the decision maker inputs the preferences on a pairwise basis, it is assumed that he or she does not know the vector of weights that characterizes the relative strength of each element with respect to a specific element at the next higher level in the hierarchy. Thus, after obtaining the pairwise judgments as in Table 20-1, the next step is the computation of a vector of priorities (or weighting of the elements in the matrix). In terms of matrix algebra, this consists of calculating the "principal vector" (eigenvector) of the matrix, and then normalizing it to sum to 1.0 or 100%. Standard programs are available for computing the principal vector of a matrix.

TABLE 20-1 Matrix of Paired Comparisons (Including Decimal Equivalents) for Attributes

With respect to the "best overall automated system"	A	B	C	D	E	Decimal equivalents				
						A	B	C	D	E
A. CIMS goals	1	1/3	5	6	5	1	0.333	5	6	5
B. Net present worth	3	1	6	7	6	3	1	6	7	6
C. Serviceability	1/5	1/6	1	3	1	0.20	0.167	1	3	1
D. Management/ engineering effort	1/6	1/7	1/3	1	1/4	0.167	0.143	0.333	1	0.25
E. Riskiness, lack of	1/5	1/6	1	4	1	0.20	0.167	1	4	1
						$\Sigma =$ 4.567	1.810	13.333	21	13.25

> The following is an approximation method for determining the principal vector, which is thought to provide sufficiently close results for most applications:[6] Divide the elements of each column by the sum of the column (i.e., normalize the column), then add the elements in each resulting row, and divide this sum by the number of elements in the row.

Table 20-2 shows the normalized matrix obtained by dividing each element in Table 20-1 by the sum of its respective column. Finally, row entries in the last two columns of Table 20-2 are comprised by the sum of the five elements in the row and the average of those row elements (principal vector), respectively. The results (principal vector) are that the attributes have the following approximate priority weights:

A.	CIMS goals	0.288
B.	Net present worth	0.489
C.	Serviceability	0.086
D.	Management/engineering effort	0.041
E.	Riskiness, lack of	0.096
		$\Sigma = \overline{1.000}$

TABLE 20-2   Normalized Matrix of Paired Comparisons and Calculations of Priority Weights (Approximate Attribute Weights)

	A	B	C	D	E	Row Σ	Average $= \Sigma/5$
A	0.219	0.184	0.375	0.286	0.377	1.441	0.288
B	0.657	0.553	0.450	0.333	0.453	2.446	0.489
C	0.044	0.092	0.075	0.143	0.076	0.430	0.086
D	0.036	0.079	0.025	0.048	0.019	0.207	0.041
E	0.044	0.092	0.075	0.190	0.074	0.477	0.096
	$\Sigma = \overline{1.000}$	$\overline{1.000}$	$\overline{1.000}$	$\overline{1.000}$	$\overline{1.000}$		$\overline{1.000}$

20.4 Determining the Relative Standing of Alternatives with Respect to Attributes

The next stage of the AHP is to determine the principal vector (priorities) of each of the alternatives with respect to each of the attributes to which they relate in the next higher level in Fig. 20-3. Typically, this is accomplished using the pairwise comparison process as was demonstrated in the previous section. However, it is also feasible to use quantified performance data (when available) to compute the priority weights of alternatives with respect to attributes. We will demonstrate the use of both methods in this section.

---

[6] The approximation method discussed in this chapter can be in error by as much as 25% when the comparison matrix is inconsistent. A better approximation is to raise the matrix to a large power until it converges.

20.4.1 Using Subjective Judgments to Prioritize Alternatives with Respect to Attributes

We will use the pairwise comparison process to determine the priority weights of alternatives with respect to the following attributes: A. CIMS goals, C. Service-ability, and E. Riskiness, lack of. We illustrate this only with respect to the first attribute, *CIMS goals*. Table 20-3 shows illustrative pairwise comparisons, and Table 20-4 shows subsequent calculations paralleling those in Table 20-2, resulting in the approximate principal vector (priority weights that are often descriptively called "evaluation ratings") with respect to CIMS goals. The results are

Alternative	Priority weight
P-1	0.21
P-2	0.55
P-3	0.24
	$\Sigma = \overline{1.00}$

Table 20-5 summarizes the results of evaluating the alternatives with respect to attributes A, C, and E. Note that the uppermost results are those for which calculations were discussed previously and shown in Tables 20-3 and 20-4. Results for attributes C and E are presented without showing the computational details.

20.4.2 Using Performance Data to Prioritize Alternatives with Respect to Attributes

When performance data are available, it is possible to rate alternatives with respect to attributes by using this data instead of the pairwise comparison process.

TABLE 20-3 Matrix of Paired Comparison Results for Alternatives (with Respect to CIMS Goals)

	P-1	P-2	P-3	P-1	P-2	P-3
P-1	1	1/3	1	1	0.333	1
P-2	3	1	2	3	1	2
P-3	1	1/2	1	1	0.500	1
				$\Sigma = \overline{5}$	$\overline{1.833}$	$\overline{4}$

TABLE 20-4 Normalized Matrix and Priority Weights for Alternatives (with Respect to CIMS Goals)

	P-1	P-2	P-3	Σ	$\Sigma/3$
P-1	0.20	0.18	0.25	0.63	0.21
P-2	0.60	0.55	0.50	1.65	0.55
P-3	0.20	0.27	0.25	0.72	0.24
	$\Sigma = \overline{1.00}$	$\overline{1.00}$	$\overline{1.00}$		$\overline{1.00}$

TABLE 20-5 Summary of Paired Comparisons and Resulting Priority
Weights for Alternatives with Respect to Selected Attributes

		P-1	P-2	P-3	Priority weights	Consistency index (CI)*	Consistency ratio (C.R.)*
A. CIMS goals	P-1	1	1/3	1	0.21		
	P-2	3	1	2	0.55	0.01	0.02
	P-3	1	1/2	1	$\underline{0.24}$		
					$\Sigma = 1.00$		
C. Serviceability	P-1	1	2	2	0.50		
					0.25		
	P-2	1/2	1	1	0.25	0.00	0.00
	P-3	1/2	1	1	$\underline{0.25}$		
					$\Sigma = 1.00$		
E. Riskiness,	P-1	1	3	4	0.62		
lack of					0.24		
	P-2	1/3	1	2	0.14	0.01	0.02
	P-3	1/4	1/2	1	$\underline{0.14}$		
					$\Sigma = 1.00$		

* The relevance and computation of the consistency index and the consistency ratio will be
discussed in Section 20.5.

Consider the attributes B. Net present worth and D. Management/engineering effort. In
this example, the numerical performance of each alternative with respect to these
attributes has been estimated. Table 20-6 summarizes the performance data (these data
were originally presented in Table 19-2). For each alternative, the net present worth
has been estimated in dollars and management/engineering effort has been estimated
in terms of "hours required." Notice that in the case of Net present worth, "higher" values are better, whereas "lower" values are better for Management/engineering effort.

Let us use the performance data to rate the three alternatives with respect to net
present worth. When higher values of a performance measure are "better," a one-step
normalization process of the data is utilized. To determine the priority weights for
alternatives with respect to net present worth, we simply divide each alternative's net
present worth by the sum of the net present worth of all alternatives. Doing so results
in the following priority weights for alternatives with respect to positive-valued net
present worth.

$$P-1 \quad \$300M/(\$300M + \$350M + \$400M) = 0.29$$
$$P-2 \quad \$350M/(\$300M + \$350M + \$400M) = 0.33$$
$$P-3 \quad \$400M/(\$300M + \$350M + \$400M) = \underline{0.38}$$
$$\Sigma = 1.00$$

TABLE 20-6 Estimated Alternative Performance Data for Selected Alternatives

		Alternative		
Attribute	Unit of measure	P-1	P-2	P-3
B. Net present worth	Dollars	300M	350M	400M
D. Management/engineering effort	Hours required	500	700	1100

Note that this normalization process results in P-3 having the highest weight with respect to net present worth and P-1 having the lowest weight. This is consistent with the fact that P-3 had the highest estimated net present worth while P-1 had the lowest.

We will also use performance data to obtain the vector of weights for the attribute Management/engineering effort. The performance measure used was "hours required," thus a low estimated value is desirable. When a lower performance value is "better," a two-step normalization process is utilized. The first step is to compute the ratio of the best (smallest) performance value to each alternative's performance value. The second step is to normalize these ratios such that they sum to one. This process results in the following alternative priority weights with respect to Management/engineering effort.

	Ratio	Normalized
P-1	500/500 = 1	0.46
P-2	500/700 = 0.7143	0.33
P-3	500/1100 = 0.4545	0.21
		$\Sigma = 1.00$

Note that once again, the rank order of alternative performance is retained in the normalized weights (i.e., P-1 has the most weight while P-3 has the least weight). The priority weights for alternatives with respect to attributes B and D are summarized in Table 20-7.

The use of performance data to obtain priority weights is promoted by the Auto-Man software package.[7] The advantage of this approach is its objectivity (subjective judgments are minimized). However, it is important to note that this method assumes that a linear relationship exists between a performance value and its relative weight (e.g., $50 is twice as good as $25). If a linear relationship cannot be assumed for a given attribute, then a pairwise comparison process using a nine-point scale should be used.

TABLE 20-7   Priority Weights for Alternatives with Respect to Quantified Attributes

	B. Net present worth	D. Management/ engineering effort
P-1	0.29	0.46
P-2	0.33	0.33
P-3	0.38	0.21
	$\Sigma = 1.00$	1.00

20.5 Determining the Consistency of Judgments

One of the strengths of the AHP is its ability to measure the degree of consistency present in the subjective judgments made by the decision maker. Judgmental consistency is concerned with the transitivity of preference in the pairwise comparison matrices. Consider the case in which attribute A is judged to be twice as important as attribute B and attribute B is judged to be twice as important as attribute C. Perfect cardinal consistency would then require that attribute A be judged four times as

---

[7] Stephen F. Weber, *AutoMan: Decision Support Software for Automated Manufacturing Investments* (U. S. Department of Commerce, National Institute of Standards and Technology, NISTIR 89-4166, 1989).

important as attribute C. Saaty developed a method (the AHP) by which one can measure the magnitude of departure from perfect consistency. The AHP includes both a local measure of consistency for individual comparison matrices and a global measure of consistency for the entire decision problem. We will demonstrate the calculation of the local measure of consistency for the automation alternatives example. The computation of the global measure will be discussed in Section 20.6.1.

The local consistency ratio (C.R.) is an approximate mathematical indicator, or guide, of the consistency of pairwise comparisons. It is a function of what is called the "maximum eigenvalue" and size of the matrix (called a "consistency index"), which is then compared against similar values if the pairwise comparisons had been merely random (called a "random index"). If the ratio of the consistency index to the random index (called a "consistency ratio") is no greater than 0.1, Saaty suggests the consistency is generally quite acceptable for pragmatic purposes. We will demonstrate the computation of a consistency ratio using the matrix of comparisons given in Table 20-1.

The first step is to multiply the matrix of comparisons, call it matrix $[A]$, by the principal vector or priority weights (right-hand column of Table 20-2) $[B]$ to get a new vector $[C]$.

$$
\begin{array}{ccccc}
& & [A] & & \\
1 & 0.33 & 5 & 6 & 5 \\
3 & 1 & 6 & 7 & 6 \\
0.2 & 0.167 & 1 & 3 & 1 \\
0.167 & 0.143 & 0.333 & 1 & 0.25 \\
0.2 & 0.167 & 1 & 4 & 1
\end{array}
\times
\begin{array}{c}
[B] \\
0.288 \\
0.489 \\
0.086 \\
0.041 \\
0.096
\end{array}
=
\begin{array}{c}
[C] \\
1.607 \\
2.732 \\
0.444 \\
0.212 \\
0.485
\end{array}
$$

Next, divide each element in vector $[C]$ by its corresponding element in vector $[B]$ to find a new vector $[D]$.

$$
[D] = \left| \frac{1.607}{0.288} \quad \frac{2.732}{0.489} \quad \frac{0.444}{0.086} \quad \frac{0.212}{0.041} \quad \frac{0.485}{0.096} \right|
$$

$$
= \left| 5.58 \quad 5.59 \quad 5.16 \quad 5.17 \quad 5.05 \right|
$$

Now, average the numbers in vector $[D]$. This is an approximation of what is called the "maximum eigenvalue," denoted by λ_{max}.

$$
\lambda_{max} = \frac{5.58 + 5.59 + 5.16 + 5.17 + 5.05}{5} = 5.31
$$

The consistency index (CI) for a matrix of size N is given by the formula

$$
CI = \frac{\lambda_{max} - N}{N - 1} = \frac{5.31 - 5}{5 - 1} = 0.08.
$$

Random indexes (RI) for various matrix sizes, N, have been approximated by Saaty (based on large numbers of simulation runs) as

N	1	2	3	4	5	6	7	8	9	10	11	...
RI	0.00	0.00	0.58	0.90	1.12	1.24	1.32	1.41	1.45	1.49	1.51	...

For the example above, the RI = 1.12. The consistency ratio (C.R.) can now be calculated using the relationship

$$\text{C.R.} = \frac{\text{CI}}{\text{RI}} = \frac{0.08}{1.12} = 0.07.$$

Based on Saaty's empirical suggestion that a C.R. = 0.10 is acceptable, we would conclude that the foregoing pairwise comparisons to obtain priority vectors are reasonably consistent.

20.5.1 Calculation for Alternative Comparisons with Respect to CIMS Goals (Table 20-3).

The calculation of a consistency ratio for alternatives is directly parallel to that for attributes as given above. For the pairwise comparisons in Table 20-3 (alternatives with respect to CIMS goals), the following calculations are shown without explanation.

$$\begin{matrix} [A] & & & [B] & [C] \end{matrix}$$

$$\begin{vmatrix} 1 & 0.333 & 1 \\ 3 & 1 & 2 \\ 1 & 0.5 & 1 \end{vmatrix} \times \begin{vmatrix} 0.21 \\ 0.55 \\ 0.24 \end{vmatrix} = \begin{vmatrix} 0.633 \\ 1.660 \\ 0.725 \end{vmatrix}$$

$$[D] = \begin{vmatrix} \dfrac{0.633}{0.21} & \dfrac{1.660}{0.55} & \dfrac{0.725}{0.24} \end{vmatrix} = \begin{vmatrix} 3.01 & 3.02 & 3.02 \end{vmatrix}$$

$$\lambda_{\max} = \frac{3.01 + 3.02 + 3.02}{3} = 3.02$$

$$\text{CI} = \frac{3.02 - 3}{3 - 1} = 0.01$$

$$\text{C.R.} = \frac{0.01}{0.58} = 0.02 \quad (0.58 \text{ is from bottom of p. 493})$$

The consistency ratios for the comparison matrices of alternatives with respect to attributes A, C, and E are displayed in Table 20-5. Recall that performance data were used to compute the priority vectors of alternatives with respect to attributes B and D. When performance data are used, the consistency index and the consistency ratio are equal to zero.

20.6 Determining the Overall Priority Weights of Alternatives

Table 20-8 summarizes all priority weights, in a form that is convenient for calculation of the final result, which is the vector of overall priority weights of alternatives. The following summarizes the content in "weighted evaluation" terminology (from Chapter 19):

- The attribute weights (from Table 20-2) are given in the top row in the body of Table 20-8.
- The evaluation ratings regarding how well each alternative meets each attribute (from Table 20-5 and Table 20-7) are given in the last three rows of the body of Table 20-8.

TABLE 20-8 Summary of Priority Weights Labeled as Attribute Weights, Evaluation Ratings,[a] and Weighted Evaluations

		Attribute				
	A: CIMS goals	B: Net present worth	C: Serviceability	D: Management/ engineering effort	E: Riskiness, lack of	Alternative priority weights[b]
Attribute weights	0.288	0.489	0.086	0.041	0.096	
Alternative						
P-1	0.21	0.29	0.50	0.46	0.62	0.324
P-2	0.55	0.33	0.25	0.33	0.24	0.378
P-3	0.24	0.38	0.25	0.21	0.14	0.298

[a] Evaluation ratings in body of matrix.
[b] Also called alternative "weighted evaluations" in Chapter 19.

- The weighted evaluation calculated results for each alternative are given in the right-hand column of Table 20-8.

The weighted evaluation for each alternative can be obtained by multiplying the matrix of evaluation ratings by the vector of priority weights and summing over all attributes. Expressed in conventional mathematical notation, we have

weighted evaluation for alternative $k = \Sigma_{\text{all } i \text{ attributes}}$
(priority weight$_i$ × evaluation rating$_{ik}$).

Thus, the priority weight for alternative P-1 is given by

$$0.288(0.21) + 0.489(0.29) + 0.086(0.5) + 0.041(0.46) + 0.096(0.62) = 0.324,$$

which is shown in the right-hand column of Table 20-8. Thus, alternative P-2 (with a priority weight or weighted evaluation of 0.378) is indicated to be more desirable than either alternative P-1 or P-3 (with priority weights of 0.324 and 0.298, respectively).

20.7 Added Explanation Regarding Example

Another way to show the structure of the automation alternatives example problem and the results of all priority weights (from Tables 20-2, 20-5, and 20-7) is given in Fig. 20-4. Using the results as displayed in Fig. 20-4, one can calculate the priority weight (weighted evaluation) for any alternative merely by summing the products of weights for all pathways leading to that alternative. Thus, for alternative P-1, the weighted evaluation would be $0.288(0.21) + 0.489(0.29) + 0.086(0.5) + 0.041(0.46) + 0.096(0.62) = 0.324$.

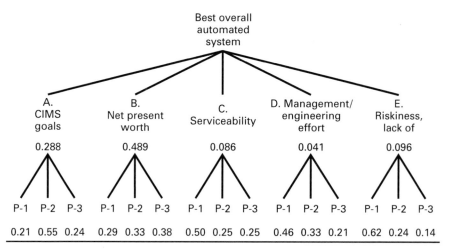

Figure 20-4   Decision hierarchy and priority weight results for the automation alternatives example.

20.7.1 The Global Consistency Ratio: C.R.H.

The measurement of consistency can be applied to the entire decision hierarchy. The global consistency ratio of the hierarchy (C.R.H.) is obtained by taking the ratio of an aggregate consistency index, M, for the entire three-level hierarchy to an aggregate random index, $\overline{M}$. These quantities are computed in the following manner.

$$M = \text{second-level CI} + \begin{vmatrix} \text{vector of} \\ \text{second-level} \\ \text{priority weights} \end{vmatrix} \times \begin{vmatrix} \text{vector of} \\ \text{third-level} \\ \text{CIs} \end{vmatrix}$$

$$\overline{M} = \text{second-level RI} + \begin{vmatrix} \text{vector of} \\ \text{second-level} \\ \text{priority weights} \end{vmatrix} \times \begin{vmatrix} \text{vector of} \\ \text{third-level} \\ \text{RIs} \end{vmatrix}$$

If the ratio of these values (C.R.H.) is no greater than 0.10, then the consistency of the hierarchy is generally acceptable. We will demonstrate the computation of the global consistency ratio for the automation alternatives example.

To compute M, we multiply the priority vector of the second-level attributes by the consistency indices of the third-level comparison matrices and add this result to the consistency index of the second-level comparison matrix. Recall that the consistency index for the attribute comparisons was equal to 0.08 (page 493). The consistency indices for the subjective comparisons were given in Table 20-5. The consistency index for both attributes B and D (where performance data were used) is equal to zero.

$$M = 0.08 + \begin{vmatrix} 0.288 & 0.489 & 0.086 & 0.041 & 0.096 \end{vmatrix} \times \begin{vmatrix} 0.01 \\ 0.00 \\ 0.00 \\ 0.00 \\ 0.01 \end{vmatrix} = 0.08$$

To compute $\overline{M}$, we multiply the vector of attribute weights by the vector of RIs corresponding to the size of the alternative comparison matrices and add to the RI corresponding to the size of the attribute comparison matrix.

$$\overline{M} = 1.12 + \begin{vmatrix} 0.288 & 0.489 & 0.086 & 0.041 & 0.096 \end{vmatrix} \times \begin{vmatrix} 0.58 \\ 0.58 \\ 0.58 \\ 0.58 \\ 0.58 \end{vmatrix} = 1.7$$

The global consistency ratio (C.R.H.) can now be calculated by using the relationship

$$\text{C.R.H.} = M / \overline{M} = \frac{0.08}{1.7} = 0.05.$$

Based on Saaty's empirical suggestion that a C.R.H. ≤ 0.10 is acceptable, we would conclude that the foregoing pairwise comparisons to obtain the overall alternative weights are reasonably consistent. This result is expected because we have already determined that each individual comparison matrix is consistent.

20.7.2 What to Do in the Presence of Inconsistency

If the consistency ratio of an individual matrix or the entire hierarchy is found to be unacceptable, the decision maker should review the judgments made and look for intransitivity. Often the process of assigning ratings on the nine-point scale is facilitated by first rank ordering the elements in the matrix from most important to least important. Rank ordering the elements in the matrix also highlights intransitivities (ratings in the upper diagonal of the matrix should then be greater than or equal to 1 and increase in value across the rows of the matrix). If the source of inconsistency is not apparent or cannot be resolved satisfactorily through the reestimation of preferences, it is possible that there is a problem with the hierarchical formulation of the problem. It may be that the attributes and/or alternatives being compared are not independent or that they are not directly comparable on a one-to-nine scale.

20.8 Hierarchies with More Than Three Levels

In complex decision problems it is likely that the decision hierarchy will consist of more than three levels. The purpose of this section is to illustrate the AHP computations for a four-level hierarchy. We will once again consider the three automation alternatives: P-1, P-2, and P-3. Figure 20-5 shows the decision elements for this problem arranged in a four-level hierarchy. The fourth level was created by associating three subattributes with the second-level attribute *CIMS goals*. In this example, the attribute CIMS goals can be fully described by the subattributes Improve product quality, Reduce inventory, and Improve manufacturing flexibility.

20.8.1 Determining the Relative Weights of Attributes and Alternatives

The basic steps of the AHP methodology remain the same regardless of the number of hierarchical levels. Pairwise comparisons must be made (or performance data used) to determine the relative weights of decision elements at one level of the hierarchy with respect to each decision element at the next-higher level of the hierarchy. In the previous three-level example, we first determined the weights of the attributes with respect to the focus and then determined the relative weights of the alternatives with respect to each attribute. The same general procedure applies to the four-level hierarchy, *except* we must perform the intermediate determination of subattribute weights with respect to the higher-level attribute CIMS goals.

The comparison of attributes with respect to the overall goal remains unchanged from the previous example. The results of this process are repeated in Table 20-9. The next step is to determine the relative weights of the third-level subattributes with respect to CIMS goals. The pairwise comparison matrix and resulting priority vector of weights are shown in Table 20-10. The subattribute Improve product quality has been judged to be the most important of the three goals.

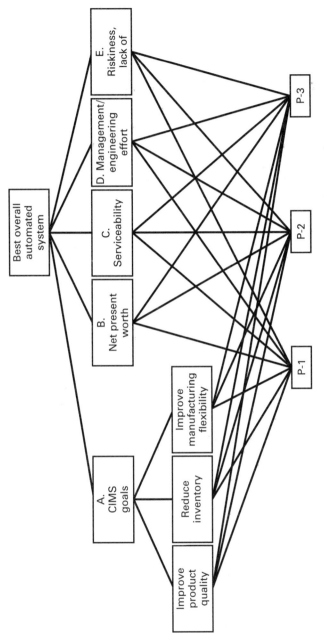

Figure 20-5 Four-level decision hierarchy for the automation alternatives example.

TABLE 20-9 Matrix of Paired Comparisons and Priority Weights for Attributes

With respect to the "best overall automated system"	A	B	C	D	E	Priority weight	Consistency
A. CIMS goals	1	1/3	5	6	5	0.288	
B. Net present worth	3	1	6	7	6	0.489	CI = 0.08
C. Serviceability	1/5	1/6	1	3	1	0.086	
D. Management/	1/6	1/7	1/3	1	1/4	0.041	C.R. = 0.07
engineering effort	1/5	1/6	1	4	1	0.096	
E. Riskiness, lack of						$\Sigma = 1.000$	

TABLE 20-10 Matrix of Paired Comparisons and Priority Weights for SubAttributes

With respect to "CIMS goals"	A1	A2	A3	Priority weight	Consistency
A1. Improve product quality	1	2	5	0.59	CI = 0.003
A2. Reduce inventory	1/2	1	2	0.28	
A3. Improve manufacturing flexibility	1/5	1/2	1	0.13	C.R. = 0.005
				$\Sigma = 1.00$	

TABLE 20-11 Priority Weights of Alternatives with Respect to Selected Attributes

	Attribute			
	B	C	D	E
P-1	0.29	0.50	0.46	0.62
P-2	0.33	0.25	0.33	0.24
P-3	0.38	0.25	0.21	0.14
CI	0.00	0.00	0.00	0.01
C.R.	0.00	0.00	0.00	0.02

The last step is to rate the alternatives with respect to the higher-level decision elements. The comparison matrices with respect to attributes C and E were previously shown in Table 20-5. Performance data were used to rate the alternatives with respect to attributes B and D, and the results were given in Table 20-7. The vectors of alternative weights with respect to these attributes are summarized in Table 20-11. The relative weights of alternatives with respect to CIMS goals are indirectly determined by determining the weights of the alternatives with respect to each of the three subattributes that together define CIMS goals. Table 20-12 shows the resulting comparison matrices and priority weights.

Before the above weights are used to compute the overall priority of alternatives, we need to check the consistency of the judgments made. As we will see in the next section, the weights shown in Table 20-12 are *not* the weights we will use in the final analysis.

TABLE 20-12 Summary of Paired Comparisons and Resulting Priority
Weights for Alternatives with Respect to Subattributes

		P-1	P-2	P-3	Priority weights	Consistency index (CI)	Consistency ratio (C.R.)
A1. Improve	P-1	1	1/2	1	0.26		
product quality	P-2	2	1	1/9	0.21	0.550	0.948
	P-3	1	9	1	0.53		
					$\Sigma = 1.00$		
A2. Reduce	P-1	1	1/2	1	0.24		
inventory	P-2	2	1	3	0.55	0.009	0.016
	P-3	1	1/3	1	0.21		
					$\Sigma = 1.00$		
A3. Improve	P-1	1	1/8	1/3	0.08		
manufacturing	P-2	8	1	3	0.68	0.001	0.002
flexibility	P-3	3	1/3	1	0.24		
					$\Sigma = 1.00$		

20.8.2 Determining the Consistency of the Comparisons

The consistency ratios of individual comparison matrices are computed as
described in Section 20.5 and are displayed next to each comparison matrix in Table
20-12. Note that the consistency ratio for the comparison of alternatives with respect
to the subattribute Improve product quality was calculated to be 0.948. This value
exceeds the empirical upper limit of 0.10. Thus we would conclude that the judgments
made were inconsistent. Before completing the AHP analysis, it is important to
resolve any detected inconsistencies.

To resolve the inconsistency problem, let us look more closely at the comparison matrix with respect to Improve product quality (shown in Table 20-12). In the first
row and second column of the comparison matrix, the value 1/2 indicates that alternatives P-2 is twice as preferred as P-1. The 1 in the third column indicates that P-1 is
equally preferred to P-3 with respect to improving product quality. Based on these
judgments, it would be reasonable to conclude that alternative P-2 should also be
twice as preferred as P-3. However, looking at the comparison matrix, P-3 has been
judged to be nine times more preferred as P-2. This is the source of the inconsistency
detected by the consistency ratio. To resolve the problem, we must decide if P-2 is
twice as preferred as P-1 (and therefore P-3) or if P-3 (and therefore P-1) is nine times
as preferred as P-2.

For this example, we will assume that P-2 is twice as preferred as both P-1 and
P-3. The revised comparison matrix and resulting priority weights are given in Table
20-13. The impact of not resolving the inconsistency problem on the overall ranking
of alternatives will be discussed in the next section.

20.8.3 Determining the Overall Weights of Alternatives

Figure 20-6 summarizes the results of the analysis in a format analogous to
Table 20-8. Using the results as displayed in this format, the overall priority weight of

TABLE 20-13   Revised Paired Comparisons and Resulting Priority
Weights for Alternatives with Respect to Subattributes

		P-1	P-2	P-3	Priority weights	Consistency index (CI)	Consistency ratio (C.R.)
A1. Improve	P-1	1	1/2	1	0.25		
product quality	P-2	2	1	2	0.50	0.000	0.000
	P-3	1	1/2	1	0.25		
					$\Sigma = 1.00$		
A2. Reduce	P-1	1	1/2	1	0.24		
inventory	P-2	2	1	3	0.55	0.009	0.016
	P-3	1	1/3	1	0.21		
					$\Sigma = 1.00$		
A3. Improve	P-1	1	1/8	1/3	0.08		
manufacturing	P-2	8	1	3	0.68	0.001	0.002
flexibility	P-3	3	1/3	1	0.24		
					$\Sigma = 1.00$		

an alternative is computed by summing the product of weights for all branches that
include the alternative. Thus, for alternative P-1, the priority weight would be

$$\text{P-1 weight} = 0.25(0.59)(0.288) + 0.24(0.28)(0.288) + 0.08(0.13)(0.288)$$
$$+ 0.29(0.489) + 0.50(0.086) + 0.46(0.041) + 0.62(0.096)$$
$$= 0.328.$$

The overall priority weights for the three alternatives are given in Fig. 20-6. The
results indicate that alternative P-2 is more desirable than either P-1 or P-3. Note that
similar results were obtained when we used a three-level hierarchy.

     Different results would have been obtained if the inconsistency discussed in
Section 20.8.2 had been left unresolved. If we had used the priority weights shown in
Table 20-12 (instead of the revised weights in Table 20-13), the overall weights for the
alternatives would have been

$$\text{P-1} = 0.330,$$
$$\text{P-2} = 0.325,$$
$$\text{P-3} = 0.345.$$

These results erroneously indicate that alternative P-3 is slightly more desirable than
either P-1 or P-2.

20.8.4 The Global Consistency Measure

     The method described in Section 20.7.1 for computing the global consistency
measure was for use with a three-level hierarchy. However, we can generalize this
procedure and apply it to a four-level (or more) hierarchy in the following manner.

     The first step of this procedure is to isolate the portion of the hierarchy related to
the second-level attribute CIMS goals (see Fig. 20-7). This attribute, its subattributes,
and the alternatives make up a three-level hierarchy. We can apply the procedure
described in Section 20.7.1 to compute an aggregate consistency index, M, for the
CIMS goals attribute. Doing so yields $M = 0.006$. This value can now be used in

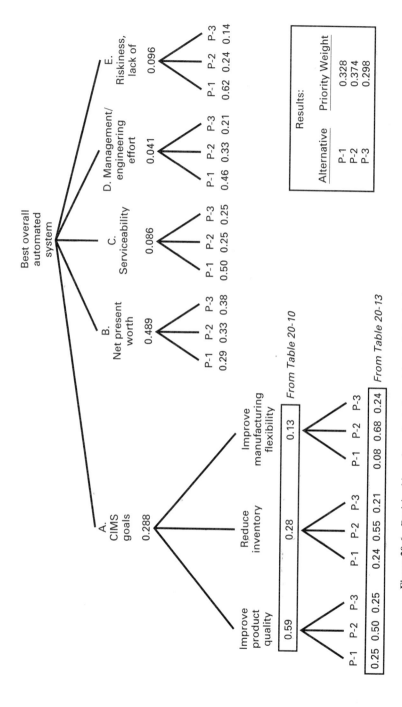

Figure 20-6 Decision hierarchy and priority weight results for the four-level automation alternatives example.

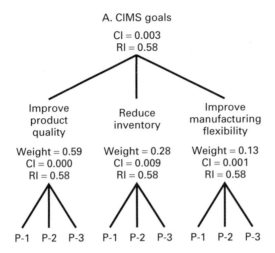

Figure 20-7 Three-level subset of decision hierarchy for calculation of an aggregate consistency index for CIMS goals.

conjunction with the CIs of the other second-level attributes to obtain the C.R.H. for the entire decision problem (see Fig. 20-8).

Now that we have a consistency index associated with each second-level attribute, we can again use the procedure described in Section 20.7.1, to compute the C.R.H. for the entire hierarchy. The results are

$$M = 0.083, \qquad \overline{M} = 1.7,$$
$$\text{C.R.H.} = M / \overline{M} = 0.049.$$

Based on the C.R.H. value, we conclude that our overall consistency is acceptable.

20.9 Problems of Which to Be Aware When Applying the AHP

Though widely used to solve decision problems, the AHP is not without its critics. A number of theoretical complaints have surfaced in the literature, especially when the AHP is applied to capital investment decision problems. The intent of this section is to make the reader aware of these issues when applying the AHP to a decision problem.

20.9.1 Vagueness of the Questioning Procedure

One of the major criticisms of the AHP is the manner in which the attribute weights are elicited and assessed. The decision maker is asked questions such as, "Which is more important, good gas mileage or low maintenance, and by how much?" The crucial observation is that such questions are meaningless. A clear interpretation of the question would require knowing what value or range of gas mileage is being compared to what level or range of maintenance.

The conclusion that has been drawn is that the decision maker must be thinking of some average quantities when making such judgments. Otherwise it would not be

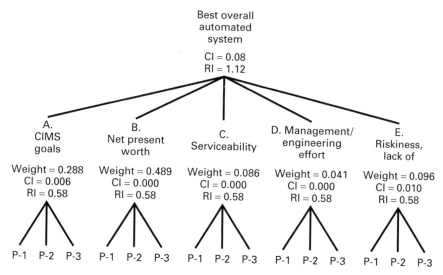

Figure 20-8   Composite three-level hierarchy for computation of the global consistency measure (C.R.H.).

reasonable to make such a judgment. Boucher and MacStravic[8] demonstrate that unless the average quantity interpretation is assumed, the AHP cannot be relied on to give results that are consistent with a simple monetary interpretation of the priority weights. In their paper, the implied dollar values of the alternatives are compared, and it is shown that the ranking given by the AHP is not consistent with the ranking obtained using the dollar values. Thus, it is important that the decision maker be aware of what is being compared when making judgments of relative importance.

20.9.2 Rank Reversal

Another well-documented problem with the AHP is the problem of *rank reversal*.[9] Rank reversal is when the introduction of a new alternative reverses the rankings of previously evaluated alternatives. This phenomenon is attributed to the fact that attribute weights are assessed independently of the alternatives under consideration. Thus, there is no relationship between the weights of the alternatives with respect to attributes and the overall attribute weights. If the anomaly of rank reversal occurs, it is because the principle of hierarchic composition (Section 20.2) is being violated. To avoid the problem of rank reversal, it is important that great care be exercised when constructing the decision hierarchy. Not only should the decision attributes be independent of each other, but the attributes should be independent of the alternatives being considered.

---

[8] Thomas O. Boucher and Elin L. MacStravic, "Multiattribute Evaluation Within a Present Worth Framework and Its Relation to the Analytic Hierarchy Process," *The Engineering Economist* 37, no. 1 (1991):1–32.

[9] Valerie Belton and Tony Gear, "On a Short-Coming of Saaty's Method of Analytic Hierarchies," *Omega* 11, no. 3 (1983):228–230; James S. Dyer, "Remarks on the Analytic Hierarchy Process," *Management Science* 36, no. 3 (1990):249–258; S. R. Watson and A. N. S. Freeling, "Assessing Attribute Weights," *Omega* 10, no. 6 (1982):582–583.

20.9.3 Aggregating Benefits and Costs

In many applications of the AHP, two hierarchies are used to evaluate alternatives, a benefits hierarchy and a cost hierarchy. The use of two hierarchies results in two measures associated with each alternative, a benefits priority vector and a cost priority vector. To aggregate these measures, Saaty[10] recommends computing the benefit–cost ratio for each alternative. The "best" alternative is then identified as the one having the highest benefit–cost ratio.

This procedure has been criticized by Bernhard and Canada[11] for failing to yield optimal results for the general case. A correct analysis needs to consider incremental benefits and costs. In addition, one needs to consider the relative preference of benefits versus costs. They conclude that an analysis not including such information can lead a decision maker to the wrong conclusion, and they illustrate this point via a numerical example.

A second problem to be aware of is the general arbitrariness inherent in any labeling procedure with regard to whether a negative feature is called a cost or a "disbenefit," or whether a positive feature is labeled a benefit or "negative cost." Bernhard and Canada show that the results obtained using Saaty's benefit–cost ratio method are sensitive to how the benefits and cost hierarchies are defined. However, this problem may be alleviated through the use of an incremental benefit–cost analysis.

20.10 Other AHP-Based Tools for Multicriteria Decision Problems

The success of the AHP has prompted researchers to develop new multicriteria decision-making methodologies that incorporate some of the features of the AHP. The use of a hierarchy to structure the decision problem and a pairwise comparison process to weight decision elements has appeared in newer methodologies. What makes these methodologies unique are their interpretation of the weights and how the weights are used to select the "best" alternative.

Putrus[12] developed a model specifically for the justification of CIM systems. A specific hierarchy was developed that identifies the important attributes of CIM systems, and pairwise comparisons are used to rate the elements of the hierarchy. These weights are then aggregated to measure total benefits and total risk. The decision to invest in CIM is based on the ratio of total benefits to total risk.

MacStravic and Boucher[13] developed a multicriteria capital investment justification technique that incorporates the finer features of the AHP but overcomes many of its perceived deficiencies. A hierarchical framework was developed for manufacturing investment decisions. A pairwise comparison process is used to weight deci-

---

[10] Thomas L. Saaty, *The Analytic Hierarchy Process* (New York: McGraw-Hill Book Company, 1980).

[11] Richard H. Bernhard and John R. Canada, "Some Problems in Using Benefit/Cost Ratios with the Analytic Hierarchy Process," *The Engineering Economist* 36, no. 1 (1990):56–65.

[12] Robert Putrus, "Non-Traditional Approach in Justifying Computer Integrated Manufacturing Systems," *AUTOFACT '89* (Oct. 30–Nov. 2, 1989), Detroit, MI.

[13] Elin L. MacStravic and Thomas O. Boucher, "Users Manual: NCIC Decision Support Software for Investment in Advanced Manufacturing Technology," Rutgers University Center for Advanced Food Technology (Aug. 1991).

sion elements. Unlike the AHP, the weights are then converted into dollar amounts. This allows the "best" alternative to be identified via traditional economic analysis techniques (e.g., present worth).

20.11 Summary

Commonly claimed benefits of the AHP are as follows:

1. It is relatively simple to use and understand.
2. It necessitates the construction of a hierarchy of attributes, subattributes, alternatives, and so on, which facilitates communication of the problem and the recommended solution(s).
3. It provides a unique means of quantifying judgmental consistency.

The AHP does provide remarkable versatility and power in structuring and analyzing complex multiattribute decision problems.

REFERENCES

ARBEL, A., AND A. SEIDMANN, "Performance Evaluation of Flexible Manufacturing Systems," *IEEE Transactions on Systems, Man, and Cybernetics* SMC-14, no. 4 (July–Aug. 1984):606–617.

BELTON, V., AND T. GEAR, "On a Short-Coming of Saaty's Method of Analytic Hierarchies," *Omega* 11, no. 3 (1983):228–230.

BERNHARD, R. H., AND J. R. CANADA, "Some Problems in Using Benefit/Cost Ratios with the Analytic Hierarchy Process," *The Engineering Economist* 36, no. 1 (1990):56–65.

BOUCHER, T. O., AND E. L. MACSTRAVIC, "Multiattribute Evaluation Within a Present Worth Framework and Its Relation to the Analytic Process," *The Engineering Economist* 37, no. 1 (1991):1–32.

DYER, J. S., "Remarks on the Analytic Hierarchy Process," *Management Science* 36, no. 3 (1990):249–258.

FORMAN, E. H., T. L. SAATY, M. A. SELLY, AND R. WALDRON, *Expert Choice,* McLean, VA: Decision Support Software, Inc., 1983.

MACSTRAVIC, E. L., AND T. O. BOUCHER, "Users Manual: NCIC Decision Support Software for Investment in Advanced Manufacturing Technology," Rutgers University Center for Advanced Food Technology (Aug. 1991).

PUTRUS, R., "Non-Traditional Approach in Justifying Computer Integrated Manufacturing Systems," *AUTOFACT '89* (Oct. 30–Nov. 2, 1989), Detroit, MI.

SAATY, T. L., *The Analytic Hierarchy Process.* New York: McGraw-Hill Book Company, 1980.

SAATY, T. L., *Decision Making for Leaders.* Belmont, CA: Wadsworth Publishing Company, Inc., 1982.

WATSON, S. R. AND A. N. S. FREELING, "Assessing Attribute Weights," *Omega* 10, no. 6 (1982): 582–583.

WEBER, S. F., *AutoMan: Decision Support Software for Automated Manufacturing Investments.* U. S. Department of Commerce, National Institute of Standards and Technology, NISTIR 89-4166, 1989.

PROBLEMS

20-1. Three alternatives are under consideration to improve certain manufacturing operations. The following attributes are considered important to the selection of the best alternative: Reduce product lead time, Improve product quality, and Maximize net present worth.

a. Given the following matrix of comparisons, compute the priority weights for attributes with respect to the focus.

	A	B	C
A: Reduce product lead time	1	1/5	1/9
B: Improve product quality	5	1	1/2
C: Maximize net present worth	9	2	1

b. Use the following performance data to rate the three alternatives with respect to the decision attributes.

		Performance of		
Attribute	Unit of measure	Alt. 1	Alt. 2	Alt. 3
Reduce product lead time	Product lead time in days	8 days	10 days	7 days
Improve product quality	% of nondefective products	97%	99%	95%
Maximize net present worth	Dollars	500M	600M	700M

c. Based on the priority weights computed in (a) and (b), which alternative would you recommend?

20-2. The method presented in this chapter for computing priority weights is an approximation. The priority vector shown below was computed via a better approximation method (the matrix was raised to successively larger powers until it converged). Use the method discussed in this chapter to approximate the priority vector. Comment on any differences.

	A	B	C	D	E	Priority weights
A	1	7	8	7	9	0.6143
B	1/7	1	7	7	7	0.2402
C	1/8	1/7	1	2	3	0.0657
D	1/7	1/7	1/2	1	2	0.0476
E	1/9	1/7	1/3	1/2	1	0.0322

20-3. Compute the consistency ratio for the following matrix of comparisons. Comment on the source(s) of inconsistency (if any).

	A	B	C	D	E	Priority weights
A	1	7	8	7	9	0.6143
B	1/7	1	7	7	7	0.2402
C	1/8	1/7	1	2	3	0.0657
D	1/7	1/7	1/2	1	2	0.0476
E	1/9	1/7	1/3	1/2	1	0.0322

20-4. An engineer expresses his preferences between major attributes with respect to his focus objective of career satisfaction as follows:

	(M)	(W)	(F)
Money (M)	1	2	1/3
Work (W)	1/2	1	1
Family (F)	3	1	1

His evaluation of each of job types A, B, and C are expressed as follows with respect to M, W, and F:

(M)	A	B	C		(W)	A	B	C		(F)	A	B	C
A	1	4	7		A	1	1/2	1		A	1	2	5
B	1/4	1	2		B	2	1	1/3		B	1/2	1	2
C	1/7	1/2	1		C	1	3	1		C	1/5	1/2	1

a. Find the priority weights for all matrices, and for the job types.

b. Find consistency ratios for all matrices. What are your conclusions from this and part (a)?

20-5. It is desired to determine which of three alternatives is best for a high-volume coil winding operation using the AHP method. The alternatives are

X: Upgrade existing machine

Y: New semiautomatic machine

Z: Fully automatic machine

The attributes important to the decision are   The comparison matrices are given below:

N: Net present worth

R: Risk

F: Future capability

M: Management requirements

a. Determine priority weights for each matrix and for the alternatives.

b. Determine consistency ratios for all matrices. What are your conclusions?

c. Show your results of part (a) in bar graph (histogram) form. Incorporate the results of part (b) in whatever way seems useful.

Using the classical 1 to 9 scale, the following express degrees of preference, with respect to overall goal:

	N	R	F	M
N	1	3	6	7
R		1	5	6
F			1	3
M				1

With respect to:

N : net present worth			
	X	Y	Z
X	1	1/5	1/7
Y		1	1/3
Z			1

R : risk			
	X	Y	Z
X	1	1/6	1
Y		1	6
Z			1

With respect to:

F : future capability			
	X	Y	Z
X	1	1/6	1/7
Y		1	1/3
Z			1

M : management requirements			
	X	Y	Z
X	1	1/5	1
Y		1	5
Z			1

20-6. You have just been transferred! Your company has offered two alternatives for a new position with them, and you decide to stay with your present employer. Use AHP to select your new job. (Now compare results with problem 19-11 results.)

	Alternative	
Attribute	Dry Gulch	Rapid Creek
Salary	$40,000	$35,000
Proximity to relatives	2,000 miles	500 miles
Promotion potential	Fair	Good
Commuting time (per day)	40 minutes	90 minutes

20-7. Select a personal decision problem to which you can relate fairly readily (such as alternative jobs, cars, housing, travel, etc.) and for which you have not previously calculated attribute weights using the AHP methodology. Identify two or three alternatives that might be compared (for reference). Then identify the four most important attributes that are, in your mind, reasonably distinct and independent. Rank order the alternatives, placing the most important first, and so on.

a. Weight the attributes and normalize the total to 100 points as exemplified by the *traditional* weighted evaluation method. Do a few checks for consistency until you are satisfied that those weights reasonably reflect your approximate weights. Do not alter based on any later thinking or results.

b. Weight the attributes by use of the AHP method of pairwise comparisons based on Saaty's ratio preference/importance factors of 1 to 9. Calculate your C.R. (consistency ratio). If it is below 0.10, great—otherwise, go back and reask yourself some or all pairwise comparison questions and recalculate your weights until the C.R. ≤ 0.10.

c. Repeat part (b) except use Canada's ratio preference/importance factors of 1 to 3. The following reflects scales for parts (b) and (c).

If x(row) is … as y(column)	Saaty's factors		Canada's factors	
	No.	1/No.	No.	1/No.
Equally important/preferred	1	1	1	1
Weakly more important/preferred	3	1/3 = 0.33	1.5	1/1.5 = 0.67
Strongly more important/preferred	5	1/5 = 0.20	2	1/2 = 0.50
Very strongly more important/preferred	7	1/7 = 0.14	2.5	1/2.5 = 0.40
Absolutely more important/preferred	9	1/9 = 0.11	3	1/3 = 0.33

d. What can you conclude about weighting results using Saaty's versus Canada's factors regarding closeness to "traditional" (in part (a))?

20-8. Use AHP to select a multiple attribute decision model from this set: weighted evaluation (WE, Chapter 19), Brown-Gibson (Chapter 19), and AHP (Chapter 20). Your overall goal (focus) is to be satisfied that you've done a good job with the treatment of nonmonetary criteria in a multimillion-dollar plant expansion. Your recommendation will be to employ one of the three models noted above in this important evaluation. Your audience is the firm's Board of Directors, which consists of two engineers, three attorneys and three MBAs. Remember, your job is to select the most appropriate model. Select or create five attributes from those listed below.

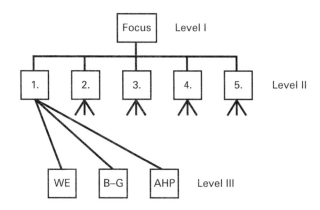

(1) *Theoretically sound.* The model must be internally and logically consistent with the decision maker's value system in the choice situation.

(2) *Credible.* The user must have confidence in the model's recommendation(s). The model must be accurate.

(3) *Verifiable.* Objectivity (lack of bias), consistency of results, and explicitness of the process are key to replicability and verification.

(4) *Comprehensive.* Can all relevant aspects of the choice situation be handled by the model?

(5) *Reasonable data requirements.* Will obtaining data to support the model cost more than the value of (hopefully) improved decisions? If decision making is not improved, the value of information is negligible.

(6) *Assumptions made explicit.* To be credible, the underlying assumptions of the model must be known to the user.

(7) *Uncertainties (riskiness) explicitly treated.* Sources of variation in a model's recommendation(s) must be identified, quantified, and eventually managed (assuming that the model is a realistic portrayal of the decision situation).

(8) *Effects of time taken into account.* Sequential decisions over time are common to continuous improvement programs and ought to be adequately included in the model.

(9) *Suitable to a group process.* Complex decision opportunities span many traditional functional areas of the organization, and a multiattribute model must accommodate consensus-driven decision making.

(10) *Communication ease.* This is what it all comes down to: If the user doesn't understand the model or its findings, she or he will not use it. The multiattribute model chosen must be one that supports decision-making activity because of its usefulness.

20-9. Formulate the following situation at the Consumer Products Company as an AHP problem.

Question:	If the advertising budget for Brand X is doubled, what will happen to sales (and market share)?
Response:	Market share will change −2% to +10%; current production capacity is available to respond to resulting sales at a +5% market share (or less).
Actions:	Anticipate increase and build inventory at distribution centers.
	Wait and see—if market share increases above +5%, subcontract business to a foreign competitor.
	Wait and see—if market share increases above +5%, build a second production line in an existing building.
Investment decision:	If the advertising budget is doubled, how can the second production line be justified? What information is required, and how would it be used to evaluate this situation?

20-10. Refer to Table 20-12 and confirm that the consistency ratio for Improve product quality is correct (0.948). Explain in your own words where the inconsistency lies in the matrix.

20-11. Compute the global consistency ratio (C.R.H.) for Problem 20-5.

20-12. Research Question. Investigate the properties of these four types of measurement scales: nominal, ordinal, interval, and ratio. Which measurement scale is implied by the AHP? What can you say about the results of the AHP if these scale properties are not present in the data being used in the analysis?

APPENDIX A

Tables of Discrete Compounding Interest Factors (for Various Common Values of i from ¼% to 25%)

TABLE A-1 Discrete Compounding; $i = 1/4\%$

	Single payment		Uniform series				
	Compound amount factor	Present worth factor	Compound amount factor	Present worth factor	Sinking fund factor	Capital recovery factor	
	To find F given P	To find P given F	To find F given A	To find P given A	To find A given F	To find A given P	
N	F/P	P/F	F/A	P/A	A/F	A/P	N
1	1.0025	0.9975	1.0000	0.9975	1.0000	1.0025	1
2	1.0050	0.9950	2.0025	1.9925	0.4994	0.5019	2
3	1.0075	0.9925	3.0075	2.9851	0.3325	0.3350	3
4	1.0100	0.9901	4.0150	3.9751	0.2491	0.2516	4
5	1.0126	0.9876	5.0250	4.9627	0.1990	0.2015	5
6	1.0151	0.9851	6.0376	5.9478	0.1656	0.1681	6
7	1.0176	0.9827	7.0527	6.9305	0.1418	0.1443	7
8	1.0202	0.9802	8.0703	7.9107	0.1239	0.1264	8
9	1.0227	0.9778	9.0905	8.8885	0.1100	0.1125	9
10	1.0253	0.9753	10.1132	9.8638	0.0989	0.1014	10
11	1.0278	0.9729	11.1385	10.8367	0.0898	0.0923	11
12	1.0304	0.9705	12.1663	11.8072	0.0822	0.0847	12
13	1.0330	0.9681	13.1967	12.7753	0.0758	0.0783	13
14	1.0356	0.9656	14.2297	13.7409	0.0703	0.0728	14
15	1.0382	0.9632	15.2653	14.7041	0.0655	0.0680	15
16	1.0408	0.9608	16.3035	15.6650	0.0613	0.0638	16
17	1.0434	0.9584	17.3442	16.6234	0.0577	0.0602	17
18	1.0460	0.9561	18.3876	17.5795	0.0544	0.0569	18
19	1.0486	0.9537	19.4335	18.5331	0.0515	0.0540	19
20	1.0512	0.9513	20.4821	19.4844	0.0488	0.0513	20
21	1.0538	0.9489	21.5333	20.4333	0.0464	0.0489	21
22	1.0565	0.9466	22.5872	21.3799	0.0443	0.0468	22
23	1.0591	0.9442	23.6436	22.3241	0.0423	0.0448	23
24	1.0618	0.9418	24.7027	23.2659	0.0405	0.0430	24
25	1.0644	0.9395	25.7645	24.2054	0.0388	0.0413	25
26	1.0671	0.9371	26.8289	25.1425	0.0373	0.0398	26
27	1.0697	0.9438	27.8959	26.0773	0.0358	0.0383	27
28	1.0724	0.9325	28.9657	27.0098	0.0345	0.0370	28
29	1.0751	0.9302	30.0381	27.9399	0.0333	0.0358	29
30	1.0778	0.9278	31.1132	28.8678	0.0321	0.0346	30
35	1.0913	0.9163	36.5291	33.4723	0.0274	0.0299	35
40	1.1050	0.9050	42.0130	38.0197	0.0238	0.0263	40
45	1.1189	0.8937	47.5659	42.5107	0.0210	0.0235	45
50	1.1330	0.8826	53.1884	46.9460	0.0188	0.0213	50
55	1.1472	0.8717	58.8817	51.3262	0.0170	0.0195	55
60	1.1616	0.8609	64.6464	55.6521	0.0155	0.0180	60
65	1.1762	0.8502	70.4836	59.9244	0.0142	0.0167	65
70	1.1910	0.8396	76.3941	64.1436	0.0131	0.0156	70
75	1.2059	0.8292	82.3788	68.3105	0.0121	0.0146	75
80	1.2211	0.8189	88.4388	72.4257	0.0113	0.0138	80
85	1.2364	0.8088	94.5748	76.490	0.0106	0.0131	85
90	1.2520	0.7987	100.788	80.504	0.0099	0.0124	90
95	1.2677	0.7888	107.079	84.467	0.0093	0.0118	95
100	1.2836	0.7790	113.45	88.382	0.0088	0.0113	100
∞				400.0		0.0025	∞

TABLE A-2 Discrete Compounding; $i = 1/2\%$

	Single payment		Uniform series				
	Compound amount factor	Present worth factor	Compound amount factor	Present worth factor	Sinking fund factor	Capital recovery factor	
	To find F given P	To find P given F	To find F given A	To find P given A	To find A given F	To find A given P	
N	F/P	P/F	F/A	P/A	A/F	A/P	N
1	1.0050	0.9950	1.0000	0.9950	1.0000	1.0050	1
2	1.0100	0.9901	2.0050	1.9851	0.4988	0.5038	2
3	1.0151	0.9851	3.0150	2.9702	0.3317	0.3367	3
4	1.0202	0.9802	4.0301	3.9505	0.2481	0.2531	4
5	1.0253	0.9754	5.0502	4.9259	0.1980	0.2030	5
6	1.0304	0.9705	6.0755	5.8964	0.1646	0.1696	6
7	1.0355	0.9657	7.1059	6.8621	0.1407	0.1457	7
8	1.0407	0.9609	8.1414	7.8229	0.1228	0.1278	8
9	1.0459	0.9561	9.1821	8.7790	0.1089	0.1139	9
10	1.0511	0.9513	10.2280	9.7304	0.0978	0.1028	10
11	1.0564	0.9466	11.2791	10.6770	0.0887	0.0937	11
12	1.0617	0.9419	12.3355	11.6189	0.0811	0.0861	12
13	1.0670	0.9372	13.3972	12.5561	0.0746	0.0796	13
14	1.0723	0.9326	14.4642	13.4887	0.0691	0.0741	14
15	1.0777	0.9279	15.5365	14.4166	0.0644	0.0694	15
16	1.0831	0.9233	16.6142	15.3399	0.0602	0.0652	16
17	1.0885	0.9187	17.6973	16.2586	0.0565	0.0615	17
18	1.0939	0.9141	18.7857	17.1727	0.0532	0.0582	18
19	1.0994	0.9096	19.8797	18.0823	0.0503	0.0553	19
20	1.1049	0.9051	20.9791	18.9874	0.0477	0.0527	20
21	1.1104	0.9006	22.0839	19.8879	0.0453	0.0503	21
22	1.1160	0.8961	23.1944	20.7840	0.0431	0.0481	22
23	1.1216	0.8916	24.3103	21.6756	0.0411	0.0461	23
24	1.1272	0.8872	25.4319	22.5628	0.0393	0.0443	24
25	1.1328	0.8828	26.5590	23.4456	0.0377	0.0427	25
26	1.1385	0.8784	27.6918	24.3240	0.0361	0.0411	26
27	1.1442	0.8740	28.8303	25.1980	0.0347	0.0397	27
28	1.1499	0.8697	29.9744	26.0676	0.0334	0.0384	28
29	1.1556	0.8653	31.1243	26.9330	0.0321	0.0371	29
30	1.1614	0.8610	32.2799	27.7940	0.0310	0.0360	30
35	1.1907	0.8398	38.1453	32.0353	0.0262	0.0312	35
40	1.2208	0.8191	44.1587	36.1721	0.0226	0.0276	40
45	1.2516	0.7990	50.3240	40.2071	0.0199	0.0249	45
50	1.2832	0.7793	56.6450	44.1427	0.0177	0.0227	50
55	1.3156	0.7601	63.1256	47.9813	0.0158	0.0208	55
60	1.3488	0.7414	69.7698	51.7254	0.0143	0.0193	60
65	1.3829	0.7231	76.5818	55.3773	0.0131	0.0181	65
70	1.4178	0.7053	83.5658	58.9393	0.0120	0.0170	70
75	1.4536	0.6879	90.7262	62.4135	0.0110	0.0160	75
80	1.4903	0.6710	98.0674	65.8022	0.0102	0.0152	80
85	1.5280	0.6545	105.594	69.107	0.0095	0.0145	85
90	1.5666	0.6383	113.311	72.331	0.0088	0.0138	90
95	1.6061	0.6226	121.22	75.475	0.0082	0.0132	95
100	1.6467	0.6073	129.33	78.542	0.0077	0.0127	100
∞				200.0		0.0050	∞

	Single payment		Uniform series				
	Compound amount factor	Present worth factor	Compound amount factor	Present worth factor	Sinking fund factor	Capital recovery factor	
N	To find F given P F/P	To find P given F P/F	To find F given A F/A	To find P given A P/A	To find A given F A/F	To find A given P A/P	N
1	1.0075	0.9926	1.0000	0.9926	1.0000	1.0075	1
2	1.0151	0.9852	2.0075	1.9777	0.4981	0.5056	2
3	1.0227	0.9778	3.0226	2.9556	0.3308	0.3383	3
4	1.0303	0.9706	4.0452	3.9261	0.2472	0.2547	4
5	1.0381	0.9633	5.0756	4.8894	0.1970	0.2045	5
6	1.0459	0.9562	6.1136	5.8456	0.1636	0.1711	6
7	1.0537	0.9490	7.1595	6.7946	0.1397	0.1472	7
8	1.0616	0.9420	8.2132	7.7366	0.1218	0.1293	8
9	1.0696	0.9350	9.2748	8.6716	0.1078	0.1153	9
10	1.0776	0.9280	10.3443	9.5996	0.0967	0.1042	10
11	1.0857	0.9211	11.4219	10.5207	0.0876	0.0951	11
12	1.0938	0.9142	12.5076	11.4349	0.0800	0.0875	12
13	1.1020	0.9074	13.6014	12.3423	0.0735	0.0810	13
14	1.1103	0.9007	14.7034	13.2430	0.0680	0.0755	14
15	1.1186	0.8940	15.8136	14.1370	0.0632	0.0707	15
16	1.1270	0.8873	16.9322	15.0243	0.0591	0.0666	16
17	1.1354	0.8807	18.0592	15.9050	0.0554	0.0629	17
18	1.1440	0.8742	19.1947	16.7791	0.0521	0.0596	18
19	1.1525	0.8676	20.3386	17.6468	0.0492	0.0567	19
20	1.1612	0.8612	21.4912	18.5080	0.0465	0.0540	20
21	1.1699	0.8548	22.6523	19.3628	0.0441	0.0516	21
22	1.1787	0.8484	23.8222	20.2112	0.0420	0.0495	22
23	1.1875	0.8421	25.0009	21.0533	0.0400	0.0475	23
24	1.1964	0.8358	26.1884	21.8891	0.0382	0.0457	24
25	1.2054	0.8296	27.3848	22.7187	0.0365	0.0440	25
26	1.2144	0.8234	28.5902	23.5421	0.0350	0.0425	26
27	1.2235	0.8173	29.8046	24.3594	0.0336	0.0411	27
28	1.2327	0.8112	31.0282	25.1707	0.0322	0.0397	28
29	1.2420	0.8052	32.2609	25.9758	0.0310	0.0385	29
30	1.2513	0.7992	33.5028	26.7750	0.0298	0.0373	30
35	1.2989	0.7699	39.8537	30.6826	0.0251	0.0326	35
40	1.3483	0.7416	46.4464	34.4469	0.0215	0.0290	40
45	1.3997	0.7145	53.2900	38.0731	0.0188	0.0263	45
50	1.4530	0.6883	60.3941	41.5664	0.0166	0.0241	50
55	1.5083	0.6630	67.7686	44.9315	0.0148	0.0223	55
60	1.5657	0.6387	75.4239	48.1733	0.0133	0.0208	60
65	1.6253	0.6153	83.3706	51.2962	0.0120	0.0195	65
70	1.6871	0.5927	91.6198	54.3045	0.0109	0.0184	70
75	1.7514	0.5710	100.183	57.2026	0.0100	0.0175	75
80	1.8180	0.5500	109.072	59.9943	0.0092	0.0167	80
85	1.8872	0.5299	118.300	62.6837	0.0085	0.0160	85
90	1.9591	0.5104	127.879	65.275	0.0078	0.0153	90
95	2.0337	0.4917	137.822	67.770	0.0073	0.0148	95
100	2.1111	0.4737	148.14	70.175	0.0068	0.0143	100
∞				133.333		0.0075	∞

TABLE A-4 Discrete Compounding; i = 1%

	Single payment		Uniform series				
	Compound amount factor	Present worth factor	Compound amount factor	Present worth factor	Sinking fund factor	Capital recovery factor	
	To find F given P	To find P given F	To find F given A	To find P given A	To find A given F	To find A given P	
N	F/P	P/F	F/A	P/A	A/F	A/P	N
1	1.0100	0.9901	1.0000	0.9901	1.0000	1.0100	1
2	1.0201	0.9803	2.0100	1.9704	0.4975	0.5075	2
3	1.0303	0.9706	3.0301	2.9410	0.3300	0.3400	3
4	1.0406	0.9610	4.0604	3.9020	0.2463	0.2563	4
5	1.0510	0.9515	5.1010	4.8534	0.1960	0.2060	5
6	1.0615	0.9420	6.1520	5.7955	0.1625	0.1725	6
7	1.0721	0.9327	7.2135	6.7282	0.1386	0.1486	7
8	1.0829	0.9235	8.2857	7.6517	0.1207	0.1307	8
9	1.0937	0.9143	9.3685	8.5660	0.1067	0.1167	9
10	1.1046	0.9053	10.4622	9.4713	0.0956	0.1056	10
11	1.1157	0.8963	11.5668	10.3676	0.0865	0.0965	11
12	1.1268	0.8874	12.6825	11.2551	0.0788	0.0888	12
13	1.1381	0.8787	13.8093	12.1337	0.0724	0.0824	13
14	1.1495	0.8700	14.9474	13.0037	0.0669	0.0769	14
15	1.1610	0.8613	16.0969	13.8650	0.0621	0.0721	15
16	1.1726	0.8528	17.2578	14.7178	0.0579	0.0679	16
17	1.1843	0.8444	18.4304	15.5622	0.0543	0.0643	17
18	1.1961	0.8360	19.6147	16.3982	0.0510	0.0610	18
19	1.2081	0.8277	20.8109	17.2260	0.0481	0.0581	19
20	1.2202	0.8195	22.0190	18.0455	0.0454	0.0554	20
21	1.2324	0.8114	23.2391	18.8570	0.0430	0.0530	21
22	1.2447	0.8034	24.4715	19.6603	0.0409	0.0509	22
23	1.2572	0.7954	25.7162	20.4558	0.0389	0.0489	23
24	1.2697	0.7876	26.9734	21.2434	0.0371	0.0471	24
25	1.2824	0.7798	28.2431	22.0231	0.0354	0.0454	25
26	1.2953	0.7720	29.5256	22.7952	0.0339	0.0439	26
27	1.3082	0.7644	30.8208	23.5596	0.0324	0.0424	27
28	1.3213	0.7568	32.1290	24.3164	0.0311	0.0411	28
29	1.3345	0.7493	33.4503	25.0657	0.0299	0.0399	29
30	1.3478	0.7419	34.7848	25.8077	0.0287	0.0387	30
35	1.4166	0.7059	41.6602	29.4085	0.0240	0.0340	35
40	1.4889	0.6717	48.8863	32.8346	0.0205	0.0305	40
45	1.5648	0.6391	56.4809	36.0945	0.0177	0.0277	45
50	1.6446	0.6080	64.4630	39.1961	0.0155	0.0255	50
55	1.7285	0.5785	72.8523	42.1471	0.0137	0.0237	55
60	1.8167	0.5505	81.6695	44.9550	0.0122	0.0222	60
65	1.9094	0.5237	90.9364	47.6265	0.0110	0.0210	65
70	2.0068	0.4983	100.676	50.1684	0.0099	0.0199	70
75	2.1091	0.4741	110.912	52.5870	0.0090	0.0190	75
80	2.2167	0.4511	121.671	54.8881	0.0082	0.0182	80
85	2.3298	0.4292	132.979	57.0776	0.0075	0.0175	85
90	2.4486	0.4084	144.86	59.161	0.0069	0.0169	90
95	2.5735	0.3886	157.35	61.143	0.0064	0.0164	95
100	2.7048	0.3697	170.48	63.029	0.0059	0.0159	100
∞				100.000		0.0100	∞

TABLE A-5 Discrete Compounding; $i = 1\text{-}1/2\%$

	Single payment		Uniform series				
	Compound amount factor	Present worth factor	Compound amount factor	Present worth factor	Sinking fund factor	Capital recovery factor	
	To find F given P	To find P given F	To find F given A	To find P given A	To find A given F	To find A given P	
N	F/P	P/F	F/A	P/A	A/F	A/P	N
1	1.0150	0.9852	1.0000	0.9852	1.0000	1.0150	1
2	1.0302	0.9707	2.0150	1.9559	0.4963	0.5113	2
3	1.0457	0.9563	3.0452	2.9122	0.3284	0.3434	3
4	1.0614	0.9422	4.0909	3.8544	0.2444	0.2594	4
5	1.0773	0.9283	5.1523	4.7826	0.1941	0.2091	5
6	1.0934	0.9145	6.2295	5.6972	0.1605	0.1755	6
7	1.1098	0.9010	7.3230	6.5982	0.1366	0.1516	7
8	1.1265	0.8877	8.4328	7.4859	0.1186	0.1336	8
9	1.1434	0.8746	9.5593	8.3605	0.1046	0.1196	9
10	1.1605	0.8617	10.7027	9.2222	0.0934	0.1084	10
11	1.1779	0.8489	11.8632	10.0711	0.0843	0.0993	11
12	1.1956	0.8364	13.0412	10.9075	0.0767	0.0917	12
13	1.2136	0.8240	14.2368	11.7315	0.0702	0.0852	13
14	1.2318	0.8118	15.4504	12.5434	0.0647	0.0797	14
15	1.2502	0.7999	16.6821	13.3432	0.0599	0.0749	15
16	1.2690	0.7880	17.9323	14.1312	0.0558	0.0708	16
17	1.2880	0.7764	19.2013	14.9076	0.0521	0.0671	17
18	1.3073	0.7649	20.4893	15.6725	0.0488	0.0638	18
19	1.3270	0.7536	21.7967	16.4261	0.0459	0.0609	19
20	1.3469	0.7425	23.1236	17.1686	0.0432	0.0582	20
21	1.3671	0.7315	24.4705	17.9001	0.0409	0.0559	21
22	1.3876	0.7207	25.8375	18.6208	0.0387	0.0537	22
23	1.4084	0.7100	27.2251	19.3308	0.0367	0.0517	23
24	1.4295	0.6995	28.6335	20.0304	0.0349	0.0499	24
25	1.4509	0.6892	30.0630	20.7196	0.0333	0.0483	25
26	1.4727	0.6790	31.5139	21.3986	0.0317	0.0467	26
27	1.4948	0.6690	32.9866	22.0676	0.0303	0.0453	27
28	1.5172	0.6591	34.4814	22.7267	0.0290	0.0440	28
29	1.5400	0.6494	35.9986	23.3761	0.0278	0.0428	29
30	1.5631	0.6398	37.5386	24.0158	0.0266	0.0416	30
35	1.6839	0.5939	45.5920	27.0756	0.0219	0.0369	35
40	1.8140	0.5513	54.2676	29.9158	0.0184	0.0334	40
45	1.9542	0.5117	63.6141	32.5523	0.0157	0.0307	45
50	2.1052	0.4750	73.6827	34.9997	0.0136	0.0286	50
55	2.2679	0.4409	84.5294	37.2714	0.0118	0.0268	55
60	2.4432	0.4093	96.2145	39.3802	0.0104	0.0254	60
65	2.6320	0.3799	108.803	41.3378	0.0092	0.0242	65
70	2.8355	0.3527	122.364	43.1548	0.0082	0.0232	70
75	3.0546	0.3274	136.97	44.8416	0.0073	0.0223	75
80	3.2907	0.3039	152.71	46.4073	0.0065	0.0215	80
85	3.5450	0.2821	169.66	47.8607	0.0059	0.0209	85
90	3.8189	0.2619	187.93	49.2098	0.0053	0.0203	90
95	4.1141	0.2431	207.61	50.4622	0.0048	0.0198	95
100	4.4320	0.2256	228.80	51.6247	0.0044	0.0194	100
∞				66.667		0.0150	∞

TABLE A-6   Discrete Compounding; $i = 2\%$

	Single payment		Uniform series				
	Compound amount factor	Present worth factor	Compound amount factor	Present worth factor	Sinking fund factor	Capital recovery factor	
	To find F given P	To find P given F	To find F given A	To find P given A	To find A given F	To find A given P	
N	F/P	P/F	F/A	P/A	A/F	A/P	N
1	1.0200	0.9804	1.0000	0.9804	1.0000	1.0200	1
2	1.0404	0.9612	2.0200	1.9416	0.4950	0.5150	2
3	1.0612	0.9423	3.0604	2.8839	0.3268	0.3468	3
4	1.0824	0.9238	4.1216	3.8077	0.2426	0.2626	4
5	1.1041	0.9057	5.2040	4.7135	0.1922	0.2122	5
6	1.1262	0.8880	6.3081	5.6014	0.1585	0.1785	6
7	1.1487	0.8706	7.4343	6.4720	0.1345	0.1545	7
8	1.1717	0.8535	8.5830	7.3255	0.1165	0.1365	8
9	1.1951	0.8368	9.7546	8.1622	0.1025	0.1225	9
10	1.2190	0.8203	10.9497	8.9826	0.0913	0.1113	10
11	1.2434	0.8043	12.1687	9.7868	0.0822	0.1022	11
12	1.2682	0.7885	13.4121	10.5753	0.0746	0.0946	12
13	1.2936	0.7730	14.6803	11.3484	0.0681	0.0881	13
14	1.3195	0.7579	15.9739	12.1062	0.0626	0.0826	14
15	1.3459	0.7430	17.2934	12.8493	0.0578	0.0778	15
16	1.3728	0.7284	18.6393	13.5777	0.0537	0.0737	16
17	1.4002	0.7142	20.0121	14.2919	0.0500	0.0700	17
18	1.4282	0.7002	21.4123	14.9920	0.0467	0.0667	18
19	1.4568	0.6864	22.8405	15.6785	0.0438	0.0638	19
20	1.4859	0.6730	24.2974	16.3514	0.0412	0.0612	20
21	1.5157	0.6598	25.7833	17.0112	0.0388	0.0588	21
22	1.5460	0.6468	27.2990	17.6580	0.0366	0.0566	22
23	1.5769	0.6342	28.8449	18.2922	0.0347	0.0547	23
24	1.6084	0.6217	30.4218	18.9139	0.0329	0.0529	24
25	1.6406	0.6095	32.0303	19.5234	0.0312	0.0512	25
26	1.6734	0.5976	33.6709	20.1210	0.0297	0.0497	26
27	1.7069	0.5859	35.3443	20.7069	0.0283	0.0483	27
28	1.7410	0.5744	37.0512	21.2813	0.0270	0.0470	28
29	1.7758	0.5631	38.7922	21.8444	0.0258	0.0458	29
30	1.8114	0.5521	40.5681	22.3964	0.0246	0.0446	30
35	1.9999	0.5000	49.9944	24.9986	0.0200	0.0400	35
40	2.2080	0.4529	60.4019	27.3555	0.0166	0.0366	40
45	2.4379	0.4102	71.8927	29.4902	0.0139	0.0339	45
50	2.6916	0.3715	84.5793	31.4236	0.0118	0.0318	50
55	2.9717	0.3365	98.5864	33.1748	0.0101	0.0301	55
60	3.2810	0.3048	114.051	34.7609	0.0088	0.0288	60
65	3.6225	0.2761	131.126	36.1975	0.0076	0.0276	65
70	3.9996	0.2500	149.978	37.4986	0.0067	0.0267	70
75	4.4158	0.2265	170.792	38.6771	0.0059	0.0259	75
80	4.8754	0.2051	193.772	39.7445	0.0052	0.0252	80
85	5.3829	0.1858	219.144	40.7113	0.0046	0.0246	85
90	5.9431	0.1683	247.16	41.5869	0.0040	0.0240	90
95	6.5617	0.1524	278.08	42.3800	0.0036	0.0236	95
100	7.2446	0.1380	312.23	43.0983	0.0032	0.0232	100
∞				50.0000		0.0200	∞

TABLE A-7 Discrete Compounding; $i = 3\%$

	Single payment		Uniform series				
	Compound amount factor	Present worth factor	Compound amount factor	Present worth factor	Sinking fund factor	Capital recovery factor	
N	To find F given P F/P	To find P given F P/F	To find F given A F/A	To find P given A P/A	To find A given F A/F	To find A given P A/P	N
1	1.0300	0.9709	1.0000	0.9709	1.0000	1.0300	1
2	1.0609	0.9426	2.0300	1.9135	0.4926	0.5226	2
3	1.0927	0.9151	3.0909	2.8286	0.3235	0.3535	3
4	1.1255	0.8885	4.1836	3.7171	0.2390	0.2690	4
5	1.1593	0.8626	5.3091	4.5797	0.1884	0.2184	5
6	1.1941	0.8375	6.4684	5.4172	0.1546	0.1846	6
7	1.2299	0.8131	7.6625	6.2303	0.1305	0.1605	7
8	1.2668	0.7894	8.8923	7.0197	0.1125	0.1425	8
9	1.3048	0.7664	10.1591	7.7861	0.0984	0.1284	9
10	1.3439	0.7441	11.4639	8.5302	0.0872	0.1172	10
11	1.3842	0.7224	12.8078	9.2526	0.0781	0.1081	11
12	1.4258	0.7014	14.1920	9.9540	0.0705	0.1005	12
13	1.4685	0.6810	15.6178	10.6349	0.0640	0.0940	13
14	1.5126	0.6611	17.0863	11.2961	0.0585	0.0885	14
15	1.5580	0.6419	18.5989	11.9379	0.0538	0.0838	15
16	1.6047	0.6232	20.1569	12.5611	0.0496	0.0796	16
17	1.6528	0.6050	21.7616	13.1661	0.0460	0.0760	17
18	1.7024	0.5874	23.4144	13.7535	0.0427	0.0727	18
19	1.7535	0.5703	25.1168	14.3238	0.0398	0.0698	19
20	1.8061	0.5537	26.8703	14.8775	0.0372	0.0672	20
21	1.8603	0.5375	28.6765	15.4150	0.0349	0.0649	21
22	1.9161	0.5219	30.5367	15.9369	0.0327	0.0627	22
23	1.9736	0.5067	32.4528	16.4436	0.0308	0.0608	23
24	2.0328	0.4919	34.4264	16.9355	0.0290	0.0590	24
25	2.0938	0.4776	36.4592	17.4131	0.0274	0.0574	25
26	2.1566	0.4637	38.5530	17.8768	0.0259	0.0559	26
27	2.2213	0.4502	40.7096	18.3270	0.0246	0.0546	27
28	2.2879	0.4371	42.9309	18.7641	0.0233	0.0533	28
29	2.3566	0.4243	45.2188	19.1884	0.0221	0.0521	29
30	2.4273	0.4120	47.5754	19.6004	0.0210	0.0510	30
35	2.8139	0.3554	60.4620	21.4872	0.0165	0.0465	35
40	3.2620	0.3066	75.4012	23.1148	0.0133	0.0433	40
45	3.7816	0.2644	92.7197	24.5187	0.0108	0.0408	45
50	4.3839	0.2281	112.797	25.7298	0.0089	0.0389	50
55	5.0821	0.1968	136.071	26.7744	0.0073	0.0373	55
60	5.8916	0.1697	163.053	27.6756	0.0061	0.0361	60
65	6.8300	0.1464	194.332	28.4529	0.0051	0.0351	65
70	7.9178	0.1263	230.594	29.1234	0.0043	0.0343	70
75	9.1789	0.1089	272.630	29.7018	0.0037	0.0337	75
80	10.6409	0.0940	321.362	30.2008	0.0031	0.0331	80
85	12.3357	0.0811	377.856	30.6311	0.0026	0.0326	85
90	14.3004	0.0699	443.35	31.0024	0.0023	0.0323	90
95	16.5781	0.0603	519.27	31.3227	0.0019	0.0319	95
100	19.2186	0.0520	607.29	31.5989	0.0016	0.0316	100
∞				33.3333		0.0300	∞

TABLE A-8   Discrete Compounding; $i = 4\%$

	Single payment		Uniform series				
	Compound amount factor	Present worth factor	Compound amount factor	Present worth factor	Sinking fund factor	Capital recovery factor	
N	To find F given P F/P	To find P given F P/F	To find F given A F/A	To find P given A P/A	To find A given F A/F	To find A given P A/P	N
1	1.0400	0.9615	1.0000	0.9615	1.0000	1.0400	1
2	1.0816	0.9246	2.0400	1.8861	0.4902	0.5302	2
3	1.1249	0.8890	3.1216	2.7751	0.3203	0.3603	3
4	1.1699	0.8548	4.2465	3.6299	0.2355	0.2755	4
5	1.2167	0.8219	5.4163	4.4518	0.1846	0.2246	5
6	1.2653	0.7903	6.6330	5.2421	0.1508	0.1908	6
7	1.3159	0.7599	7.8983	6.0021	0.1266	0.1666	7
8	1.3686	0.7307	9.2142	6.7327	0.1085	0.1485	8
9	1.4233	0.7026	10.5828	7.4353	0.0945	0.1345	9
10	1.4802	0.6756	12.0061	8.1109	0.0833	0.1233	10
11	1.5395	0.6496	13.4863	8.7605	0.0741	0.1141	11
12	1.6010	0.6246	15.0258	9.3851	0.0666	0.1066	12
13	1.6651	0.6006	16.6268	9.9856	0.0601	0.1001	13
14	1.7317	0.5775	18.2919	10.5631	0.0547	0.0947	14
15	1.8009	0.5553	20.0236	11.1184	0.0499	0.0899	15
16	1.8730	0.5339	21.8245	11.6523	0.0458	0.0858	16
17	1.9479	0.5134	23.6975	12.1657	0.0422	0.0822	17
18	2.0258	0.4936	25.6454	12.6593	0.0390	0.0790	18
19	2.1068	0.4746	27.6712	13.1339	0.0361	0.0761	19
20	2.1911	0.4564	29.7781	13.5903	0.0336	0.0736	20
21	2.2788	0.4388	31.9692	14.0292	0.0313	0.0713	21
22	2.3699	0.4220	34.2480	14.4511	0.0292	0.0692	22
23	2.4647	0.4057	36.6179	14.8568	0.0273	0.0673	23
24	2.5633	0.3901	39.0826	15.2470	0.0256	0.0656	24
25	2.6658	0.3751	41.6459	15.6221	0.0240	0.0640	25
26	2.7725	0.3607	44.3117	15.9828	0.0226	0.0626	26
27	2.8834	0.3468	47.0842	16.3296	0.0212	0.0612	27
28	2.9987	0.3335	49.9676	16.6631	0.0200	0.0600	28
29	3.1187	0.3207	52.9663	16.9837	0.0189	0.0589	29
30	3.2434	0.3083	56.0849	17.2920	0.0178	0.0578	30
35	3.9461	0.2534	73.6522	18.6646	0.0136	0.0536	35
40	4.8010	0.2083	95.0255	19.7928	0.0105	0.0505	40
45	5.8412	0.1712	121.029	20.7200	0.0083	0.0483	45
50	7.1067	0.1407	152.667	21.4822	0.0066	0.0466	50
55	8.6464	0.1157	191.159	22.1086	0.0052	0.0452	55
60	10.5196	0.0951	237.991	22.6235	0.0042	0.0442	60
65	12.7987	0.0781	294.968	23.0467	0.0034	0.0434	65
70	15.5716	0.0642	364.290	23.3945	0.0027	0.0427	70
75	18.9452	0.0528	448.631	23.6804	0.0022	0.0422	75
80	23.0498	0.0434	551.245	23.9154	0.0018	0.0418	80
85	28.0436	0.0357	676.090	24.1085	0.0015	0.0415	85
90	34.1193	0.0293	827.98	24.2673	0.0012	0.0412	90
95	41.5113	0.0241	1012.78	24.3978	0.0010	0.0410	95
100	50.5049	0.0198	1237.62	24.5050	0.0008	0.0408	100
∞				25.0000		0.0400	∞

TABLE A-9 Discrete Compounding; $i = 5\%$

	Single payment		Uniform series				
	Compound amount factor	Present worth factor	Compound amount factor	Present worth factor	Sinking fund factor	Capital recovery factor	
	To find F given P	To find P given F	To find F given A	To find P given A	To find A given F	To find A given P	
N	F/P	P/F	F/A	P/A	A/F	A/P	N
1	1.0500	0.9524	1.0000	0.9524	1.0000	1.0500	1
2	1.1025	0.9070	2.0500	1.8594	0.4878	0.5378	2
3	1.1576	0.8638	3.1525	2.7232	0.3172	0.3672	3
4	1.2155	0.8227	4.3101	3.5460	0.2320	0.2820	4
5	1.2763	0.7835	5.5256	4.3295	0.1810	0.2310	5
6	1.3401	0.7462	6.8019	5.0757	0.1470	0.1970	6
7	1.4071	0.7107	8.1420	5.7864	0.1228	0.1728	7
8	1.4775	0.6768	9.5491	6.4632	0.1047	0.1547	8
9	1.5513	0.6446	11.0266	7.1078	0.0907	0.1407	9
10	1.6289	0.6139	12.5779	7.7217	0.0795	0.1295	10
11	1.7103	0.5847	14.2068	8.3064	0.0704	0.1204	11
12	1.7959	0.5568	15.9171	8.8633	0.0628	0.1128	12
13	1.8856	0.5303	17.7130	9.3936	0.0565	0.1065	13
14	1.9799	0.5051	19.5986	9.8986	0.0510	0.1010	14
15	2.0789	0.4810	21.5786	10.3797	0.0463	0.0963	15
16	2.1829	0.4581	23.6575	10.8378	0.0423	0.0923	16
17	2.2920	0.4363	25.8404	11.2741	0.0387	0.0887	17
18	2.4066	0.4155	28.1324	11.6896	0.0355	0.0855	18
19	2.5269	0.3957	30.5390	12.0853	0.0327	0.0827	19
20	2.6533	0.3769	33.0659	12.4622	0.0302	0.0802	20
21	2.7860	0.3589	35.7192	12.8212	0.0280	0.0780	21
22	2.9253	0.3418	38.5052	13.1630	0.0260	0.0760	22
23	3.0715	0.3256	41.4305	13.4886	0.0241	0.0741	23
24	3.2251	0.3101	44.5020	13.7986	0.0225	0.0725	24
25	3.3864	0.2953	47.7271	14.0939	0.0210	0.0710	25
26	3.5557	0.2812	51.1134	14.3752	0.0196	0.0696	26
27	3.7335	0.2678	54.6691	14.6430	0.0183	0.0683	27
28	3.9201	0.2551	58.4026	14.8981	0.0171	0.0671	28
29	4.1161	0.2429	62.3227	15.1411	0.0160	0.0660	29
30	4.3219	0.2314	66.4388	15.3725	0.0151	0.0651	30
35	5.5160	0.1813	90.3203	16.3742	0.0111	0.0611	35
40	7.0400	0.1420	120.800	17.1591	0.0083	0.0583	40
45	8.9850	0.1113	159.700	17.7741	0.0063	0.0563	45
50	11.4674	0.0872	209.348	18.2559	0.0048	0.0548	50
55	14.6356	0.0683	272.713	18.6335	0.0037	0.0537	55
60	18.6792	0.0535	353.584	18.9293	0.0028	0.0528	60
65	23.8399	0.0419	456.798	19.1611	0.0022	0.0522	65
70	30.4264	0.0329	588.528	19.3427	0.0017	0.0517	70
75	38.8327	0.0258	756.653	19.4850	0.0013	0.0513	75
80	49.5614	0.0202	971.228	19.5965	0.0010	0.0510	80
85	63.2543	0.0158	1245.09	19.6838	0.0008	0.0508	85
90	80.7303	0.0124	1594.61	19.7523	0.0006	0.0506	90
95	103.035	0.0097	2040.69	19.8059	0.0005	0.0505	95
100	131.501	0.0076	2610.02	19.8479	0.0004	0.0504	100
∞				20.0000		0.0500	∞

	Single payment		Uniform series				
	Compound amount factor	Present worth factor	Compound amount factor	Present worth factor	Sinking fund factor	Capital recovery factor	
N	To find F given P F/P	To find P given F P/F	To find F given A F/A	To find P given A P/A	To find A given F A/F	To find A given P A/P	N
1	1.0600	0.9434	1.0000	0.9434	1.0000	1.0600	1
2	1.1236	0.8900	2.0600	1.8334	0.4854	0.5454	2
3	1.1910	0.8396	3.1836	2.6730	0.3141	0.3741	3
4	1.2625	0.7921	4.3746	3.4651	0.2286	0.2886	4
5	1.3382	0.7473	5.6371	4.2124	0.1774	0.2374	5
6	1.4185	0.7050	6.9753	4.9173	0.1434	0.2034	6
7	1.5036	0.6651	8.3938	5.5824	0.1191	0.1791	7
8	1.5938	0.6274	9.8975	6.2098	0.1010	0.1610	8
9	1.6895	0.5919	11.4913	6.8017	0.0870	0.1470	9
10	1.7908	0.5584	13.1808	7.3601	0.0759	0.1359	10
11	1.8983	0.5268	14.9716	7.8869	0.0668	0.1268	11
12	2.0122	0.4970	16.8699	8.3838	0.0593	0.1193	12
13	2.1329	0.4688	18.8821	8.8527	0.0530	0.1130	13
14	2.2609	0.4423	21.0151	9.2950	0.0476	0.1076	14
15	2.3966	0.4173	23.2760	9.7122	0.0430	0.1030	15
16	2.5404	0.3936	25.6725	10.1059	0.0390	0.0990	16
17	2.6928	0.3714	28.2129	10.4773	0.0354	0.0954	17
18	2.8543	0.3503	30.9056	10.8276	0.0324	0.0924	18
19	3.0256	0.3305	33.7600	11.1581	0.0296	0.0896	19
20	3.2071	0.3118	36.7856	11.4699	0.0272	0.0872	20
21	3.3996	0.2942	39.9927	11.7641	0.0250	0.0850	21
22	3.6035	0.2775	43.3923	12.0416	0.0230	0.0830	22
23	3.8197	0.2618	46.9958	12.3034	0.0213	0.0813	23
24	4.0489	0.2470	50.8155	12.5504	0.0197	0.0797	24
25	4.2919	0.2330	54.8645	12.7834	0.0182	0.0782	25
26	4.5494	0.2198	59.1563	13.0032	0.0169	0.0769	26
27	4.8223	0.2074	63.7057	13.2105	0.0157	0.0757	27
28	5.1117	0.1956	68.5281	13.4062	0.0146	0.0746	28
29	5.4184	0.1846	73.6397	13.5907	0.0136	0.0736	29
30	5.7435	0.1741	79.0581	13.7648	0.0126	0.0726	30
35	7.6861	0.1301	111.435	14.4982	0.0090	0.0690	35
40	10.2857	0.0972	154.762	15.0463	0.0065	0.0665	40
45	13.7646	0.0727	212.743	15.4558	0.0047	0.0647	45
50	18.4201	0.0543	290.336	15.7619	0.0034	0.0634	50
55	24.6503	0.0406	394.172	15.9905	0.0025	0.0625	55
60	32.9876	0.0303	533.128	16.1614	0.0019	0.0619	60
65	44.1449	0.0227	719.082	16.2891	0.0014	0.0614	65
70	59.0758	0.0169	967.931	16.3845	0.0010	0.0610	70
75	79.0568	0.0126	1300.95	16.4558	0.0008	0.0608	75
80	105.796	0.0095	1746.60	16.5091	0.0006	0.0606	80
85	141.579	0.0071	2342.98	16.5489	0.0004	0.0604	85
90	189.464	0.0053	3141.07	16.5787	0.0003	0.0603	90
95	253.546	0.0039	4209.10	16.6009	0.0002	0.0602	95
100	339.301	0.0029	5638.36	16.6175	0.0002	0.0602	100
∞				16.6667		0.0600	∞

	Single payment		Uniform series				
	Compound amount factor	Present worth factor	Compound amount factor	Present worth factor	Sinking fund factor	Capital recovery factor	
N	To find F given P F/P	To find P given F P/F	To find F given A F/A	To find P given A P/A	To find A given F A/F	To find A given P A/P	N
1	1.0700	0.9346	1.0000	0.9346	1.0000	1.0700	1
2	1.1449	0.8734	2.0700	1.8080	0.4831	0.5531	2
3	1.2250	0.8163	3.2149	2.6243	0.3111	0.3811	3
4	1.3108	0.7629	4.4399	3.3872	0.2252	0.2952	4
5	1.4026	0.7130	5.7507	4.1002	0.1739	0.2439	5
6	1.5007	0.6663	7.1533	4.7665	0.1398	0.2098	6
7	1.6058	0.6227	8.6540	5.3893	0.1156	0.1856	7
8	1.7182	0.5820	10.2598	5.9713	0.0975	0.1675	8
9	1.8385	0.5439	11.9780	6.5152	0.0835	0.1535	9
10	1.9672	0.5083	13.8164	7.0236	0.0724	0.1424	10
11	2.1049	0.4751	15.7836	7.4987	0.0634	0.1334	11
12	2.2522	0.4440	17.8884	7.9427	0.0559	0.1259	12
13	2.4098	0.4150	20.1406	8.3576	0.0497	0.1197	13
14	2.5785	0.3878	22.5505	8.7455	0.0443	0.1143	14
15	2.7590	0.3624	25.1290	9.1079	0.0398	0.1098	15
16	2.9522	0.3387	27.8880	9.4466	0.0359	0.1059	16
17	3.1588	0.3166	30.8402	9.7632	0.0324	0.1024	17
18	3.3799	0.2959	33.9990	10.0591	0.0294	0.0994	18
19	3.6165	0.2765	37.3790	10.3356	0.0268	0.0968	19
20	3.8697	0.2584	40.9955	10.5940	0.0244	0.0944	20
21	4.1406	0.2415	44.8652	10.8355	0.0223	0.0923	21
22	4.4304	0.2257	49.0057	11.0612	0.0204	0.0904	22
23	4.7405	0.2109	53.4361	11.2722	0.0187	0.0887	23
24	5.0724	0.1971	58.1766	11.4693	0.0172	0.0872	24
25	5.4274	0.1842	63.2490	11.6536	0.0158	0.0858	25
26	5.8074	0.1722	68.6764	11.8258	0.0146	0.0846	26
27	6.2139	0.1609	74.4838	11.9867	0.0134	0.0834	27
28	6.6488	0.1504	80.6977	12.1371	0.0124	0.0824	28
29	7.1143	0.1406	87.3465	12.2777	0.0114	0.0814	29
30	7.6123	0.1314	94.4607	12.4090	0.0106	0.0806	30
35	10.6766	0.0937	138.237	12.9477	0.0072	0.0772	35
40	14.9744	0.0668	199.635	13.3317	0.0050	0.0750	40
45	21.0024	0.0476	285.749	13.6055	0.0035	0.0735	45
50	29.4570	0.0339	406.529	13.8007	0.0025	0.0725	50
55	41.3150	0.0242	575.928	13.9399	0.0017	0.0717	55
60	57.9464	0.0173	813.520	14.0392	0.0012	0.0712	60
65	81.2728	0.0123	1146.75	14.1099	0.0009	0.0709	65
70	113.989	0.0088	1614.13	14.1604	0.0006	0.0706	70
75	159.876	0.0063	2269.66	14.1964	0.0004	0.0704	75
80	224.234	0.0045	3189.06	14.2220	0.0003	0.0703	80
85	314.500	0.0032	4478.57	14.2403	0.0002	0.0702	85
90	441.102	0.0023	6287.18	14.2533	0.0002	0.0702	90
95	618.669	0.0016	8823.85	14.2626	0.0001	0.0701	95
100	867.715	0.0012	12381.7	14.2693	[a]	0.0701	100

[a] Less than 0.0001.

TABLE A-12 Discrete Compounding; $i = 8\%$

	Single payment		Uniform series				
	Compound amount factor	Present worth factor	Compound amount factor	Present worth factor	Sinking fund factor	Capital recovery factor	
	To find F given P	To find P given F	To find F given A	To find P given A	To find A given F	To find A given P	
N	F/P	P/F	F/A	P/A	A/F	A/P	N
1	1.0800	0.9259	1.0000	0.9259	1.0000	1.0800	1
2	1.1664	0.8573	2.0800	1.7833	0.4808	0.5608	2
3	1.2597	0.7938	3.2464	2.5771	0.3080	0.3880	3
4	1.3605	0.7350	4.5061	3.3121	0.2219	0.3019	4
5	1.4693	0.6806	5.8666	3.9927	0.1705	0.2505	5
6	1.5869	0.6302	7.3359	4.6229	0.1363	0.2163	6
7	1.7138	0.5835	8.9228	5.2064	0.1121	0.1921	7
8	1.8509	0.5403	10.6366	5.7466	0.0940	0.1740	8
9	1.9990	0.5002	12.4876	6.2469	0.0801	0.1601	9
10	2.1589	0.4632	14.4866	6.7101	0.0690	0.1490	10
11	2.3316	0.4289	16.6455	7.1390	0.0601	0.1401	11
12	2.5182	0.3971	18.9771	7.5361	0.0527	0.1327	12
13	2.7196	0.3677	21.4953	7.9038	0.0465	0.1265	13
14	2.9372	0.3405	24.2149	8.2442	0.0413	0.1213	14
15	3.1722	0.3152	27.1521	8.5595	0.0368	0.1168	15
16	3.4259	0.2919	30.3243	8.8514	0.0330	0.1130	16
17	3.7000	0.2703	33.7502	9.1216	0.0296	0.1096	17
18	3.9960	0.2502	37.4502	9.3719	0.0267	0.1067	18
19	4.3157	0.2317	41.4463	9.6036	0.0241	0.1041	19
20	4.6610	0.2145	45.7620	9.8181	0.0219	0.1019	20
21	5.0338	0.1987	50.4229	10.0168	0.0198	0.0998	21
22	5.4365	0.1839	55.4567	10.2007	0.0180	0.0980	22
23	5.8715	0.1703	60.8933	10.3711	0.0164	0.0964	23
24	6.3412	0.1577	66.7647	10.5288	0.0150	0.0950	24
25	6.8485	0.1460	73.1059	10.6748	0.0137	0.0937	25
26	7.3964	0.1352	79.9544	10.8100	0.0125	0.0925	26
27	7.9881	0.1252	87.3507	10.9352	0.0114	0.0914	27
28	8.6271	0.1159	95.3388	11.0511	0.0105	0.0905	28
29	9.3173	0.1073	103.966	11.1584	0.0096	0.0896	29
30	10.0627	0.0994	113.283	11.2578	0.0088	0.0888	30
35	14.7853	0.0676	172.317	11.6546	0.0058	0.0858	35
40	21.7245	0.0460	259.056	11.9246	0.0039	0.0839	40
45	31.9204	0.0313	386.506	12.1084	0.0026	0.0826	45
50	46.9016	0.0213	573.770	12.2335	0.0017	0.0817	50
55	68.9138	0.0145	848.923	12.3186	0.0012	0.0812	55
60	101.257	0.0099	1253.21	12.376(	0.0008	0.0808	60
65	148.780	0.0067	1847.25	12.4160	0.0005	0.0805	65
70	218.606	0.0046	2720.08	12.4428	0.0004	0.0804	70
75	321.204	0.0031	4002.55	12.4611	0.0002	0.0802	75
80	471.955	0.0021	5886.93	12.4735	0.0002	0.0802	80
85	693.456	0.0014	8655.71	12.4820	0.0001	0.0801	85
90	1018.92	0.0010	12723.9	12.4877	a	0.0801	90
95	1497.12	0.0007	18071.5	12.4917	a	0.0801	95
100	2199.76	0.0005	27484.5	12.4943	a	0.0800	100
∞				12.5000		0.0800	∞

^{a} Less than 0.0001.

TABLE A-13 Discrete Compounding; i = 9%

	Single payment		Uniform series				
	Compound amount factor	Present worth factor	Compound amount factor	Present worth factor	Sinking fund factor	Capital recovery factor	
	To find F given P	To find P given F	To find F given A	To find P given A	To find A given F	To find A given P	
N	F/P	P/F	F/A	P/A	A/F	A/P	N
1	1.0900	0.9174	1.000	0.917	1.00000	1.09000	1
2	1.1881	0.8417	2.090	1.759	0.47847	1.56847	2
3	1.2950	0.7722	3.278	2.531	0.30505	0.39505	3
4	1.4116	0.7084	4.573	3.240	0.21867	0.30867	4
5	1.5386	0.6499	5.985	3.890	0.16709	0.25709	5
6	1.6771	0.5963	7.523	4.486	0.13292	0.22292	6
7	1.8280	0.5470	9.200	5.033	0.10869	0.19869	7
8	1.9926	0.5019	11.028	5.535	0.09067	0.18067	8
9	2.1719	0.4604	13.021	5.995	0.07680	0.16680	9
10	2.3674	0.4224	15.193	6.418	0.06582	0.15582	10
11	2.5804	0.3875	17.560	6.805	0.05695	0.14695	11
12	2.8127	0.3555	20.141	7.161	0.04965	0.13965	12
13	3.0658	0.3262	22.953	7.487	0.04357	0.13357	13
14	3.3417	0.2992	26.019	7.786	0.03843	0.12843	14
15	3.6425	0.2745	29.361	8.061	0.03406	0.12406	15
16	3.9703	0.2519	33.003	8.313	0.03030	0.12030	16
17	4.3276	0.2311	36.974	8.544	0.02705	0.11705	17
18	4.7171	0.2120	41.301	8.756	0.02421	0.11421	18
19	5.1417	0.1945	46.018	8.950	0.02173	0.11173	19
20	5.6044	0.1784	51.160	9.129	0.01955	0.10955	20
21	6.1088	0.1637	56.765	9.292	0.01762	0.10762	21
22	6.6586	0.1502	62.873	9.442	0.01590	0.10590	22
23	7.2579	0.1378	69.532	9.580	0.01438	0.10438	23
24	7.9111	0.1264	76.790	9.707	0.01302	0.10302	24
25	8.6231	0.1160	84.701	9.823	0.01181	0.10181	25
26	9.3992	0.1064	93.324	9.929	0.01072	0.10072	26
27	10.2451	0.0976	102.723	10.027	0.00973	0.09973	27
28	11.1671	0.0895	112.968	10.116	0.00885	0.09885	28
29	12.1722	0.0822	124.135	10.198	0.00806	0.09806	29
30	13.2677	0.0753	136.308	10.274	0.00734	0.09734	30
35	20.4140	0.0490	215.711	10.567	0.00464	0.09464	35
40	31.4094	0.0318	337.882	10.757	0.00296	0.09296	40
45	48.3273	0.0207	525.859	10.881	0.00190	0.09190	45
50	74.3575	0.0134	815.084	10.962	0.00123	0.09123	50
55	114.4083	0.0087	1260.092	11.014	0.00079	0.09079	55
60	176.0313	0.0057	1944.792	11.048	0.00051	0.09051	60
65	270.8460	0.0037	2998.288	11.070	0.00033	0.09033	65
70	416.7301	0.0024	4619.223	11.084	0.00022	0.09022	70
75	641.1909	0.0016	7113.232	11.094	0.00014	0.09014	75
80	986.5517	0.0010	10950.574	11.100	0.00009	0.09009	80
85	1517.9320	0.0007	16854.800	11.104	0.00006	0.09006	85
90	2235.5266	0.0004	25939.184	11.106	0.00004	0.09004	90
95	3593.4971	0.0003	39916.635	11.108	0.00003	0.09003	95
100	5529.0408	0.0002	61422.675	11.109	0.00002	0.09002	100

TABLE A-14 Discrete Compounding; $i = 10\%$

	Single payment		Uniform series				
	Compound amount factor	Present worth factor	Compound amount factor	Present worth factor	Sinking fund factor	Capital recovery factor	
	To find F given P	To find P given F	To find F given A	To find P given A	To find A given F	To find A given P	
N	F/P	P/F	F/A	P/A	A/F	A/P	N
1	1.1000	0.9091	1.0000	0.9091	1.0000	1.1000	1
2	1.2100	0.8264	2.1000	1.7355	0.4762	0.5762	2
3	1.3310	0.7513	3.3100	2.4869	0.3021	0.4021	3
4	1.4641	0.6830	4.6410	3.1699	0.2155	0.3155	4
5	1.6105	0.6209	6.1051	3.7908	0.1638	0.2638	5
6	1.7716	0.5645	7.7156	4.3553	0.1296	0.2296	6
7	1.9487	0.5132	9.4872	4.8684	0.1054	0.2054	7
8	2.1436	0.4665	11.4359	5.3349	0.0874	0.1874	8
9	2.3579	0.4241	13.5795	5.7590	0.0736	0.1736	9
10	2.5937	0.3855	15.9374	6.1446	0.0627	0.1627	10
11	2.8531	0.3505	18.5312	6.4951	0.0540	0.1540	11
12	3.1384	0.3186	21.3843	6.8137	0.0468	0.1468	12
13	3.4523	0.2897	24.5227	7.1034	0.0408	0.1408	13
14	3.7975	0.2633	27.9750	7.3667	0.0357	0.1357	14
15	4.1772	0.2394	31.7725	7.6061	0.0315	0.1315	15
16	4.5950	0.2176	35.9497	7.8237	0.0278	0.1278	16
17	5.0545	0.1978	40.5447	8.0216	0.0247	0.1247	17
18	5.5599	0.1799	45.5992	8.2014	0.0219	0.1219	18
19	6.1159	0.1635	51.1591	8.3649	0.0195	0.1195	19
20	6.7275	0.1486	57.2750	8.5136	0.0175	0.1175	20
21	7.4002	0.1351	64.0025	8.6487	0.0156	0.1156	21
22	8.1403	0.1228	71.4027	8.7715	0.0140	0.1140	22
23	8.9543	0.1117	79.5430	8.8832	0.0126	0.1126	23
24	9.8497	0.1015	88.4973	8.9847	0.0113	0.1113	24
25	10.8347	0.0923	98.3470	9.0770	0.0102	0.1102	25
26	11.9182	0.0839	109.182	9.1609	0.0092	0.1092	26
27	13.1100	0.0763	121.100	9.2372	0.0083	0.1083	27
28	14.4210	0.0693	134.210	9.3066	0.0075	0.1075	28
29	15.8631	0.0630	148.631	9.3696	0.0067	0.1067	29
30	17.4494	0.0573	164.494	9.4269	0.0061	0.1061	30
35	28.1024	0.0356	271.024	9.6442	0.0037	0.1037	35
40	45.2592	0.0221	442.592	9.7791	0.0023	0.1023	40
45	72.8904	0.0137	718.905	9.8628	0.0014	0.1014	45
50	117.391	0.0085	1163.91	9.9148	0.0009	0.1009	50
55	189.059	0.0053	1880.59	9.9471	0.0005	0.1005	55
60	304.481	0.0033	3034.81	9.9672	0.0003	0.1003	60
65	490.370	0.0020	4893.71	9.9796	0.0002	0.1002	65
70	789.746	0.0013	7887.47	9.9873	0.0001	0.1001	70
75	1271.89	0.0008	12708.9	9.9921	[a]	0.1001	75
80	2048.40	0.0005	20474.0	9.9951	[a]	0.1000	80
85	3298.97	0.0003	32979.7	9.9970	[a]	0.1000	85
90	5313.02	0.0002	53120.2	9.9981	[a]	0.1000	90
95	8556.67	0.0001	85556.7	9.9988	[a]	0.1000	95
100	13780.6	[a]	137796	9.9993	[a]	0.1000	100
∞				10.0000		0.1000	∞

[a] Less than 0.0001.

TABLE A-15 Discrete Compounding; $i = 12\%$

	Single payment		Uniform series				
	Compound amount factor	Present worth factor	Compound amount factor	Present worth factor	Sinking fund factor	Capital recovery factor	
N	To find F given P F/P	To find P given F P/F	To find F given A F/A	To find P given A P/A	To find A given F A/F	To find A given P A/P	N
1	1.1200	0.8929	1.0000	0.8929	1.0000	1.1200	1
2	1.2544	0.7972	2.1200	1.6901	0.4717	0.5917	2
3	1.4049	0.7118	3.3744	2.4018	0.2963	0.4163	3
4	1.5735	0.6355	4.7793	3.0373	0.2092	0.3292	4
5	1.7623	0.5674	6.3528	3.6048	0.1574	0.2774	5
6	1.9738	0.5066	8.1152	4.1114	0.1232	0.2432	6
7	2.2107	0.4523	10.0890	4.5638	0.0991	0.2191	7
8	2.4760	0.4039	12.2997	4.9676	0.0813	0.2013	8
9	2.7731	0.3606	14.7757	5.3282	0.0677	0.1877	9
10	3.1058	0.3220	17.5487	5.6502	0.0570	0.1770	10
11	3.4785	0.2875	20.6546	5.9377	0.0484	0.1684	11
12	3.8960	0.2567	24.1331	6.1944	0.0414	0.1614	12
13	4.3635	0.2292	28.0291	6.4235	0.0357	0.1557	13
14	4.8871	0.2046	32.3926	6.6282	0.0309	0.1509	14
15	5.4736	0.1827	37.2797	6.8109	0.0268	0.1468	15
16	6.1304	0.1631	42.7533	6.9740	0.0234	0.1434	16
17	6.8660	0.1456	48.8837	7.1196	0.0205	0.1405	17
18	7.6900	0.1300	55.7497	7.2497	0.0179	0.1379	18
19	8.6128	0.1161	63.4397	7.3658	0.0158	0.1358	19
20	9.6463	0.1037	72.0524	7.4694	0.0139	0.1339	20
21	10.8038	0.0926	81.6987	7.5620	0.0122	0.1322	21
22	12.1003	0.0826	92.5026	7.6446	0.0108	0.1308	22
23	13.5523	0.0738	104.603	7.7184	0.0096	0.1296	23
24	15.1786	0.0659	118.155	7.7843	0.0085	0.1285	24
25	17.0001	0.0588	133.334	7.8431	0.0075	0.1275	25
26	19.0401	0.0525	150.334	7.8957	0.0067	0.1267	26
27	21.3249	0.0469	169.374	7.9426	0.0059	0.1259	27
28	23.8839	0.0419	190.699	7.9844	0.0052	0.1252	28
29	26.7499	0.0374	214.583	8.0218	0.0047	0.1247	29
30	29.9599	0.0334	241.333	8.0552	0.0041	0.1241	30
35	52.7996	0.0189	431.663	8.1755	0.0023	0.1223	35
40	93.0509	0.0107	767.091	8.2438	0.0013	0.1213	40
45	163.988	0.0061	1358.23	8.2825	0.0007	0.1207	45
50	289.002	0.0035	2400.02	8.3045	0.0004	0.1204	50
55	509.320	0.0020	4236.00	8.3170	0.0002	0.1202	55
60	897.596	0.0011	7471.63	8.3240	0.0001	0.1201	60
65	1581.87	0.0006	13173.9	8.3281	[a]	0.1201	65
70	2787.80	0.0004	23223.3	8.3303	[a]	0.1200	70
75	4913.05	0.0002	40933.8	8.3316	[a]	0.1200	75
80	8658.47	0.0001	72145.6	8.3324	[a]	0.1200	80
∞				8.333		0.1200	∞

[a] Less than 0.0001.

TABLE A-16 Discrete Compounding; i = 15%

	Single payment		Uniform series				
	Compound amount factor	Present worth factor	Compound amount factor	Present worth factor	Sinking fund factor	Capital recovery factor	
	To find F given P	To find P given F	To find F given A	To find P given A	To find A given F	To find A given P	
N	F/P	P/F	F/A	P/A	A/F	A/P	N
1	1.1500	0.8696	1.0000	0.8696	1.0000	1.1500	1
2	1.3225	0.7561	2.1500	1.6257	0.4651	0.6151	2
3	1.5209	0.6575	3.4725	2.2832	0.2880	0.4380	3
4	1.7490	0.5718	4.9934	2.8550	0.2003	0.3503	4
5	2.0114	0.4972	6.7424	3.3522	0.1483	0.2983	5
6	2.3131	0.4323	8.7537	3.7845	0.1142	0.2642	6
7	2.6600	0.3759	11.0668	4.1604	0.0904	0.2404	7
8	3.0590	0.3269	13.7268	4.4873	0.0729	0.2229	8
9	3.5179	0.2843	16.7858	4.7716	0.0596	0.2096	9
10	4.0456	0.2472	20.3037	5.0188	0.0493	0.1993	10
11	4.6524	0.2149	24.3493	5.2337	0.0411	0.1911	11
12	5.3502	0.1869	29.0017	5.4206	0.0345	0.1845	12
13	6.1528	0.1625	34.3519	5.5831	0.0291	0.1791	13
14	7.0757	0.1413	40.5047	5.7245	0.0247	0.1747	14
15	8.1371	0.1229	47.5804	5.8474	0.0210	0.1710	15
16	9.3576	0.1069	55.7175	5.9542	0.0179	0.1679	16
17	10.7613	0.0929	65.0751	6.0472	0.0154	0.1654	17
18	12.3755	0.0808	75.8363	6.1280	0.0132	0.1632	18
19	14.2318	0.0703	88.2118	6.1982	0.0113	0.1613	19
20	16.3665	0.0611	102.444	6.2593	0.0098	0.1598	20
21	18.8215	0.0531	118.810	6.3125	0.0084	0.1584	21
22	21.6447	0.0462	137.632	6.3587	0.0073	0.1573	22
23	24.8915	0.0402	159.276	6.3988	0.0063	0.1563	23
24	28.6252	0.0349	184.168	6.4338	0.0054	0.1554	24
25	32.9189	0.0304	212.793	6.4641	0.0047	0.1547	25
26	37.8568	0.0264	245.712	6.4906	0.0041	0.1541	26
27	43.5353	0.0230	283.569	6.5135	0.0035	0.1535	27
28	50.0656	0.0200	327.104	6.5335	0.0031	0.1531	28
29	57.5754	0.0174	377.170	6.5509	0.0027	0.1527	29
30	66.2118	0.0151	434.745	6.5660	0.0023	0.1523	30
35	133.176	0.0075	881.170	6.6166	0.0011	0.1511	35
40	267.863	0.0037	1779.09	6.6418	0.0006	0.1506	40
45	538.769	0.0019	3585.13	6.6543	0.0003	0.1503	45
50	1083.66	0.0009	7217.71	6.6605	0.0001	0.1501	50
55	2179.62	0.0005	14524.1	6.6636	a	0.1501	55
60	4384.00	0.0002	29220.0	6.6651	a	0.1500	60
65	8817.78	0.0001	58778.5	6.6659	a	0.1500	65
70	17735.7	a	118231	6.6663	a	0.1500	70
75	35672.8	a	237812	6.6665	a	0.1500	75
80	71750.8	a	478332	6.6666	a	0.1500	80
∞				6.667		0.1500	∞

a Less than 0.0001.

TABLE A-17 Discrete Compounding; $i = 18\%$

	Single payment		Uniform series				
	Compound amount factor	Present worth factor	Compound amount factor	Present worth factor	Sinking fund factor	Capital recovery factor	
N	To find F given P F/P	To find P given F P/F	To find F given A F/A	To find P given A P/A	To find A given F A/F	To find A given P A/P	N
1	1.1800	0.8475	1.000	0.847	1.00000	1.18000	1
2	1.3924	0.7182	2.180	1.566	0.45872	0.63872	2
3	1.6430	0.6086	3.572	2.174	0.27992	0.45992	3
4	1.9388	0.5158	5.215	2.690	0.19174	0.37174	4
5	2.2878	0.4371	7.154	3.127	0.13978	0.31978	5
6	2.6996	0.3704	9.442	3.498	0.10591	0.28591	6
7	3.1855	0.3139	12.142	3.812	0.08236	0.26236	7
8	3.7589	0.2660	15.327	4.078	0.06524	0.24524	8
9	4.4355	0.2255	19.086	4.303	0.05239	0.23239	9
10	5.2338	0.1911	23.521	4.494	0.04251	0.22251	10
11	6.1759	0.1619	28.755	4.656	0.03478	0.21478	11
12	7.2876	0.1372	34.931	4.793	0.02863	0.20863	12
13	8.5994	0.1163	42.219	4.910	0.02369	0.20369	13
14	10.1472	0.0985	50.818	5.008	0.01968	0.19968	14
15	11.9737	0.0835	60.965	5.092	0.01640	0.19640	15
16	14.1290	0.0708	72.939	5.162	0.01371	0.19371	16
17	16.6722	0.0600	87.068	5.222	0.01149	0.19149	17
18	19.6733	0.0508	103.740	5.273	0.00964	0.18964	18
19	23.2144	0.0431	123.414	5.316	0.00810	0.18810	19
20	27.3930	0.0365	146.628	5.353	0.00682	0.18682	20
21	32.3238	0.0309	174.021	5.384	0.00575	0.18575	21
22	38.1421	0.0262	206.345	5.410	0.00485	0.18485	22
23	45.0076	0.0222	244.487	5.432	0.00409	0.18409	23
24	53.1090	0.0188	289.494	5.451	0.00345	0.18345	24
25	62.6686	0.0160	342.603	5.467	0.00292	0.18292	25
26	73.9490	0.0135	405.272	5.480	0.00247	0.18247	26
27	87.2598	0.0115	479.221	5.492	0.00209	0.18209	27
28	102.9665	0.0097	566.481	5.502	0.00177	0.18177	28
29	121.5005	0.0082	669.447	5.510	0.00149	0.18149	29
30	143.3706	0.0070	790.948	5.517	0.00126	0.18126	30
31	169.1774	0.0059	934.319	5.523	0.00107	0.18107	31
32	199.6293	0.0050	1103.496	5.528	0.00091	0.18091	32
33	235.5625	0.0042	1303.125	5.532	0.00077	0.18077	33
34	277.9638	0.0036	1538.688	5.536	0.00065	0.18065	34
35	327.9973	0.0030	1816.652	5.539	0.00055	0.18055	35
40	750.3783	0.0013	4163.213	5.548	0.00024	0.18024	40
45	1716.6839	0.0006	9531.577	5.552	0.00010	0.18010	45
50	3927.3569	0.0003	21813.094	5.554	a	0.18005	50
∞				5.556	a	0.18000	∞

a Less than 0.0001.

TABLE A-18 Discrete Compounding; $i = 20\%$

	Single payment		Uniform series				
	Compound amount factor	Present worth factor	Compound amount factor	Present worth factor	Sinking fund factor	Capital recovery factor	
	To find F given P	To find P given F	To find F given A	To find P given A	To find A given F	To find A given P	
N	F/P	P/F	F/A	P/A	A/F	A/P	N
1	1.2000	0.8333	1.0000	0.8333	1.0000	1.2000	1
2	1.4400	0.6944	2.2000	1.5278	0.4545	0.6545	2
3	1.7280	0.5787	3.6400	2.1065	0.2747	0.4747	3
4	2.0736	0.4823	5.3680	2.5887	0.1863	0.3863	4
5	2.4883	0.4019	7.4416	2.9906	0.1344	0.3344	5
6	2.9860	0.3349	9.9299	3.3255	0.1007	0.3007	6
7	3.5832	0.2791	12.9159	3.6046	0.0774	0.2774	7
8	4.2998	0.2326	16.4991	3.8372	0.0606	0.2606	8
9	5.1598	0.1938	20.7989	4.0310	0.0481	0.2481	9
10	6.1917	0.1615	25.9587	4.1925	0.0385	0.2385	10
11	7.4301	0.1346	32.1504	4.3271	0.0311	0.2311	11
12	8.9161	0.1122	39.5805	4.4392	0.0253	0.2253	12
13	10.6993	0.0935	48.4966	4.5327	0.0206	0.2206	13
14	12.8392	0.0779	59.1959	4.6106	0.0169	0.2169	14
15	15.4070	0.0649	72.0351	4.6755	0.0139	0.2139	15
16	18.4884	0.0541	87.4421	4.7296	0.0114	0.2114	16
17	22.1861	0.0451	105.931	4.7746	0.0094	0.2094	17
18	26.6233	0.0376	128.117	4.8122	0.0078	0.2078	18
19	31.9480	0.0313	154.740	4.8435	0.0065	0.2065	19
20	38.3376	0.0261	186.688	4.8696	0.0054	0.2054	20
21	46.0051	0.0217	225.026	4.8913	0.0044	0.2044	21
22	55.2061	0.0181	271.031	4.9094	0.0037	0.2037	22
23	66.2474	0.0151	326.237	4.9245	0.0031	0.2031	23
24	79.4968	0.0126	392.484	4.9371	0.0025	0.2025	24
25	95.3962	0.0105	471.981	4.9476	0.0021	0.2021	25
26	114.475	0.0087	567.377	4.9563	0.0018	0.2018	26
27	137.371	0.0073	681.853	4.9636	0.0015	0.2015	27
28	164.845	0.0061	819.223	4.9697	0.0012	0.2012	28
29	197.814	0.0051	984.068	4.9747	0.0010	0.2010	29
30	237.376	0.0042	1181.88	4.9789	0.0008	0.2008	30
35	590.668	0.0017	2948.34	4.9915	0.0003	0.2003	35
40	1469.77	0.0007	7343.85	4.9966	0.0001	0.2001	40
45	3657.26	0.0003	18281.3	4.9986	a	0.2001	45
50	9100.43	0.0001	45497.2	4.9995	a	0.2000	50
55	22644.8	a	113219	4.9998	a	0.2000	55
60	56347.5	a	281732	4.9999	a	0.2000	60
∞				5.0000		0.2000	∞

a Less than 0.0001.

TABLE A-19 Discrete Compounding; $i = 25\%$

	Single payment		Uniform series				
	Compound amount factor	Present worth factor	Compound amount factor	Present worth factor	Sinking fund factor	Capital recovery factor	
	To find F given P	To find P given F	To find F given A	To find P given A	To find A given F	To find A given P	
N	F/P	P/F	F/A	P/A	A/F	A/P	N
1	1.2500	0.8000	1.0000	0.8000	1.0000	1.2500	1
2	1.5625	0.6400	2.2500	1.4400	0.4444	0.6944	2
3	1.9531	0.5120	3.8125	1.9520	0.2623	0.5123	3
4	2.4414	0.4096	5.7656	2.3616	0.1734	0.4234	4
5	3.0518	0.3277	8.2070	2.6893	0.1218	0.3718	5
6	3.8147	0.2621	11.2588	2.9514	0.0888	0.3388	6
7	4.7684	0.2097	15.0735	3.1611	0.0663	0.3163	7
8	5.9605	0.1678	19.8419	3.3289	0.0504	0.3004	8
9	7.4506	0.1342	25.8023	3.4631	0.0388	0.2888	9
10	9.3132	0.1074	33.2529	3.5705	0.0301	0.2801	10
11	11.6415	0.0859	42.5661	3.6564	0.0235	0.2735	11
12	14.5519	0.0687	54.2077	3.7251	0.0184	0.2684	12
13	18.1899	0.0550	68.7596	3.7801	0.0145	0.2645	13
14	22.7374	0.0440	86.9495	3.8241	0.0115	0.2615	14
15	28.4217	0.0352	109.687	3.8593	0.0091	0.2591	15
16	35.5271	0.0281	138.109	3.8874	0.0072	0.2572	16
17	44.4089	0.0225	173.636	3.9099	0.0058	0.2558	17
18	55.5112	0.0180	218.045	3.9279	0.0046	0.2546	18
19	69.3889	0.0144	273.556	3.9424	0.0037	0.2537	19
20	86.7362	0.0115	342.945	3.9539	0.0029	0.2529	20
21	108.420	0.0092	429.681	3.9631	0.0023	0.2523	21
22	135.525	0.0074	538.101	3.9705	0.0019	0.2519	22
23	169.407	0.0059	673.626	3.9764	0.0015	0.2515	23
24	211.758	0.0047	843.033	3.9811	0.0012	0.2512	24
25	264.698	0.0038	1054.79	3.9849	0.0009	0.2509	25
26	330.872	0.0030	1319.49	3.9879	0.0008	0.2508	26
27	413.590	0.0024	1650.36	3.9903	0.0006	0.2506	27
28	516.988	0.0019	2063.95	3.9923	0.0005	0.2505	28
29	646.235	0.0015	2580.94	3.9938	0.0004	0.2504	29
30	807.794	0.0012	3227.17	3.9950	0.0003	0.2503	30
35	2465.19	0.0004	9856.76	3.9984	0.0001	0.2501	35
40	7523.16	0.0001	30088.7	3.9995	a	0.2500	40
45	22958.9	a	91831.5	3.9998	a	0.2500	45
50	70064.9	a	280256	3.9999	a	0.2500	50
∞				4.0000		0.2500	∞

a Less than 0.0001.

TABLE A-20 Gradient to Present Worth Conversion Factor for Discrete Compounding (to Find P, Given G)

$$(P/G, i\%, N) = \frac{1}{i}\left[\frac{(1+i)^N - 1}{i(1+i)^N} - \frac{N}{(1+i)^N}\right]$$

n	1%	2%	4%	6%	8%	10%	12%	15%	20%	25%	n
1	0.00	0.00	0.00	0.00	0.00	0.00	0.00	0.00	0.00	0.00	1
2	0.98	0.96	0.92	0.89	0.86	0.83	0.80	0.76	0.69	0.64	2
3	2.92	2.85	2.70	2.57	2.45	2.33	2.22	2.07	1.85	1.66	3
4	5.80	5.62	5.27	4.95	4.65	4.38	4.13	3.79	3.30	2.89	4
5	9.61	9.24	8.55	7.93	7.37	6.86	6.40	5.78	4.91	4.20	5
6	14.32	13.68	12.50	11.46	10.52	9.68	8.93	7.94	6.58	5.51	6
7	19.92	18.90	17.07	15.45	14.02	12.76	11.64	10.19	8.26	6.77	7
8	26.38	24.88	22.18	19.84	17.81	16.03	14.47	12.48	9.88	7.95	8
9	33.69	31.57	27.80	24.58	21.81	19.42	17.36	14.75	11.43	9.02	9
10	41.84	38.95	33.88	29.60	25.98	22.89	20.25	16.98	12.89	9.99	10
11	50.80	47.00	40.38	34.87	30.27	26.40	23.13	19.13	14.23	10.85	11
12	60.57	55.67	47.25	40.34	34.63	29.90	25.95	21.18	15.47	11.60	12
15	94.48	85.20	69.74	57.55	47.89	40.15	33.92	26.69	18.51	13.33	15
20	165.46	144.60	111.56	87.23	69.09	55.41	44.97	33.58	21.74	14.89	20
25	252.89	214.26	156.10	115.97	87.80	67.70	53.10	38.03	23.43	15.56	25
30	355.00	291.72	201.06	142.36	103.46	77.08	58.78	40.75	24.26	15.83	30
35	470.15	374.88	244.88	165.74	116.09	83.99	62.61	42.36	24.66	15.94	35
40	596.85	461.99	286.53	185.96	126.04	88.95	65.12	43.28	24.85	15.98	40
45	733.70	551.56	325.40	203.11	133.73	92.45	66.73	43.81	24.93	15.99	45
50	879.41	642.36	361.16	217.46	139.59	94.89	67.76	44.10	24.97	16.00	50
60	1192.80	823.70	423.00	239.04	147.30	97.70	68.81	44.34	24.99	—	60
70	1528.64	999.83	472.48	253.33	151.53	98.99	69.21	44.42	—	—	70
80	1879.87	1166.79	511.12	262.55	153.80	99.56	69.36	44.47	—	—	80
90	2240.55	1322.17	540.77	268.39	154.99	99.81	—	—	—	—	90
100	2605.76	1464.75	563.12	272.05	155.61	99.92	—	—	—	—	100

TABLE A-21 Gradient to Uniform Series Conversion Factor for Discrete Compounding (to Find A, Given G)

$$(A/G, i\%, N) = \left[\frac{1}{i} - \frac{N}{(1+i)^N - 1} \right]$$

n	1%	2%	4%	6%	8%	10%	12%	15%	20%	25%	n
1	0.0001	0.0000	0.0000	0.0000	0.0000	0.0000	0.0000	0.0000	0.0000	0.0000	1
2	0.4974	0.4950	0.4902	0.4854	0.4808	0.4762	0.4717	0.4651	0.4545	0.4444	2
3	0.9932	0.9868	0.9739	0.9612	0.9487	0.9366	0.9246	0.9071	0.8791	0.8525	3
4	1.4874	1.4752	1.4510	1.4272	1.4040	1.3812	1.3589	1.3263	1.2742	1.2249	4
5	1.9799	1.9604	1.9216	1.8836	1.8465	1.8101	1.7746	1.7228	1.6405	1.5631	5
6	2.4708	2.4422	2.3857	2.3304	2.2763	2.2236	2.1720	2.0972	1.9788	1.8683	6
7	2.9600	2.9208	2.8433	2.7676	2.6937	2.6216	2.5515	2.4498	2.2902	2.1424	7
8	3.4476	3.3961	3.2944	3.1952	3.0985	3.0045	2.9131	2.7813	2.5756	2.3872	8
9	3.9335	3.8680	3.7391	3.6133	3.4910	3.3724	3.2574	3.0922	2.8364	2.6048	9
10	4.4177	4.3367	4.1773	4.0220	3.8713	3.7255	3.5847	3.3832	3.0739	2.7971	10
11	4.9003	4.8021	4.6090	4.4213	4.2395	4.0641	3.8953	3.6549	3.2893	2.9663	11
12	5.3813	5.2642	5.0343	4.8113	4.5957	4.3884	4.1897	3.9082	3.4841	3.1145	12
15	6.8141	6.6309	6.2721	5.9260	5.5945	5.2789	4.9803	4.5650	3.9588	3.4530	15
20	9.1692	8.8433	8.2091	7.6051	7.0369	6.5081	6.0202	5.3651	4.4643	3.7667	20
25	11.4829	10.9744	9.9925	9.0722	8.2254	7.4580	6.7708	5.8834	4.7352	3.9052	25
30	13.7555	13.0251	11.6274	10.3422	9.1897	8.1762	7.2974	6.2066	4.8731	3.9628	30
35	15.9869	14.9961	13.1198	11.4319	9.9611	8.7086	7.6577	6.4019	4.9406	3.9858	35
40	18.1774	16.8885	14.4765	12.3590	10.5699	9.0962	7.8988	6.5168	4.9728	3.9947	40
45	20.3271	18.7033	15.7047	13.1413	11.0447	9.3740	8.0572	6.5830	4.9877	3.9980	45
50	22.4362	20.4420	16.8122	13.7964	11.4107	9.5704	8.1597	6.6205	4.9945	3.9993	50
60	26.5331	23.6961	18.6972	14.7909	11.9015	9.8023	8.2664	6.6530	4.9989	—	60
70	30.4701	26.6632	20.1961	15.4613	12.1783	9.9113	8.3082	6.6627	—	—	70
80	34.2490	29.3572	21.3718	15.9033	12.3301	9.9609	8.3241	6.6656	—	—	80
90	37.8723	31.7929	22.2826	16.1891	12.4116	9.9831	—	—	—	—	90
100	41.3424	33.9863	22.9800	16.3711	12.4545	9.9927	—	—	—	—	100

APPENDIX B

Tables of Continuous Compounding Interest Factors (for Various Common Values of r from 2% to 25%)

TABLE B-1   Continuous Compounding; $r = 2\%$

	Discrete flows				Continuous flows		
	Single payment		Uniform series		Uniform series		
	Compound amount factor	Present worth factor	Compound amount factor	Present worth factor	Compound amount factor	Present worth factor	
	To find F given P	To find P given F	To find F given A	To find P given A	To find F given $\overline{A}$	To find P given $\overline{A}$	
N	F/P	P/F	F/A	P/A	$F/\overline{A}$	$P/\overline{A}$	N
1	1.0202	0.9802	1.0000	0.9802	1.0101	0.9901	1
2	1.0408	0.9608	2.0202	1.9410	2.0405	1.9605	2
3	1.0618	0.9418	3.0610	2.8828	3.0918	2.9118	3
4	1.0833	0.9231	4.1228	3.8059	4.1644	3.8442	4
5	1.1052	0.9048	5.2061	4.7107	5.2585	4.7581	5
6	1.1275	0.8869	6.3113	5.5976	6.3748	5.6540	6
7	1.1503	0.8694	7.4388	6.4670	7.5137	6.5321	7
8	1.1735	0.8521	8.5891	7.3191	8.6755	7.3928	8
9	1.1972	0.8353	9.7626	8.1544	9.8609	8.2365	9
10	1.2214	0.8187	10.9598	8.9731	11.0701	9.0635	10
11	1.2461	0.8025	12.1812	9.7756	12.3038	9.8741	11
12	1.2712	0.7866	13.4273	10.5623	13.5625	10.6686	12
13	1.2969	0.7711	14.6985	11.3333	14.8465	11.4474	13
14	1.3231	0.7558	15.9955	12.0891	16.1565	12.2108	14
15	1.3499	0.7408	17.3186	12.8299	17.4929	12.9591	15
16	1.3771	0.7261	18.6685	13.5561	18.8564	13.6925	16
17	1.4049	0.7118	20.0456	14.2678	20.2474	14.4115	17
18	1.4333	0.6977	21.4505	14.9655	21.6665	15.1162	18
19	1.4623	0.6839	22.8839	15.6494	23.1142	15.8069	19
20	1.4918	0.6703	24.3461	16.3197	24.5912	16.4840	20
21	1.5220	0.6570	25.8380	16.9768	26.0981	17.1477	21
22	1.5527	0.6440	27.3599	17.6208	27.6354	17.7982	22
23	1.5841	0.6313	28.9126	18.2521	29.2037	18.4358	23
24	1.6161	0.6188	30.4967	18.8709	30.8037	19.0608	24
25	1.6487	0.6065	32.1128	19.4774	32.4361	19.6735	25
26	1.6820	0.5945	33.7615	20.0719	34.1014	20.2740	26
27	1.7160	0.5827	35.4435	20.6547	35.8003	20.8626	27
28	1.7507	0.5712	37.1595	21.2259	37.5336	21.4395	28
29	1.7860	0.5599	38.9102	21.7858	39.3019	22.0051	29
30	1.8221	0.5488	40.6962	22.3346	41.1059	22.5594	30
35	2.0138	0.4966	50.1824	24.9199	50.6876	25.1707	35
40	2.2255	0.4493	60.6663	27.2591	61.2770	27.5336	40
45	2.4596	0.4066	72.2528	29.3758	72.9802	29.6715	45
50	2.7183	0.3679	85.0578	31.2910	85.9141	31.6060	50
55	3.0042	0.3329	99.2096	33.0240	100.208	33.3564	55
60	3.3201	0.3012	114.850	34.5921	116.006	34.9403	60
65	3.6693	0.2725	132.135	36.0109	133.465	36.3734	65
70	4.0552	0.2466	151.238	37.2947	152.760	37.6702	70
75	4.4817	0.2231	172.349	38.4564	174.084	38.8435	75
80	4.9530	0.2019	195.682	39.5075	197.652	39.9052	80
85	5.4739	0.1827	221.468	40.4585	223.697	40.8658	85
90	6.0496	0.1653	249.966	41.3191	252.482	41.7351	90
95	6.6859	0.1496	281.461	42.0978	284.295	42.5216	95
100	7.3891	0.1353	316.269	42.8023	319.453	43.2332	100

	Discrete flows				Continuous flows		
	Single payment		Uniform series		Uniform series		
	Compound amount factor	Present worth factor	Compound amount factor	Present worth factor	Compound amount factor	Present worth factor	
	To find F given P	To find P given F	To find F given A	To find P given A	To find F given $\overline{A}$	To find P given $\overline{A}$	
N	F/P	P/F	F/A	P/A	$F/\overline{A}$	$P/\overline{A}$	N
1	1.0513	0.9512	1.0000	0.9512	1.0254	0.9754	1
2	1.1052	0.9048	2.0513	1.8561	2.1034	1.9033	2
3	1.1618	0.8607	3.1564	2.7168	3.2367	2.7858	3
4	1.2214	0.8187	4.3183	3.5355	4.4281	3.6254	4
5	1.2840	0.7788	5.5397	4.3143	5.6805	4.4240	5
6	1.3499	0.7408	6.8237	5.0551	6.9972	5.1836	6
7	1.4191	0.7047	8.1736	5.7598	8.3814	5.9062	7
8	1.4918	0.6703	9.5926	6.4301	9.8365	6.5936	8
9	1.5683	0.6376	11.0845	7.0678	11.3662	7.2474	9
10	1.6487	0.6065	12.6528	7.6743	12.9744	7.8694	10
11	1.7333	0.5769	14.3015	8.2512	14.6651	8.4610	11
12	1.8221	0.5488	16.0347	8.8001	16.4424	9.0238	12
13	1.9155	0.5220	17.8569	9.3221	18.3108	9.5591	13
14	2.0138	0.4966	19.7724	9.8187	20.2751	10.0683	14
15	2.1170	0.4724	21.7862	10.2911	22.3400	10.5527	15
16	2.2255	0.4493	23.9032	10.7404	24.5108	11.0134	16
17	2.3396	0.4274	26.1287	11.1678	26.7929	11.4517	17
18	2.4596	0.4066	28.4683	11.5744	29.1921	11.8686	18
19	2.5857	0.3867	30.9279	11.9611	31.7142	12.2652	19
20	2.7183	0.3679	33.5137	12.3290	34.3656	12.6424	20
21	2.8577	0.3499	36.2319	12.6789	37.1530	13.0012	21
22	3.0042	0.3329	39.0896	13.0118	40.0833	13.3426	22
23	3.1582	0.3166	42.0938	13.3284	43.1639	13.6673	23
24	3.3201	0.3012	45.2519	13.6296	46.4023	13.9761	24
25	3.4903	0.2865	48.5721	13.9161	49.8069	14.2699	25
26	3.6693	0.2725	52.0624	14.1887	53.3859	14.5494	26
27	3.8574	0.2592	55.7317	14.4479	57.1485	14.8152	27
28	4.0552	0.2466	59.5891	14.6945	61.1040	15.0681	28
29	4.2631	0.2346	63.6443	14.9291	65.2623	15.3086	29
30	4.4817	0.2231	67.9074	15.1522	69.6338	15.5374	30
35	5.7546	0.1738	92.7346	16.1149	95.0921	16.5245	35
40	7.3891	0.1353	124.613	16.8646	127.781	17.2933	40
45	9.4877	0.1054	165.546	17.4484	169.755	17.8920	45
50	12.1825	0.0821	218.105	17.9032	223.650	18.3583	50
55	15.6426	0.0639	285.592	18.2573	292.853	18.7214	55
60	20.0855	0.0498	372.247	18.5331	381.711	19.0043	60
65	25.7903	0.0388	483.515	18.7479	495.807	19.2245	65
70	33.1155	0.0302	626.385	18.9152	642.309	19.3961	70
75	42.5211	0.0235	809.834	19.0455	830.422	19.5296	75
80	54.5981	0.0183	1045.39	19.1469	1071.963	19.6337	80
85	70.1054	0.0143	1347.84	19.2260	1382.108	19.7147	85
90	90.0171	0.0111	1736.20	19.2875	1780.342	19.7778	90
95	115.584	0.0087	2234.87	19.3354	2291.686	19.8270	95
100	148.413	0.0067	2875.17	19.3727	2948.263	19.8652	100

TABLE B-3   Continuous Compounding; $r = 10\%$

	Discrete flows				Continuous flows		
	Single payment		Uniform series		Uniform series		
	Compound amount factor	Present worth factor	Compound amount factor	Present worth factor	Compound amount factor	Present worth factor	
N	To find F given P F/P	To find P given F P/F	To find F given A F/A	To find P given A P/A	To find F given $\overline{A}$ $F/\overline{A}$	To find P given $\overline{A}$ $P/\overline{A}$	N
1	1.1052	0.9048	1.0000	0.9048	1.0517	0.9516	1
2	1.2214	0.8187	2.1052	1.7236	2.2140	1.8127	2
3	1.3499	0.7408	3.3266	2.4644	3.4986	2.5918	3
4	1.4918	0.6703	4.6764	3.1347	4.9182	3.2968	4
5	1.6487	0.6065	6.1683	3.7412	6.4872	3.9347	5
6	1.8221	0.5488	7.8170	4.2900	8.2212	4.5119	6
7	2.0138	0.4966	9.6391	4.7866	10.1375	5.0341	7
8	2.2255	0.4493	11.6528	5.2360	12.2554	5.5067	8
9	2.4596	0.4066	13.8784	5.6425	14.5960	5.9343	9
10	2.7183	0.3679	16.3380	6.0104	17.1828	6.3212	10
11	3.0042	0.3329	19.0563	6.3433	20.0417	6.6713	11
12	3.3201	0.3012	22.0604	6.6445	23.2012	6.9881	12
13	3.6693	0.2725	25.3806	6.9170	26.6930	7.2747	13
14	4.0552	0.2466	29.0499	7.1636	30.5520	7.5340	14
15	4.4817	0.2231	33.1051	7.3867	34.8169	7.7687	15
16	4.9530	0.2019	37.5867	7.5886	39.5303	7.9810	16
17	5.4739	0.1827	42.5398	7.7713	44.7395	8.1732	17
18	6.0496	0.1653	48.0137	7.9366	50.4965	8.3470	18
19	6.6859	0.1496	54.0634	8.0862	56.8589	8.5043	19
20	7.3891	0.1353	60.7493	8.2215	63.8906	8.6466	20
21	8.1662	0.1225	68.1383	8.3440	71.6617	8.7754	21
22	9.0250	0.1108	76.3045	8.4548	80.2501	8.8920	22
23	9.9742	0.1003	85.3295	8.5550	89.7418	8.9974	23
24	11.0232	0.0907	95.3037	8.6458	100.232	9.0928	24
25	12.1825	0.0821	106.327	8.7278	111.825	9.1791	25
26	13.4637	0.0743	118.509	8.8021	124.637	9.2573	26
27	14.8797	0.0672	131.973	8.8693	138.797	9.3279	27
28	16.4446	0.0608	146.853	8.9301	154.446	9.3919	28
29	18.1741	0.0550	163.298	8.9852	171.741	9.4498	29
30	20.0855	0.0498	181.472	9.0349	190.855	9.5021	30
35	33.1155	0.0302	305.364	9.2212	321.154	9.6980	35
40	54.5981	0.0183	509.629	9.3342	535.982	9.8168	40
45	90.0171	0.0111	846.404	9.4027	890.171	9.8889	45
50	148.413	0.0067	1401.65	9.4443	1474.13	9.9326	50
55	244.692	0.0041	2317.10	9.4695	2436.92	9.9591	55
60	403.429	0.0025	3826.43	9.4848	4024.29	9.9752	60
65	665.142	0.0015	6314.88	9.4940	6641.42	9.9850	65
70	1096.63	0.0009	10417.6	9.4997	10956.3	9.9909	70
75	1808.04	0.0006	17182.0	9.5031	18070.7	9.9945	75
80	2980.96	0.0003	28334.4	9.5051	29799.6	9.9966	80
85	4914.77	0.0002	46721.7	9.5064	49137.7	9.9980	85
90	8103.08	0.0001	77037.3	9.5072	81020.8	9.9988	90
95	13359.7	[a]	127019	9.5076	133587	9.9993	95
100	22026.5	[a]	209425	9.5079	220255	9.9995	100

[a] Less than 0.0001.

TABLE B-4 Continuous Compounding; $r = 15\%$

	Discrete flows				Continuous flows		
	Single payment		Uniform series		Uniform series		
	Compound amount factor	Present worth factor	Compound amount factor	Present worth factor	Compound amount factor	Present worth factor	
	To find F given P	To find P given F	To find F given A	To find P given A	To find F given $\overline{A}$	To find P given $\overline{A}$	
N	F/P	P/F	F/A	P/A	$F/\overline{A}$	$P/\overline{A}$	N
1	1.1618	0.8607	1.0000	0.8607	1.0789	0.9286	1
2	1.3499	0.7408	2.1618	1.6015	2.3324	1.7279	2
3	1.5683	0.6376	3.5117	2.2392	3.7887	2.4158	3
4	1.8221	0.5488	5.0800	2.7880	5.4808	3.0079	4
5	2.1170	0.4724	6.9021	3.2603	7.4467	3.5176	5
6	2.4596	0.4066	9.0191	3.6669	9.7307	3.9562	6
7	2.8577	0.3499	11.4787	4.0168	12.3843	4.3337	7
8	3.3201	0.3012	14.3364	4.3180	15.4674	4.6587	8
9	3.8574	0.2592	17.6565	4.5773	19.0495	4.9384	9
10	4.4817	0.2231	21.5139	4.8004	23.2113	5.1791	10
11	5.2070	0.1920	25.9956	4.9925	28.0465	5.3863	11
12	6.0496	0.1653	31.2026	5.1578	33.6643	5.5647	12
13	7.0287	0.1423	37.2522	5.3000	40.1913	5.7182	13
14	8.1662	0.1225	44.2809	5.4225	47.7745	5.8503	14
15	9.4877	0.1054	52.4471	5.5279	56.5849	5.9640	15
16	11.0232	0.0907	61.9348	5.6186	66.8212	6.0619	16
17	12.8071	0.0781	72.9580	5.6967	78.7140	6.1461	17
18	14.8797	0.0672	85.7651	5.7639	92.5315	6.2186	18
19	17.2878	0.0578	100.645	5.8217	108.585	6.2810	19
20	20.0855	0.0498	117.933	5.8715	127.237	6.3348	20
21	23.3361	0.0429	138.018	5.9144	148.907	6.3810	21
22	27.1126	0.0369	161.354	5.9513	174.084	6.4208	22
23	31.5004	0.0317	188.467	5.9830	203.336	6.4550	23
24	36.5982	0.0273	219.967	6.0103	237.322	6.4845	24
25	42.5211	0.0235	256.565	6.0338	276.807	6.5099	25
26	49.4024	0.0202	299.087	6.0541	322.683	6.5317	26
27	57.3975	0.0174	348.489	6.0715	375.983	6.5505	27
28	66.6863	0.0150	405.886	6.0865	437.909	6.5667	28
29	77.4785	0.0129	472.573	6.0994	509.856	6.5806	29
30	90.0171	0.0111	550.051	6.1105	593.448	6.5926	30
35	190.566	0.0052	1171.36	6.1467	1263.78	6.6317	35
40	403.429	0.0025	2486.67	6.1638	2682.86	6.6501	40
45	854.059	0.0012	5271.19	6.1719	5687.06	6.6589	45
50	1808.04	0.0006	11166.0	6.1757	12046.9	6.6630	50
55	3827.63	0.0003	23645.3	6.1775	25510.8	6.6649	55
60	8103.08	0.0001	50064.1	6.1784	54013.9	6.6658	60
65	17154.2	a	105993	6.1788	114355	6.6663	65
70	36315.5	a	224393	6.1790	242097	6.6665	70
75	76879.9	a	475047	6.1791	512526	6.6666	75
80	162755	a	1005680	6.1791	1085030	6.6666	80

[a] Less than 0.0001.

TABLE B-5 Continuous Compounding; r = 20%

	Discrete flows				Continuous flows		
	Single payment		Uniform series		Uniform series		
	Compound amount factor	Present worth factor	Compound amount factor	Present worth factor	Compound amount factor	Present worth factor	
	To find F given P	To find P given F	To find F given A	To find P given A	To find F given $\overline{A}$	To find P given $\overline{A}$	
N	F/P	P/F	F/A	P/A	F/$\overline{A}$	P/$\overline{A}$	N
1	1.2214	0.8187	1.0000	0.8187	1.1070	0.9063	1
2	1.4918	0.6703	2.2214	1.4891	2.4591	1.6484	2
3	1.8221	0.5488	3.7132	2.0379	4.1106	2.2559	3
4	2.2255	0.4493	5.5353	2.4872	6.1277	2.7534	4
5	2.7183	0.3679	7.7609	2.8551	8.5914	3.1606	5
6	3.3201	0.3012	10.4792	3.1563	11.6006	3.4940	6
7	4.0552	0.2466	13.7993	3.4029	15.2760	3.7670	7
8	4.9530	0.2019	17.8545	3.6048	19.7652	3.9905	8
9	6.0496	0.1653	22.8075	3.7701	25.2482	4.1735	9
10	7.3891	0.1353	28.8572	3.9054	31.9453	4.3233	10
11	9.0250	0.1108	36.2462	4.0162	40.1251	4.4460	11
12	11.0232	0.0907	45.2712	4.1069	50.1159	4.5464	12
13	13.4637	0.0743	56.2944	4.1812	62.3187	4.6286	13
14	16.4446	0.0608	69.7581	4.2420	77.2232	4.6959	14
15	20.0855	0.0498	86.2028	4.2918	95.4277	4.7511	15
16	24.5325	0.0408	106.288	4.3325	117.663	4.7962	16
17	29.9641	0.0334	130.821	4.3659	144.820	4.8331	17
18	36.5982	0.0273	160.785	4.3932	177.991	4.8634	18
19	44.7012	0.0224	197.383	4.4156	218.506	4.8881	19
20	54.5981	0.0183	242.084	4.4339	267.991	4.9084	20
21	66.6863	0.0150	296.682	4.4489	328.432	4.9250	21
22	81.4509	0.0123	363.369	4.4612	402.254	4.9386	22
23	99.4843	0.0101	444.820	4.4713	492.422	4.9497	23
24	121.510	0.0082	544.304	4.4795	602.552	4.9589	24
25	148.413	0.0067	665.814	4.4862	737.066	4.9663	25
26	181.272	0.0055	814.227	4.4917	901.361	4.9724	26
27	221.406	0.0045	995.500	4.4963	1102.03	4.9774	27
28	270.426	0.0037	1216.91	4.5000	1347.13	4.9815	28
29	330.299	0.0030	1487.33	4.5030	1646.50	4.9849	29
30	403.429	0.0025	1817.63	4.5055	2012.14	4.9876	30
35	1096.63	0.0009	4948.60	4.5125	5478.17	4.9954	35
40	2980.96	0.0003	13459.4	4.5151	14899.8	4.9983	40
45	8103.08	0.0001	36594.3	4.5161	40510.4	4.9994	45
50	22026.5	a	99481.4	4.5165	110127	4.9998	50
55	59874.1	a	270426	4.5166	299366	4.9999	55
60	162755	a	735103	4.5166	813769	5.0000	60

[a] Less than 0.0001.

TABLE B-6 Continuous Compounding; $r = 25\%$

	Discrete flows				Continuous flows		
	Single payment		Uniform series		Uniform series		
	Compound amount factor	Present worth factor	Compound amount factor	Present worth factor	Compound amount factor	Present worth factor	
N	To find F given P F/P	To find P given F P/F	To find F given A F/A	To find P given A P/A	To find F given $\overline{A}$ $F/\overline{A}$	To find P given $\overline{A}$ $P/\overline{A}$	N
1	1.2840	0.7788	1.0000	0.7788	1.1361	0.8848	1
2	1.6487	0.6065	2.2840	1.3853	2.5949	1.5739	2
3	2.1170	0.4724	3.9327	1.8577	4.4680	2.1105	3
4	2.7183	0.3679	6.0497	2.2256	6.8731	2.5285	4
5	3.4903	0.2865	8.7680	2.5121	9.9614	2.8540	5
6	4.4817	0.2231	12.2584	2.7352	13.9268	3.1075	6
7	5.7546	0.1738	16.7401	2.9090	19.0184	3.3049	7
8	7.3891	0.1353	22.4947	3.0443	25.5562	3.4587	8
9	9.4877	0.1054	29.8837	3.1497	33.9509	3.5784	9
10	12.1825	0.0821	39.3715	3.2318	44.7300	3.6717	10
11	15.6426	0.0639	51.5539	3.2957	58.5705	3.7443	11
12	20.0855	0.0498	67.1966	3.3455	76.3421	3.8009	12
13	25.7903	0.0388	87.2821	3.3843	99.1614	3.8449	13
14	33.1155	0.0302	113.073	3.4145	128.462	3.8792	14
15	42.5211	0.0235	146.188	3.4380	166.084	3.9059	15
16	54.5982	0.0183	188.709	3.4563	214.393	3.9267	16
17	70.1054	0.0143	243.307	3.4706	276.422	3.9429	17
18	90.0171	0.0111	313.413	3.4817	356.068	3.9556	18
19	115.584	0.0087	403.430	3.4904	458.337	3.9654	19
20	148.413	0.0067	519.014	3.4971	589.653	3.9730	20
21	190.566	0.0052	667.427	3.5023	758.265	3.9790	21
22	244.692	0.0041	857.993	3.5064	974.768	3.9837	22
23	314.191	0.0032	1102.69	3.5096	1252.76	3.9873	23
24	403.429	0.0025	1416.88	3.5121	1609.72	3.9901	24
25	518.013	0.0019	1820.30	3.5140	2068.05	3.9923	25
26	665.142	0.0015	2338.31	3.5155	2656.57	3.9940	26
27	854.059	0.0012	3003.46	3.5167	3412.23	3.9953	27
28	1096.63	0.0009	3857.52	3.5176	4382.53	3.9964	28
29	1408.10	0.0007	4954.15	3.5183	5628.42	3.9972	29
30	1808.04	0.0006	6362.26	3.5189	7228.17	3.9978	30
35	6310.69	0.0002	22215.2	3.5203	25238.8	3.9994	35
40	22026.5	[a]	77547.5	3.5207	88101.9	3.9998	40
45	76879.9	[a]	270676	3.5208	307516	3.9999	45
50	268337	[a]	944762	3.5208	1073350	4.0000	50

[a] Less than 0.0001.

APPENDIX C

Table of Random Numbers[*]

48867	33971	29678	13151	56644	49193	93469	43252	14006	47173
32267	69746	00113	51336	36551	56310	85793	53453	09744	64346
27345	03196	33877	35032	98054	48358	21788	98862	67491	42221
55753	05256	51557	90419	40716	64589	90398	37070	78318	02918
93124	50675	04507	44001	06365	77897	84566	99600	67985	49133
98658	86583	97433	10733	80495	62709	61357	66903	76730	79355
68216	94830	41248	50712	46878	87317	80545	31484	03195	14755
17901	30815	78360	78260	67866	42304	07293	61290	61301	04815
88124	21868	14942	25893	72695	56231	18918	72534	86737	77792
83464	36749	22336	50443	83576	19238	91730	39507	22717	94719
91310	99003	25704	55581	00729	22024	61319	66162	20933	67713
32739	38352	91256	77744	75080	01492	90984	63090	53087	41301
07751	66724	03290	56386	06070	67105	64219	48192	70478	84722
55228	64156	90480	97774	08055	04435	26999	42039	16589	06757
89013	51781	81116	24383	95569	97247	44437	36293	29967	16088
51828	81819	81038	89146	39192	89470	76331	56420	14527	34828
59783	85454	93327	06078	64924	07271	77563	92710	42183	12380
80267	47103	90556	16128	41490	07996	78454	47929	81586	67024
82919	44210	61607	93001	26314	26865	26714	43793	94937	28439
77019	77417	19466	14967	75521	49967	74065	09746	27881	01070
66225	61832	06242	40093	40800	76849	29929	18988	10888	40344
98534	12777	84601	56336	00034	85939	32438	09549	01855	40550
63175	70789	51345	43723	06995	11186	38615	56646	54320	39632
92362	73011	09115	78303	38901	58107	95366	17226	74626	78208
61831	44794	65079	97130	94289	73502	04857	68855	47045	06309
42502	01646	88493	48207	01283	16474	08864	68322	92454	19287
89733	86230	04903	55015	11811	98185	32014	84761	80926	14509
01336	66633	26015	66768	24846	00321	73118	15082	13549	41335
72623	56083	65799	88934	87274	19417	84897	90877	76472	52145
74004	68388	04090	35239	49379	04456	07642	68642	01026	43810
09388	54633	27684	47117	67583	42496	20703	68579	65883	10729
51771	92019	39791	60400	08585	60680	28841	09921	00520	73135
69796	30304	79836	20631	10743	00246	24979	35707	75283	39211
98417	33403	63448	90462	91645	24919	73609	26663	09380	30515
56150	18324	43011	02660	86574	86097	49399	21249	90380	94375
76199	75692	09063	72999	94672	69128	39046	15379	98450	09159
74978	98693	21433	34676	97603	48534	59205	66265	03561	83075
85769	92530	04407	53725	96963	19395	16193	51018	70333	12094
63819	65669	38960	74631	39650	39419	93707	61365	46302	26134
18892	43143	19619	43200	49613	50904	73502	19519	11667	53294
32855	17190	61587	80411	22827	38852	51952	47785	34952	93574
29435	96277	53583	92804	05027	19736	54918	66396	96547	00351
36211	67263	82064	41624	49826	17566	02476	79368	28831	02805
73514	00176	41638	01420	31850	41380	11643	06787	09011	88924
90895	93099	27850	29423	98693	71762	39928	35268	59359	20674
69719	90656	62186	50435	77015	29661	94698	56057	04388	33381
94982	81453	87162	28248	37921	21143	62673	81224	38972	92988
84136	04221	72790	04719	34914	95609	88695	60180	58790	12802
58515	80581	88442	65727	72121	40481	06001	13159	55324	93591
20681	59164	75797	08928	68381	12616	97487	84803	92457	88847

[*] Reproduced with permission from the Rand Corporation, *A Million Random Numbers*. (New York: The Free Press, 1955).

Table of Random Normal Deviates*

1.102	− .944	.401	.226	1.396	−1.030	−1.723	− .368	2.170	.393
.148	−1.140	.492	−1.210	− .998	.573	.893	− .855	−2.209	− .267
2.372	1.353	− .900	− .554	− .343	.470	−1.033	−1.026	2.172	.195
− .145	.466	.854	− .282	−1.504	.431	− .060	.952	− .343	.735
.104	.732	.604	− .016	− .266	1.372	− .925	−1.594	−2.004	1.925
1.419	−1.853	− .347	.155	−1.078	.623	− .024	.498	.466	.049
.069	− .411	− .661	− .037	.703	.532	− .177	.395	− .278	.240
.797	.488	−1.070	− .721	−1.412	− .976	−1.953	− .206	1.848	.632
− .393	− .351	.222	.557	−1.094	1.403	.173	− .113	.806	.939
− .874	−1.336	.523	.848	.304	− .202	−1.279	.501	.396	.859
.125	−1.170	− .192	1.387	2.291	− .959	.090	1.031	.180	−1.389
−1.091	− .649	− .514	− .232	−1.198	.822	.240	.951	−1.736	.270
2.304	.481	− .987	−1.222	.549	−1.056	.277	− .919	.148	1.517
− .961	2.057	− .546	− .896	.165	− .343	.696	.628	− .929	− .965
− .783	.854	− .139	1.087	.515	− .876	− .448	.485	.589	− .804
.487	.557	.327	1.280	−1.731	− .339	.295	− .724	.720	.331
− .299	.979	− .924	− .649	.574	1.407	− .292	− .775	− .511	.026
1.831	− .937	−1.321	−1.734	1.677	−1.393	−1.187	− .079	− .181	− .844
.243	.466	−1.330	1.078	−1.102	1.123	− .421	− .674	2.951	− .743
−2.181	−1.854	−1.059	− .478	−1.119	.272	− .800	.841	− .061	2.261
.154	− .333	1.011	−1.565	1.261	.776	1.130	1.552	− .563	.558
−1.065	1.610	.463	.062	− .086	.021	1.633	1.788	.480	2.824
1.083	− .760	− .012	.183	.155	.676	−1.315	.067	.213	2.380
.615	− .594	− .028	− .506	− .054	3.173	.817	.210	1.699	1.950
.178	− .500	1.100	1.613	1.048	2.323	− .174	− .033	2.220	− .661
− .507	−1.273	.596	.690	−1.724	−1.689	.163	− .199	− .450	.244
.362	− .588	−1.386	.072	.778	− .591	.365	.465	2.472	1.049
.775	1.546	.217	−1.012	.778	.246	1.055	1.071	.447	− .585
.818	.561	−1.024	2.105	− .868	.060	− .385	1.089	.017	− .873
.014	.240	− .632	− .225	− .844	.448	1.651	1.423	.425	.252
−1.236	−1.045	−1.628	.687	.983	− .840	−1.835	−1.864	1.327	− .408
− .567	−1.161	.010	− .853	.111	1.145	1.015	.056	.141	1.471
.278	−1.783	.170	− .358	.705	− .054	1.098	.707	− .585	− .305
− .959	− .497	.688	− .268	−1.431	− .791	− .727	.958	.237	.092
1.249	.037	.497	.579	− .227	.860	.349	2.355	2.184	−1.744
− .915	− .164	−1.166	1.529	.008	.636	−1.080	− .688	2.444	−1.316
.132	2.809	−1.918	−1.083	− .642	− .179	.339	.637	.063	− .079
− .156	−1.664	1.140	.295	1.086	−2.546	− .002	− .672	.205	− .039
.538	−1.143	− .390	.165	− .160	.457	−1.307	.273	− .670	− .988
.027	− .057	.742	− .149	− .801	1.702	− .346	− .053	.892	−1.181
.023	.423	1.051	− .831	− .325	− .795	−1.129	− .287	.172	− .793
− .196	−1.457	1.060	.557	− .190	− .891	− .768	.282	−1.432	− .447
.133	.577	− .332	−1.932	.220	.189	−1.521	.896	− .781	− .899
.020	− .217	− .856	.605	.072	.520	1.222	− .181	− .266	−1.222
1.405	1.065	1.350	1.353	−2.289	−1.003	.375	−1.621	−1.126	.937
.178	−1.237	− .520	− .603	−1.615	− .358	.605	− .407	−2.579	−1.811
−1.438	.104	−1.821	− .390	− .630	1.294	1.470	.991	− .355	−1.285
1.768	− .175	− .450	.915	− .221	− .019	1.864	.038	.058	1.212
.099	1.076	2.348	−1.550	.458	.147	−1.223	.994	−1.657	1.264
.951	.252	−1.261	− .963	.221	− .036	− .395	− .252	−1.379	1.885

*Reproduced with permission from the Rand Corporation, *A Million Random Numbers*. (New York: The Free Press, 1955).

APPENDIX E

The Standardized Normal Distribution Function,[*] $F(S)$

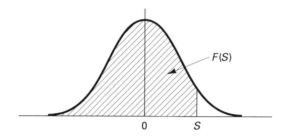

s	0.00	0.01	0.02	0.03	0.04	0.05	0.06	0.07	0.08	0.09
0.0	0.5000	0.5040	0.5080	0.5120	0.5160	0.5199	0.5239	0.5279	0.5319	0.5359
0.1	0.5398	0.5438	0.5478	0.5517	0.5557	0.5596	0.5636	0.5675	0.5714	0.5753
0.2	0.5793	0.5832	0.5871	0.5910	0.5948	0.5987	0.6026	0.6064	0.6103	0.6141
0.3	0.6179	0.6217	0.6255	0.6293	0.6331	0.6368	0.6406	0.6443	0.6480	0.6517
0.4	0.6554	0.6591	0.6628	0.6664	0.6700	0.6736	0.6772	0.6808	0.6844	0.6879
0.5	0.6915	0.6950	0.6985	0.7019	0.7054	0.7088	0.7123	0.7157	0.7190	0.7224
0.6	0.7257	0.7291	0.7324	0.7357	0.7389	0.7422	0.7454	0.7486	0.7517	0.7549
0.7	0.7580	0.7611	0.7642	0.7673	0.7703	0.7734	0.7764	0.7794	0.7823	0.7852
0.8	0.7881	0.7910	0.7939	0.7967	0.7995	0.8023	0.8051	0.8078	0.8106	0.8133
0.9	0.8159	0.8186	0.8212	0.8238	0.8264	0.8289	0.8315	0.8340	0.8365	0.8389
1.0	0.8413	0.8438	0.8461	0.8485	0.8508	0.8531	0.8554	0.8577	0.8599	0.8621
1.1	0.8643	0.8665	0.8686	0.8708	0.8729	0.8749	0.8770	0.8790	0.8810	0.8830
1.2	0.8849	0.8869	0.8888	0.8907	0.8925	0.8944	0.8962	0.8980	0.8997	0.90147
1.3	0.90320	0.90490	0.90658	0.90824	0.90988	0.91149	0.91309	0.91466	0.91621	0.91774
1.4	0.91924	0.92073	0.92220	0.92364	0.92507	0.92647	0.92785	0.92922	0.93056	0.93189
1.5	0.93319	0.93448	0.93574	0.93699	0.93822	0.93943	0.94062	0.94179	0.94295	0.94408
1.6	0.94520	0.94630	0.94738	0.94845	0.94950	0.95053	0.95154	0.95254	0.95352	0.95449
1.7	0.95543	0.95637	0.95728	0.95818	0.95907	0.95994	0.96080	0.96164	0.96246	0.96327
1.8	0.96407	0.96485	0.96562	0.96638	0.96712	0.96784	0.96856	0.96926	0.96995	0.97062
1.9	0.97128	0.97193	0.97257	0.97320	0.97381	0.97441	0.97500	0.97558	0.97615	0.97670
2.0	0.97725	0.97778	0.97831	0.97882	0.97932	0.97982	0.98030	0.98077	0.98124	0.98169
2.1	0.98214	0.98257	0.98300	0.98341	0.98382	0.98422	0.98461	0.98500	0.98537	0.98574
2.2	0.98610	0.98645	0.98679	0.98713	0.98745	0.98778	0.98809	0.98840	0.98870	0.98899
2.3	0.98928	0.98956	0.98983	0.9^20097	0.9^20358	0.9^20613	0.9^20863	0.9^21106	0.9^21344	0.9^21576
2.4	0.9^21802	0.9^22024	0.9^22240	0.9^22451	0.9^22656	0.9^22857	0.9^23053	0.9^23244	0.9^23431	0.9^23613
2.5	0.9^23790	0.9^23963	0.9^24132	0.9^24297	0.9^24457	0.9^24614	0.9^24766	0.9^24915	0.9^25060	0.9^25201
3.0	0.9^28650	0.9^28649	0.9^28736	0.9^28777	0.9^28817	0.9^28856	0.9^28893	0.9^28930	0.9^28965	0.9^28999
3.5	0.9^37674	0.9^37759	0.9^37842	0.9^37922	0.9^37999	0.9^38074	0.9^38146	0.9^38215	0.9^38282	0.9^38347
4.0	0.9^46833	0.9^46964	0.9^47090	0.9^47211	0.9^47327	0.9^47439	0.9^47546	0.9^47649	0.9^47748	0.9^47843

For example: $F(2.41) = 0.9^22024 = 0.992024$

[*] Reprinted from A. Hald, *Statistical Tables and Formulas* (New York: John Wiley & Sons, Inc., 1952), by permission of the publisher.

Bibliography

PART I: ARTICLES

ABERNATHY, W. J., AND K. WAYNE, "Limits of the Learning Curve," *Harvard Business Review* (Sept.–Oct. 1974):109–119.

AGGARWAL, R., "Corporate Use of Sophisticated Capital Budgeting Techniques," *Interfaces* no. 2 (April 1980):31–34.

BALDWIN, C. Y., AND R. S. RUBACK, "Inflation Uncertainty, and Investment," *Journal of Finance* no. 3 (July 1986):657–669.

BALDWIN, C. Y., "How Capital Budgeting Deters Innovation—And What to Do About It," *Research Technology Management* no. 6 (November 1991):39–45.

BAUMOL, W. J., AND R. E. QUANDT, "Investment and Discount Rates Under Capital Budgeting—A Programming Approach," *Economic Journal* 75, no. 298 (June 1965):317–329.

BENNETT, E. D., AND J. A. HENDRICKS, "Justifying the Acquisition of Automated Equipment," *Management Accounting* (July 1987):39–46.

BERNHARD, R. H., "Discount Methods for Expenditure Evaluation—A Clarification of Their Assumptions," *Journal of Industrial Engineering* 18, no. 1 (Jan.–Feb. 1962):19–27.

BERNHARD, R. H., "A Comprehensive Comparison and Critique of Discounting Indices Proposed for Capital Investment Evaluation," *The Engineering Economist* 16, no. 3 (Spring 1971):157–186.

BERNHARD, R. H., "Income, Wealth Base and Rate of Return Implications of Alternative Project Evaluation Criteria," *The Engineering Economist* 38, no. 3 (Spring 1993).

COOK, T. J., AND RIZZUTO, R. J., "Capital Budgeting Practices for R&D: A Survey and Analysis of Business Week's R&D Scoreboard," *The Engineering Economist* 34, no. 4 (Summer 1989).

DEBONDT, W. F. M., AND A. K. MAKHIJA, "Throwing Good Money After Bad? Nuclear Power Plant Investment Decisions and the Relevance of Sunk Costs," *Journal of Economic Behavior and Organization* no. 2 (Sept. 1988):173–199.

DONALDSON, G., "Strategic Hurdle Rates for Capital Investments," *Harvard Business Review* no. 2 (Mar.–Apr. 1972):50–58.

DREYFUS, S. E., "A Generalized Equipment Replacement Study," *Journal of the Society for Industrial and Applied Mathematics* 8, no. 3 (1960):425–435.

DUDLEY, C. L., JR., "A Note on Reinvestment Assumptions in Choosing Between Net Present Value and Internal Rate of Return," *Journal of Finance* (Sept. 1972):907–915.

FARRINGTON, E. J., "Capital Budgeting Practices of Non-Industrial Firms," *The Engineering Economist* 31, no. 4 (1986):293–302.

FREMGEN, J. M., "Capital Budgeting Practices: A Survey," *Management Accounting* (May 1973):19–25.

GURNANI, C., "Capital Budgeting: Theory and Practice," *The Engineering Economist* 24, no. 1 (Fall 1984).

HAKA, S. F., L. A. GORDON, AND G. E. PINCHES, "Sophisticated Capital Budgeting Selection Techniques and Firm Performance," *Accounting Review* no. 4 (Oct. 1985):651–669.

HAYS, R. H., AND D. A. GARVIN, "Managing if Tomorrow Mattered," *Harvard Business Review* no. 3 (May–June 1982):70–79.

HIRSHLEIFER, J., "On the Theory of Optimal Investment Decision," *Journal of Political Economy* 66 (Aug. 1958).

ISTVAN, D. F., "The Economic Evaluation of Capital Expenditures," *Journal of Business* (Jan. 1961):45–51.

KAPLAN, R., "The Evolution of Management Accounting," *Accounting Review* no. 3 (July 1984):390–418.

KAPLAN, R. S., "Must CIM be Justified by Faith Alone?" *Harvard Business Review* (Mar.–Apr. 1986):67–75.

KESTER, W. C., "Today's Options for Tomorrow's Growth," *Harvard Business Review* (Mar.–Apr. 1984):153–160.

KLAMMER, T., "Empirical Evidence of the Adoption of Sophisticated Capital Budgeting Techniques," *Journal of Business* (Oct. 1972):387–397.

KNIGHT, C. F., "Emerson Electric: Consistent Profits, Consistently," *Harvard Business Review* no. 1 (Jan.–Feb. 1992):57–69.

KULONDA, D. J., "Replacement Analysis with Unequal Lives," *The Engineering Economist* 23, no. 3 (Spring 1978):171–179.

LOHMANN, J. R., "The IRR, NPV and the Fallacy of the Reinvestment Rate Assumptions," *The Engineering Economist* 33, no. 4 (Summer 1988).

LORIE, J., AND L. J. SAVAGE, "Three Problems in Rationing Capital," *Journal of Business* XXVIII (1955):229.

LUTZ, R. P., AND H. A. CROWLES, "Estimation Deviations: Their Effect Upon the Benefit–Cost Ratio," *The Engineering Economist* 16, no. 1 (Fall 1971).

MEREDITH, J. R., AND M. M. HILL, "Justifying New Manufacturing Systems: A Managerial Approach," *Sloan Management Review* 28, no. 4 (Summer 1987):49–61.

MILLER, M. H., AND F. MODIGLIANI, "Cost of Capital to Electric Utility Industry," *American Economic Review* (June 1966):333–391.

MUKHERJEE, T. K., AND G. V. HENDERSON, "The Capital Budgeting Process: Theory and Practice," *Interfaces* 17, no. 2 (Mar.–Apr. 1987).

PARK, C. S., AND Y. SON, "An Economic Evaluation Model for Advanced Manufacturing Systems," *The Engineering Economist* 34, no. 1 (1988):1–26.

PETTY, W. J., D. F. SCOTT, AND M. M. BIRD, "The Capital Expenditure Decision-Making Process of Large Corporations," *The Engineering Economist* 20, no. 3 (Spring 1975):159–172.

PIKE, R., "Do Sophisticated Capital Budgeting Approaches Improve Investment Decision-Making Effectiveness?" *The Engineering Economist* (Winter 1989):149–161.

RAPPAPORT, A., AND R. TAGGART, "Evaluation of Capital Expenditure Proposals Under Inflation," *Financial Management* no. 1 (Spring 1982):5–13.

SARNAT, M., AND H. LEVY, "The Relationship of Rules of Thumb to the Internal Rate of Return: A Restatement and Generalization," *Journal of Finance* (1969):479–490.

SHARP, W. F., "A Simplified Model for Portfolio Analysis," *Management Science* 9, no. 2 (Jan. 1963).

TEICHROEW, D., A. A. ROBICHEK, AND M. MONTALBANO, "Mathematical Analysis of Rates of Return Under Certainty," *Management Science* 11, no. 3 (Jan. 1965):395–403.

TEICHROEW, D., A. A. ROBICHEK, AND M. MONTALBANO, "An Analysis of Criteria for Investment and Financing Decisions Under Certainty," *Management Science* 12, no. 3 (Nov. 1965): 151–179.

THOMPSON, H. E., "Mathematical Programming, the Capital Asset Pricing Model, and Capital Budgeting of Interrelated Projects," *Journal of Finance* no. 1 (March 1976):125–131.

WEINGARTNER, H. M., "Capital Budgeting of Interrelated Projects: Survey and Synthesis," *Management Science* (Mar. 1966):485–516.

WILNER, N., B. KOCH, AND T. KLAMMER, "Justification of High Technology Capital Investment—An Empirical Study," *The Engineering Economist* 37, no. 4 (Summer 1992).

PART I: BOOKS

AMERICAN TELEPHONE AND TELEGRAPH COMPANY, *Engineering Economy: A Manager's Guide to Decision Making*, 3d. ed. New York: McGraw-Hill, 1977.

AU, T., AND T. P. AU, *Engineering Economics for Capital Investment Analysis.* Boston: Allyn and Bacon, 1992.

BARISH, N. N., AND S. KAPLAN, *Economic Analysis for Engineering and Managerial Decision Making,* 2d. ed. New York: McGraw-Hill, 1978.

CLARK, F. D., AND A. B. LORENZONI, *Applied Cost Engineering.* New York: Marcel Dekker, Inc., 1978.

DEAN, J. W., JR., *Deciding to Innovate: How Firms Justify Advanced Technology.* Cambridge, ✓ MA: Ballinger, 1988.

DEAN, J., *Capital Budgeting.* New York: Columbia University Press, 1951.

GORDON, M. J., *The Cost of Capital to a Public Utility.* East Lansing, MI: Michigan State University, 1974.

KEENEY, R. L., AND H. RAIFFA, *Decisions with Multiple Objectives; Preferences and Value* ✓ *Tradeoffs.* New York: John Wiley and Sons, 1976.

LEVY, H., AND M. SARNAT, *Portfolio and Investment Selection: Theory and Practice.* Englewood Cliffs, NJ: Prentice-Hall, 1984.

MARKOWITZ, H., *Portfolio Selection.* New Haven, CT: Yale University Press, 1959.

NEWNAN, D. G., *Engineering Economic Analysis,* 3d. ed. San Jose, CA: Engineering Press, 1988.

OAKFORD, R. V., *Capital Budgeting.* New York: The Ronald Press Company, 1970.

O'NEIL, J. N., *Construction Cost Estimating for Project Control.* Englewood Cliffs, NJ: Prentice-Hall, 1982.

OSTWALD, P. F., *Cost Estimating,* 3d. ed. Englewood Cliffs, NJ: Prentice-Hall, 1989.

PARK, C. S., AND G. P. SHARPE-BETTE, *Advanced Engineering Economics.* New York: John Wiley & Sons, Inc., 1990.

QUINN, G. D., AND J. C. WINGINTON, *Analyzing Capital Expenditures: Private and Public Perspectives.* Homewood, IL: Richard D. Irwin, 1981.

RUEGG, R. T., AND H. E. MARSHALL, *Building Economics: Theory and Practice.* New York: Van Nostrand Reinhold, 1990.

SMITH, G. W., *Engineering Economy: Analysis of Capital Expenditures,* 4th ed. Ames, IA: Iowa State University Press, 1987.

STEWART, R. D., *Cost Estimating.* New York: John Wiley & Sons, 1982.

WEINGARTNER, H. M., *Mathematical Programming and the Analysis of Capital Budgeting Problems.* Englewood Cliffs, NJ: Prentice-Hall, 1963.

PART II AND III: ARTICLES

AGGARWAL, R., AND L. SOENEN, "Project Exit Value as a Measure of Flexibility and Risk Exposure," *The Engineering Economist* 35, no. 1 (Fall 1989):39–54.

ANKROM, R. K., "Top-Level Approach to the Foreign Exchange Problem," *Harvard Business Review* (July–Aug. 1974).

ANTLE, R., AND G. EPPEN, "Capital Rationing and Organizational Slack in Capital Budgeting," *Management Science* 31, no. 2 (Feb. 1985):163–174.

BAUMOL, W. J., "An Expected Gain-Confidence Limit Criteria for Portfolio Selection," *Management Science* 10, no. 1 (Oct. 1963):174–182.

BELL, D. E., "Disappointment in Decision Making Under Uncertainty," *Operations Research* 33, no. 1 (Jan.–Feb. 1985):1–27.

BERNHARD, R. H., "Mathematical Programming Models for Capital Budgeting—A Survey, Generalization and Critique," *Journal of Financial and Quantitative Analysis* (June 1969): 111–158.

BIERMAN, H., JR., AND W. R. HAUSMAN, "The Resolution of Investment Uncertainty Through Time," *Management Science* 18, no. 12 (1972):B:654–B:662.

BLUME, M. E., "On the Assessment of Risk," *Journal of Finance* 26 (March 1971):1–10.

BONINI, C. P., "Comment on Formulating Correlated Cash Flow Streams," *Engineering Economist* 20 (Spring 1975):209–214.

BONINI, C. P., "Risk Evaluation of Investment Projects," *Omega* 3, no. 6 (1975):735–750.

BOOTH, L., "The Influence of Productive Technology on Risk and Cost of Capital," *Journal of Financial and Quantitative Analysis* 26, no. 1 (March 1991):109–127.

BUSSEY, L. E., AND G. T. STEVENS, JR., "Formulating Correlated Cash Flow Streams," *The Engineering Economist* 18 (Fall 1972):1–30.

BUTLER, J. S., AND B. SCHACHTER, "The Investment Decision: Estimation Risk and Risk Adjusted Discount Rates," *Financial Management* 18, no. 4 (Winter 1989):13–22.

DEMERS, M., "Investment Under Uncertainty, Irreversibility and the Arrival of Information over Time," *Review of Economic Studies* 58, no. 2 (April 1991):333–350.

EVANS, D. A., "Investment Decision Making Under Uncertainty: Potential Environmental/Social Impacts of New Products," *The Engineering Economist* 32, no. 4 (Summer 1987).

EVANS, J. R., "Sensitivity Analysis in Decision Theory," *Decision Science* 15 (Spring 1984): 239–247.

EVERETT, M. D., "A Simplified Guide to Capital Investment Risk Analysis," *Planning Review* 14 (July 1986):32–36.

FALKNER, C. H., AND S. BENHAJLA, "Multi-Attribute Decision Models in the Justification of CIM Systems," *The Engineering Economist* 35, no. 2 (Winter 1990):91–114.

FISHBURN, P. C., "Foundations of Decision Analysis," *Management Science* 35, no. 4 (April 1989):387–405.

FRASER, J. M., AND R. P. FLYNN, "A New Method to Teach Multi-Attribute Utility Assessment," *The Engineering Economist* 36, no. 1 (Fall 1990):11–20.

HAYES, R. H., "Incorporating Risk Aversion into Risk Analysis," *The Engineering Economist* 20 (Winter 1975):99–121.

HERTZ, D. B., "Risk Analysis in Capital Investment," *Harvard Business Review* 42 (1964):95–106.

HILLIER, F. S., "Derivation of Probabilistic Information for the Evaluation of Risky Investments," *Management Science* 2, no. 3 (1963):485–489.

HILLIER, F. S., "A Basic Model for Capital Budgeting of Risky Interrelated Projects," *The Engineering Economist* 17 (Oct.–Nov. 1971):1–30.

HIRSHLEIFER, J., "Investment Decision Under Uncertainty: Choice Theoretic Approaches," *Quarterly Journal of Economics* 79 (1965):509–536.

HIRSHLEIFER, J., AND J. G. RILEY, "The Analytics of Uncertainty and Information—An Expository Survey," *Journal of Economic Literature* 42, no. 4 (Dec. 1979):1375–1421.

HODDER, J., AND H. RIGGS, "Pitfalls in Evaluating Risky Assets," *Harvard Business Review* 53 (Jan.–Feb. 1985):128–135.

HULL, J. C., "The Input to and Output from Risk Evaluation Models," *European Journal of Operational Research* 1 (Nov. 1977):368–375.

JENSEN, R. E., "Capital Budgeting Under Risk and Inflation," *Advances in Accounting* 3 (1986): 255–279.

KRYZANOWSKI, L., P. LUSZTIG, AND B. SCHWAB, "Monte Carlo Simulation and Capital Expenditure Decisions—A Case Study," *The Engineering Economist* 18 (Fall 1972):31–48.

MACCRIMMON, K. R., AND D. A. WEHRUNG, "Characteristics of Risk Taking Executives," *Management Science* 36, no. 4 (April 1990):422–435.

MACHINA, M. J., "Decision-Making in the Presence of Risk," *Science* 236 (May 1987): 537–543.

MAGEE, J. F., "How to Use Decision Trees in Capital Investment," *Harvard Business Review* 42 (Sept.–Oct. 1964):79–96.

MAO, J. C. T., AND J. F. HELLIWELL, "Investment Decision Under Uncertainty: Theory and Practice," *Journal of Finance* 24, no. 2 (May 1969):323–338.

MARCH, G. J., AND Z. SHAPIRA, "Managerial Perspectives on Risk and Risk Taking," *Management Science* 33, no. 11 (Nov. 1987):1404–1418.

MARCH, G. J., AND Z. SHAPIRA, "Managerial Perspectives on Risk and Risk Taking," *Management Science* 33, no. 11 (Nov. 1987):1404–1418.

PARK, C. S., AND THUESEN, G. J., "Combining the Concepts of Uncertainty Resolution and Project Balance for Capital Allocation Decisions," *The Engineering Economist* 24, no. 2 (Winter 1979).

ROBICHEK, A. A., "Interpreting the Results of Risk Analysis," *Journal of Finance* 30 (Dec. 1975):1384–1386.

ROBICHEK, A. A., AND S. C. MYERS, "Conceptual Problems in the Use of Risk-Adjusted Discount Rates," *Journal of Finance* 21 (Dec. 1966):727.

SICK, G. A., "Certainty Equivalent Approach to Capital Budgeting," *Financial Management* 15, no. 5 (Winter 1986):23–32.

SON, Y. K., "Simulation-Based Manufacturing Accounting for Modern Management," *Journal of Manufacturing Systems* 12, no. 5 (1993):417–427.

STOBAUGH, R. B., "How to Analyze Foreign Investment Climates," *Harvard Business Review* (Sept.–Oct. 1969):100–108.

SURESH, N. C., "Towards an Integrated Evaluation of Flexible Automation Investments," *International Journal of Production Research* 28, no. 9 (1990):1657–1672.

SWALM, R. O., "Utility Theory—Insights into Risk Taking," *Harvard Business Review* (Nov.–Dec. 1966):123.

VAN HORNE, J. C., "The Analysis of Uncertainty Resolution in Capital Budgeting for New Products," *Management Science* (April 1969).

WAGLE, B., "A Statistical Analysis of Risk in Capital Investment Projects," *Operational Research Quarterly* 18 (1967):13–33.

WHISLER, W. D., "Sensitivity Analysis of Rates of Return," *Journal of Finance* 31 (Mar. 1976): 63–70.

PARTS II AND III: BOOKS

AGGARWAL, R., *Capital Budgeting Under Uncertainty.* Englewood Cliffs, NJ: Prentice-Hall, 1993.

BIERMAN, H., AND S. SMIDT, *The Capital Budgeting Decision: Economic Analysis of Investment Projects.* New York: Macmillan Publishing Company, 1993.

BUSSEY, L. E., AND T. G. ESCHENBACH, *The Economic Analysis of Industrial Projects.* Englewood Cliffs, NJ: Prentice-Hall, 1992.

CANADA, J. R., AND W. G. SULLIVAN, *Economic and Multi-Attribute Evaluation of Advanced Manufacturing Systems.* Englewood Cliffs, NJ: Prentice-Hall, 1989.

COOPER, R., AND R. S. KAPLAN, *The Design of Cost Management Systems: Text, Cases and Readings.* Englewood Cliffs, NJ: Prentice-Hall, 1991.

HERTZ, D. B., AND H. THOMAS, *Risk Analysis and Its Applications.* New York: John Wiley & Sons, 1983.

HERTZ, D. B., AND H. THOMAS, *Practical Risk Analysis: An Approach Through Case Histories.* New York: John Wiley & Sons, 1984.

HILLIER, F. S., *The Evaluation of Risky Interrelated Investments.* Amsterdam: North-Holland, 1971.

MACCRIMMON, K. R., AND D. A. WEHRUNG, *Take Risks: The Management of Uncertainty.* New York: Free Press, 1986.

PARK, C. S., AND G. P. SHARPE-BETTE, *Advanced Engineering Economics.* New York: John Wiley & Sons, 1990.

WINKLER, R. L., *Introduction to Bayesian Inference and Decision.* New York: Holt, Rinehart and Winston, 1972.

ZELENY, M., *Multiple Criteria Decision Making.* New York: McGraw-Hill, 1982.

Index

T

U

Types of Risk

- PM Integration   - Cost
                     - Time
- Scope   - Risk
- Quality
                     - Human Resources
- Contract/Procurement - Communications

Process –
   Identification
   Assessment
   Response +
   Documentation

$$IRR - PW \text{ cash in} = PW \text{ cash out}$$

or

$$PW \text{ of net cash flow} = 0$$

$$AW \text{ of cost} = AW \text{ of Benefits}$$

Break Even

$$AW \text{ of cost} = AW \text{ of Benefits}$$

$$= Capital \ Recovery - \cancel{p \ S}$$

$$(P-S) * (A/P, 8\%, ?) + S * (8\%) = Annual \ receipts - disbursements$$

or

Net ~~profit~~ cash flow

Inital Inv   Salvage

ERR

$$FW \text{ of investment} = Accumulation \text{ of reinvested return at } i_0$$

(Present, not annual)

— last yr — there, salvage there

$$PW (ATCF) - 167$$

TABLE 2-1 Summarization of Discrete Compound Interest Factors and Symbols

To find	Given	Multiply "Given" by factor below	Factor name	Factor functional symbol	Example (answer for i = 5%) (Note: All uniform series problems assume end-of-period payments.)
F	P	$(1+i)^N$	Single sum compound amount	$(F/P, i\%, N)$	A firm borrows $1,000 for 5 years. How much must it repay in a lump sum at the end of the fifth year? *Ans.*: $1,276
P	F	$\dfrac{(1+i)^N - 1}{i(1+i)^N}$	Single sum present worth	$(P/F, i\%, N)$	A company desires to have $1,000 8 years from now. What amount is needed now to provide for it? *Ans.*: $676.80
P	A	$\dfrac{(1+i)^N - 1}{i(1+i)^N}$	Uniform series present worth	$(P/A, i\%, N)$	How much should be deposited in a fund to provide for 5 annual withdrawals of $100 each? First withdrawal 1 year after deposit. *Ans.*: $432.95
A	P	$\dfrac{(1+i)^N - 1}{i(1+i)^N}$	Capital recovery	$(A/P, i\%, N)$	What is the size of 10 equal annual payments to repay a loan of $1,000? First payment 1 year after receiving loan. *Ans.*: $129.50
F	A	$\dfrac{(1+i)^N - 1}{i(1+i)^N}$	Uniform series compound amount	$(F/A, i\%, N)$	If 4 annual deposits of $2,000 each are placed in an account, how much money has accumulated immediately after the last deposit? *Ans.*: $8,620
A	F	$\dfrac{(1+i)^N - 1}{i(1+i)^N}$	Sinking fund	$(A/F, i\%, N)$	How much should be deposited each year in an account in order to accumulate $10,000 at the time of the fifth annual deposit? *Ans.*: $1,810

Key: i = Interest rate per interest period
N = Number of interest periods

A = Uniform series amount
F = Future worth

P = Present worth